Advances in
Human Aspects
of Aviation

Advances in Human Factors and Ergonomics Series

Series Editors

Gavriel Salvendy

Professor Emeritus
School of Industrial Engineering
Purdue University

Chair Professor & Head
Dept. of Industrial Engineering
Tsinghua Univ., P.R. China

Waldemar Karwowski

Professor & Chair
Industrial Engineering and
Management Systems
University of Central Florida
Orlando, Florida, U.S.A.

3rd International Conference on Applied Human Factors and Ergonomics (AHFE) 2010

Advances in Applied Digital Human Modeling
Vincent G. Duffy

Advances in Cognitive Ergonomics
David Kaber and Guy Boy

Advances in Cross-Cultural Decision Making
Dylan D. Schmorrow and Denise M. Nicholson

Advances in Ergonomics Modeling and Usability Evaluation
Halimahtun Khalid, Alan Hedge, and Tareq Z. Ahram

Advances in Human Factors and Ergonomics in Healthcare
Vincent G. Duffy

Advances in Human Factors, Ergonomics, and Safety in Manufacturing and Service Industries
Waldemar Karwowski and Gavriel Salvendy

Advances in Occupational, Social, and Organizational Ergonomics
Peter Vink and Jussi Kantola

Advances in Understanding Human Performance: Neuroergonomics, Human Factors Design, and Special Populations
Tadeusz Marek, Waldemar Karwowski, and Valerie Rice

4th International Conference on Applied Human Factors and Ergonomics (AHFE) 2012

Advances in Affective and Pleasurable Design
Yong Gu Ji

Advances in Applied Human Modeling and Simulation
Vincent G. Duffy

Advances in Cognitive Engineering and Neuroergonomics
Kay M. Stanney and Kelly S. Hale

Advances in Design for Cross-Cultural Activities Part I
Dylan D. Schmorrow and Denise M. Nicholson

Advances in Design for Cross-Cultural Activities Part II
Denise M. Nicholson and Dylan D. Schmorrow

Advances in Ergonomics in Manufacturing
Stefan Trzcielinski and Waldemar Karwowski

Advances in Human Aspects of Aviation
Steven J. Landry

Advances in Human Aspects of Healthcare
Vincent G. Duffy

Advances in Human Aspects of Road and Rail Transportation
Neville A. Stanton

Advances in Human Factors and Ergonomics, 2012-14 Volume Set:
Proceedings of the 4th AHFE Conference 21-25 July 2012
Gavriel Salvendy and Waldemar Karwowski

Advances in the Human Side of Service Engineering
James C. Spohrer and Louis E. Freund

Advances in Physical Ergonomics and Safety
Tareq Z. Ahram and Waldemar Karwowski

Advances in Social and Organizational Factors
Peter Vink

Advances in Usability Evaluation Part I
Marcelo M. Soares and Francisco Rebelo

Advances in Usability Evaluation Part II
Francisco Rebelo and Marcelo M. Soares

Advances in
Human Aspects
of Aviation

CRC Press
Taylor & Francis Group
6000 Broken Sound Parkway NW, Suite 300
Boca Raton, FL 33487-2742

© 2013 by Taylor & Francis Group, LLC
CRC Press is an imprint of Taylor & Francis Group, an Informa business

No claim to original U.S. Government works

Printed in the United States of America on acid-free paper
Version Date: 20120529

International Standard Book Number: 978-1-4398-7116-4 (Hardback)

Visit the Taylor & Francis Web site at
http://www.taylorandfrancis.com

and the CRC Press Web site at
http://www.crcpress.com

Table of Contents

viii

Section III: Human Performance in Aviation

Section VI: Modeling

Section VIII: Cognition and Workload

Section IX: Experiments and Evaluation

Preface

This book brings together the most recent human factors work in the aviation domain, including empirical research, human performance and other types of modeling, analysis, and development. Since the very earliest years of aviation, it was clear that human factors were critical to the success and safety of the system. As aviation has matured, the system has become extremely complex. Although largely devoid of automation, air traffic control is expected to undergo major changes in the next several decades as air traffic modernization efforts are enacted to vastly increase the capacity of the system. The flight deck of modern commercial airlines has become highly automated and computerized, a trend that is expected to continue. The issues facing engineers, scientists, and other practitioners of human factors are becoming more challenging and more critical.

In keeping with a system that is vast in its scope and reach, the chapters in this book cover a wide range of topics. The chapters are organized into several sections:

I: Flight Deck Interfaces
II: Air Traffic Control Interfaces and Operations
III: Human Performance
IV: Automation and Function Allocation
V: The Next Generation Air Transportation System (NextGen)
VI: Modeling
VII: Safety and Ethics
VIII: Cognition and Workload
IX: Experiments and Evaluations

The common theme across these sections is that they deal with the intersection of the human and the system. Moreover, many of the chapter topics cross section boundaries, for instance by focusing on function allocation in NextGen or on the safety benefits of a tower controller tool. This is in keeping with the systemic nature of the problems facing human factors experts in aviation – it is becoming increasingly important to view problems not as isolated issues that can be extracted from the system environment, but as embedded issues that can only be understood as a part of an overall system.

Section I and II discuss interface and operations issues from the perspectives of pilots and air traffic controllers, respectively. Current modernization efforts suggest that major changes are coming to flight decks and to air traffic controller systems and duties. Section III contains a number of chapters that discuss specific human performance issues, studied from within the context of the air transportation system. Section IV contains chapters that look into specific issues related to automation and the delineation of function between automation and human within the current and future system. Section V collects a number of chapters specifically on the U.S. air traffic modernization effort, called NextGen. In section VI, a number of diverse modeling perspectives and methods are presented. Safety and ethics have always been driving factors for change in the air transportation system, and section VII gathers together several chapters on this topic. Aviation has also been a common application area for the study of specific aspects of cognition and workload, and several chapters on this topic are presented in section VIII. Lastly, in section IX the reader will find several chapters on the latest empirical research and evaluation within the air transportation domain.

These chapters were all contributed, solicited, and/or reviewed by the members of the technical program board. Therefore, the majority of the credit for the usefulness of this book goes to those people, who are listed below.

S. Verma, USA	B. Willems, USA
K. Vu, USA	H. Davison Reynolds, USA
E. Rantanen, USA	A. Majumdar, UK
J. Mercer, USA	T. von Thaden, USA
A. Sebok, USA	B. Hooey, USA
A. Alexander, USA	M. Draper, USA
C. Samms, USA	N. McDonald, Ireland
E. Rovira, USA	B. Gore, USA

The book should be of use to aviation human factors professionals, researchers, and students. It documents the latest research in this area across a broad range of topics. Readers should be able to not only take specific information from each chapter, but should also be able to detect broad themes and trends in this field. It is our sincere hope that this volume is read and referenced many times and for many years in the future, thereby demonstrating its value to those in the field of aviation human factors.

March 2012

Steven J. Landry
Purdue University
USA

Editor

Section I

Flight Deck Interfaces

Towards an Improved Pilot-Vehicle Interface for Highly Automated Aircraft: Evaluation of the Haptic Flight Control System

Paul Schutte, Kenneth Goodrich

NASA Langley Research Center
Hampton, VA 23681
Contact: Paul.C.Schutte@NASA.gov

Ralph Williams

Analytical Mechanics Associates, Inc.
Hampton, VA 23666

ABSTRACT

The control automation and interaction paradigm (e.g., manual, autopilot, flight management system) used on virtually all large highly automated aircraft has long been an exemplar of breakdowns in human factors and human-centered design. An alternative paradigm is the Haptic Flight Control System (HFCS) that is part of NASA Langley Research Center's Naturalistic Flight Deck Concept. The HFCS uses only stick and throttle for easily and intuitively controlling the actual flight of the aircraft without losing any of the efficiency and operational benefits of the current paradigm. Initial prototypes of the HFCS are being evaluated and this paper describes one such evaluation. In this evaluation we examined claims regarding improved situation awareness, appropriate workload, graceful degradation, and improved pilot acceptance.

Twenty-four instrument-rated pilots were instructed to plan and fly four different flights in a fictitious airspace using a moderate fidelity desktop simulation. Three different flight control paradigms were tested: Manual control, Full

Automation control, and a simplified version of the HFCS. Dependent variables included both subjective (questionnaire) and objective (SAGAT) measures of situation awareness, workload (NASA-TLX), secondary task performance, time to recognize automation failures, and pilot preference (questionnaire).

The results showed a statistically significant advantage for the HFCS in a number of measures. Results that were not statistically significant still favored the HFCS. The results suggest that the HFCS does offer an attractive and viable alternative to the tactical components of today's FMS/autopilot control system. The paper describes further studies that are planned to continue to evaluate the HFCS.

1   INTRODUCTION

There have been many criticisms of the current flight management/autoflight system in modern aircraft (Weiner, 1998). In 1996, the FAA determined that a major problem with the modern flight deck was the disconnect between the pilot and the autoflight systems in the aircraft (Abbott, et. al, 1996). This disconnect led to instances where pilots would remark, "Why is the aircraft doing that?" or "What's it doing now?" There are many ways to frame and consider these problems in terms of human factors. The perspective considered in this paper is one of languages and interfaces. After framing the problem in this way, the paper briefly describes the Haptic Flight Control System (HFCS) and the potential benefits of the HFCS. Next, it describes a simulation study that evaluated some of these benefits.

1.1   Languages and interfaces

There are three 'languages' that the pilot uses to command the modern highly automated aircraft. The first is manipulating the control surfaces and propulsion systems via the control inceptors, that is, stick, rudder and throttles. Examples of the commands are pitch up, bank left, and increase thrust. The mapping of the commands onto the actions on the control inceptors is intuitive. Pull back to pitch up, turn right to bank right, push the throttles forward for more thrust.

The second language is based on direction and speed. The commands in this language are heading commands, airspeed commands, altitude and altitude rate commands. These commands are usually made through the knobs and dials on the autoflight system.

The third language is one of earth-referenced locations and clock time. It is by far the richest language. The pilot can specify an approach into Reagan National Airport, a departure from Heathrow, a specific jetway in the airspace system, or a waypoint by its latitude and longitude. The commands are given to the FMS cockpit display unit through an alphanumeric keypad with single purpose and multi-use buttons. The language has an elaborate and rigid syntax.

The pilot of a modern aircraft is faced with three different languages, using three different input devices and all three different systems can be controlling the aircraft in some combination at the same time. Given the limitations of the human pilot, this design is a recipe for confusion, high memory workload, high display requirements, and miscommunication. This error-proneness is further exacerbated by the ability to preprogram route changes and arm modes – allowing for the passage of time between error commission (e.g., when the pilot sets a command) and the manifestation of that error (e.g., when the aircraft actually performs that command).

We have developed a single flight control system that speaks all three languages and uses a single intuitive interface – it's called the Haptic-Multimodal Flight Control System.

1.2   The Haptic-Multimodal Flight Control System

In the Haptic-Multimodal Flight Control System (HFCS), the pilot issues commands to the aircraft solely through the stick and throttle (Schutte, et al, 2007, Goodrich, et al, 2011). The HFCS can best be described as a point and shoot interface. All of the route information currently in the FMS (e.g., waypoints and airway structure) is presented on the Primary Flight Display (PFD) and the Map Display (MD). The pilot points the aircraft at one of these features, selects it, pulls the trigger on the stick and the automation then flies the aircraft according to the procedure in the database. To fly to a heading, the pilot simply turns the aircraft to that heading and pulls the trigger. Since no published procedure is selected, the automation holds that heading. If the aircraft is pointing at multiple features, the pilot can cycle through them and pull the trigger when the feature he wants is selected. If the pilot wants to arrive at the next waypoint at a certain time, he points the aircraft at that waypoint, pulls the trigger, then moves the throttle while watching the predicted arrival time to that waypoint (displayed on the PFD and MD). He adjusts the speed until it is at the desired arrival time, and pulls the trigger on the throttle. Even non-published commands such as fly parallel to a jetway can be easily performed. Nearly every tactical command that can be given to the current stick/autoflight/FMS combination can be accomplished through the two inceptors.

The common elements between the pilot and the automation with regard to the control of the aircraft are the back-driven stick and throttle. When the automation moves the aircraft, the pilot can see and feel what the automation is doing by watching the controls and/or having a hand on the inceptor. When the pilot moves the aircraft using the inceptors, the automation is aware of what the pilot is doing.

One feature that current automation has but is missing from HFCS is the ability to preprogram an entire route and then 'set it and forget it'. While the pilot can plan the route on a separate planning device and display this plan on the PFD and MD, the automation will not fly the entire route – rather the pilot has to make all coarse turns and altitude changes. The automation will accomplish the individual features with all the efficiency of current automation, but the pilot must be in the loop whenever major changes in the aircraft's trajectory are made. This lack of preprogramming provides a benefit from a human factors perspective. Humans become complacent with reasonably reliable pre-programmed automation. But one of their primary roles is to monitor the mission progress and the automation. In order to effectively monitor over long durations they need to be engaged in the task at regular intervals. Having the pilot perform the simple task of pointing the aircraft to the next goal can provide just such engagement. In the HFCS, the pilot will never ask what the aircraft is doing, because the pilot is the one that just commanded it to do it.

One potential Achilles Heel for the HFCS is that the pilot must remember to reengage and make the next move. Humans in general have a very difficult time with this prospective memory – easily forgetting or becoming distracted. Automation, however does not suffer from this problem. So the automation is used to prompt the pilot (at various levels of interruption) that a move will soon need to be made, is pending, or has been missed. Thus if the HFCS is coupled to a jetway and is approaching a 'fork in the road' or decision point, the pilot will be given alerting cues. If the pilot has entered a plan (using a separate planning device), these cues can be more specific to the plan.

At first glance, the single-point-of-control concept can appear to be a technological step backwards. In a modern flight deck, the pilot rarely has to touch the control inceptor. In fact, the use of the control inceptor to fly the aircraft is virtually the hallmark of the unautomated aircraft. The HFCS appears to be increasing pilot workload back to the levels of manual flight before automation. The purpose of this study is to explore how pilots feel about this new control concept and to see how it affects their workload and their situation awareness.

2   METHOD

2.1   Participants

Twenty-four general aviation pilots were used as subjects. The subjects were all right-handed males with no color blindness. They were VFR-certified but not IFR-certified. Each had more than 50 hours and fewer that 300 hours total flying time and they had logged between 12 and 48 hours in the 12 months prior to the experiment.

2.2   Apparatus

The experiment used a PC-based flight simulation environment (Figure 1). The aircraft model used was a simplified version of a deHaviland Dash-8. The control inceptor was an active, force feedback side arm controller made by Stirling Dynamics, Ltd. There were no throttles (speed was automatically controlled in all conditions) or rudder pedals. There were two main displays, a Primary Flight Display (PFD) and a Map Display (MD). In addition to the MD, there was a planning interface that was implemented on a tablet PC to the subject's right. Three large out-the-window displays and a laptop computer were used to provide secondary tasking.

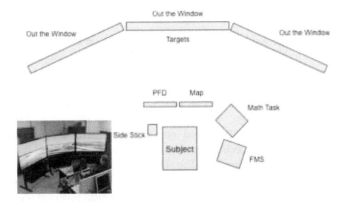

Figure -1: PC-based simulator

2.3   Subject Task

A test run for this experiment occurred in three phases. (1) Planning the route on paper, (2) entering the route into the electronic planner, and (3) flying that route using the simulator.

Planning the route on paper - For this experiment, we created an airspace over a fictitious island consisting of seven airports. Each airport had one or two runways. The terminal area was defined by pathways that were used for departures and approaches. Aircraft could fly from airport to airport using 18 airways connected by eight waypoints. The prescribed flight altitude for all airways was 2500 ft. Each subject an origin and a destination along with necessary charts and airway instructions. The subject then highlighted the route on the paper charts.

Entering the route into the electronic planner - The subject would take his paper map over to the electronic planner. He entered the route by 'connecting the dots' on the map. The route was then reviewed by both the subject and the experimenter. Then it was transmitted to the flight control and display system to be presented on the display (for all conditions) and used by the HFCS and Full Automation conditions to provide automated control.

Flying that route using the simulator - The simulator was initialized to a fixed point along the route and was paused. The subject was responsible for flying the aircraft to its destination along the route that he had planned and entered. At different points in the simulation, the subject was requested to make one of three changes: a strategic change in the route, which consisted of him making a change to the flight plan but required no change in his flight trajectory at that time; a tactical change that required him to fly a route parallel to the planned route (supposedly to fly by traffic); and a parallel runway change on approach. The run was always terminated before touch down.

2.4 Experimental Protocol

Over a two-day period, each subject spent 12 hours in the study. The first half day consisted of training on all tasks. The subject flew the simulation in the Manual mode (i.e., no automation assistance) in order to become acquainted with the simulator handling and performance. In the second half, the subject was trained on the first automation condition (Manual, HFCS, Full Automation). He was given one of four routes to plan/fly. The conditions and routes (see below) were counterbalanced. The subject then filled out a post run questionnaire. After a break, he was trained on another condition, given another route to plan and fly. Again, he was given the post run questionnaire. The subject left for the day.

The subject returned the next morning and was allowed to fly the simulator in the Manual mode to refresh his familiarity. Then the subject received training on the last condition and flew another run with this condition and route. After completing the post run questionnaire, the subject was given a break. The subject was then given one last route to plan and fly. One of the three automation conditions was assigned (counterbalanced). After the post run questionnaire, the subject was given a post experiment questionnaire and released. Thus each subject saw all three conditions at least once and saw one condition twice.

2.5 Experimental Conditions

There were three experimental conditions: Full Automation (FA), HFCS, and Manual. In all conditions, the subject received visual flight guidance on the PFD and MD as well as aural and tactile warnings of impending turns. PFD guidance used a tunnel in the sky with rings that the aircraft had to fly through.

Full Automation: The subject had only to monitor the flight. The automation was always coupled to the flight plan. Any tactical changes (i.e., runway changes, side-step maneuvers) were made in the electronic planner. This condition represents the most automated functionality of modern aircraft.

Haptic Flight Control System: The subject could lock on or couple to any straight line route segment (planned or published) and the automation would hold the aircraft on that segment. At the end of a segment, the automation would release control of the aircraft and the pilot would have to steer the aircraft onto the next segment. Once in the proximity of the next segment, the subject pulled the trigger on the side stick and the automation coupled to the path. When uncoupled to a path, the flight control system was essentially that of the Manual condition.

Manual: In the Manual condition, the subject would hand fly the aircraft using the stick. They were instructed to follow the centerline of the route as displayed on the PFD and fly through tunnel rings. Random winds were introduced to move the aircraft off course if its attitude was not attended to by the subject.

2.6   Run Definitions

Four runs were used in the experiment. In each run the subject was given tactical and strategic changes such as might be given from Air Traffic Control. The tactical changes were introduced by having the subject either fly an offset parallel to the planned course or make a runway change while on approach. The strategic change involved a change in the flight plan that required no immediate movement of the aircraft. In addition, one run contained an automation failure where the automation was turned off with no alert. This meant that in the FA condition, the aircraft would remain in a turn (flying in circles) and in the HFCS condition, the automation could not be recoupled after the turn.

2.7   Secondary Tasks

Workload was increased for each subject by giving them two secondary tasks – a visual task and a cognitive task. The visual task required the subject to monitor the center out-the-window display and look for dots that randomly appeared and then disappeared. The dots had no operational significance. The cognitive secondary task was a simple addition of two double-digit numbers. Thus the subject could set his own pace regarding answering the math questions. The subject was told that flying the aircraft was always to be his highest priority.

2.8   Dependent Variables

The following dependent variables were recorded: Situation Awareness was recorded using SAGAT (Endsley, 1988) and subjective ratings. Workload was measured using the NASA-TLX (Hart and Staveland, 1998) and subjective ratings. Pilot Involvement was measured using subjective ratings. Secondary task performance was objectively measured. The time to recognize the automation failure was measured from the onset of the failure until the subject either verbalized the problem or corrected for the problem – whichever came first. Remaining subjective data was collected using questionnaires.

2.9   Hypotheses

Hypothesis 1: Flight Situation Awareness for the HFCS condition will be higher than that of the FA condition.

Hypothesis 2: Secondary Task Situation Awareness for the HFCS condition will be higher than that of the Manual Condition.

Hypothesis 3: Secondary Task Performance for the HFCS condition will be higher than that of the Manual condition.

Hypothesis 4: Subjective Workload for the HFCS condition will be less than that of the Manual condition.

Hypothesis 5: Automation failure will be detected sooner in the HFCS condition when compared to the FA condition.

Hypothesis 6: Subjects will prefer the HFCS condition over the Manual and the FA conditions.

3   EXPERIMENT RESULTS

While all results favored the HFCS condition and the hypotheses, only a few of those turned out to be statistically significant. All statistical tests were performed using IBM© SPSS© Statistics Version 19.0.

Hypothesis 1: Situation awareness was measured objectively using the SAGAT score. While the results favored the hypothesis (FA: Correct = 138, Incorrect = 86; HFCS: Correct = 149, Incorrect = 75), a Pearson Chi-Square test of the differences between the HFCS condition and the FA conditions was not significant $\chi^2(1, N=448) = 1.73, p=.29$.

The post-run questionnaire asked the subject to rate their awareness regarding aircraft position, aircraft heading, and the progress of the aircraft with regard to the flight plan. Table 1 shows the data. While the raw data favors the hypothesis for all three, independent t-tests demonstrated that only the difference for the progress awareness question was significant, $t(62) = 2.00, p < .05$.

Table 1: Results of Subjective Awareness Assessment

	FA		HFCS	
	M	*SD*	*M*	*SD*
Aircraft Position	13.9	4.0	15.1	3.3
Aircraft Heading	13.6	4.6	13.9	4.5
Aircraft Progress to Plan	15.3	3.8	16.9	2.4

Hypothesis 2: Comparisons between Manual and HFCS of SAGAT measurements for the secondary tasks yielded results that favored the hypothesis (HFCS: Correct = 113, Incorrect = 111; Manual: Correct = 101, Incorrect = 123). However, these results were not statistically significant $\chi^2(1, N=448) = 1.29, p=.26$.

Hypothesis 3: Again, for this hypothesis, the results all favored HFCS but not statistically, although the Math Speed scores approached significance ($t(62) = -1.85$, $p = .07$). Table 2 shows the raw data.

Table 2: Results of Secondary Task Performance

	HFCS		Manual	
	M	*SD*	*M*	*SD*
Target Accuracy (% Correct)	81.4	12.5	75.5	18.1
Math Accuracy (% Correct)	96.6	2.2	96.0	2.9
Math Speed (Question/min)	4.3	2.0	3.4	2.0

Hypothesis 4: Subjective workload was measured using the NASA-TLX workload instrument. The TLX asks the subject to rate six different workload categories (Mental, Physical, Temporal, Performance, Effort, and Frustration) using a Likert scale with 20 divisions. Table 3 shows the data. Independent t-tests were used in comparing the HFCS condition to the Manual condition. Only the Mental ($t(62) = 4.98$, $p < .01$) and Effort ($t(62) = 4.31$, $p < .01$) measures were found to be statistically significant.

Table 3: Results of NASA-TLX

	HFCS		Manual	
	M	*SD*	*M*	*SD*
Mental *(lower = better)*	12.5	4.0	16.6	2.3
Physical *(lower = better)*	8.5	3.9	8.9	4.2
Temporal *(lower = better)*	9.0	4.6	9.9	4.5
Performance *(higher = better)*	14.1	3.3	13.5	3.5
Effort *(lower = better)*	12.5	3.5	15.6	1.9
Frustration *(lower = better)*	7.5	4.0	9.3	3.9

Hypothesis 5: All subjects detected the automation failure. An independent t-test comparing the two conditions showed that the failures in the HFCS condition were detected sooner than the FA condition. This difference was statistically significant $t(14) = -2.19$, $p < .05$, (HFCS $M=26.4$ seconds, $SD = 7.44$; AUTO $M=117.0$, $SD = 116.78$).

Hypothesis 6: After experiencing all conditions, subjects were asked to choose one condition from among the three as their most preferred. They were asked this with regard to just the flying task (i.e., if there was no secondary task) and with regard to the combined task (i.e., flying and all secondary tasks). Pearson Chi-square test was used to test for significance. For both flying and combined, the subjects preferred the HFCS condition over both the Manual and the FA conditions at the .01 significance level.

Comparing the HFCS condition with the FA condition yielded $\chi^2(2, N = 24) = 6.86$, $p < .01$ (HFCS = 15, FA = 6) for just the flying task and $\chi^2(2, N = 24) = 6.76$, $p < .01$ (HFCS = 16, FA = 7) for the combined tasks.

Similarly, comparing the HFCS condition with the Manual condition yielded $\chi^2(2, N = 24) = 10.54$, $p < .01$ (HFCS = 15, Manual = 3) for just the flying task and $\chi^2(2, N = 24) = 20.49$, $p < .01$ (HFCS = 16, Manual = 1) for the combined tasks.

4   DISCUSSION OF EXPERIMENTAL RESULTS

To summarize the statistically significant findings: The HFCS condition improved flight progress situation awareness over that of FA. The HFCS condition caused less Mental workload and Effort when compared to the Manual condition. Subjects detected a failure of the automation in the HFCS condition sooner than they detected it in the FA condition. Subjects preferred the HFCS condition over both the FA and the Manual conditions when considering just flying the aircraft and when considering flying the aircraft along with secondary tasks.

It is also important to note that none of the actual data refuted the hypotheses – that is, they just did not pass the test of statistical significance. The number of SAGAT probes that could be reasonably used was relatively small (usually only three or four per run) and the distribution of answers in the different categories was not normal. Since all of the data supported the hypotheses, it may be that the sample size was not large enough for sufficient power or that the measurements were not sensitive enough.

Nonetheless, the statistically significant results themselves are impressive and they reinforce the claims of increased situation awareness, reduced workload, and high pilot preference when using the HFCS. The HFCS holds promise for ameliorating many of the human factors problems found in current automation in modern flight decks.

5   FUTURE RESEARCH

In the future, the power and sensitivity components of the dependent measures will be improved to further examine the hypotheses. Also, more mature versions of the HFCS including a more robust alerting system will be tested. Other types of failures including failures of the alerting system (that is, the subjects will not be cued to make a transition) will be tested. Usability studies will be conducted to improve the HFCS interface.

REFERENCES

Abbott, K., Slotte, S. M., & Stimson, D. K., 1996. The interface between flightcrews and modern flight deck systems. Washington D.C.: Federal Aviation Administration.

Endsley, M.R., 1988, Situation Awareness Global Assessment Technique (SAGAT). In Proceedings of the National Aerospace and Electronics Conference (NAECON) (pp 789-795). New York: IEEE.

Goodrich, K.H., Schutte, P.C. and Williams, R.A, 2011, Haptic-Multimodal Flight Control System Update, 11th AIAA Aviation Technology, Integration, and Operations Conference (ATIO), Virginia Beach, VA.

Hart, S. and Staveland, L. E., 1988,. *Human Mental Workload*, chapter Development of NASA TLX: Results of Empirical and Theoretical Research. North Holland Press, Amsterdam.

Schutte, P. C., Goodrich, K. H., et al, 2007, "The Naturalistic Flight Deck System: An Integrated System Concept for Improved Single-Pilot Operations," NASA/TM-2007-21509. National Aeronautics and Space Administration: Washington, DC.

Wiener Earl, 1998, "Cockpit Automation," In E. L. Wiener & D. C. Nagel (Eds.), *Human Factors In Aviation*, San Diego, CA: Academic Press, pp. 433-461.

CHAPTER 2

The Study of Optimum Cognition on Character Images for In-vehicle Navigation Systems

Ming-Chyuan Lin[1], Ming-Shi Chen[2], Chen-Cheng Lin[3], C. Alec Chang[4]

[1,3]Nanhua University, Taiwan; [2]TransWorld University, Taiwan;
[4]University of Missouri-Columbia, USA
minglin@mail.nhu.edu.tw; mschen@twu.edu.tw;
chencheng@mail.nhu.edu.tw; changC@missouri.edu

ABSTRACT

The progress of technology in computers has made the in-vehicle navigation system of the global positioning systems (GPS) be integrated with communication, information, entertainment and personal service products. However, if the in-vehicle navigation system (IVNS) provides inadequate user interface information, the user might have difficulty in operating the system. The graphics images in GPS systems are very attractive and cognitive to the user, but an IVNS needs the character images as an auxiliary tool to provide more accurate information with the user. Character images in current IVNSs are seldom adjustable and the colors between characters and backgrounds are not planned with high legibility or cognition. The research proposed an integrated evaluation procedure using the back-propagation neural network to recommend the optimum cognition of character images based on the investigation of user visual cognition. The portable navigation device (PND) was used for the design of character images. Computer software of Microsoft Visual Basic 6.0 was used to build an interactive and simulated navigation screen system for tested subjects to identify the most suitable and legible display. The optimum character images design alternative recommended in this research will provide designers with a reference for developing a more efficient and use-friendly operation interface of the in-vehicle navigation system.

Keywords: global positioning system, in-vehicle navigation system, back-propagation neural network, interaction interface design, character images display

1   INTRODUCTION

The enhancement of the Internet and Wi-Fi (Wireless Fidelity) on mobility, functions and capability has greatly changed the living styles of our societies. The 3C (Computers, Communications and Consumer-Electronics) products, such as cell phones, tablet PCs, PDAs (Personal Digital Assistant), PNDs (Portable Navigation Device) and electronic books become popular. In general, most 3C products have a screen/monitor to allow the user to setup, operate and display digital contents. The representation of digital contents includes graphic pictures, character/word images, voices and even dynamic images. Although the graphic pictures in digital contents can attract people immediate attention, only the character images can communicate accurate information to people (Tomioka, 2006). Therefore, the character images play an important role in digital contents, especially in-vehicle navigation systems (INVSs). The IVNS of the global positioning systems (GPS) is integrated with communication, information, entertainment and personal service. Lee and Cheng (2008) observed that using a portable navigation system such as smart phones will perform better than those of using paper maps. Lee and Cheng (2010) further observed that using the portable navigation system will obtain better efficiency on driving performance than that of using the onboard one (cell phone). Because the IVNS needs quick response to the user, the user might have difficulty in operating the system if it can not perform the adequate interface information. For example, unsuitable icons and inappropriate digital contents will confuse the users and increase the user's mental workload when driving a vehicle. In fact, the character images have been considered as an auxiliary tool in the IVNSs to help people obtain more accurate information in driving. An appropriate form and representation of digital character images on an IVNS becomes worthy of investigating issue. Mills and Weldon (1987) noted that small size of character images will have a faster reading speed than those of large size. Pastoor (1990) indicated that people prefer cool colors as backgrounds associated with non-saturated colors of character images in a digital display. Sanders and McCormick (1993) suggested that not use too many colors in a screen and avoid color discord and combinations of similar colors, red and blue, red and green and blue and green. The character images versus grounds should have a higher color difference (Sanders and McCormick, 1993). Darroch, et al. (2005) found that the aged people favor 8 pts (Points, 1 point = 0.35mm) to 12 pts of characters. As to reading magazines or books, the size 9 to 10 pts of characters will be most suitable (2005). Lee, et al, (2008) further stressed that electronic E-paper displays need greater illumination and greater character size to enhance search speed and recognition accuracy. However, the character images in current IVNSs are seldom adjustable, the ranges between words and lines are simultaneously scaled and can not be adjusted based on the user preference. Besides, the colors between character images and the corresponding backgrounds are not compatibly planned with high legibility or cognition. The above deficiencies of character cognition will affect the user on reading efficiency and operating accuracy. To enhance the cognition of character images in the GPS for an IVNS, the research proposed an integrated evaluation procedure incorporated with the concept of back-propagation neural network to search for the optimum cognition of character images. The portable navigation device (PND) was used as an example to help explore the feasibility of the design of character images.

2   THE DEVELOPMENT PROCEDURE

The research will focus on identifying the most suitable parameter combination of character images for the PND interface design. The parameters that are identified for the research include: (1) colors between the character image and the corresponding background, (2) row distances between lines of characters, (3) distances between characters, (4) sizes of characters and (5) fonts of characters. The implement of the research will be based on an interactive experiment that the above parameters are adjusted according to the choice of the background color. The experimental data that obtained from the test subjects are then forwarded to the training process of back-propagation neural network to establish an optimum recommendation inference network for the representation of character images on an in-vehicle navigation system. Note that the construction of an interactive interface for experiments can be used to control the PND simulation display and adjust parameters according to the visual preference of test subjects.

The framework of the research procedure is divided into three stages: (1) definition of experimental parameters and simulated display, (2) conduction of the experimental design, (3) application of back-propagation neural network. A four pronged procedure for the Stage 1 includes (1) definition of representational interface for experimental uses and results, (2) questionnaire of character color planning and clustering analysis, (3) determination of threshold values of character parameter attributes and (4) identification of a simulated PND character pattern. In Stage 2, the conduction procedure includes (1) Establishment of the experimental interface of simulated navigation display, (2) development of experimental programming, (3) experiment of PND character optimum representation and (4) induction of experimental data analysis. As to Stage 3, the procedure includes (1) training of the back-propagation neural network, (2) simulation of the back-propagation neural network and (3) representation of the optimum display.

3   DEFINITION OF EXPERIMENTAL PARAMETERS AND SIMULATED DISPLAY

The research first deals with the definition of two display interfaces for further experiment design: (1) choice of an appropriate navigation display pattern from current navigation systems and (2) transformation of the chosen pattern into a simulated navigation display. The research then proceeds to the data collection regarding the color combination between character images and the corresponding backgrounds. Some lower legibility color combinations are eliminated based on the results of a legibility questionnaire. The retained data of the color combinations are categorized with the concept of hierarchical clustering analysis. The best legibility of color combination on each category is selected as a representative color combination for experiment. The research also investigates character image representation patterns from current PND displays and induces the most optimum areas of character images representation for an experimental PND display. Note that the threshold values of character parameter attributes are then identified to determine the minimum and the maximum adjustable values of character images for experiment.

Current in-vehicle navigation systems have at most four layers of interface. Each layer involves a variety of functions. A total of 61 GPS functional items including common and individual items from four brands (Garmin, Mio, TomTom and Holux) of 67 marketed in-vehicle navigation systems were collected. According to the investigation of navigation displays from four brands of PND displays, the current navigation displays are classified into three types (1) simultaneous representation of simulated scene and electronic map and uniformly distributed representation of character images, (2) representation of electronic map on the middle and character images on the top and bottom and (3) representation of electronic map on the left and character images on the middle, upper right and lower left. Because Type 2 of the navigation display is frequently used by the four brands of products (about 46 product samples from 67 totally collected product samples; 46/67=0.68.7), the research then chose Type 2 as a reference for the development of an experimental simulation display. A representative Type 2 of the navigation display is shown in Figure 1.

Figure 1 An example of Type 2 navigation display

To develop a simulated navigation system for the experiment to recommend the optimum representation of character images and the corresponding background colors, the research needs to determine row distances between lines of characters,

distances between characters, sizes of characters and fonts of characters. About 48 pairs of color combinations related to character images and the corresponding backgrounds are designed as a questionnaire and distributed to the tests subjects to evaluate the respective legibility. The result of 42 valid questionnaires indicated 26 pairs of color combinations having higher legibility that were selected for the experiment consideration. Because 26 pairs of color combinations are still relatively too many, the research proceeded with a questionnaire that is based on five evaluation criteria (1) personal preference, (2) appropriateness of color scheme, (3) aesthetic sense of color scheme, (4) personal view custom and (5) comfort of reading. The hierarchical clustering method is applied to group the 26 pairs of color combinations based on 37 valid questionnaires (Jobson, 1992; Sharma, 1996). Note that the statistical software SPSS 12.0 is used and 10 groups of color combinations are determined as illustrated in Table 1.

Table 1 Groups of color combinations of character image and corresponding background

Cluster	No.	FT-r	FT-g	FT-b	BG-r	BG-g	BG-b	Score	Match colors
1	40	239	155	160	255	0	255	3.86	
1*	41	252	6	130	121	251	6	4.00	
1	28	126	2	128	250	128	3	3.93	
1	3	3	7	127	124	244	9	3.64	中文 ABC
1	26	128	0	0	0	252	251	3.93	
1	19	2	256	3	1	3	123	3.93	
1	20	4	252	2	121	2	2	3.71	
2*	13	0	1	2	250	251	123	5.86	中文 ABC
3*	6	6	3	127	244	254	251	6.43	中文 ABC
4*	11	3	5	2	2	255	3	4.29	中文 ABC
4	36	121	120	249	255	255	4	3.50	
5*	2	133	6	3	5	129	247	5.00	中文 ABC
5	25	124	2	5	1	125	124	3.57	
6*	16	0	0	252	126	123	124	4.57	中文 ABC
7	46	255	130	127	3	131	2	5.07	中文 ABC
7*	14	0	0	122	251	122	120	5.14	
8	39	129	250	129	128	0	4	5.07	
8*	48	252	251	131	4	132	255	5.29	中文 ABC
8	23	0	253	251	0	0	0	5.07	
9	9	129	9	11	52	197	242	4.64	中文 ABC
9*	8	8	4	3	245	251	7	5.71	
10	7	4	7	1	251	132	3	5.43	
10	18	2	252	2	0	0	0	5.29	
10	17	0	126	249	250	252	3	3.71	中文 ABC
10	21	3	252	1	255	255	255	3.71	
10*	4	4	248	5	11	20	7	5.50	

In Table 1, the items of BG-r, BG-g and BG-b denote RGB values of background colors, while the items of FT-r, FT-g and FT-b denote RGB values of character colors. To define adjustable ranges of character image blocks for an interactive interface of the simulated in-navigation display, the research collected 30 samples regarding their threshold values of row distances between lines of characters, distances between characters and sizes of characters including PNDs, cell phones, MP4 (Media player 4), PDAs (Personal Digital Assistant), magazines and Web pages. Based on the investigation result, the research determined the threshold values for the respective character parameter attributes as illustrated in Table 2. Based on 46 Type 2 product samples, the research chose 8 blocks for the simulated character image experiment as shown in Figure 2. In Figure 2, eight blocks of character image representation are named as: (A) emergency information block, (B) area location block, (C) words of "expected arriving time," (D) digital numbers block indicating expected arriving time, (E) main functional menu block, (F) current driving road name block, (G) words of "turning distance" and (H) digital numbers block indicating remaining meters/feet of the turning distance.

Table 2 Determination of threshold values for character parameters of simulated display

Threshold	Unit	Point (pt) max	min	Millimeter (mm) max	min
Line spacing		25.0	1.0	8.81	0.35
Character spacing		22.0	0.4	7.70	0.13
Font size		42.0	5.0	14.78	1.76

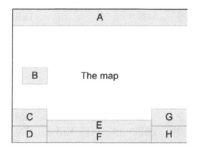

Figure 2 Block allocation of character images on an experimental simulation display

4   CONDUCTION OF THE EXPERIMENTAL DESIGN

To do an optimum display experiment for the evaluation of the PND character images, the research designed an interface including two screens for the operation of test subjects. The interface allows the test subject to make self adjustments based on his or her viewing preference and legibility. When making an adjustment from the computer screen, the test subject can view the simulated navigation display simultaneously in response to the adjustment. Note that the research used the computer software Microsoft Visual basic 6.0 to build the interface of the experiment and Microsoft Excel to store experimental data files. As mentioned before, the performance of the experiment consisted of four parts: (1) setup the experimental interface, (2) development of experimental programming, (3) experiment of PND character optimum representation and (4) induction of experimental data analysis.

Based on the chosen experimental simulation display shown in Figure 2, the research developed a simulated picture as well as a frame appearance on the simulated navigation display to obtain a lifelike reality of the display as shown in Figure 3. As to the adjustable control interface of the navigation display, the research used Microsoft Visual basic as a programming language. There are three interfaces: (1) homepage and information about tested subjects, (2) experimental operation, and (3) data recording.

Figure 3 Navigation display of the experiment simulation

1. Interface of homepage and information about tested subjects

It is the first display of the experiment interface. The display showed the research title and PND explanation on the top and 15 questions including basic data of tested subjects, vision condition and PND related concepts on the bottom.

2. Interface of experimental operation

When the tested subject finishes answering the questions on the first page, the system will proceed to the experiment operation display. The interface of the experimental operation includes four adjustable control areas: (1) selection area of character image blocks, (2) selection area of character fonts, (3) selection area of colors of character image and background and (4) selection area of character parameter attribute values. The developed interface is shown in Figure 4.

3. Interface of data recording

When the tested subject finishes selection and adjustment of the experiment operations, he or she proceeds to push the button of "viewing experimental data" on the bottom of Figure 4 and get into the interface of data recording. In the interface of data recording, the display has 8 blocks (A, B, ..., H) and each block records five adjusted data: font of character, color planning, character size, distance between rows and distance between characters. When the tested subject completes the 8 block data, he or she can then pushes the button of "data storage" to store the experimental data. Note that the storage, retrieval and update of experimental data are directly linked with the software of Microsoft Excel.

To help identify optimum representation of PND character images, the research developed a simulated driving environment associated with the simulated navigation display. The experiment apparatuses include a notebook computer associated with an additional screen. The conceptual layout is shown in Figure 5.

Experiment of PND character optimum representation			

PND layout		Font	

A		Chinese	English
B Please select the block		細明體	Arial
C	G	標楷體	Times New Roman
D E F	H	正黑體	Calibri

Character color planning

中文Chinese	中文Chinese	中文Chinese	中文Chinese
中文Chinese	中文Chinese	中文Chinese	中文Chinese
中文Chinese	中文Chinese		

Character parameter attributes

Font size		24	pt
Line spacing		45.8	mm
Character spacing		3.92	mm

Figure 4 Interface of experimental operation

The two screens have different functions for different stages and need to switch the displays to and fro to complete one cycle of experiment. A total of 48 college students including 25 males and 23 female who have basic capability of operating a computer and experience of driving a vehicle are considered as tested subjects. Note that the tested subjects have different educational fields, vision conditions, experience of driving and use of PNDs. In conducting the experiment, the tested subject first enters the experiment operation of left screen and at the moment the right screen maintains a static navigation display. When the tested subject finish adjusting the operational controls, the left screen will be changed to a simulated road driving display and the right screen will be activated to a dynamic navigation display. The experimental data including 8 blocks of attribute values of five character parameters that each tested subject adjusted are simultaneously recorded in the tables of Microsoft Excel 2007 software. These data will be forwarded to the back-propagation neural network training and simulation to search for optimum character image representation.

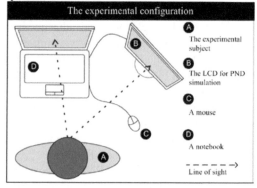

Figure 5 A conceptual diagram of experiment layout

5 APPLICATION OF THE BACK-PROPAGATION NEURAL NETWORK

Because the concept of neural network can be a multi-input and multi-output problem solving system, the research uses the back-propagation neural network to infer the optimum character image representation. To fit for the back-propagation neural network analysis, the collected data are normalized to the values between 0 and 1. In building the model of back-propagation neural network, the research considers 14 characteristic data of 48 tested subjects and 5 parameter attributes of character images as training inputs and outputs, respectively. As such, the designed back-propagation neural network model has 14 input neurons, 5 neurons in an output layer and 10 neurons in a hidden layer. Because here are 8 blocks of character images representation in the simulated navigation display, the research develops 8 respective network models for training. Note that the research uses the software MATLAB 7.0 to construct and train the back-propagation neural network.

To explore the feasibility of using back-propagation neural network in predicting the optimum representation of character images for the in-vehicle navigation system, the research chooses two user groups based on the inputted data of 48 tested subjects to process the simulation. The first sample group is defined as 25 to 30 year-old males; while the second sample group is defined as 20-25 year-old females. After finish the simulation process, the system will indicate the optimum attribute values of the character parameters according to the termination rule. However, the optimum attribute values obtained from the simulation are normalized values and will need to be transformed into the original type of values. The recommended character image representation for group 1 is shown in Figure 6.

Figure 6 Recommendation of optimum character representation for group 1 users

6 CONCLUSIONS

The research proposed an integrated approach of using the concept of neural network in the in-vehicle navigation system to help determine the optimum representation of character images. 67 marketed portable navigation devices were collected and analyzed to induce the most popular navigation display for further experiment and simulation. The research also collected 48 pairs of color combinations related to character images and the corresponding backgrounds and remained 26 pairs. The 26 pairs of color combinations are classified into 10 categories. The threshold values of three character parameters including row distances between lines of characters, distances between characters and sizes of characters are determined. To ensure the experiment is in a realistic scene, an experimental environment was built to simulate driving condition and in-vehicle navigation operation. The data collected from the experiment were then forwarded to the process of training of the back-propagation neural network. The data of two groups of tested subjects were chosen to the simulation of back-propagation neural network. Finally, the recommendations of the optimum representation of character images for two groups of tested subjects were presented.

It is expected that the results will provide designers with an effective procedure on building an optimum representation of character images in an in-vehicle navigation system for specific user groups.

ACKNOWLEDGMENTS

The authors are grateful to the National Science Council, Taiwan for supporting this research under the grant number NSC99-2221-E-343-008-MY3

REFERENCES

Darroch, I., J. Goodman, S. A. Brewster, and P. D. Gray. 2005. The Effect of Age and Font Size on Reading Text on Handheld Computers, *In Proceedings of IFIP Interact 2005*, Springer Perlin, Heidelberg, German, 253-266.

Jobson, D. 1992. *Applied Multivariate Data Analysis Volume II : Categorical and Multivariate Methods*, Springer-Verlag, New York.

Lee, et al. 2008. Effect of character size and lighting on legibility of electronic papers, *Displays*, 29: 10-17.

Lee, W.-C. and B.-W. Cheng. 2008. Effects of using a portable navigation system and paper map in real driving, *Accident Analysis and Prevention*, 40: 303-308.

Lee, W.-C. and B.-W. Cheng. 2010. Comparison of portable and onboard navigation system for the effects in real driving, *Safety Science*, 48: 1421-1426.

Mills, C. B. and L. J. Weldon. 1987. Reading text from computer screens, *ACM Computing Surveys*, 19(4): 329-358.

Pastoor, S. 1990. Legibility and subjective preference for color combinations in text, *Human Factors*, 32(2): 157-171.

Sanders, M. S. and E. J. McCormick. 1993. *Human Factors in Engineering and Design*, McGraw-Hill, 7th edition, New York.

Sharma, S. 1996. *Applied Multivariate Techniques*, John Wiley & Sons, Inc., New York.

Tomioka, K. 2006. Study on legibility of characters for the elderly-effects of character display modes, *Journal of Physiological Anthropology*, 26(2): 159-164.

CHAPTER 3

Designing an Adaptive Flight Deck Display for Trajectory-Based Operations in NextGen

Sylvain BRUNI, Andy CHANG, Alan CARLIN,
Leah SWANSON, and Stephanie PRATT

Aptima, Inc.
Woburn, MA, USA
sbruni@aptima.com

ABSTRACT

In the NextGen environment, pilots will be required to make more complex and strategic decisions than are required in current-day operations, such as self-separation management or route replanning for weather avoidance. However, no effective flight deck tool exists to support such activities. To address this gap, Aptima is developing a Trajectory-Based Operations Adaptive Information Display (TBO-AID), which is aimed at providing real-time, mission-critical, and time-sensitive decisional information. TBO-AID will incorporate innovative display techniques that specifically address (1) the unique information needs associated with conducting 4D operations; (2) the uncertainty and risks associated with weather and aircrafts being equipped, or not, with NextGen technologies; (3) the potential advantages gained through multimodal information presentation; and (4) the need for model-based, situationally-aware display adaptation to support information processing and decision making. TBO-AID is expected to increase the safety and efficiency of air traffic self-management and navigational deconfliction in NextGen operations.

Keywords: adaptive automation, model-based visualization, multimodal interfaces, uncertainty visualization, NextGen, trajectory-based operations

1 THE CHALLENGES OF TRAJECTORY-BASED OPERATIONS IN NEXTGEN

Managing trajectory-based operations (TBO) encompasses one of the primary means to enable airspace efficiency in the Next Generation Air Transportation System (NextGen). The vision of NextGen is one in which pilots will be responsible for following 4-dimensional (4D) trajectories while maintaining separation from other aircrafts and avoiding weather. Ongoing research focuses heavily on the infrastructure and procedures required to conduct 4D TBO. However, there is currently a lack of effective flight deck displays that support pilots who will be faced with the challenge of making more complex and strategic decisions than are required in current-day operations.

In response to this challenge, we created a Trajectory-Based Operations Adaptive Information Display (TBO-AID), a flight deck decision-support system for pilots to conduct 4D-trajectory management, including self-separation and weather avoidance. TBO-AID was build using innovative display techniques to specifically address the following system requirements and design mandates:

1. TBO-AID must fulfill the unique **information needs and requirements** associated with conducting 4D operations;

2. TBO-AID must provide means to **adaptively mitigate the uncertainty and risks** associated with weather and aircrafts equipped with and without NextGen technologies ("mixed-equipage environments");

3. TBO-AID must leverage the advantages gained through **multimodal information presentation**;

4. TBO-AID must take advantage of **model-based, situationally-aware display adaptation** to support information processing, decision-making.

TBO-AID is expected to provide increased safety and efficiency in air traffic self-management and navigational deconfliction in NextGen operations. Additionally, TBO-AID will help reach a number of NASA milestones related to assessing NextGen concepts for improved and novel visual, aural/speech, and multimodal interface capabilities to support 4D-TBO (NASA, 2008), as well as several JPDO-identified operational improvements (JPDO, 2008). It is anticipated that TBO-AID can equip both legacy and new commercial and military aircrafts, as well as flight deck suites, and full flight simulators. TBO-AID can also be utilized in a number of military domains dependent on advanced air traffic management.

2 DESIGN METHODS

Four principled design methods were employed to build TBO-AID and each specifically targeted one of the four system-level requirements and design mandates identified previously:

1. **Scenario-based iterative design**: leveraging subject-matter expertise through knowledge elicitation sessions, we developed a NextGen use case scenario highlighting the requirements associated with 4D-TBO, particularly in the context of weather events and mixed-equipage constraints;

2. **Adaptive graphical user interface design**: we designed a visual language for communicating uncertainty and risk to the pilot, and structured the visual display to permit efficient and effective adaptation of information display and automated support;

3. **Multimodal interface design**: we explored and designed innovative, multimodal interface concepts merging visual, aural, and tactile interactions; and

4. **Adaptive automation through model-based visualizations**: in order to drive display adaptation with respect to the situation and environment, we furthered the development of and integrated models of hazard and operator state that capture the context of operations and output model-driven information for TBO-AID and its embedded automation.

3   SCENARIO-BASED ITERATIVE DESIGN

In order to ground the development of TBO-AID operationally, Aptima implemented the early stages of the hybrid cognitive task analysis process (Tappan et al., 2011), a methodology that enables system designers to define requirements for systems that do not exist in any shape or form. Specifically, Aptima held knowledge elicitation sessions with a subject matter expert (SME), a commercial airline pilot, in order to develop a use case scenario that would include examples of the expected states and behaviors of a pilot—future user of TBO-AID—, and would describe expectations for the TBO-AID interface and its automated assistance technology. From this use case scenario, we extracted the functional, information and interaction requirements that allowed for graphical user interface design.

During the knowledge elicitation sessions, our SME provided such insights as the expected level of workload associated with various phases of TBO flights, appropriate speeds and altitudes at various waypoints en route to the runway, a realistic schedule of events given a weather conflict and a traffic conflict and the operational objectives relevant to 4D trajectory management. From this information, we developed a narrative that outlined the specific, envisioned interactions between the pilot and the TBO-AID tool, in order to maintain safe separation from the weather and traffic hazards, and meet operational objectives.

The resulting use case is based on an approach course in the Dallas-Fort Worth (DFW) terminal airspace, at night. Ownship, flight US438, and another aircraft, flight UA434, are flying the approach course into DFW. A third aircraft, flight AA152, is transitioning through the airspace without any intention to land at DFW. Flight AA152 does not include on-board Automatic Dependent Surveillance-

Broadcast (or ADS-B) technology, therefore its exact location in space is not known at all times. The ownship pilot must therefore rely on intermittent radar data to know or project AA152's location. In addition, two convective weather cells exist in the airspace. According to the pre-planned 4D trajectories of all NextGen-equipped aircrafts in the area, and to the weather data currently available, two upcoming conflicts need to be addressed by the pilot (Figure 1): a loss of separation (LOS) will occur between ownship and UA434 around the GIBBI waypoint, and ownship trajectory will intersect with one of the weather cells after the SILER waypoint. These conflicts are expected to delay the arrival of ownship at DFW.

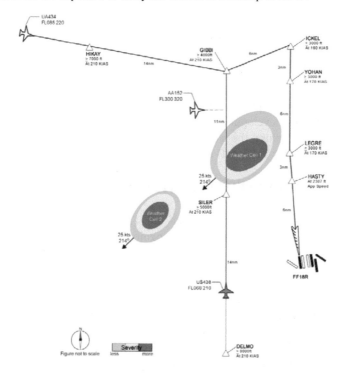

Figure 1 Map-based representation of the TBO-AID use case, showing two future issues with the scheduled 4D flight path of ownship: weather conflict between the SILER and GIBBI waypoints, and loss of separation with UA434 around the GIBBI waypoint.

To resolve these conflicts, the pilot will rely on the TBO-AID interface, which will be proactive in analyzing the current and projected states of the terminal area, in determining the hazard state of the conflict situations, in anticipating the pilot's state (in terms of workload, fatigue, etc...), and in providing multimodal support to propose feasible and safe courses of action to the pilot. The TBO-AID modes of intervention are presented in the subsequent sections.

The design of TBO-AID consisted in iterative cycles leveraging this use case: at each step, the design was checked against this narrative and its requirements to ensure that design choices satisfied the operational needs.

4   ADAPTIVE GRAPHICAL USER INTERFACE DESIGN

The core function of TBO-AID is to provide proactive support to the pilot, based on the current situation and projected environmental conditions. To achieve this function we opted for designing a display which adapts itself based on context. This adaptation occurs at two levels: first in terms of what and how information is displayed to the pilot, and second in terms of how much automated support TBO-AID provides. Based on existing capabilities developed for NASA in previous work (Carlin et al., 2010b) which we enhanced in the course of the design of TBO-AID (see "6 Adaptive automation through model-based visualizations"), we focused the development of the interface on defining a visual language for communicating uncertainty and risk to the pilot, and on structuring the visual display so TBO-AID provides the right information at the right time, based on four levels of alerts - advisory, caution, warning, and directive- and four stages of automation support (Carlin et al., 2010b).

Figure 2 highlights a few examples of display elements communicating uncertainty in TBO-AID: the uncertainty of location for aircrafts not equipped with ADS-B is represented by growing circles around the aircraft icon, the uncertainty linked to their speed and trajectory is coded using circles with diminishing intensity (both in terms of thickness and shade of grey), and the uncertainty on weather cells is displayed with a checkered pattern. The TBO-AID visual language also displays risk, with the traditional green/yellow/red gradient for increasing risk, and a specific coding of each alert level: in order to highlight the source of the alert, TBO-AID displays a colored outline around the icon of the element in the display at the origin of the alert. Advisory is coded with a white outline, caution with a yellow outline, warning with a steady red outline and directive with a blinking red outline.

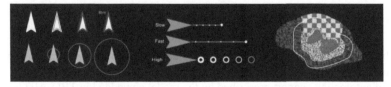

Figure 2 Examples of the visual language used to portray information and uncertainty in TBO-AID.

Figure 3 illustrates the overall structure of the TBO-AID visual interface, composed of four main elements: a library of preset geo-visualizations on the left, a 4D geo-visualization of the airspace in the center, with a timeline of waypoints and hazards at the bottom, and an automation support panel on the right side.

The 4D geo-visualization is an augmentation of the 3-dimensional cockpit display of traffic information (3D-CDTI) created at NASA Ames, which displays, in real-time, ADS-B, radar and weather data, as well as projected data (such as expected trajectories of aircrafts or displacement of weather cells). The pilot can manipulate this representation of the airspace by panning, tilting, and zooming, in order to build better situation awareness and a better mental model of the airspace and hazards. Above the 4D geo-visualization is basic flight information about ownship such as ground speed, altitude, wind speed and heading, ownship heading, and time and distance information about the next waypoint. Below the 4D geo-visualization is a timeline with markers indicating when waypoints will be reached, and when projected hazards are likely to occur. These markers are color-coded to indicate the alert level, as specified by the TBO-AID visual language and MIL-STD-411F, "Aircrew Station Alerting Systems" standard. As the alert levels increases over time, these colors will change accordingly.

The left side of the screen includes dynamic preset views, showing a top, side, and isometric angle. Each view dynamically updates to show real time information, and can be used in conjunction to mimic horizontal and vertical situation displays in the cockpit. Selecting a preset view will place it in the center of the screen, enlarged.

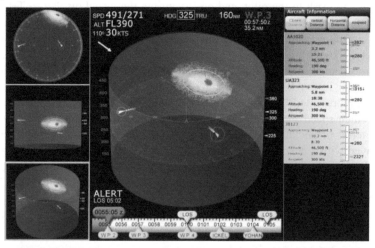

Figure 3 The TBO-AID visual display.

The panel to the right displays context-based and decision-oriented support information for the pilot (Figure 4). This panel is driven by the level of alert and stage of automation support prescribed by the model embedded in TBO-AID (see "6 Adaptive automation through model-based visualizations"). Four stages have been designed to match the four stages of information procession (Parasuraman et al., 2000). In Stage 1, the panel provides the pilots with important information about other entities in the airspace, such as an aircraft's call sign, altitude, heading, airspeed, and some information about its destination (i.e., waypoint name, distance, and time of arrival). A visualization of the altitudes based on an absolute scale can also be shown to give the pilot a graphical representation of this information. These aircrafts can be sorted by closest overall distance to ownship, or closest vertical/horizontal distance using the three large buttons along the top. In Stage 2, the aircraft at the origin of a future LOS is brought to the top of the stack and surrounded by a yellow border to indicate a caution-level status. At this stage, the automation performs some level of integration on the information to aid the pilot's mental processing. This can take the form of using ownship information as a frame of reference, and displaying information in relative terms, such as relative altitude, bearing, and vertical speed. In Stage 3, the automation suggests potential alternative routes, and provides high level decision criteria to aid the pilot in making the most appropriate selection, based on their criteria of interest, such as resulting time delays, fuel consumption, and the number of future potential conflicts when rerouting. Each reroute alternative also provides detailed instructions of each step in the new path, such as heading, airspeed, altitude, and duration. In the highest stage of automation, the information panel presents an automation-selected reroute plan to be implemented by the pilot. We refer to this stage of automation as Stage 3.5, since in a typical Stage 4, the automation would take over and implement the maneuver on its own (a level which we are not providing in TBO-AID).

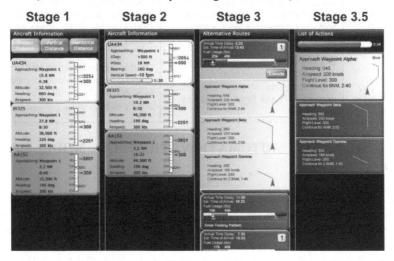

Figure 4 Examples of contextual information displayed for each stage of automation.

5   MULTIMODAL INTERFACE DESIGN

Beyond the visual display described in the previous section, TBO-AID was designed to incorporate two additional modalities to convey information and alerts to the pilot: as a starting point, we opted to include aural and tactile feedback as means to augment the visual display and cue the pilot for impending hazards. The aural and tactile feedback signals were designed to be sent respectively as beeps through the pilot's headset, and as vibrotactile pulses through a small wearable actuator, that could fit in a wristwatch, in a pocket, or on the pilot's seat or harness.

Additionally, a multimodal alerting strategy was defined, to modulate which mode would be involved, and how each mode would deliver the alerts to the pilot, based on alert level and selected stage of automation. As an initial implementation, the alert levels and stages of automation have been directly mapped 1-to-1 to multimodal interventions, and the strategy increases the alert intensity with increased urgency, in order to guide the pilot's attention and focus to the most important information to mitigate the hazard, as summarized in Table 1.

Table 1 The multimodal alerting strategy to cue pilots for varying alert levels

Alert Level	Auditory	Tactile
Advisory (Stage 1)	Single tone to indicate the presence of new information	None
Caution (Stage 2)	Double tone to indicate presence of integrated and processed information	Double vibration pulse to supplement auditory signal
Warning (Stage 3)	Double tone with increasing pitch (Hellier et al., 1993) as a cue to verbal command "select a new route" (Li Wei et al., 2009)	Double vibration pulse that correlates with the auditory signal (Salzer et al., 2011; Jones et al., 2008)
Directive (Stage 3.5)	Double tone with increasing pitch and faster period (Hellier et al., 1993) as a cue to upcoming verbal command "bank right" (Li Wei et al., 2009)	Double vibration pulse that correlates with the auditory double tone cue (Salzer et al., 2011; Jones et al., 2008)

6   ADAPTIVE AUTOMATION THROUGH MODEL-BASED VISUALIZATIONS

In order to drive display adaptation with respect to the situation and environment, TBO-AID was designed to integrate models of hazard and user state in order to select the best possible combination of visual, aural and tactile information for the pilot. Specifically, TBO-AID includes a planner that selects the best possible stage of automation (Parasuraman et al., 2000; Galster, 2003) for the automated support. The planner was created as an extension of the framework of (Carlin et al., 2010a) and (Carlin et al., 2010b) developed for NASA to derive alerting policies in the cockpit for NextGen environments.

Figure 5 summarizes the architecture of the adaptive automation. Proceeding from left to right, multiple hazards exist in the environment and result in alerts on the flight deck. The hazards themselves, and the sensor alerting systems, are external to the TBO-AID system. The alerts are issued to an Integrated System-User Model, where they are treated as evidence: from this evidence, a "state estimation module" estimates the actual hazards using a Dynamic Bayesian Network (DBN).

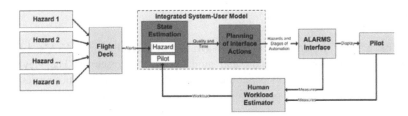

Figure 5 System architecture for the model-based adaptive automation.

This hazard state is combined with the estimated state of the pilot (described in the next section), to form a complete state estimate. This pilot and hazard estimate is fed into a "planning" module, which recommends the stage of automation for addressing the hazards. This recommendation is fed into an "interface" module that defines what information should be communicated to the pilot, and how.

During the design of TBO-AID, two critical enhancements to the model were made. First, the "hazard state" model was instantiated specifically to include weather and loss of separation hazards in the pilot's environment, as well as the uncertainty for both types of hazards.

Second, the "human workload estimator" module was augmented to enable the possibility to capture direct measurements from the pilot. Indeed, we identified and adapted the architecture of the "human workload estimator" module for a set of measures TBO-AID could capture in a future cycle of development, and leverage to enhance its ability to estimate pilot state. Such measures include physiological measures (such as pupillary response, blink rate, heart rate, or galvanic skin response), system-based measures (such as the number and complexity of pilot tasks, the number of tasks that compete for time or resources), and subjective measures (such as, for each pilot task, mental demand, physical demand, temporal demand, performance, difficulty, or frustration). Each of these measures is assigned a linear weight, which yields a single scalar representing an estimate of workload. This scalar is a parameter used by the "planning" module (Carlin et al., 2010a).

7   CONCLUSION AND FUTURE WORK

The development of TBO-AID resulted in the creation of several artifacts, including an operationally realistic use case scenario, a context-based model of environmental hazards, operator state, uncertainty, and alternative routes, and an adaptive, multimodal interface that provides decision-support capabilities to pilots for 4D-TBO management from the flight deck. The graphical interface of TBO-AID was implemented as a proof-of-concept on a tactile tablet and demonstrated at NASA Langley Research Center. Future work will include the integration of TBO-AID in a larger simulation system where the context-based models will be receiving simulator data. This demonstration will enable the showcasing of TBO-AID's ability to adapt, in real-time, the visual, aural and tactile displays, and the recommendations provided to the pilot.

32

ACKNOWLEDGMENTS

The authors would like to thank Dr. Kara Latorella of NASA's Langley Research Center for her support and guidance in this research. The views and conclusions presented in this paper are those of the authors and do not represent an official opinion, expressed or implied, of Langley Research Center or NASA.

REFERENCES

Carlin, A., Alexander, A. L., and Schurr, N. (2010a). Modeling pilot state in next generation aircraft alert systems. Presented at the MODSIM World 2010 Conference & Expo.

Carlin, A., Schurr, N., and Marecki, J. (2010b). ALARMS: Alerting and Reasoning Management System for Next Generation Aircraft Hazards. Proceedings of the 26th Conference on Uncertainty in Artificial Intelligence (UAI 2010). P. Grünwald and P. Spirtes (Editors). AUAI Press.

Galster, S. (2003). An examination of complex human-machine system performance under multiple levels and stages of automation. WPAFB, OH: Technical Report No. AFRL-HE-WP-TR- 2003-0149.

Hellier, E. J., Edworth, J. and Dennis, I. (1993). Improving auditory warning design: Quantifying and predicting the effects of different warning parameters on perceived urgency. *Human Factors, 35*(4), 693-706.

Jones, L. A. and Sarter, N. B. (2008). Tactile displays: Guidance for their design and application. *Human Factors, 50*(1), 90-111.

JPDO (2008). NextGen Integrated Work Plan - Appendix I - Operational Improvements (available at www.jpdo.gov/iwp/IWP_V1.0_Appendix_I_OIs.pdf)

Li Wei, C. & Wei Liang Kenny, C. (2009). An evaluation of icon, tone, and speech cues for detecting anomalies and auditory communications for onboard military vehicle displays. *HFES Annual Meeting Proceedings, 53*(17), 1076-1080.

NASA (2008). Airspace Systems Program, NextGen-Airspace Project, FY2009 Project Plan (available at www.aeronautics.nasa.gov/pdf/airspace_508.pdf)

Parasuraman, R., Sheridan, T. B., and Wickens, C. D. (2000). A model of types and levels of human interaction with automation. *IEEE Transactions on Systems, Man, and Cybernetics— Part A: Systems and Humans*, 30, 286–297.

Salzer, Y., Oron-Gilad, T., Ronen, A. and Oarmet, Y. (2011). Vibrotactile "on thigh" alerting system in the cockpit. *Human Factors, 53*(2), 118-131.

Tappan, J.M., Pitman, D.J., Cummings, M.L., and Miglianico, D. (2011). Display Requirements for an Interactive Rail Scheduling Display, *HCI International Conference*, Orlando, FL.

CHAPTER 4

Enhanced Audio for NextGen Flight Decks

Robert W. Koteskey[1,2], Shu-Chieh Wu[1,2], Vernol Battiste[1,2],
Elizabeth M. Wenzel[1], Joel Lachter[1,2], Durand R. Begault[1], Jimmy Nguyen[3],
Michael S. Politowicz[4], and Walter W. Johnson[1]
NASA Ames Research Center[1], San Jose State University[2], California State
University Long Beach[3], University of Michigan, Ann Arbor[4]
Robert.W.Koteskey@NASA.gov

ABSTRACT

Implementation of the Next Generation Air Transportation System (NextGen) will require shifting more roles to the flight deck. The proposed tools and displays for facilitating these added tasks primarily deliver information through visual means. This saturates an already loaded channel while perhaps underutilizing the auditory modality. This paper describes audio enhancements we have developed to compliment NextGen tools and displays, and reports on preliminary observations from a simulation incorporating these enhancements. Pilots were generally receptive to the broad concept, but opinions diverged regarding specific features, suggesting potential for this strategy, and that user defined settings may be important.

Keywords: NextGen, aurally enhanced flight deck, CDTI, synthetic voice

1   INTRODUCTION

It is envisioned in the Next Generation Air Transportation System (NextGen) that flight decks will take on some of the responsibilities traditionally associated with Air Traffic Control (ATC). Specifically, Air Traffic Management (ATM), once under the sole purview of ATC, will be increasingly integrated into the flight deck as the automation necessary to implement NextGen becomes operational. This shift in roles and responsibility on the flight deck will drive a need for new equipage and capabilities. For example, pilots will manage new tools and ATM procedures such as conflict detection and resolution (CD&R), and arrival interval management. They will also have the responsibility to reroute themselves around weather and other hazards. While emerging technologies would allow these new tasks to be entirely controlled on the flight deck, these new roles will likely impose additional challenges for pilots whose existing responsibilities are already demanding.

A primary purpose of this simulation study was to explore the use of audio technologies to mitigate some of the burdens placed on the NextGen flight deck. In the following sections, we will first discuss our view of some specific challenges posed by the new role of the NextGen pilot, followed by a summary of possible mitigations for these challenges through the use of sound as an information source. We will then discuss our concept of the aurally enhanced NextGen flight deck, followed by a description of a simulation that implemented these new technologies in a NextGen operational environment. We conclude with a general discussion of pilots' experiences with these technologies and implications for future development.

2   NEXTGEN FLIGHT DECK CHALLENGES

2.1   Delegation of ATM Responsibilities

One challenge faced by the NextGen flight deck concerns responsibility for separation. In the current National Airspace System (NAS), aircraft separation and sequencing is the exclusive purview of ATC. Pilots today have need of these services provided by the controller, and are obliged by regulation to use them. There is no display on the flight deck capable of facilitating these tasks, nor are today's pilots trained to accomplish them except in the narrow sense of maintaining visual, out-the-window separation in the vicinity of airports.

Under a number of proposed new applications (oceanic in-trail procedures – ITP, closely spaced parallel approach operations – CSPA, and flight deck interval management – FIM), limited delegation of separation responsibility will be placed on flight crews to maintain a safe distance from other aircraft. The implementation of this shared responsibility will require greater predictability which will be provided through the use of Trajectory Based operations (TBO), tighter required navigation performance (RNP), flight deck interval management tools, and onboard conflict detection and resolution (CD&R) tools, to name a few.

2.2   Increased Use of Data Link Communications

A related challenge arising from the delegation of separation responsibility is the expected increase in the use of Controller Pilot Data Link Communications (CPDLC). This increased use of data link will be driven by saturation of existing VHF radio networks as both the number of aircraft increases, and the number of coordination tasks between each flight deck and ATC increases. Data link communication offers a viable solution for this frequency congestion, but has some possible disadvantages. As data communications increase, the information available through an open-circuit with human talkers is eliminated. This potentially disrupts one of the primary means through which pilots obtain situational awareness (SA).

Indeed, in the results of several surveys administered to pilots on party line information, Pritchett, Hansman and Midkiff (1995) found that traffic and weather information obtained from radio party line communications was rated critical for maintaining SA. Pilots' subjective assessment of the importance of party line information was confirmed in a recent study by Boehm-Davis, Gee, Baker, and Medina-Mora (2010), who found that the use of data link not only reduced SA but also increased workload in some cases. Boehm-Davis et al.'s results again attest to the importance of party line information to pilots and suggest the need to replace this information where data link is implemented.

3   SOUND AS AN INFORMATION SOURCE

There is a large body of research that indicates multi-modal presentation, specifically aural enhancement of visual displays may be effective in improving human performance in applied settings (for a review, see Proctor & Vu 2010). For example, Ho and Spence (2005) assessed auditory cues as a means for capturing a driver's attention to relevant events. Auditory cues were presented either as semantically meaningful verbal cues (i.e., spoken words) or directional sonifications (i.e., the sound of squealing brakes). They found improved reaction time and accuracy in drivers' responses when aural cues were added to simulated driving scenarios. Ho and Spence attributed the effect to the ability of the aural cues to help orient visual attention to critical visual stimuli.

Additionally, three-dimensional aural augmentation of visual displays has been shown to successfully focus user attention to specific tasks and has proven to improve reaction times in locating visual targets. Begault (1993) showed a 2.2 second improvement in out-the-window visual acquisition of targets when target presentation was augmented by spatial 3D audio. Tannen, Nelson, Bolia, et al. (2000) found a significant reduction in head movement when visual search targets were accompanied by spatially corresponding audio cues. Veltman, Oving and Bronkhorst (2004) reported similar improvements in performance during pursuit and intercept tasks conducted in a simulated fighter cockpit.

In addition to its innately spatial nature, the sound itself can also convey information about the world. A small pebble hits the water with a splash; a large rock hits the water with a thump. Graver (1986) used the term "auditory icons" to describe those kinds of informative sounds. Graver suggested that the principles behind auditory icons could be used to provide supplemental, descriptive information about computer generated objects rather the arbitrary and metaphorical relations often employed in designing sound alerts in computer interfaces.

In environments involving multiple speech sources, Ericson, Brungart and Simpson (2004) found that subjects were better able to attend to, and differentiate content if different voices were spatially separated. They also found notable effects by differentiating the multiple speakers by gender and by sound level.

4   AN AURALLY ENHANCED NEXTGEN FLIGHT DECK

The aforementioned research shows that sounds when used appropriately can not only deliver messages without disrupting ongoing visual processing but also convey spatial information. On NextGen advanced flight deck displays where traffic and aircraft system information abounds, we believe pilots can benefit greatly from carefully designed auditory enhancements to visual stimuli. In the following we describe several proposed NextGen flight deck tools and the types of audio enhancements that could potentially improve the efficiency of their use. In addition, we describe a potential solution to the loss of party line information through the use of synthetic speech.

4.1   Advanced NextGen Flight Deck Tools

The platform on which we implement proposed NextGen flight deck tools and audio enhancements is the Three-Dimensional Cockpit Situation Display (3D CSD), an advanced version of a cockpit display of traffic information (CDTI). This experimental prototype was developed by the Flight Deck Display Research Laboratory at NASA Ames Research Center (Granada et al. 2005). The CSD provides displays of traffic, weather, and flight trajectory information. Several tools supporting ATM capabilities are provided. Among them are the Route Assessment Tool (RAT), a graphical tool used for in-flight trajectory modifications, and a Conflict Detection and Alerting (CD&A) tool which graphically alerts the crew to conflicts with other aircraft. In conjunction with the RAT, the CD&A tool assists the crew in finding conflict free routes. The CSD also has a Spacing Tool that provides automation (with a graphical user interface) for performing interval management. Using the tool a pilot can select a lead aircraft that is then followed on the arrival. The tool calculates a speed which is delivered to the aircraft autothrottle. In addition to the 3D CSD, an enhanced data link interface was provided to allow data link to be used for all communications.

4.2   Proposed Audio Enhancements

Figure 1 illustrates our proposed audio enhancements to the CSD tools and how they are designed to be delivered to the crew. These enhancements fall into the five categories summarized below:

Audio Display Scheme

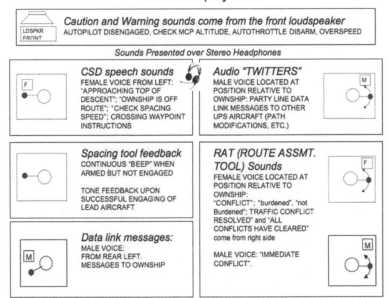

Figure 1 Auditory enhancements to CSD tools and data communications including delivery mode.

1. **Audio feedback on tool usage**: In a multi-tasking environment, users can be distracted and forget to resume interrupted tasks. In those cases, providing aural feedback with tool usage to signify step completion or incompletion can help users stay on task (e.g., Brewster 1998). We propose adding such feedback to the spacing tool which requires the pilot to perform a sequence of actions on multiple objects to complete the task.

2. **Spatially localized voice messages accompanying visual alerts**: Previous research suggests that spatially localized sounds can help users detect and orient attention toward visual events in the world (Begault 1993; Ho & Spence 2005). We propose to capitalize on this finding by adding spatially localized voice messages to traffic conflict alerts that have only been represented visually in the past. Synthetic voice messages carrying information on the conflicts are presented along with visual alerts. They are heard in the direction of the intruding aircraft in the physical airspace. The degree of urgency is conveyed by the gender of the voice: female for non-urgent and male for urgent.

3. **Voice reminders of procedural compliance**: Maintaining compliance with flight procedures often requires pilots to routinely monitor numerical changes in different parts of the displays (e.g., distance from waypoints). In emergencies, routine monitoring may be disrupted leading to procedure noncompliance that will not be corrected until visual monitoring is resumed. We propose to use synthesized voice messages as reminders to

nominal and off-nominal indications so that pilots can be notified of noncompliance as soon as it occurs even in the absence of monitoring.

4. **Voice augmentation for data link messages**: It can be anticipated that increasing the use of data link messages will lead to more head down time for the crew. This disadvantage of data link usage can be potentially remedied by narrating DataText information from uplinked data messages.

5. **Voice replacement of party line information**: We propose the use of an aural enhancement that we call audio "twitter" to replace the loss of party line information due to the use of data comm. The idea is for the system to generate synthetic voice messages approximating radio transmissions that might occur as the result of changes in the states of one aircraft (e.g. a course change, or the start of a descent), and to allow pilots of other aircraft to monitor those changes through a subscription-based mechanism. In theory, the selection of which aircraft to monitor could be initiated by the pilot or pre-determined by the airline or other stakeholders. As an initial demonstration, we propose to select only ownship's lead aircraft in an arrival interval management pair, and all other aircraft ahead on the arrival that fall within certain range limits. Audio feeds are triggered when a subscribed aircraft changes route (as in a weather deviation or a conflict resolution) or begins descent.

5   SIMULATION

To demonstrate the utility of these proposed audio enhancements, we implemented them in a research simulator which incorporated the CSD display with the aforementioned flight deck tools. We contrasted this *enhanced audio* environment with a *minimal audio* environment representing the current-day flight deck aural environment. In both audio environments the simulator modeled other standard transport aircraft controls and displays.

5.1   Method

5.1.1   Participants

Ten transport category aircraft pilots familiar with Flight Management Computer (FMC) operations and glass cockpit procedures participated in the simulation and were compensated for their time at a rate of $25/hr.

5.1.2   Apparatus

The simulation utilized a mid-fidelity fixed-based research simulator consisting of a two person cab configured as an advanced transport category flight deck. Four 50 inch plasma displays presented out-of-window views of weather and traffic. Most flight deck controls and auto-flight capabilities were modeled to provide an immersive and physically realistic environment to the participants. Pilots occupied the left (captain) seat of the simulator and used an autopilot mode control panel (MCP), a flight management system (FMS) with a control display unit (CDU), and a computer mouse to interact with simulation software and displays.

A real-time signal processing engine referred to as CrewSound supported the generation of audio and voice cues (Begault et al. 2010). Hardware included a fully configurable 24-channel audio interface (MOTU 24 I/O core system PCIe), multiple loudspeakers, and supra-aural stereo aviation headsets with active noise cancellation and a customized push-to-talk capability (Sennheiser HMEC46-BV-K). Software included a custom graphical user interface enabling up to 24 channels of synthesized speech messages and/or non-speech alerts.

Airspace and traffic for the simulation was generated using the Multi-Aircraft Control System (MACS) software (Prevot 2002). This software package allows creation of an accurate three dimensional airspace model which can be populated by a variety of controllable simulated aircraft.

5.1.3   Design and Scenarios

Eight 20-minute traffic scenarios, four in each of the two audio conditions were presented to allow pilots to experience these enhancements in a variety of situations. Pilots flew aircraft arriving into Louisville International Airport (SDF). Each scenario was a 20 minute segment of a nominal arrival trajectory that began in the en route environment approximately 90 minutes west of SDF at a cruising altitude of 33000 to 35000 feet. The route included a planned optimal profile descent (OPD) into the airport. Convective weather cells were placed at a location about 150 NM from the starting point along route. To maximize the opportunity for exposure to a variety of audio events during different phases of flight, each of the four scenarios in a given audio condition started at a different point on the nominal trajectory (150 NM from weather, 50 NM from weather, passing abeam weather to the north or south, and past the weather near the top-of-descent). The two audio conditions were counterbalanced across participants.

Planned traffic conflicts were engineered to occur with the experimental aircraft at specific times during a particular scenario, generated either by the simulation software or by confederate pseudo-pilots according to scripted timing. Confederate pseudo-pilots also handled additional air traffic to bring the total traffic load up to approximately 1.5 times of current day traffic.

5.1.4   Procedures

The simulation was conducted during ten, daylong sessions each lasting eight hours. The pilots were instructed that their primary responsibilities included guiding their simulated aircraft though an OPD, maintaining a temporal interval from an assigned lead aircraft, and avoiding other aircraft and hazardous weather exclusively through use of onboard tools (i.e. without contacting ATC). A confederate "co-pilot" was present to aid the subject pilot as necessary.

In both audio conditions, the pilot received an arrival clearance message that included a scheduled time of arrival, a speed profile, and an interval management clearance related to an assigned lead aircraft. During pilot initiated route changes for weather or traffic conflicts, communication with ATC was discouraged.

After completing the scenario runs, pilots completed a 79-question post-simulation questionnaire that asked them to use a 5-point scale to rate a variety of aspects of the simulation such as training, displays, tools, controls and tasks, as well as provide written comments on these topics. Space for written comment was provided for feedback on areas not directly queried.

Finally, pilots also participated in a twenty to forty minute verbal de-briefing. One interview was lost due to recording equipment malfunction.

6   OBSERVATIONS AND DISCUSSION

Due to the wide range of audio enhancements and the areas to which they were applied, it was difficult to access their utility individually. However, opinions expressed by the participants through their questionnaire responses offer a glimpse of how well these audio enhancements facilitated the use of various tools and compliance with procedures as a whole. Ratings from the questionnaire showed that participants, in general, had positive impressions of the auditory display and the types of messages that were presented. The use of synthesized text-to-speech messages for alerting, merging and spacing, data link, and situational awareness ("twitter") messages was rated highly (4.4/5). Participants expressed clear preference for the text-plus-speech display over the text-only one. Participates also responded positively to the general quality and acceptability of the acoustic simulation environment and audio presentation (4.3/5). One area that did not fare well and perhaps requires further refinement in its implementation was the use of spatial location cues (3.8/5).

Opinions volunteered in written comments echoed rating responses, favoring the concept of enhanced voice information and audio cueing. A typical respondent stated: "Some radio chatter can help with w/x avoidance and situational awareness." However, there were exceptions to this general trend. Specifically, the overlap of audio cues and recurring alerts in this simulation were mentioned as areas needing improvement, with one subject stating: "I would prefer only one message at a time."

While a majority of opinions favored the enhanced audio concept in general, there was not clear consensus on what form that should take. Divergent views of participants emerged during conversations during post-simulation debriefing sessions. Seven of the ten subjects seemed to agree with the notion that some form of audio enhancement will be necessary to augment the advanced NextGen flight deck. Two participants specifically liked the "twitter" information that was received from other arrival aircraft. One participant listed the audio feedback paired with the spacing tool and the RAT as the best of the enhancements, stating that it saved him from making sequencing mistakes with those tools on more than one occasion.

However, two subjects preferred to just use advanced visual display to gain and maintain situation awareness. Their comments suggested that too much audio was actually detrimental to their level of performance. Both stated that, to a great extent, the advanced aspects of the CSD made up for any audio information they were potentially missing by not having the open radio circuit available. They acknowledged that audio enhancements would be needed, but that the ones they were exposed to in this study may not have been ideal for them.

This divergence of opinion concerning the utility of specific features, and in two cases the entire concept as presented, suggests that a high degree of user-selected flexibility in flight deck audio presentation may be necessary. As with current flight deck visual displays, users may need the ability to fit the audio "display" to their situation. This idea is borne out by the answers volunteered when the subjects were asked about a future ability to filter or select what they could listen to with the "twitter" function. Eight of the subjects felt that customizing the audio presentation to their own preferences and needs would be highly desirable. Four expanded their answer to include the ability to adjust features for the entire audio suite.

Parallels to existing visual alerting systems, such as EICAS (Engine Indication and Crew Alerting System), were drawn by two participants. Both the logical hierarchy, and the ability to cancel and recall an alert or synthesized voice message were listed as features that would be desired as future audio enhancements. In addition, two comments suggest that concurrent and integrated audio and visual alerts were quite helpful when presented correctly. The visual flashing of a target on the CSD while receiving an audio "twitter" about that aircraft was suggested as a good example of this type of integration during the current study.

When asked about audio features the subjects would like to have in the future, pilots suggested that they would like information about changes in airport status, changes in expected routing, and changes in enroute weather. Additionally they listed other considerations such as holding or diversions in progress a their airports of interest. It was further suggested that these aural notifications would be particularly useful for information that was beyond the range of the visual display.

6.3   Final Thoughts

The present simulation represents a preliminary effort in exploring the use of auditory technologies to aid pilots in managing their new roles and responsibilities on the NextGen flight deck. Subjective evaluations from the pilots showed that they were receptive to the audio enhancements in general and found many of the individual features to be highly useful. While pilots preferred the enhanced audio concept in general, their evaluations of specific tools varied widely. Their diverse preferences suggest that future audio environments can benefit from allowing user-customized settings to meet individual needs and preferences in varying situations.

ACKNOWLEDGMENTS

This work was supported by NASA's Aviation Safety Program, Integrated

Intelligent Flight Deck Technologies (IIFDT) Project and Airspace Systems Program, Concepts and Technology Development Project. Jimmy Nguyen, now at Gulfstream, was supported in part by a NASA cooperative agreement (NNX09AU66A) with the Center for Human Factors in Advanced Aeronautics Technologies at California State University – Long Beach, awarded as a part of NASA Group 5 University Research Center Program by NASA's Office of Education. Michael Politowicz, now at GE Aviation Systems, was supported by an internship from NASA's Undergraduate Student Research Program. We thank Jonathan Luk, Summer Brandt, Sarah Ligda, Dominic Wong, George Lawton, Tom Quinonez, and Patrick Cravalho for their help as well.

REFERENCES

Begault, D.R. et al., 2010. Applying Spatial Audio to Human Interfaces: 25 Years of NASA Experience. In *Proceedings of the 40th International Conference: Spatial Audio: Sense the Sound of Space*. New York: Audio Engineering Society. Available at: http://www.aes.org/e-lib/browse.cfm?elib=15546.

Begault, D.R., 1993. Head-Up Auditory Displays for Traffic Collision Avoidance System Advisories: A Preliminary Investigation. *Human Factors: The Journal of the Human Factors and Ergonomics Society*, 35(4), pp.707–717.

Boehm-Davis, D.A. et al., 2010. Effect of Party Line Loss and Delivery Format on Crew Performance and Workload. In *Human Factors and Ergonomics Society Annual Meeting Proceedings*. pp. 126–130.

Brewster, S.A., 1998. Sonically-enhanced drag and drop. In *Proceedings of the International Conference on Auditory Display*. pp. 1–7.

Ericson, M.A., Brungart, D.S. & Simpson, B.D., 2004. Factors That Influence Intelligibility in Multitalker Speech Displays. *The International Journal of Aviation Psychology*, 14(3), pp.313–334.

Gaver, W.W., 1986. Auditory icons: Using sound in computer interfaces. *Human-Computer Interaction*, 2(2), pp.167–177.

Granada, S. et al., 2005. Development and Integration of a Human-Centered Volumetric Cockpit Situation Display for Distributed Air-Ground Operations. In International Symposium on Aviation Psychology.

Ho, C. & Spence, C., 2005. Assessing the effectiveness of various auditory cues in capturing a driver's visual attention. *Journal of experimental psychology: Applied*, 11(3), p.157.

Prevot, T., 2002. Exploring the many perspectives of distributed air traffic management: The Multi Aircraft Control System MACS. In *Proceedings of the HCI-Aero*. pp. 149–154.

Pritchett, A., Midkiff, A. & Hansman, R.J., 1995. Party line information use studies and implications for ATC datalink communications. In *Digital Avionics Systems Conference, 1995., 14th DASC*. pp. 26–31.

Proctor, R.W. & Vu, K.-P.L., 2010. Cumulative Knowledge and Progress in Human Factors. In *Annual Review of Psychology*. Palo Alto: Annual Reviews, pp. 623–651.

Tannen, R.S. et al., 2000. Adaptive integration of head-coupled multi-sensory displays for target localization. In *Proceedings of the Human Factors and Ergonomics Society Annual Meeting*. pp. 77–80.

Veltman, J.A., Oving, A.B. & Bronkhorst, A.W., 2004. 3-D Audio in the Fighter Cockpit Improves Task Performance. *The International Journal of Aviation Psychology*, 14(3), pp.239–256.

CHAPTER 5

Aviation Display Dynamics and Flight Domain in Pilot Perceptions of Display Clutter

James T. Naylor[1], David B. Kaber[2], Sang-Hwan Kim[3], Guk-Ho Gil[2] and Carl Pankok[2]

[1]U.S. Army, Technology Applications Program Office,
Ft. Eustis, VA 23604, USA
[2]Edwards P. Fitts Department of Industrial and Systems Engineering,
North Carolina State University, Raleigh, NC, 27695-7906, USA
[3]Department of Industrial and Manufacturing Systems Engineering,
University of Michigan-Dearborn, Dearborn, MI, 48128-1491, USA

ABSTRACT

This research involved a meta-analysis of data sets from three prior empirical studies on pilot perceptions of clutter in aviation displays. The objectives were to assess the affects of cockpit display dynamics and aviation domain (fixed vs. vertical takeoff and landing) on perceived clutter based on overall ratings and a multi-dimensional score collected during simulated flight tasks. Comparisons were made of observations on pilots with similar levels of experience within domain using displays (head-down or head-up) with similar features in similar segments of landing approach scenarios. Results revealed both display dynamics and the flight domain to be significant in perceived clutter and the need for such factors to be incorporated in models and measures of clutter. The findings also emphasize the need for designers to consider how display features influence human information processing and how specific visual characteristics of displays may lead to perceptions of clutter and potential performance problems.

Keywords: aviation human factors, cockpit display design, perceived clutter, pilot performance

1   RESEARCH OVERVIEW AND MOTIVATION

The present study involved a meta-analysis of multiple data sets generated by a three-year program of NASA-sponsored research aimed at developing measures and models of aviation display clutter (MMADC). During the first year (Y1), four expert test pilots evaluated static images of head-up cockpit displays (HUD), as a basis for prototyping a comprehensive subjective measure of clutter. They provided overall clutter ratings (CRs) as well as ratings of specific displays characteristics/qualities (e.g., density, redundancy, salience, etc.). A collection of six different rating scales resulted from this study targeting those qualities pilots considered to be most influential in perceived clutter (Kaber et al., 2007). During the second year (Y2), the multi-dimensional subjective measurement technique was applied in a high-fidelity Boeing (B)-757 flight simulator with a sample of 24 current commercial airline pilots with varied levels of experience in order to assess the sensitivity, diagnosticity and reliability of the measure in evaluating various display conditions. Clutter scores (CSs) were determined for each test HUD and pilot flight control performance was recorded. The relationship of perceived clutter to performance was examined (Kaber et al., 2008). The Year 3 (Y3) study was conducted using a high-fidelity vertical take-off & landing (VTOL) aircraft simulator with head-down displays (HDDs). Sixteen test subjects with varying levels of flight experience participated. The intent of this study was to assess the validity of the previously developed clutter measure for application to similar cockpit display concepts but under different flight scenarios and with a different vehicle type (Kaber et al., 2009). Table 1 summarizes the research methods used across the studies.

No direct comparison has previously been made of the results of these studies. Since the display configurations and flight scenarios were similar for the Y1 and Y2 studies, there was an opportunity to assess the effect of display dynamics on pilot perceptions of clutter. In addition, since comparable displays were used across the Y2 and Y3 studies, there was an opportunity to assess the effect of aviation domain on pilot perceptions of clutter. It was expected that this research would identify additional aviation system and task factors that might be influential in pilot perceptions of display clutter. Developing comprehensive measures and models of display clutter is important to the aircraft conceptual design phase in order to identify display configurations and task conditions under which confusion of visual imagery might degrade the pilot performance. One previous study found that visual display information overload can lead to increased pilot visual search times (Wickens et al., 2004). In addition, other work has revealed a significant correlation between the perception of clutter and pilot performance (Kim et al., 2011), specifically increased cockpit display clutter was found to have a negative impact on pilot flight control (localizer deviations, glide slope deviations, etc.).

Table 1 Research Methods Used Across the three MMADC studies

	Y1 Study	Y2 Study	Y3 Study
Display Dynamics	Static HUD	Dynamic HUD	Dynamic HDD
Aircraft Type	Fixed-wing	Fixed-wing	VTOL
Participants	4 experts pilots	24 pilots (in total): - 6 pilots for each of low, medium, and high flight experience -6 expert pilots with HUD experience	14 pilots (novice in VTOL)
Flight Scenario	Instrument Landing System (ILS) approach into Reno, NV (KRNO), consisting of level flight and descent.	ILS approach into KRNO, consisting of level flight and descent.	ILS approach, consisting of descent and hover phases.
Independent Variables	Display features	Display features & conditions (high, med, low clutter), flight workload, and flight segment	Display conditions (high, med, low clutter)
Dependent Variables	CR	CR, CS, Flight performance, Workload	CR, CS, Flight performance

1.1 Defining and Measuring Clutter

Prior research has defined clutter as, "an unintended consequence of a displaying imagery that obscures or confuses other information or is irrelevant to the task at hand" (Kaber et al., 2008). Different approaches have been defined for clutter measurement, such as quantification of target-to-background contrast (Aviram & Rotman, 2000) with validation based on human subjective clutter ratings. Ewing et al. (2006) and Rotman et al. (1994) focused on measurement of local display density as an indicator of clutter and related this to human performance in visual search. However, other measures have been based on the degree of global display clutter (Rosenholtz et al., 2005). The main limitation of these methods is that they simplify the concept of clutter to one or two physical display characteristics. This limitation motivated the development and assessment of the multi-dimensional measure by Kaber et al. (2007) or clutter score.

Of particular relevance to the present study, Kaber et al. (2007) found pilot subjective assessments of display dynamics to be useful in describing perceptions of clutter. This relationship may be attributable to increased difficulty of visual search with dynamic versus static features (Beijer et al., 2004), general dynamic display complexity (Xing, 2004), and the likelihood of overlap of visual objects in dynamic displays (Wang et al., 2001). Kaber et al. (2008) also identified the potential for perceived display clutter to be influenced by differences in pilot flight task workload, which may vary from one domain to another.

1.2   Research Hypotheses

In general, it was expected that introduction of dynamics in display imagery in the Y2 study caused pilots to perceive an increase in display clutter due to a greater potential for overlap or confusion of features, as compared with Y1 (Hypothesis 1 (H1)). This hypothesis was tested by comparing the CRs among static (Y1) vs. dynamic (Y2) displays. Higher mean CRs were expected for Y2. Due to consistency in the aviation domain between the Y1 and Y2 studies, observed differences in CRs among display conditions (within year) were expected to remain consistent (H2). This hypothesis was tested by comparing the CRs for common displays across the studies as well as looking for common patterns in post-hoc groupings of the various display conditions based on mean CRs among years. The difference in aviation domain among the Y2 and Y3 studies was expected to alter pilot goal states and tasks such that display features would affect CSs differently between the years (H3). This hypothesis was tested by looking for a significant interaction effect of study year and display on clutter scores. It was also expected (H4) that the Y3 test scenario (VTOL for novice fixed-wing pilots) would pose greater task difficulty and generate higher mean CSs vs. the Y2 test scenario (fixed-wing craft with low to medium experience pilots). This hypothesis was tested by comparing the CSs among years with expectation of a higher Y3 mean. As a result of the difficulty of the Y3 flight task, pilots were expected to be more sensitive to display feature and clutter differences in Y3, as compared to Y2 (H5). This hypothesis was tested by comparing the effect of display condition on CSs and CRs within year. We also examined post-hoc groupings of display conditions in terms of the responses for Y2 and Y3 and looked for any significant differences.

2   METHODOLOGY

2.1   Pilots and Display Conditions

In the Y1 study, total flight hours of the subjects ranged from 1300 to 9000 (M=5325 hours). In order to make the comparison with the Y2 study results, only observations on the captains with 15 or more yrs. of experience or more in commercial operations from the Y2 sample were used in this meta-analysis. Those pilots also had some advanced HUD experience and total flight hours ranged from 7500 to 32000 (M=16842 hours). The Y3 pilots were also experts in fixed-wing flight with all being first officers with over 15 years of line experience and a mean of 14000 total flight hours. However, 12 of the pilots were novices to the VTOL domain and their perceptions of HDD clutter were compared with those of the low and medium experience Y2 pilots (n=12). The mean total flight hours for these two groups was 5958.7.

As for the display conditions investigated in the various studies, the HUDs and HDDs presented similar primary flight display configurations and the HDDs included a navigation display at bottom. There was a common set of information features toggled "on" or "off" in all displays including: 1) synthetic vision system (SVS) imagery, which presented a wire-frame representation of a terrain model (see Fig. 1a); 2) enhanced vision system (EVS) imagery, which presented forward-looking infrared sensor imagery of terrain previously collected on actual approaches into KRNO (Fig. 1b); 3) flight pathway guidance (a tunnel or highway-in-the-sky display), represented by four sets of crows-feet outlining boxes along the flight path in HUD/HDD image (Fig. 1c); and 4) visual meteorological condition (VMC) symbology (i.e., pitch ladder, ground speed, selected speed, airspeed tape, speed bug, altitude tape, selected altitude, baro setting, and runway outline (depending on the approach segment)). Figure 1d shows the HUD with the VMC symbology toggled "on" (the "off" setting represented an instrument meteorological condition (IMC) mode of operation).

Seven display conditions were matched between the Y2 and Y3 studies in terms of specific features presented to pilots with similar characteristics in similar phases of approach scenarios (described below) during the experiment. Table 2 summarizes the display conditions and variable names used in the Y1/Y2/Y3 experiments. The categorizations of "high", "medium" and "low" clutter were based on measured and predicted expert pilot ratings collected in the Y1 study.

Table 2 Display Configurations Tested Across the three MMADC Studies

	Label	EVS	SVS	Tunnel	Used in Y1	Used in Y2	Used in Y3
High Clutter (HC)	HC1	On	On	On	Yes	Yes	Yes
	HC2	On	Off	On	No	No	Yes
	HC3	On	Off	Off	Yes	Yes*	Yes
Medium Clutter (MC)	MC1	On	On	Off	Yes	Yes	Yes
	MC2	Off	On	On	No	Yes*	Yes
	MC3	Off	Off	On	No	Yes	Yes
Low Clutter (LC)	LC1	Off	On	Off	No	Yes*	Yes
	LC2	Off	Off	Off	Yes	Yes	Yes

(* - indicates VMC flight symbology was "off".)

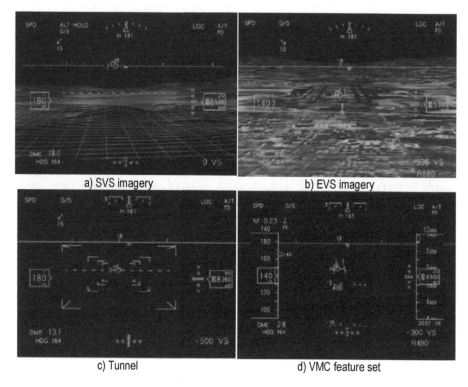

Figure 1. Examples of HUD/HDD features manipulated in experimental studies.

2.2 Scenarios and Procedures

With respect to the flight scenarios, the landing approach presented in Y1 and Y2 (runway 16R at KRNO) was identical. However, the Y1 pilots were provided a verbal description of aircraft flight information and saw static images of different HUD configurations at various points along the approach while the Y2 pilots flew the simulator and experienced different HUD configurations in different segments. Figure 2 shows the three segments of the approach with the first involving level flight from an initial approach fix (IAF) to glideslope (G/S) intercept. The pilot's primary task was to maintain the localizer course before intercepting the G/S, all while slowing the aircraft from the initial airspeed (210 KIAS) to a final approach speed (138 KIAS) and extending the landing gear and flaps. In the second segment, the pilot's task was to maintain the localizer course while flying at the correct rate of descent to stay on the G/S. The pilot's task was the same in third segment but they were also expected to attend to the out-of-cockpit view, as the aircraft neared decision height. The segment continued until the pilot made the decision to land or perform a missed approach at the appropriate decision height for the HUD condition. The simulator was halted at the end of each of these segments, and pilots provided subjective workload and HUD clutter assessments. (We note here that the results of the Y2 study revealed no significant effect of flight segment on CRs; therefore, the pilot ratings for comparable segments of the approach across the Y1 and Y2 studies were aggregated for the present analysis.)

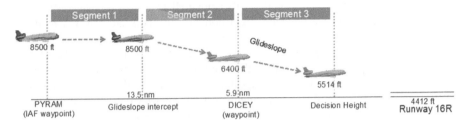

Figure 2. Segments of landing approach scenario for Y1 and Y2 studies.

The Y3 scenario was also a precision ILS approach but to an unlit desert landing site at night under IMC with a 500-foot overcast cloud ceiling. This approach was divided into two segments, including: 1) a descent and decelerating approach to a hover 250 feet above a touchdown point; followed by 2) a vertical descent to termination. The entire approach required constant deceleration of the aircraft and a corresponding decrease in rate of descent in order to remain on G/S. Since the vertical descent to landing was dissimilar from any portion of the Y2 approach, only data collected during the first segment of Y3 trials was used in the Y2/Y3 comparisons. As in the Y2 study, the simulator was halted at the end of each of segment and pilots provided HDD clutter assessments.

2.3   Clutter Measures

For the CRs (clutter ratings), pilots used a uni-dimensional scale ranging from 0-20 points. The CSs (clutter scores) were determined using the multidimensional index developed by Kaber et al. (2008), including six subdimensions (Redundancy, Colorfulness, Salience, Dynamics, Variability and Density). Scores were computed as a rank-weighted sum of ratings on the individual subdimensions. Ranks were determined based on pairwise comparisons of the dimensions of clutter that pilots made at the outset of the Y2 and Y3 experiments. They were asked to identify those qualities that they thought would contribute most to perceptions of clutter with the target displays under the defined flight scenarios. The resulting scores ranged from 0-100%.

3   RESULTS AND INFERENCES

Regarding the influence of display dynamics on pilot perceptions of clutter, ANOVA results revealed a significant main effect of study year (Y1 vs. Y2) $(F(1,72)=7.4751, p=0.0079)$ on CRs. However, counter to expectation (H1), the mean CR for Y2 (12.0 of 20 points) was lower than Y1 (13.8). In fact, the CRs in Y2 were lower for all but one of the four display conditions. Display dynamics did not increase clutter; it appeared that trend information conveyed by the display dynamics effectively supported pilot task performance and reduced perceptions of clutter.

With respect to our expectation that CRs among HUDs would remain consistent from Y1 to Y2 due to the common flight domain (i.e., H2), an ANOVA model including display condition, study year and the interaction of these predictors revealed significant main effects (display: $F(3,72)=17.3584$, $P<0.0001$; year: $F(1,72)=7.4751$, $p=0.0079$) as well as the interaction ($F(3,72)=4.1741$, $p=0.0088$). The interaction was counter our expectation and indicated dynamics led to differences in perception of clutter among the various HUDs within each study year. Post-hoc analyses using Tukey's tests also revealed differences in the groups of display conditions, in terms of mean, among the study years. In Y1, the ratings for display HC1 and HC3 were significantly different ($p<0.05$) from MC1 and LC2 (baseline display), and MC1 also differed from LC2. In Y2, there were no significant differences in ratings between display conditions. This was due greater variability in ratings and less difference between means for each condition. It appeared that ratings for the HC1 display were reduced due to the tunnel drawing pilot attention away from other "cluttered" areas of the display, and the dynamic information content of the HC3 HUD drew visual attention away from anomalous EVS returns that might otherwise have inflated ratings. It is also possible that the familiarity of the flight task in Y2 led to smaller differences in CRs, as compared to Y1.

As for our expectation that differences in the flight domain from Y2 to Y3 would lead to differences in CSs among displays, an ANOVA model including display, study year and the interaction of these terms revealed significant main and interaction effects. The statistical significance of the interaction ($F(6,114.5)=7.1725$, $p<0.0001$) supported our expectation.

Our hypothesis that higher task difficulty in the Y3 study for the novice VTOL pilots would lead to higher mean CSs (H4) was also supported by the significance of the year term in the ANOVA model ($F(1,21.3)=8.8760$, $p=0.0071$). The average CS for Y3 was 55.5 whereas the average score for Y2 was 46.8. The pilots in the Y2 study had some relevant experience in the test domain and were likely able to use a pattern-matching approach to HUD information processing, typical of experts in knowledge-based task performance. Such behavior would logically generate different perceptions display clutter than sequential attention to individual display features typical of novice task performance, as in Y3. The same pattern of results was observed for the CRs among the two studies (Y2 and Y3), save the year effect.

On our expectation that pilots would be more sensitive to display feature manipulations and clutter in the Y3 study due to higher task difficulty (H5), ANOVAs on the CSs responses within each study year revealed the display condition to be significant (Y2: $F(6,105)=8.1493$, $p<0.0001$; Y3: $F(6,76)=8.7054$, $p<0.0001$). A full statistical model, including the interaction of display and year, also revealed a significant effect on CS. This suggested some inconsistency in differences among CSs for displays from one year to the next. However, post-hoc analyses using Tukey's tests revealed that all but two display conditions fell into exactly the same CS groups across years. Therefore, there were only limited differences in pilot sensitivities to clutter in Y3 vs. Y2. These differences appeared to be due to two specific display features. In the Y2 study, a common feature of those HUDs yielding "low clutter" ratings was the absence of VMC symbology. In these displays there was a lower occurrence of iconic and graphical feature overlap due to the reduced symbol set. Since the pilots in Y2 had domain-relevant experience, it is likely they instinctively looked to the Instrument Flight Rules flight symbology for status information. Being perceived as task irrelevant, the presence of the HUD VMC symbology would have increased perceptions of clutter. Secondly, in the Y3 study, those HDDs with the highest clutter ratings included the SVS feature. The wire-frame was densely depicted in the upper-portion of the display, as compared to the Y2 HUDs, and created greater overlap with the most significant flight information. Pilots also seemed to have a greater sense of clutter when the wire-frame appeared higher above the artificial horizon line. Again, the pattern of results on the CRs for the two studies mimicked the pattern of results on CSs.

4   CONCLUSIONS

The overarching goal of the three years of the MMADC experimentation was to develop a multidimensional measure of display clutter for aviation displays. The purpose of the present study was to provide additional insight into influential factors in perceptions of clutter by contrasting pilot assessments of comparable displays under static vs. dynamic presentation conditions as well as different flight domains. It appears that the primary effect of dynamics is trend information that provides additional context for pilots and largely mitigates effects of clutter observed under static display conditions. The analysis of flight domain effects generally supported the use of the clutter measurement method developed by Kaber et al. (2009) across domains. The present meta-analysis revealed the multidimensional CS to be sensitive to pilot goal states and tasks within aviation domains as well as relevant experience level to task.

The main limitation of the present research is that only two potential factors in pilot perceptions of display clutter were examined. There are likely many other factors, particularly related to pilot demographics and background, that may be critical in the sense of display information density, feature redundancy, etc. In addition, some alternate formats of the display features manipulated in the MMADC studies have recently been tested and have shown favorable results. For example, photo-realistic SVS formats with various shading techniques and color schemes have been implemented in aircraft cockpits (Schnell et al., 2009) and have proved superior to the wire-frame format assessed here in terms of pilot task performance. These newer display feature formats should use this model and measure of perceived clutter for consistent comparisons with existing HUD/HDD features.

ACKNOWLEDGMENTS

This research was supported by NASA Grants NNL06AA21A and NNX09AN72A. The technical monitors were Lance Prinzel and Randy Bailey. The opinions expressed are those of the authors and do not necessarily reflect the views of NASA.

REFERENCES

Aviram, A., & Rotman, S. (2000). Evaluating human detection performance of targets and false alarms, using a statistical texture image metric. *Optical Engineering*, 39, 2285-2295.

Beijer, D., Smiley, A., & Eizenman, M. (2004). Driver and vehicle simulation, human performance, and information systems for highways; railroad safety; and visualization in transportation. In *Transportation Research Record*. National Research Council

Ewing, G., Woodruff, C., & Vickers, D. (2006). Effects of 'local' clutter on human target detection. *Spatial Vision*, 19(1), 37-60.

Kaber, D. B., Alexander, A., Kaufmann, K., Kim, S-H., Naylor, J. T. & Entin, E. (2009). *Testing and validation of a psychophysically defined metric of display clutter* (Final Report: NASA Langley Research Center (LaRC) Grant #NNL06AA21A). Hampton, VA: NASA LaRC.

Kaber, D., Alexander, A., Stelzer, E., Kim, S.-H., Kaufmann, K., Cowley, J., Hsiang, S. & Bailey, N. (2007). *Testing and validation of a psychophysically defined metric of clutter* (Annual Report: NASA LaRC Grant #NNL06AA21A). Hampton, VA: NASA.

Kaber, D. B., Kim, S-H., Kaufmann, K., Veil, T., Alexander, A., Selzer, E., Hsiang, S., & Bailey, N. (2008). *Modeling the effects of HUD visual properties, pilot experience, and flight scenario on a multi-dimensional measure of clutter* (NASA LaRC Grant Number NNL06AA21A). Hampton, VA: NASA LaRC.

Kim, S-H., Prinzel, L., Kaber, D., Alexander, A., Stelzer, E, Kaufmann, K. & Veil, T. (2011). Multidimensional measure of display clutter and pilot performance for advanced head-up display. *Aviation, Space, and Environmental Medicine*, 82(11), 1013-1022.

Rosenholtz, R., Li, Y., Mansfield, J., & Jin, Z. (2005). Feature congestion: A measure of display clutter. In *Proceedings of CHI 2005* (pp. 761-770). Portland: ACM.

Rotman, S., Tidhar, G., & Kowalczyk, M. (1994). Clutter metrics for target detection systems. *IEEE Transactions on Aerospace and Electronic Systems*, 30(1), 81-90.

Schnell, T., Keller, M., & Etherington, T. (2009). Trade-offs in synthetic vision system display resolution, field of regard, terrain data density, texture, and shading during off-path operations. The International Journal of Aviation Psychology, 19(1), 33-48.

Wang, C., Griebel, S., Brandstein, M., & Hsu, B. (2001). Real-time automated video and audio capture with multiple cameras and microphones. *Journal of VLSI Signal Processing Systems for Signal Image and Video Technology*, 28(1), 81-99.

Wickens, C., Alexander, A., Ambinder, M., & Martens, M. (2004). The role of highlighting in visual search through maps. *Spatial Vision*, 17(4-5), 373-388.

Xing, J. (2004). *Measure of Information Complexity and the Implications for Automation Design*. Oklahoma City, OK: Federal Aviation Administration, Civil Aeromedical Inst.

CHAPTER 6

Effectiveness of 3D Cockpit Displays of Weather on Pilot Performance in NextGen Airspace

Gregory A. Morales, Tiana Higham, Vyha Ho, Kim-Phuong L. Vu and Thomas Z. Strybel

California State University, Long Beach
Long Beach, CA
gregory.morales1@gmail.com

Vernol Battiste, Joel Lachter, and Walter Johnson

NASA Ames Research Center
Moffett Field, in California

ABSTRACT

The Next Generation Air Transportation System (NextGen) is a continuing program to improve safety and efficiency in the National Airspace System. Weather and workload are two major challenges that must be addressed to meet the goals of NextGen. In this simulation, pilots flew in three plausible concepts of operation for separation assurance, using two types of weather displays (current day radar and 3D NEXRAD). Performance data of pilots maneuvering past weather is presented. When using NEXRAD pilots began maneuvers to avoid weather earlier, passed weather faster, and required fewer flight plan changes compared to conditions using current day radar. As such, NEXRAD may be a more efficient tool for strategic weather avoidance. Weather avoidance performance was also affected by operating concept for separation assurance.

Keywords: 3D Display, NEXRAD, NextGen, Weather avoidance

1.0   INTRODUCTION

The Next Generation Air Transportation System (NextGen), established in 2003 by act of congress, is an ongoing program of improvements and upgrades to the National Airspace System (NAS) for enhancing the safety, reliability and efficiency of air transportation (FAA, 2011). To achieve these goals all components of the NAS, including air traffic management operations, airports, air operations centers and flight decks are being overhauled. NextGen changes are being rolled out incrementally, based in part on the users' willingness to equip aircraft with the required technologies. Many of these changes are aimed at eliminating barriers to improved efficiency in the current day system. This paper is focused on two such barriers, weather and operator workload.

Weather is a major source of aviation accidents and delays.   It has been estimated that during 2009, 65.7% of NAS delays were weather related (RITA, 2012). Weather reduces traffic flow over pre-defined corner posts, increases air traffic controller (ATC) workload and creates significant flight delays.   It is generally assumed that providing more precise weather information in the cockpit will reduce delays and improve safety in NextGen if weather information is formatted based on operator needs (e.g., Berringer & Schvaneveldt, 2002). However, even if flight decks are provided with more precise weather information, the anticipated shift from clearance-based to trajectory-based operations (TBOs) creates another problem: Under clearance-based operations, pilots make imprecise requests for weather deviations. With TBO, pilots can fly negotiated flight paths necessary for full performance-based navigation, taking both operator preferences and optimal airspace system performance into consideration.   However, there has been limited discussion on how weather avoidance decisions will be made and executed in the context of performing TBO, and how weather displays should be designed to support these operations.

Operator workload is another bottleneck currently limiting NAS capacity. NextGen strategies for reducing ATC workload include new automation tools, Data Comm, and changes in operating concepts.   In NextGen, responsibility for separation assurance may be allocated either to automated agents or to appropriately equipped flight decks in addition to ATC. The final function allocation solution is still being debated.   Nevertheless, if pilots become responsible for maintaining separation from traffic, it is essential that flight decks be equipped with cockpit displays of traffic information (CDTIs) that promote awareness of nearby traffic without increasing pilot workload (e.g., Comerford, 2004).

Solving the weather and workload problems cannot be achieved in isolation: The methods used by pilots for avoiding weather may depend to some extent on the concept of operation for separation assurance. Presently, pilots request vectors for maneuvering around weather (e.g., "request 20 degrees left for weather") that must be cleared by ATC.   When a request is made, ATC must determine if the request is free of traffic conflicts.   If a conflict is detected, ATC must deny the request, propose another vector to the flight deck, or move the other traffic involved in the conflict.   If pilots assume separation assurance responsibility, flight-deck-initiated weather deviations also must be conflict free, and CDTIs that integrate weather and traffic information would be necessary. In the simulation reported here, we examined the effectiveness of ground- and airborne-based weather radar integrated with a CDTI under three plausible NextGen concepts of operation for separation assurance: pilot primary (pilots responsible for solving conflicts with ownship), ATC primary (ATC responsible for most traffic separation), or automation primary (a ground-based automated agent responsible for most traffic separation).

1.1   NEXRAD vs. Radar Weather

Traditionally pilots receive weather information via onboard radar. This system is based on reflections from a weather source such as precipitation. Current radar technology has a limited effective coverage, with a forward azimuth of 120 degrees and a range of 150 miles or less, based on transmitter power and line of sight. Line of sight limitations are caused by obstacles and the curvature of the earth. Moreover, the incoming signal can be attenuated when the energy of the radar is not completely reflected back to the aircraft because of the intensity of the rain from a cell. This means that pilots may not see the other side of large storm cells. Another source of weather information is NEXRAD. NEXRAD is a system of multiple overlapping ground-based radars. Under most circumstances the NEXRAD system has relatively unlimited coverage. NEXRAD information can be provided to the flight deck by satellite-based systems such as SiriusXM Radio. However, since NEXRAD data can be delayed 6-7 minutes following an actual radar scan, it has been suggested that both NEXRAD and radar displays be installed in the cockpit. NEXRAD would be used for long-range strategic planning, whereas onboard radar with up-to-the-minute information would be used for short-term tactical weather avoidance as the aircraft approaches the weather cells (Beringer & Ball, 2004).

Research on pilot weather avoidance using NEXRAD displays has shown that pilots often use this information for tactical planning, despite the fact that the information may be outdated. Pilot performance with NEXRAD displays is also affected by the resolution of the display. Beringer & Ball (2004), in an investigation of general aviation pilots, found that pilots spent more time looking at higher resolution images, and pilots viewing low resolution images deferred their decisions longer than pilots using high resolution images. The usefulness of NEXRAD also depends on the display format. For example, Yuchnovicz et al., (2001) found no benefit of NEXRAD with a display that did not include ownship position relative to the weather. In a follow up study, Novaczek et al. (2001) showed that the addition of ownship information reduced pilot workload and improved situation awareness. Wu et al. (2011) evaluated three methods of showing ownship relationship to weather, Pulse, Slider and Route, in a part task investigation of weather avoidance using predictive weather information. Pilots modified flight plan trajectories to avoid weather when necessary. Although no one method produced superior weather avoidance performance, pilot preference for each display was based on the type weather encounter scenarios.

Research on displays that integrate weather with traffic information is sparse, and limited by the displays used. Wickens (Boyer & Wickens, 1994; O'Brien & Wickens, 1997) showed that the integration of weather and traffic in one display produced biases in avoidance maneuvers. When 2D and 3D displays were compared, Wickens showed that pilots created more efficient routes with 2D displays. Moreover, displays that integrated weather and traffic produced increased use of horizontal changes compared with altitude changes. Comerford (2004) suggested however, that the displays used by Wickens did not provide the pilot with a method of assessing flight plan changes before they were executed. Comerford developed recommendations for integrated traffic and weather displays and suggested that pilots need a "what if" tool for assessing potential avoidance maneuvers. Comerford also noted that for these integrated displays of traffic and weather, the source of the hazard must be clearly identified because the consequences of encountering weather and traffic are different. Mistakenly entering into weather *may* lead to catastrophe; encountering traffic *will* be catastrophic.

In the present paper, we report on a test of weather avoidance efficiency when using a 3D CDTI that integrated weather depictions based on either simulated NEXRAD or airborne data sources (i.e., weather radar), with traffic information. The CDTI also had a Route Assessment Tool (RAT) that allowed pilots to evaluate and modify their current flight plan using a graphical user interface to avoid weather, and, under some operating concepts, maintain separation from traffic. We were interested in whether the source and format of weather information would interact with changes in responsibility for conflict detection and resolution between pilots, ATCs, and an automated, ground-based conflict resolution agent. Pilots flew desktop simulators in 12, 90-minute scenarios in which they had to deviate from their planned trajectory route to avoid weather, perform interval management operations, and execute a continuous descent approach (CDA) under one of three plausible NextGen concepts of operation for separation assurance. Here, we report on weather avoidance; for information regarding other outcomes of this distributed simulation see, for example, Strybel et al., 2010; Vu et al., 2012; and Johnson et al., 2010.

2.0 METHOD

The study was conducted over a two-week period. Due to equipment failures, only data from the second week will be reported. Seven experimental pilots were tested during the second week. Four of the pilots were at the professional rank of captain and three were first officers. The participants reported no prior experience with interval management operations, but three of the pilots had experience flying continuous descent arrivals (CDAs).

2.1   Simulation Configuration

The simulation environment was managed by the Multiple Aircraft Control System (MACS) software (Prevot, 2002). A distributed simulation environment was used. Experimental pilots, and the simulation hub, were located at NASA Ames' Flight Deck Display Research Laboratory (FDDRL). ATCs were located in the Center for Human Factors in Advanced Aeronautic Technologies (CHAAT) at California State University Long Beach. CHAAT also hosted confederate or pseudopilot stations, which managed additional aircraft in the simulation. Additional pseudopilot stations were located at the Systems Engineering Research Laboratory (SERL) at California State University Northridge, and the Human Integrated Systems Engineering Laboratory (HISEL) at Purdue University.

The simulated airspace spanned Kansas City (ZKC) and Indianapolis Centers (ZID). Traffic in each sector was based on real traffic feeds, but traffic density was increased to produce 3X traffic density environment, as might be encountered in NextGen airspaces. Aircraft populating the simulation were designated as either AFR (Autonomous Flight Rules) or IFR (Instrument Flight Rules). AFR aircraft had flight decks equipped with Data Com, path replanning tools, and sometimes conflict detection tools. AFR aircraft pilots were sometimes responsible for ensuring their own separation from other aircraft, and sometimes were managed by a ground based automated agent that provided separation. IFR aircraft only had Data Com, and were always managed by a human controller. All experimental pilots flew simulated AFR desktop stations. Pseudopilots flew all IFR and the remaining AFR aircraft populating the airspace.

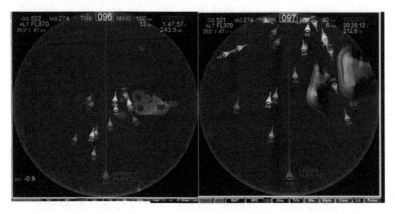

Figure 1. Current-day and 3D-NEXRAD weather displayed on the CDTI.

Several advanced tools were available to pilots and controllers for managing traffic separation and weather avoidance. Interval management (see Ho et. al, 2010, Johnson et. al, 2010 for descriptions of the interval management task) and weather avoidance tasks were supported by a simulated CDTI (Granada, Dao, Wong, Johnson, Battiste, 2005). Embedded in both the MACS ATC display and, during the pilot and ATC responsible conditions, in the CDTI, were tools for conflict detection, and resolution (Erzberger 2006). The Route Assessment Tool (RAT) for pilots and the Trial Planner for air traffic controllers allow for creation of flight plan changes. When these tools are coupled with the Conflict Probe, they allow the operator to determine whether proposed route changes are conflict free. Pilots can also use the RAT for weather avoidance. The CDTI also displayed graphical weather information as shown in Figure 1. On half of the scenarios, the weather was based on NEXRAD weather data and displayed as a 3-dimensional image based on the orientation of the CDTI. For the remaining scenarios, weather was displayed in two dimensions based on current day airborne weather radar systems. The current day radar emulation was designed with an adjustable ±5 degree tilt and range of 150nm. No gain control was provided.

The auto-resolver tool was also used as a ground-based traffic management system in some conditions. The auto-resolver agent was responsible for autonomously data linking conflict resolutions to AFR flight decks in Concepts 2 and 3. The agent could not move aircraft around weather. Under different concepts of operation, responsibility for traffic separation variously was allocated to pilots, ATCs or the automated agent (Table 1).

2.2 Concepts of Operation

Three plausible NextGen concepts of operation were tested as shown in Table 1. In Concept 1, AFR pilots had the capability, responsibility, and authority for avoiding weather and separating their Ownship from other aircraft using the advanced traffic separation tools provided. Pilots made route modifications for traffic and weather avoidance and executed them; they did not have to datalink route modifications to a controller for approval. AFR pilots were given voice frequencies to monitor but were told that they would not receive clearances from the ATC unless they returned responsibility for interval management to the ATC. During the CDA, AFR pilots were instructed not to use the auto-resolver or RAT tools; the controller was to monitor them for conflicting traffic.

Table 1. Concepts of Operation for Separation Assurance

	1:Pilot Primary	2: ATC Primary	3:Automation Primary
Separation Responsibility			
Pilot	All conflicts with ownship	None - conflict information displayed	None – no conflict information
ATC	IFR-IFR conflicts	AFR-IFR, IFR-IFR conflicts	IFR-IFR conflicts
Automation	None	AFR-AFR conflicts	AFR-AFR, AFR-IFR conflicts
AFR-IFR Conflict Rules	AFR moves	IFR moves	AFR moves

In Concept 2, AFR pilots were responsible for avoiding weather but not avoiding traffic. All flight plan changes were downlinked to the ATC for approval. AFR pilots could use the advanced tools to generate a conflict resolution, but it had to be approved by the controller. In Concept 2, the ATC was responsible for resolving IFR-IFR and IFR-AFR conflicts. For IFR-AFR conflicts, the controller was to move the IFR aircraft. Conflicts between AFR-AFR aircraft were the responsibility of the auto-resolver agent. In Concept 3, conflict detection and resolution tools were not available to the pilots. The Route Assessment Tool was available for weather avoidance or other routing requests, but it was not coupled with the conflict resolution tools. The human controller was assigned responsibility for resolving IFR-IFR conflicts and the auto-resolver agent was allocated responsibility for resolving IFR-AFR and AFR-AFR conflicts. For IFR-AFR conflicts, the AFR aircraft was burdened to move unless the aircraft was on CDA.

2.3   Procedures

Twelve, 90-minute scenarios, were run, four replications of each concept of operation shown in Table 1. All scenarios were based on arrivals into Louisville Standiford International Airport (SDF), performing interval management while circumventing en route convective weather. For each concept of operation, simulated cockpits were equipped with the 3D NEXRAD weather on two scenarios and current day radar on two scenarios. Four static weather scenarios varying in number and size of cells were used within each operating concept. Within the first 10 minutes of the scenario, pilots were assigned a lead aircraft and given spacing instructions. The weather cells were encountered early, and therefore avoided early, in the scenario. On-line workload/situation awareness probes were also used. See Strybel et al., 2010; Ligda et al.2010 for description and results.

Across the two-week simulation, operating concept, weather display and weather complexity were completely crossed in a mixed-factorial design. Data from week one was not analyzed due to equipment malfunctions. Therefore, Week 2 weather avoidance data was averaged across all 4 weather scenarios to eliminate potential confounds. Measures of weather avoidance performance consisted of 1) the duration from when the experimental aircraft entered the scenario west of the weather cells until when the experimental aircraft cleared the eastern-most point of the weather cluster (time around weather), 2) the time from scenario start until the time when the pilot executed the first flight plan change for weather (time until first flight plan change), and 3) the number of flight plan changes made to circumvent the weather cells (number flight plan changes). For each dependent measure a two-factor repeated measures ANOVA was run with the factors operating concept and weather information.

3.0   RESULTS

Figure 2 shows the mean time around weather for each display and operating concept. Significant main effects of display ($F(1, 6)= 9.36$; $p=.02$) and operating concept ($F(2,12)= 6.30$; $p=.01$) were obtained; the interaction was nonsignficant. On average pilots took 32.5 min (SEM=.52 min) to avoid weather when using

NEXRAD displays and 35.3 min (SEM=.72 min) when using radar displays. Weather was circumvented fastest in the ATC primary condition (M=31.8 min; SEM=.76 min) compared with pilot primary (M=34.6 min; SEM=.69 min) and automation primary (M=35.2 min; SEM=.85 min) conditions.

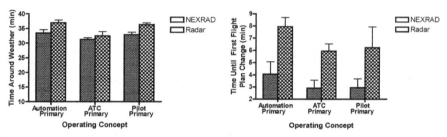

Figure 2. Mean time around weather for display type and operating concept.

Figure 3. Mean time until the first flight plan change for weather avoidance.

Figure 3 shows the average time interval from the beginning of the scenario until the first flight plan change was made for weather avoidance. Only the main effect of display type was significant, $F(1,6)= 9.03$; $p=.02$. On average pilots using the NEXRAD display made initial changes for weather earlier in the scenario (and farther from the weather) compared with the radar display condition (NEXRAD: M=3.31 min; radar: M=6.7 min).

Figure 4 shows the number of flight plan changes made to avoid weather. Although on average fewer flight plan changes were required with NEXRAD weather displays, the effect of display was nonsignficant ($p=.12$). However, a marginally significant interaction between display type and operating concept was obtained, $F(2, 12)= 3.42$; $p=.07$. As shown in Figure 3, for automation and ATC primary conditions, more flight plan changes were needed to avoid weather when using radar than NEXRAD. However, for pilot primary condition, the number of flight plan changes to maneuver around weather was roughly equivalent.

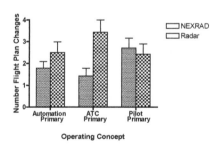

Figure 4. Average number flight plan changes to pass weather by display and operating concept.

4.0 DISCUSSION

This preliminary investigation of two weather displays under three concepts of operation for separation assurance showed the advantage of NEXRAD weather information in the cockpit. With 3D NEXRAD weather integrated with traffic, pilots in our scenarios began maneuvering around weather sooner, reached the other side of the weather faster and required fewer flight plan changes compared with current-day radar information. The fact that pilots began maneuvers around weather earlier in the scenario suggests that NEXRAD may be more efficient as a strategic tool for avoiding weather, as suggested by Beringer & Ball (2004). However, as the weather in the present simulation was static, both the benefits and costs of NEXRAD information cannot be fully assessed. These results might indicate that the 3D format of NEXRAD was responsible for earlier planning.

The concept of operation for separation assurance did affect some aspects of weather avoidance. Pilots made the other side of weather fastest when ATC was responsible for resolving traffic conflicts. This may be because when ATC was primary, conflicts between AFR and IFR aircraft required that the IFR aircraft be moved. Therefore, ATCs may have cleared the airspace of potential conflicts near weather, thus making it easier for pilots to circumvent the weather cells. Note that in this condition, pilots were able to see traffic and downlink suggested conflict resolution maneuvers, but were not responsible for resolving conflicts. On the other hand, when pilots were responsible for separating themselves from traffic, the number of flight plan changes was roughly equivalent for both weather displays, suggesting that avoiding weather and traffic changed the task.

ACKNOWLEDGEMENTS

The simulation described in the paper was supported in part by NASA cooperative agreement NNA06CN30A, *Metrics for Situation Awareness, Workload, and Performance in Separation Assurance Systems* (Walter Johnson Technical Monitor). Preparation of this paper was supported by NASA cooperative agreement NNX09AU66A, *Group 5 University Research Center: Center for the Human Factors in Advanced Aeronautics Technologies* (Brenda Collins Technical Monitor).

REFERENCES

Beringer, D.B. and J.D. Ball. 2004. The Effects of NexRad Graphical Data Resolution and Direct Weather Viewing on Pilots' Judgments of Weather Severity and Their Willingness to Continue a Flight. (DOT/FAA/AM-04/5). Washington, DC: Federal Aviation Administration.

Beringer, D. B., & R. W. Schvaneveldt. 2002. Priorities of weather information in various phases of flight. *Paper presented at the Human Factors and Ergonomics Society Meetings*, Baltimore, October.

Boyer, S.B. and C.D. Wickens. 1994. 3D weather displays for aircraft cockpits. (Tech report ARL-94-11/NASA-94-4).

Comerford, D. A. 2004. Recommendations for a Cockpit Display that Integrates Weather Information with Traffic Information. *NASA Technical Memorandum* 2004-212830.

Durso, F., K. Bleckley, and A. Dattel. 2006. Does situation awareness add to the validity of cognitive tests? *Human Factors*, 48, 721–733.

Erzberger, H., 2006. Automated conflict resolution for air traffic control. *Proceedings of the 25th International Congress of the Aeronautical Sciences (ICAS)*.

Federal Aviation Administration (2011). *FAA's NextGen Implementation Plan, March 2011*. Federal Aviation Administration.

Granada, S., Dao, A. Q., Wong, D., Johnson, W. W., & Battiste, V. 2005. Development and integration of a human-centered volumetric cockpit display for distributed air-ground operations. *Proceedings of the 12th International Symposium on Aviation Psychology*.

Johnson, W. W., N. Ho, V. Battiste, K. L. Vu, J. B. Lachter, S. V. Ligda, Q. Dao, and P. Martin. 2010. Management of continuous descent approach during interval management operation, *Digital Avionics Systems Conference (DASC), 2010 IEEE/AIAA 29th*. 4.D.4-1-4.D.4-13.

Ligda, S. V., A.-Q. V. Dao, T. Z. Strybel, K.-P. L. Vu, V. Battiste, and W. W. Johnson. 2010. Impact of conflict avoidance responsibility allocation on pilot workload in a distributed air traffic management system. *Proceedings of the Human Factors and Ergonomics Society's 54th Annual Meeting*. 55–59.

Novacek, P.F., M.A. Burgess, M.L. Heck, and A.F. Stokes. 2001. The effect of ownship information and NEXRAD resolution on pilot decision making in the use of a cockpit weather information display. Springfield, VA: NTIS, Technical Report NASA/CR-01-210845

O'Brien, J.V. and C.D. Wickens. 1997. Cockpit displays of traffic and weather information: Effects of dimension and data base integration. (Tech. Report ARL-97-3/NASA-97-1).

Prevot, T. 2002. Exploring the many perspectives of distributed air traffic management: The Multi Aircraft Control System: MACS. *International Conference on Human–Computer Interaction in Aeronautics, HCI–Aero 2002*, 23–25.

Research and Innovative Technology Administration (RITA)-U.S.Department of Transportation. *Understanding the Reporting of Causes of Flight Delays and Cancellations*. Retrieved http://www.bts.gov/help/aviation/html/understanding.html, Feb 11, 2012.

Strybel, T. Z., K. Minakata, J. Kraut, P. Bacon, V. Battiste, and W. W. Johnson. 2010. Diagnosticity of an online query technique for measuring pilot situation awareness in NextGen. *29th Digital Avionics Systems Conference*. CD ROM pp. 4. B.1-1–4.B.1-12.

Vu, K.-P. L, T. Z. Strybel, V. Battiste, J. B. Lachter, A.-Q. V. Dao, S. L. Brandt, S. V. Ligda, and W.W. Johnson. 2012: Pilot Performance in Trajectory-Based Operations Under Concepts of Operation That Vary Separation Responsibility Across Pilots, Air Traffic Controllers, and Automation. *International Journal of Human Computer Interaction* 28(2): 107-118

Wu, S.C., C. G. Duong, R. W. Koteskey, and W. W. Johnson. 2011. Designing a Flight Deck Predictive Weather Forecast Interface Supporting Trajectory-Based Operations, *Ninth USA/Europe Air Traffic Management Research and Development Seminar*, ATM2011 (Paper 132), EUROCONTROL/FAA. Berlin, Germany, June 2011.

Yuchnovicz, D.E., P.F. Novacek, M.A. Burgess, M.L. Heck, and A.F. Stokes. 2001. Use of a datalinked weather information display and effects on pilot navigation decision making in a piloted simulation study. Springfield, VA: NTIS, Technical Report NASA/CRT-2001-211047.

CHAPTER 7

A User-centered Optimization Method for Helicopter Flight Instrument Panels

Cem Alppay, Nigan Bayazit

Istanbul Technical University
Istanbul, Turkey
calppay@itu.edu.tr
bayazit@itu.edu.tr

ABSTRACT

In this paper flight instrument panels, main user-interfaces of multi-purpose civil helicopters are studied. Current arrangement approaches for instrument panels are studied and analyzed and a new user-centered design method is developed for civil rotary-wing aircraft that can be described as an interface optimization based on user opinions and preferences. An interview was conducted among 15 helicopter pilots constituted the first step of the research. In the second phase of the research, a paper prototyping method has been applied on helicopter pilots. The proposal of an optimization method of instrument panel arrangement based on user opinions is the primary output of the study. The optimum instrument panel design is , constructed depending on an approach, in which locating displays on the instrument panel using their average locational values and then making revisions over the first design, using interview findings and pilots' opinions and preferences. As the result of both interview and experimental studies; it has been observed that the principle of functional grouping of displays is the primary consideration of users. Additionally, two new concepts, "inseparability relation" and "locational value" has been proposed by the researcher.

Keywords: User centered design, user interface, helicopter instrument panel, interface optimization, locational value, inseparability relation

1.   INTRODUCTION

User interfaces, being conceptual spaces in which power and information is transferred between human and machines constitute one of the main components of man-machine systems in the context of human factors. Display panels of transport vehicles are widely used examples of user-interfaces in daily life. Display and control panels found in different modes of transportation vehicles such as automobiles, motorcycles, boats, rail vehicles, aircraft and helicopters are the main user interfaces for the operators of these vehicles. With the advancements in technological field the number of displays increases in many modes of transportation vehicles resulting with more complex display panel arrangements. Helicopters, being a complex man-machine system, are a good example for the increasing complexity of display panels.

The aim of the study is to constitute a research method to arrange and locate displays in flight instrument panels of multi-purpose civil helicopters. This research method is based on user-centered design approach with human factors perspective. Current interface arrangement approaches have been analyzed and criticized and a new flight instrument arrangement method have been proposed. The study aimed to design an optimum flight instrument panel arrangement based primarily on user opinions. An interview is conducted among civil helicopter pilots and panel arrangements from a paper prototyping study with participants of the interview constituted two main data sources of the research. Therefore, this research can be described as an user interface optimization process using user opinions and preferences.

2.   CONTROL OF HELICOPTERS AND BASIC ISSUES IN HELICOPTER INSTRUMENT PANELS

The interaction between a user and a product or system occurs over a user interface. The function of any user interface can be described to facilitate the transfer of any knowledge and force between the user and product or system necessary for the operation of the system. In the area of transportation design the main user-interface of any vehicle is the display panel, where the main interaction between the user and product takes place. The purpose of any interaction which occurs on the user-interface is to allow efficient use and control of the vehicle and to give necessary feedback to help making operational decisions of the user. User interfaces can be hardware based as well as software based systems.

As a civil transport vehicle, helicopters require a very complex control process for pilots as their operators. On the contrary of a general belief, their flying controls are extremely different than aircrafts' control principles. A helicopter pilot uses his both hands and foot to fly his vehicle. He controls the "collective pitch" with his/her left hand while handling the "cyclic". In the same time as he occupies his hands; his feet controls the ant-torque pedals in order to control the direction in which the nose of the aircraft is pointed. This situation strictly limits pilot's use of his hands and feet. Lovesey (1975) describes the helicopter pilot as an operator who works under most difficult conditions, in terms of human factors, because of large scale maneuver abilities, instability of its dynamical nature and resulting control problems. Display/control compatibility is one the basic problems which emerge in that situation. According to Lovesey (1975), if these controls could function separately from each other, the task of flying a helicopter should be a little bit more difficult from the same task of a fixed-wing aircraft. But a small change of any flight parameter of a helicopter requires that all controls have to be arranged. On the other hand Hart (1988) indicates that, although helicopter pilots' control needs and mental workloads are higher than a fixed-wing aircraft pilot's, many helicopters are equipped with flight displays designed for fixed-wing aircraft.

3.   RESEARCH AND DESIGN METHODS FOR PANEL ARRANGEMENT

Sanders and McCormick (1993) have studied display arrangements in user interfaces under the title of "arrangement of components in a physical space". In their approach they are dividing the problem in two general steps: *General location of components and Specific arrangement of components in their general location.* Sanders and McCormick (1993) define four principles to be used in steps cited above. These principles are *importance, frequency of use, functional similarity* and *sequence of use.*

In the context of aviation, early studies to design flight instrument panels and to arrange displays in aircraft cockpits began in 1930's in Great Britain. In 1937 Royal Air Force-RAF conducted a research for flight instrument panel arrangements of military aircraft which will be used during World War II and published the results of this research. Supported with visual scanning research, this basic arrangement has been refined and reached its final "T" shape. This "T" shape arrangement is still being used in contemporary fixed wing and rotary wing aircrafts (Hawkins 2002). The need of reading four flying parameters, which are attitude being the most important, speed, altitude and heading, quickly and correctly has resulted in the modification of this basic "T" arrangement. Gray and Flower (1943) have also developed a model for locating flight displays in an instrument panel. In their model all displays were grouped in a square-shaped space (1 m2) and were located in accordance with the aircraft's left-right or up-down maneuvers. On the other hand Kelley (1972) describes two approaches for the design and arrangement of display panels. These approaches are the information requirement analysis of the operator and experimental study for personal preferences. The information requirement analysis is related with the relative importance of data sources and their inter-relations. The experimental study, on the other hand, can be either conducted in a simulator or in a real environment to discover personal preferences of users.

Arrangement of components within a user interface can also be supported with data provided by methods based on visual analysis. Visual analysis is based on direct observation of user action in the process of use of a certain product or system. Chapanis (1965) describes visual analysis methods as a data basis to refine existing products or systems. Visual analysis methods can be divided in four groups: *user opinions, activity-sampling techniques, process-analysis techniques and micro-motion techniques* (Chapanis, 1965). Chanel et al. (1947) used flow-chart method to re-design the cockpit of a military aircraft Chapanis (1965) used observation and multiple-flow-chart techniques to research the operation process of landing gear system of a R5D aircraft. Fitts et al (1950) studied eye-motions using link-analysis techniques for landing procedure of aircrafts. Senol et al. (2010) use qualitative and quantitative methods to design the flight instrument panel of a multi-purpose civil helicopter. They have utilized multi-criteria decision making and card-sorting techniques during their research.

4.   METHOD

This study is basically structured on user-opinions by utilizing two different research methods: an interview study and a paper prototyping process with helicopter pilots. The number of participants in the research is limited to 15 male pilots for the interview and 10 pilots for the paper-prototyping process. Because of the very-specialized nature of helicopters the difficulty of finding participants for the study limited the number of participants.

In the interview phase of the research, the participants have been asked to evaluate, using a Likert Scale 1-5, each display to be used in the instrument panel according to interface arrangement principles proposed by Sanders and McCormick (1993) as well as their general views and opinions about instrument panels. The study resulted with quantitative and qualitative answers from participants. For the responses of all pilots to "degree of importance" and "degree of frequency of use" arithmetic averages have been calculated. The qualitative answers given for the "principle of functional similarity" have been analyzed using functional similarity matrices and display groups based on functional similarity principle have been generated by the researcher.

The paper-prototyping study, which can be described as a user-centered design technique, with the participants of the interview constituted the second phase of the research. A three-dimensional empty model of a Bell 407 helicopter has been made using Styrofoam. Pilots, who have also answered the interview, have been presented the 3-D Styrofoam model along with a set of helicopter displays drawn in a CAD environment. The panel is originally designed that the pilot seat is located at the right side. The following question "*if you were designing your own helicopter; how you would locate these displays on this empty instrument panel?*" has been directed to participants. Each participant's paper-prototyping process has been video-recorded and a detailed picture of the final arrangement has been captured. Each participant's own arrangement has been analyzed using following *general explanation of pilots own prototyping process documentation of pilot's important sentences and explanations, realistic drawing and a "letter coded" drawing of each pilot's design comparison of locational values with average value for degree of importance and frequency of use.* At this stage the concept of "**locational value**" has been proposed by the researcher in order to analyze each pilot's own display arrangement. Area 1 being the most important (each display placed in this area should have a locational value of 5) followed by Area 2, Area 3 Area 4 and Area 5 with values 4, 3, 2 and 1 consecutively. Area 1 is the closest area to the pilot where the pilot is sitting on the right side of the panel. Area 4 and 5 are most difficult areas to look for the pilot because of maximum head and eye movement requirements. In order to turn average values obtained from the interview a reading-direction relationship has been developed. This locating-logic, which can be described as "area based reading systematic", is shown in Figure 1. This logic is developed as the result of observations of pilots protocols during the paper prototyping processes. According to this systematic the display with the highest average value should be placed at position nr 1, systematic the display with the second highest average value should be placed at position nr 2 and consecutively.

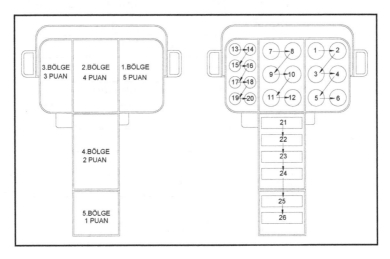

Figure 1: Instrument panel's area and their values (left) and locating logic of displays

In the last phase of the research an optimization method, based on the observations and analysis of interview and paper-prototyping data, has been developed. In this method, interview data is based on theoretical knowledge of participants while the paper-prototyping data is based on direct practical data. Therefore the paper prototyping technique has been considered as the primary source of data and the interview as the secondary source of data. The optimization method (Figure 2) is generally based on the optimization of paper prototyping data and followed by an improvement process using interview data.

5. FINDINGS

5.1 Results of interview and paper prototyping studies

Table 1 shows all pilots' arithmetic average values for evaluation of all displays during the interview. Using these average values, four different instrument panel arrangements have been designed based on:

- Average values for the degree of importance of displays (Figure 3)
- Average values for the degree of frequency of use of displays
- Functional groupings and average values for the degree of importance of displays
- Functional groupings and average values for the degree of frequency of use of displays

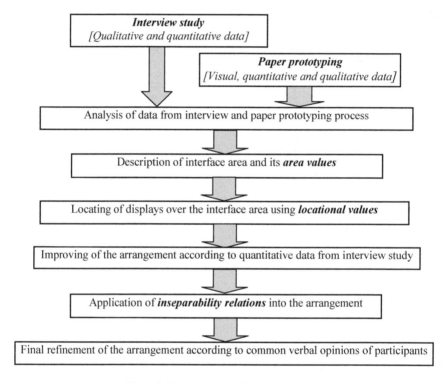

Figure 2: User-centered optimization method

In the paper- prototyping phase of the study, which can be described as the experimental phase, 10 pilots have produced their own instrument panel arrangements. The duration of the study was limited to 20 minutes for each pilot and they have been told that they are free to make their own design and also they can make any improvement and corrections during that time. An example of instrument panel arrangement as the outcome of paper prototyping process can be seen in Figure 4. In the analyzing stage, verbal expressions about general and common opinions of pilots have been documented such as:; *The most important thing for me is the frequency of use. The most frequently looked display must be placed in a position where I won't make any head movements.*; *"Engine displays must form a group.", "Rotor/turbine tachymeter must be placed in a very central point."* In that stage, each pilot's own personal evaluation during the interview study and the locational values of his paper-prototype has been compared in a graphic chart in order to explore which of the Sanders and McCormick' (1993) principles is dominant in pilots own arrangements.

Table 2 shows the comparison table of same pilot's, whom own arrangement, is seen in Figure 4, locational values with average values from interview data. In the comparison of all pilots' paper-prototype arrangements with average values from interview data it has seen that neither the degree of importance nor the degree of frequency of use is a dominant concept in display arrangements. Also, its evident that the principle of functional grouping is dominant in display arrangements. This is also supported with a visual comparison of average interview data based panel design and pilots' own design as the outcome of paper prototyping.

Table 1 : Average values of displays

Code	Display	V1	V2
[A]	Air Speed Indicator	4.3	3.9
[B]	Altimeter	4.2	3.8
[C]	Compass	3.5	2.5
[D]	Attitude Indicator	4.1	3.6
[E]	Directional Gyro	4.4	3.8
[F]	Clock	2	1.7
[G]	Turn and Bank Indicator	3.5	3.1
[H]	Vertical Speed Indicator	3.5	2.8
[I]	Transponder	2,2	1.7
[J]	NavCom	3.4	2.7
[K]	Audio Panel, Marker Receiver	2	1.7
[L]	Automatic Direction Finder	2.2	2.3
[M]	Distance Measuring Unit	2.9	2.2
[N]	GPS NavCom	4	4
[O]	Electronic Flight Information System	3.8	3.4
[P]	Fuel Gauge	4.2	3.9
[R]	Fuel Pressure Indicator	4	3.2
[S]	Transmission Oil Pressure and Heat Indicator	4	3.5
[T]	Engine Oil Pressure and Temperature Indicator	4.3	3.5
[U]	Gas Temperature Indicator	4.3	3.9
[V]	Gas Producer Tachometer	4.4	3.7
[Y]	Rotor/Engine Tachometer	4.7	4.3
[Z]	Torque Indicator	4.9	4.7

V1: Average value for degree of importance
V2: Average value for degree of frequency of use

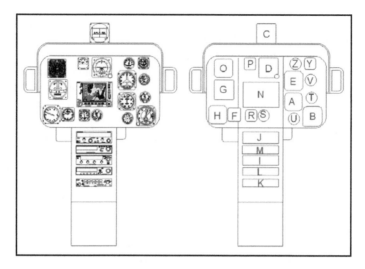

Figure 3: Panel arrangement based on average values for the degree of importance of displays

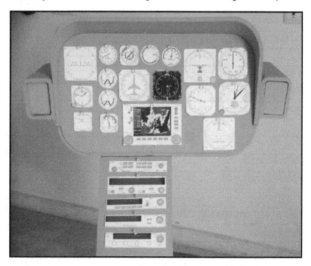

Figure 4: One of the instrument panel arrangements resulting from the paper-prototyping method

5.2　Panel arranegement resulting from the optimization method

The instrument panel's realistic and coded AutoCad drawing is also shown in Figure 5. The test and evaluation of the results of this optimization method is subject for a future study at Istanbul Technical University. This method can also be applied in the design of other product or system interfaces used or operated by specialized users or operator such city buses, trucks or yachts.

Table 2: Comparison of locational values with average values of interview data

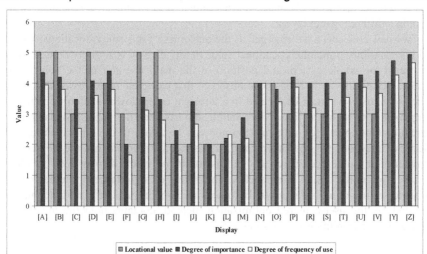

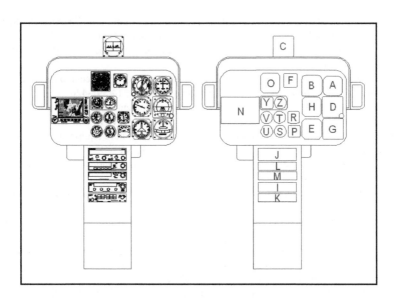

Figure 5: The instrument panel arrangement as the outcome of the optimization method

6   CONCLUSION

One of the early results of this study is the fact that the concepts of the degree of importance and the degree of frequency of use is not the primary parameter in an instrument panel arrangement. The principle of functional grouping plays the primary role in a panel layout. Any interface arrangement problem is a usability problem as well as a design problem. Therefore, the arrangement of the same display set used in this study in another instrument panel with a different shape will result in a different panel layout.

Two new concepts have emerged as the outcomes of this study: locational value and inseparability principle. Locational value of any display is strongly related with the use of the whole system and position of the main operator of the vehicle or system. During the study it has been observed that participants expressed that they prefer some displays must be located in a left-right or up-down relation to be more compatible with their function. This relation ship has been defined as inseparability principle or inseparability relation ship and can be asked to the participant as an interview question. On the other hand, it has been observed that the degree of importance or the degree of frequency of use of an individual display does not play a crucial role in the design of a panel but the degree of importance or the degree of frequency of use of a display group becomes more important. The main outcome of this study is the user-centered optimization method, contains both the locational value and inseparability principles in its structure.

REFERENCES

Channell R. C., (1947), An analysis of pilots' performances in multi-engine aircraft (R5D), Division of Bio-Mechanics, The psychological corporation, New York., NY

Chapanis A., (1965), Research techniques in human engineering, The Johns Hopkins Press, Baltimore, MD

Claman M., Kaber D. B., (2004), Application of usability evaluation techniques to aviation systems, *The international journal of aviation psychology,* 14:395-420

Fitts P. M., Jones R. E., Milton J. L., 1950, Eye movements of aircraft pilots during instrument-landing approaches, *Aeronautical engineering review,* 9:24-29.

Grey H. E., Flower S., (1943), Proposed instrument panel for future equipment, Engineering department, Pan American Airways System, New York, NY

Lovesey E.J., (1975), The helicopter, some ergonomic factors, *Applied ergonomics,* 6:139-146

Lovesey E. J, (1977), The instrument explosion – a study of aircraft cockpit instruments, *Applied Ergonomics,* 8:23-30

Hart S.G., (1988), Helicopter human factors, *Human factors in aviation,* Wiener E. L., Nagel D. C., Academic Press, New York, NY

Hawkins F. H., (2002), Human factors in flight, Ashgate Publishing Limited, Hants, UK

Kelley J. R., (1972), Display layout, *Displays and controls,* Bernotat R.K., Gartner K, P: (eds.), Swets&Zeitlinger, Amsterdam, Netherlands

Mcfarland R. A., (1946), Human factors in air transport design, McGraw Hill, New York, NY

Newman R. L., Greeley K. W., (2001), Cockpit displays: Test and evaluation, Ashgate, Publishers Ltd., Surrey, UK

Sanders M., McCormick S, (1993), Human factors in engineering and design, 7th edition, McGraw Hill, New York, NY

Senol M. B., Dagdeviren M., Cilingir C., Kurt M., 2010, Display panel design of a general utility helicopter by applying quantitative and qualitative approaches, *Human factors and ergonomics in manufacturing and service industries,* 1:73-86.

Section II

Air Traffic Control Interfaces and Operations

Evaluation of the Terminal Area Precision Scheduling and Spacing System for Near-Term NAS Application

Jane Thipphavong

NASA Ames Research Center
Moffett Field, California
Jane.Thipphavong@nasa.gov

Harry Swenson

NASA Ames Research Center
Moffett Field, California
Harry.N.Swenson@nasa.gov

Lynne Martin

San Jose State University
NASA Ames Research Center
Moffett Field, California
Lynne.Martin@nasa.gov

Paul Lin and Jimmy Nguyen

Optimal Synthesis, Inc.
Los Altos, CA
PLin@optisyn.com
JNguyen@optisyn.com

ABSTRACT

NASA has developed a capability for terminal area precision scheduling and spacing (TAPSS) to provide higher capacity and more efficiently manage arrivals during peak demand periods. This advanced technology is NASA's vision for the NextGen terminal metering capability. A set of human-in-the-loop experiments was conducted to evaluate the performance of the TAPSS system for near-term implementation. The experiments evaluated the TAPSS system under the current terminal routing infrastructure to validate operational feasibility. A second goal of the study was to measure the benefit of the Center and TRACON advisory tools to help prioritize the requirements for controller radar display enhancements. Simulation results indicate that using the TAPSS system provides benefits under current operations, supporting a 10% increase in airport throughput. Enhancements to Center decision support tools had limited impact on improving the efficiency of terminal operations, but did provide more fuel-efficient advisories to achieve scheduling conformance within 20 seconds. The TRACON controller decision support tools were found to provide the most benefit, by improving the precision in schedule conformance to within 20 seconds, reducing the number of arrivals having lateral path deviations by 50% and lowering subjective controller workload. Overall, the TAPSS system was found to successfully develop an achievable terminal arrival metering plan that was sustainable under heavy traffic demand levels and reduce the complexity of terminal operations when coupled with the use of the terminal controller advisory tools.

Keywords: terminal metering, arrival scheduling, controller decision support tools, air traffic control automation

1   INTRODUCTION

The nation's future air transportation system, known as the Next Generation Air Transportation System (or NextGen), is being designed to handle the predicted increases in traffic volume and to improve the capacity, efficiency and safety of the National Airspace System (NAS). NextGen goals include expanding the capacity of high-demand airports, while maintaining the efficiency of arriving aircraft.[1] Arrivals into high-density airports experience significant inefficiencies resulting from use of Miles-in-Trail procedures, step-down descents, and excess vectoring close to the airport. Use of these current procedures contributes to reducing airport capacity, increasing controller workload, increasing arrival delay, as well as increasing fuel burn, emissions and noise.[2]

NASA has developed a capability for terminal area precision scheduling and spacing (TAPSS) to increase the use of fuel-efficient arrival procedures during periods of traffic congestion at high-density airports. The TAPSS system is a 4-D trajectory-based strategic and tactical air traffic control (ATC) decision support tool (DST) for arrival management. In this concept as originally developed,[3] arrival aircraft are assigned optimized Area Navigation (RNAV) Standard Terminal Arrival Routes (STAR) prior to top-of-descent (TOD) with routing defined to a specific runway. The Precision Scheduler in the TAPSS system then computes an efficient schedule for these aircraft that facilitates continuous descent operations through the routing topology from TOD to landing. To meet this schedule, controllers are given a set of advisory tools to precisely control aircraft.

The TAPSS system was tested in a series of human-in-the-loop (HITL) simulations during the Fall of 2010 to evaluate the integrated performance of the precision scheduler and control tools. Results show a reduction in the complexity of terminal area operations, which in turn helps increase airport throughput without negatively impacting the environment. The performance of the TAPSS system over current operations was found to achieve up to a 10% increase in airport throughput with reduced controller workload.[3,4] The TAPSS advisory tools also resulted in aircraft maintaining continuous descent operations longer and with better scheduling conformance, under heavy traffic demand levels.[5] These previous HITL simulations explored the benefits of using the TAPSS system, with experiment assumptions made such that the operations concept could be deployable in 5-10 years. The TAPSS system, however, could provide benefits in the near-term (i.e., 3-5 years.)

This paper will focus on the results from a study performed in 2011 that evaluates the performance of the TAPSS system for near-term NAS application. The main research objective is to assess the TAPSS system under terminal routing infrastructure that more closely resembles current practices. The secondary objective of the study is to determine the incremental benefits gained when using the advisory control tools versus simpler control advisories. For comparison, metrics used to evaluate performance will be the same as previous HITL experiments. The paper is organized as follows: The next section describes the TAPSS operational concept and system. Section 3 details the experimental setup of the human-in-the-loop simulations. Results from the simulations are then discussed in section 4, which first discusses the evaluation of the TAPSS system under current procedures and then examines the benefit of each controller advisory tool. Section 5 concludes with a summary of key findings and plans for further research.

2   TERMINAL AREA PRECISION SCHEDULING AND SPACING SYSTEM OPERATIONAL CONCEPT

The TAPSS system is used for integrated arrival management between the Air Route Traffic Control Center (Center) and Terminal Radar Approach Control (TRACON) airspace. The TAPSS system consists of two major capabilities: 1) the time-based Precision Scheduler[6-7] and 2) the controller advisory tools. Arrivals are managed by the TAPSS system starting in Center airspace approximately 200 nmi from the airport. The Precision Scheduler provides the arrival sequence, scheduled times of arrival (STAs), runway assignments, and delay. Center controllers use this information to assign each arrival its RNAV STAR ending at its assigned runway. They are given an advisory tool, the Efficient Descent Advisor (EDA), to provide speed and path-stretch advisories to meet the meter fix STAs.[8-10] TRACON controllers are also given a set of advisory tools, called the Controller Managed Spacing (CMS) tools, which provide slot marker circles, speed advisories, early/late indicators, and timelines to meet STAs to meter points in the terminal area.[11-13] Flight crews fly VNAV (Vertical NAVigation) descents along the RNAV approach and follow any controller clearances.

3   EXPERIMENT DESIGN

3.1   Simulation Environment

The HITL simulations were conducted during the Fall of 2011 at NASA Ames Research Center using the Multi-Aircraft Control System (MACS) simulation platform.[14-15] MACS provides high-fidelity display emulations for air traffic controllers/managers as well as user interfaces and displays for confederate pilots, experiment managers, analysts, and observers. MACS also has flight deck capabilities that simulate current-day flight technologies that allow controllers to issue ATC clearances. The Center and TRACON controllers worked with operational emulations of radar displays. The Aeronautical Datalink and Radar Simulator (ADRS) served as a communication hub to provide the networking infrastructure that allowed the necessary information to be transferred between the precision scheduler and controller advisory tools.

3.2   Airspace

Los Angeles International Airport (LAX) arrivals were modeled using the West flow runway configuration with runways 24R and 25L runway under Instrument Meteorological Conditions (IMC). Figure 1 illustrates the STARs modeled in the simulation. The RIIVR and SEAVU STARs are used by Westbound traffic, accounting for more than 50% of the arrival traffic. These arrivals may be assigned to either 24R or 25L as determined by the Precision Scheduler runway balancing algorithms. Approximately a third of the traffic arrives on the KIMMO and SADDE STARs and only use runway 24R. The rest of the arrivals from the South are always assigned runway 25L. Arrivals into LAX currently have an aircraft mix of approximately 85% jets.

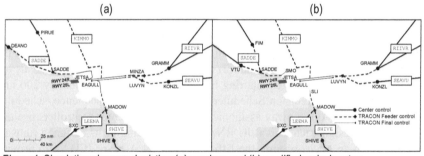

Figure 1. Simulation airspace depicting (a) previous and (b) modified arrival routes.

The simulation airspace is segregated into two main areas of control: Los Angeles Center (ZLA) and Southern California (SoCal) TRACON. Figure 1 shows the portion of the arrival route each of these areas was responsible for, along with their associated metering points. The ZLA controllers were responsible for managing each LAX arrival starting approximately 70 miles before its TOD and ending at its entry into terminal airspace located near the meter fixes. For simulation purposes, several of these sectors were combined so that three Center controllers were responsible for the Northwestern (i.e., DEANO and PIRUE), Eastern (i.e., GRAMM and KONZL) and Southern (i.e., SXC and SHIVE) STARs. Likewise, three TRACON Feeder controllers handled the next section of the route from the Northwestern (SADDE), Eastern (MINZA and LUVYN) and Southern (MADOW) arrival flows. The TRACON Feeder controller managing the Southern flows also controlled aircraft on the KIMMO STAR. The last aircraft hand-off is given to one of the two TRACON Final Controllers managing final spacing to LAX runways 24R and 25L respectively.

HITL simulations were conducted using a modified version of the terminal routing infrastructure to better model current operations. The previous route design is shown in Figure 1a, and Figure 1b illustrates the following changes:

- *The SADDE STAR starts at VTU and FIM.* The Center controller responsible for the SADDE arrivals previously used DEANO and PIRUE as the meter fixes into the terminal area. The TRACON boundary was relocated to its actual location closer to LAX, where FIM and VTU are the meter fixes.
- *The SADDE STAR ends at SMO, then arrivals are given heading 070 and the expected runway. The SHIVE and LEENA STARS ends at SLI, then arrivals are given heading 320 and the expected runway.* Previous simulations operated under the assumption that all arrival routes had complete RNAV routing directly transitioning to a Standard Instrument Approach Procedure (SIAP), which defines a series of predetermined maneuvers for the orderly transfer of an aircraft under IMC from the beginning of the initial approach to a landing. However, many published RNAV routes end with a specified heading near the TRACON boundary. Aircraft are then instructed to expect vectors onto the final approach.
- *Arrivals on the GRAMM and SEAVU STARs merge at LUVYN.* Past studies assumed independently operating runways, 24R and 25L. That is, arrivals on the RIIVR and SEAVU STARs were able to fly 'side-by-side' when landing on separate runways. Actual operational procedures prohibit such procedures and require these arrivals to be staggered by at least the wake separation minima until the final "capture box." LUVYN was used as a metering point for the GRAMM and SEAVU arrival flows. Thus, having the GRAMM and SEAVU arrivals merge at LUVYN allowed the Precision Scheduler to incorporate the necessary spacing.

3.3 Scenario

The simulation scenarios were based on current LAX traffic characteristics with approximately 60 minutes of traffic starting outside the Center boundary. Current airport arrival demand ranged from 55-72 aircraft per hour. Two scenarios were created, one with current LAX arrival-demand levels and the second with baseline arrival demand increasing 10%. For each level of demand, two variations of the scenarios were created with different call signs and start times.

3.4 Test Conditions

To investigate whether using the TAPSS system could be beneficial for the current airspace structure, simulations were conducted using the modified routes

80

that closely match today's current LAX arrival operations and compared with those using the original routing from past simulations. These test conditions are labeled 'Mixed RNAV' (i.e., Case 2) and 'Full RNAV' (i.e., Case 1) respectively. To evaluate the benefit of the controller advisory tools in the 'Mixed RNAV' condition, simulations were run with

Table 1. Experiment matrix.

Each case ran with two scenarios with demand levels: 1) Baseline and 2) Baseline +10%

		Tools Available				
		Center	TRACON			
	RNAV	EDA	Timelines	Early/Late Indicator	Slot Marker	Speed Advisory
CASE 1: All Tools	Full	✓	✓	✓	✓	✓
CASE 2: All Tools	Mixed	✓	✓	✓	✓	✓
CASE 3: TRACON All Tools	Mixed		✓	✓	✓	✓
CASE 4: TRACON Partial Tools	Mixed		✓	✓		
CASE 5: No Tools	Mixed					

different tools available for use and labeled 'All/Partial/No Tools' (i.e. Case 2-5.) accordingly. All test cases ran with scenarios having the baseline demand level and the traffic demand increased by 10%. The experimental matrix is presented in Table 1.

3.5   Controller and Pilot Procedures

Eight controllers participated simultaneously to cover all positions and had experience using the TAPSS system from prior HITL simulations. All participants were recently retired (within the previous 2 years) from either SoCal TRACON or Los Angeles Center and had an average of 20 years of ATC experience.

The Center controller responsibilities included assigning the expected runway and STAR clearance prior to TOD for each aircraft in its sector, and ensuring that the aircraft met the STA at the meter fix. Pseudo pilots verified the STAR in the aircraft FMS display panel along with the appropriate runway. The Center controllers then either followed the EDA advisories (when available in Case 1 and 2) or used their own techniques to control aircraft to meet the meter fix STA using the delay countdown timer displayed next to the aircraft symbol shown in seconds. Next, the TRACON Feeder controllers received the aircraft from the Center controller and controlled to the meter points within their sector referencing the advisory tools available. Lastly, the Feeder controllers handed off the aircraft to the appropriate TRACON Final controller responsible for proper spacing to the runway. In the 'Mixed RNAV' case, the arrivals on the SADDE STAR were given to 24R Final controller on a set heading. It was the responsibility of the 24R Final controller to determine when to turn the aircraft from its downwind leg onto final. Controllers were encouraged to use vectoring as a last resort, utilizing speed control foremost to manage the arrival traffic.

4   RESULTS

The full use of the TAPSS system (i.e., the 'All Tools' case) is evaluated under the current routing infrastructure by comparing the scenarios using the modified routes (i.e., Case 1: 'Mixed RNAV') and the original routes (i.e., Case 2: 'Full RNAV'). The Center and TRACON advisory tools are evaluated next under the 'Mixed RNAV' cases, by measuring the system performance in the absence of using a subset of the tools (i.e. Case 2-5.). For illustration purposes, the scope of this paper shows results for one scenario with its baseline demand level increased by 10%. These results are representative of the data trends observed in both variations of the scenarios used in the simulation.

4.1   Mixed RNAV Procedures

The lateral paths of all jets in the scenario are shown in Figure 2. Figures 2a and 2b show the results when using the original routes (i.e., Case 1: 'Full RNAV-All Tools') and the modified routes (i.e., Case 2: 'Mixed RNAV-All Tools') respectively. The terminal area is magnified in Figures 2c and 2d for the Full and

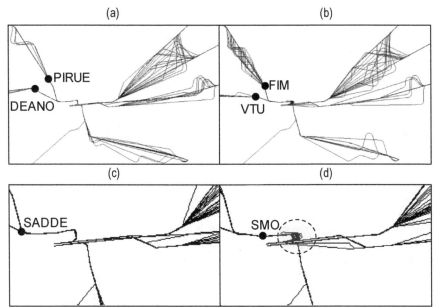

Figure 2. Lateral tracks for (a) Full and (b) Mixed RNAV-All Tools condition and corresponding magnified terminal area in (c) and (d).

Mixed RNAV cases.

Figures 2a and 2b indicate that arrivals are primarily vectored prior to the meter fixes, where the majority of the delay is absorbed at the Center level. There is noticeably more vectoring on Northwest arrival flows via SADDE in the Mixed RNAV case. Figure 2d also shows the 'tromboning' of the base leg in the Mixed RNAV case due to arrivals assigned a heading after SMO until further clearance. The throughput in both situations was found to be similar, where up to an 84 hourly arrival rate was sustained for an extended period. There are higher amounts of delay overall in the Mixed RNAV case, with larger differences in the Western arrival flows. The Mixed RNAV arrival flows via VTU and FIM have twice the amount of scheduled delays when compared to the Western flows via DEANO and PIRUE in the Full RNAV case.

The controllers were instructed to primarily use speed adjustments to absorb the scheduled delay. Excessive delay, however, may require path stretch maneuvers. Figure 3 shows the number of arrivals having flight path deviations that are more than 2.5 nm from their prescribed route in Center airspace and similarly, more than 1 nm deviation in the terminal area. Results indicate a greater number of off-route arrivals from the West (i.e. PIRUE/FIM), which is consistent with the amount of scheduled delay.

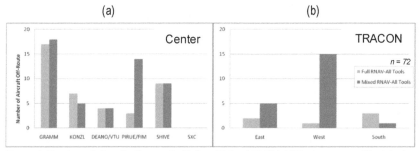

Figure 3. Number of aircraft off-route for Full and Mixed RNAV-All Tools test conditions in the (a) Center and (b) TRACON area.

The controller's ability to absorb the amount of scheduled delay (i.e., conform to the STA) is measured by examining the difference between the actual time-of-arrival (ATA) and the STA for each aircraft. The schedule conformance in the Mixed RNAV case varies in precision performance when compared to the Full RNAV case. Differences were within ±15 seconds and within the scheduling buffer used by the TMA scheduler to account for uncertainties in the system.

Workload data were collected in post-run questionnaires using the rating portion of the NASA TLX.[16] Controllers rated their level of workload on a scale from 1 "very low" to 7 "very high." The ratings were organized by the study condition and a mean was calculated for each TLX subscale. The mean ratings for

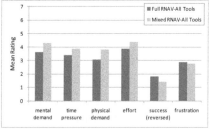

Figure 4. Mean ratings for TLX sub-scales in Full and Mixed RNAV-All Tools test conditions.

the Full and Mixed RNAV conditions are compared in Figure 4. The Mixed RNAV condition has the highest mean rating on every scale (success is reversed so the lower mean indicates higher success) except frustration. That is, participants rated the Mixed RNAV condition as having higher load on average but at the same time being more successful. However, the controller's average reported load is less than or close to 4, which is the midpoint of the scale, and so can be considered "manageable." These six pairs of ratings were compared using a Wilcoxon Signed rank test for non-parametric statistics. Participants' ratings on all the scales of the TLX were not significantly different between these two conditions.

The workload for the Final controllers was of particular interest. Their ratings were separated from the other controllers' ratings and compared. The comparison between the means of the Full and Mixed RNAV conditions is the same for the Final controllers as for the whole group of eight controllers. That is, the means for all the TLX subscales except frustration are higher, but not significantly higher, for the Mixed RNAV condition as seen in Figure 4. A point of interest is the differences between the 24R Final and the 25L Final controllers' ratings. The latter rated his/her workload lower under both conditions than the 24R Final. This could be a result of individual differences but also could indicate that the 24R Final was busier, possibly due to the complicated vectoring at the downwind turn to final from the SADDE and KIMMO arrivals. A second point of interest is the frustration scale, where both Finals consistently rated the Full RNAV condition as more frustrating than the Mixed RNAV condition. This could be due to controllers feeling more comfortable with practices that reflect current operations.

4.2   Center Advisory Tool

To evaluate the performance of EDA, three cases were run using using the TAPSS system without EDA (i.e., Case 3-5) and then compared with the Mixed RNAV-All Tools case (i.e., Case 2.) Figure 5 shows the average schedule conformance (i.e., ATA – STA) at the meter fix for each test condition. Controllers were able to meet the schedule more precisely without the use of EDA, with overall differences less than 15 seconds. The accuracy in EDA operations is limited to the corrective advisory tolerance, which was set to 20 seconds. The slight improvement in schedule confor-mance precision is possibly attributed

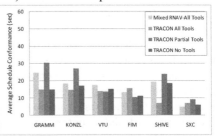

Figure 5. Average schedule conformance at each meter fix when using EDA (i.e., Mixed RNAV-All Tools) and without (i.e., all other test conditions).

to the delay countdown timer being displayed and updating in the resolution of seconds, thus allowing Center controllers to monitor performance in real-time.

When examining the number of flight path deviations that are more than 2.5 nm from their prescribed route in Center airspace, the use of EDA did increase the number slightly. This is due to the EDA tool advising path maneuvers taken at higher altitudes, which is calculated to be more fuel efficient. These path maneuvers at high altitudes will result in larger deviations because of higher ground speeds.

The average workload ratings given by Center controllers were calculated for each TLX subscale and compared in Figure 6. Center controllers reported the highest mean ratings on every scale in the Mixed RNAV-All Tools case, where they had the EDA tool available. That is, Center participants rated the condition where they had tools to use as having the highest load and feeling the least successful. However, when these

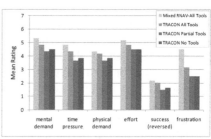

Figure 6. Mean ratings for TLX sub-scales for Center controllers.

ratings were compared using a Friedman two-way ANOVA for non-parametric statistics, there were no significant differences between them.

The mean ratings that have the largest difference were given for the frustration query, where the mean for the Mixed RNAV-All Tools condition was 4.5 and the combined rating was 2.7 for the runs where EDA was not used. Participant comments suggested that when EDA was in use, they were not able to receive an advisory for delays less than 20 seconds due to the corrective advisory tolerance set to 20 seconds. Controllers were then frustrated that the delay countdown timer was not closer to zero, and they attempted to achieve better precision in cases where EDA was not in use.

4.3   TRACON Advisory Tools

The CMS tools were also examined similarly, by comparing the Mixed RNAV-TRACON All, Partial and No Tools conditions (i.e., Case 3-5). Figure 7 shows the average schedule conformance at the terminal meter points for each test condition. The average schedule conformance improves when the CMS tools were used. Better performance is seen when the entire set of CMS tools is in use versus a subset of tools.

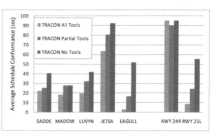

Figure 7. Average schedule conformance at each terminal metering point when using various subsets of the CMS tools.

(a)   (b)

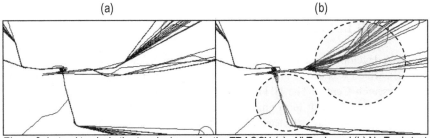

Figure 8. Lateral tracks in the terminal area for the TRACON (a) All Tools and (b) No Tools test conditions.

The lateral tracks in the terminal area for the TRACON All and No Tools case is shown in Figure 8. Although the TAPSS scheduler was used in all conditions, having performance monitoring tools reduces the variation in the lateral paths as highlighted in Figure 8. As a result, a more orderly flow is maintained in the terminal area, which facilitates high throughput to the runways.

Figure 9 shows the number of arrivals having flight path deviations that are more than 1 nm from their prescribed route in the terminal area. The No Tools condition has more arrivals off route, especially from the East side. This occurs less often when any of the CMS tools are in use.

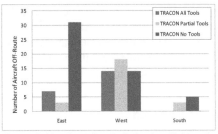

Figure 9. Number of arrivals having flight paths deviating more than 1 nm from prescribed route.

The workload analysis described in the previous section for the Center positions was repeated for the TRACON positions. The mean ratings for the TRACON controllers for the Mixed RNAV-TRACON All/Partial/No Tools conditions are compared in Figure 10. TRACON controllers reported their workload increased, on average, as the tools they had available decreased. Their highest mean workload ratings on all the TLX scales were for the No Tools condition and their lowest mean workload ratings were for the

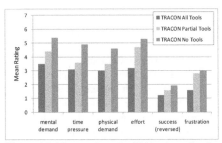

Figure 10. Mean ratings for TLX sub-scales for TRACON controllers.

TRACON All Tools condition, where the Centers did not have tools but they did. These differences are not significant for the physical demand, success or frustration scales but are significant at the P<.05 level for the mental demand, time pressure and effort scales. As an example, the mental demand ratings showed significant differences ($F(3,9) = 8.51$, p=.037), and post hoc tests indicated that the No Tools condition was reported as imposing greater mental demand than the TRACON All Tools condition. These differences among conditions may also account for the significant differences in the time pressure and the effort scales.

5 CONCLUSION

NASA developed a capability for terminal area precision scheduling and spacing (TAPSS), which was tested in a series of high-fidelity HITL simulations at NASA Ames Research Center. The HITL experiments evaluated the performance of the TAPSS system for near-term implementation by using the current-day routing infrastructure to validate the feasibility of the operational concept. The benefit of the controller advisory tools was also measured to help prioritize the requirements for controller radar display enhancements.

Simulation results indicate that using the TAPSS system provides benefits under current operations, supporting a 10% increase in airport throughput. The EDA tool had limited impact on improving the efficiency of terminal operations, but did provide more fuel-efficient advisories to achieve scheduling conformance within 20 seconds in the Center. The CMS tools were found to provide the most benefit, by improving the precision in schedule conformance to within 20 seconds, reducing the number of arrivals having lateral path deviations by 50% and lowering controller workload. Overall, the TAPSS system was found to develop an achievable arrival metering plan that was sustainable under heavy traffic demand levels, and to reduce the complexity of terminal operations when coupled with the use of the terminal controller advisory tools.

REFERENCES

1. Joint Planning and Development Office. (2010). Concept of Operations for the Next Generation Air Transportation System (Version 3.2 ed.). Washington, DC.
2. Federal Aviation Administration. (2010, May 3). Descent, Approach and Landing Benefits. Retrieved December 11, 2011, from http://www.faa.gov/nextgen/benefits/descent/
3. Swenson, H. N., Thipphavong, J., Sadovsky, A., Chen, L., Sullivan, C., & Martin, L. (2011). Design and Evaluation of the Terminal Area Precision Scheduling and Spacing System. Ninth USA/Europe Air Traffic Management Research and Development Seminar. Berlin, Germany.
4. Martin, L., Swenson, H., Thipphavong, J., Sadovsky, A., Chen, L., & Seo, Y. (2011). Effects of Scheduling and Spacing Tools on Controllers' Perceptions of the Their Load and Performance. Digitial Avionics Systems Conference. Seattle.
5. Thipphavong, J., Swenson, H., Lin, P., Seo, A., & Bagasol, L. (2011). Efficiency Benefits Using the Terminal Precision Scheduling and Spacing System. AIAA Aviation Technology Integration and Operations (ATIO) . Virginia Beach.
6. Swenson, H. N., Hoang, T., Engelland, S., Vincent, D., Sanders, T., Sanford, B., et al. (1997). Design and Operational Evaluation of the Traffic Management Advisor at the Fort Worth Air Route Traffic Control Center. 1st USA/Europe Air Traffic Management Research and Development Seminar. Saclay, France.
7. Wong, G. (2000). The Dynamic Planner: The Sequencer, Scheduler, and Runway Allocator for Air Traffic Control Automation. Moffet Field: NASA .
8. Coppenbarger, R., Dyer, G., Hayashi, M., Lanier, R., Stell, L., & Sweet, D. (2010). Development and Testing of Automation for Efficient Arrivals in Constrained Airspace. 27th International Congress of the Aeronautical Sciences. Nice, France.

9. Coppenbarger, R., Lanier, R., Sweet, D., & Dorsky, S. (2004). Design and Development of the En-Route Descent Advisor (EDA) for Conflict-Free Arrival Metering. AIAA Guidance, Navigation and Control (GNC) Conference. Providence, RI.

10. Hayashi, M., Coppenbarger, R., Sweet, D., Gaurav, N., & Dyer, G. L. (2011). Impacts of Intermediate Cruise-Altitude Advisory for Conflict-Free Continuous-Descent Arrival. AIAA Guidance, Navigation, and Control Conference. Portland, OR.

11. Callantine, T. J., & Palmer, E. A. (Sept. 21-23, 2009). Controller Advisory Tools for Efficient Arrivals in Dense Traffic Environments. 9th AIAA Aviation Technology, Integration, and Operations Conference (ATIO). Hilton Head, South Carolina.

12. Kupfer, M., Callantine, T. J., Mercer, J., Martin, L., & Palmer, E. (Aug. 2-5, 2010). Controller-Managed Spacing--A Human-In-The-Loop Simulation of Terminal-Area Operations. AIAA Guidance, Navigation, and Control Conference. Toronto, Ontario .

13. Kupfer, M., Callantine, T., Martin, L., Mercer, J., & Palmer, E. (2011). Controller Support Tools for Schedule-Based Terminal-Area Operations. Ninth USA/Europe Air Traffic Management Research and Development Seminar. Berlin, Germany.

14. Prevot, T. M. (August 2007). MACS: A Simulation Platform for Today's and Tomorrow's Air Traffic Operations. AIAA Modeling and Simulation Technologies Conference and Exhibit. Hilton Head, SC.

15. Prevot, T., Callantine, T., Lee, P., Mercer, J., Palmer, E., & Smith, N. (July 2004). Rapid Prototyping and Exploration of Advanced Air Traffic Concepts. International Conference on Computational and Engineering Science. Madeira, Portugal.

16. Hart, S. G., & Staveland, L. E. (1988). Development of the NASA-TLX (Task Load Index): Results of empirical and theoretical research. Amsterdam: P.A. Hancock and N. Meshkati.

CHAPTER 9

Data-driven Evaluation of a Flight Re-route Air Traffic Management Decision-support Tool

Lianna M. Hall, Ngaire Underhill, Yari Rodriguez, Richard DeLaura

Weather Sensing Group, MIT Lincoln Laboratory
Lexington, MA
lianna.hall@ll.mit.edu

ABSTRACT

Air traffic delays in the U.S. are problematic and often attributable to convective (thunderstorms) weather. Air traffic management is complex, dynamic, and influenced by many factors such as projected high volume of departures and uncertain forecast convective weather at airports and in the airspace. To support the complexities of making a re-route decision, which is one solution to mitigate airspace congestion, a display integrating convective weather information with departure demand predictions was prototyped jointly by MIT Lincoln Laboratory and the MITRE Corporation. The tool was deployed to twelve air traffic facilities involved in handling New York area flights for operational evaluation during the summer of 2011. Field observations, data mining and analyses were conducted under both fair and convective weather conditions. The system performance metrics chosen to evaluate the tool's effectiveness in supporting re-route decisions include predicted wheels-off error, predicted wheels-off forecast spread, and hourly departure fix demand forecast spread. The wheels-off prediction errors were near zero for half the flights across all days, but the highest 10% errors exceeded 30 minutes on convective weather days. The wheels-off forecast spread exceeded 30 minutes for 25% of forecasts on convective weather days. The hourly departure demand forecast spread was 9 flights or less for 50% of departures across all days except one. Six out of the seven days having the highest hourly departure demand forecast spreads occurred in the presence of long-lived weather impacts.

INTRODUCTION

Air traffic delays in the National Airspace System (NAS) are problematic. The New York region's airspace is highly congested, and problems originating from the region contribute to nearly three-quarters of NAS delays (Partnership for New York City, 2009). The Federal Aviation Administration (FAA) reports that 70% of delays are attributed to weather, a large portion of which are due to convective activity that is often localized and difficult to forecast (Hughes, 2011). Air traffic delays have increased during the months of the year in which convective weather predominates (Evans, 2001). Convective weather is unpredictable, greatly impacts air traffic, and adds challenges to air traffic management.

When convective weather occurs near airport terminal areas, traffic managers have difficult decisions to make to balance demand with capacity, such as choosing to delay flights, run traffic through impacted airspace, or reroute flights from impacted routes to non-impacted ones should they be available. Multiple factors are considered in departure planning, which include airspace resources such as departure fixes and jet routes. A departure fix is the first airspace location a flight passes through upon departing to a destination airport along a jet route. One departure fix may serve multiple air routes in the congested NY-area airspace. Demand includes the number of departures predicted over each fix and along each route; forecast weather impacts are considered for these resources and airports (Song, Bateman, Masalonis, & Wanke, 2009).

Departure management consists of planning and implementation activities among many stakeholders. Demand volumes over fixes and routes are considered strategically by national and regional air traffic managers to implement plans to mitigate congestion. Such plans, which may include reroutes, are subsequently implemented by regional and airport air traffic controllers in coordination with airline dispatchers and pilots (Beatty, Smith, McCoy, & Billings, 2000; Smith, Spencer, & Billings, 2007). To implement reroutes, impacted flights are identified, alternative routes are sought, sequencing of flights on the airport surface is considered, and flight plans are updated (Song et al., 2009). Air traffic managers coordinate with additional stakeholders such as pilots and airline dispatchers, but do so without the benefit of integrated tools that provide situational awareness of air traffic demand and forecasted weather impacts (Beatty et al., 2000; Song et al., 2009).

Decision-support tool components

To support the complexities of making a re-route decision, a display integrating convective weather information with departure demand predictions was prototyped jointly by MIT Lincoln Laboratory and the MITRE Corporation. The display and underlying system components are collectively called the Integrated Departure Route Planning (IDRP) tool, which provides real-time, integrated departure information (DeArmon, Jackson, Bateman, Song, & Brown, 2010). The purpose of the IDRP tool is to reduce the workload required to identify, plan, coordinate, and implement re-routes during convective weather (Masalonis et al., 2008).

The 2011 IDRP prototype is the second version of the system (DeLaura, Underhill, Hall, & Rodriguez, 2011); the first version was deployed in 2010 for a limited field study. Underlying the IDRP prototype are the Route Availability Planning Tool (RAPT) and the Corridor Integrated Weather System (CIWS), both of which are presented in Figure 1. RAPT provides 30-minute forecast weather impacts in 5-minute increments for each departure route (Robinson, DeLaura, & Underhill, 2008), and it has undergone a series of field evaluations to solidify and expand its operational concept (Davison Reynolds, DeLaura, & Robinson, 2010; Robinson et al., 2008; Robinson, DeLaura, Evans, & McGettigan, 2008). CIWS provides forecast weather on a geospatial display that allows detailed weather information to be provided (Robinson et al., 2008).

The 2011 IDRP prototype calculates departure demand forecasts using filed flight plans and current aircraft locations on the airport surface from the Airport Surveillance Detection Equipment, Model X (ASDE-X) where available. Course- and fine-grained departure forecasts across four visual components (DeLaura et al., 2011) are available; the first two are presented in Figure 1:

1. a fix list giving predicted departure demand and congestion alerts for each departure fix,
2. predictions of departure demand on each RAPT departure route,
3. a departure demand flight list that provides origin, destination, fix, flight plan, predicted departure time and RAPT status, and
4. a reroute alternative list with RAPT forecast and additional miles flown for each flight in the flight list.

The aggregate departure demand predictions, shown as totals in the right-most column of the fix list and alongside each route, provide course-grained information to support ATMs' strategic re-route planning. The aggregate departure demand predictions, shown in 15-minute bins in the fix list, along with the detailed flight and re-route lists (not shown), support fine-grained planning, coordination, and implementation of re-routes.

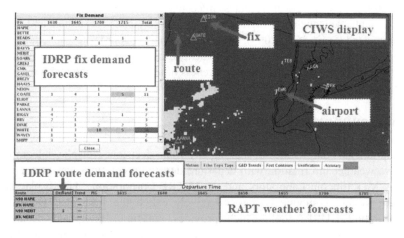

Figure 1. IDRP views for aggregate fix and route demand forecasts; NYC-area airspace.

Tool evaluation

The IDRP tool was deployed in the ZNY New York Air Route Traffic control Center (ARTCC), N90 New York Terminal Radar Approach Control (TRACON) facility, and ten other facilities for operational evaluation during the summer of 2011 (DeLaura et al., 2011). Field observations were conducted across the facilities, and data mining and analyses were performed on underlying system components, supplemented with reported flight departure (wheels-off) times and observed (true) weather impacts, to assess forecast accuracy and reliability. System stability, expressed using accuracy (correctness) and precision (consistency, reliability) measures, may impact the adoption or value of a decision-support tool. Disuse, an underutilization of automation, may result from an unstable system that can cause users to distrust the information (Parasuraman & Riley, 1997; Lee & See, 2004). Conversely, misuse, an overreliance of automation, may occur if users accept the information literally without regard for its reliability; in effect, causing an over trust of the information (Parasuraman et al., 1997; Lee et al., 2004; Smith, McCoy, & Layton, 1997).

METHODS

The tool's effectiveness in supporting re-route decisions was assessed using IDRP departure demand forecast issuances that were updated at a frequency of once every minute. The forecast performance metrics were evaluated separately for fair and convective weather days, as operations differ in the presence of convective weather. Departures from five major New York area airports were included in the analysis: Newark Liberty International, NJ (EWR); LaGuardia, NY (LGA); John F. Kennedy International, NY (JFK); Teterboro, NJ (TEB); White Plains, NY (HPN). Three of these airports (EWR, LGA, and JFK) have ASDE-X systems that provide information about aircraft location on the airport surface and in the immediate airspace. Forecast issuances for twenty-three hours of each day, ranging from midnight to 11:00pm local time (04:00Z to 02:59Z the next day), were included in the analysis.

Three forecast performance metrics were defined for the evaluation: predicted wheels-off error, predicted wheels-off spread, and predicted fix demand spread. The IDRP flights included in the predicted wheels-off analyses were limited by two additional criteria. First, the flights must correlate to reported Aviation System Performance Metrics (ASPM) flights for actual wheels-off times. Second, the flights must not have been rerouted, because their departure times may have significant changes due to the additional coordination required to implement reroutes. The predicted wheels-off forecasts were evaluated with respect to a planning horizon, which represents a decision maker's point of view for the time during which proactive reroute planning occurs to avoid convective weather and congestion. For the evaluation, a 30-minute planning horizon was used to align with the current RAPT status forecast limit. Figure 2 illustrates the forecasted

92

wheels-off issuances, and identifies the relationship between the forecasted wheels-off times and the planning horizon.

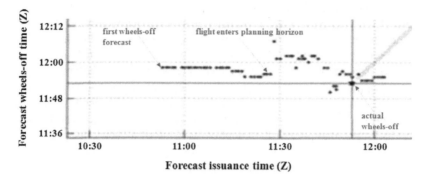

Figure 2. The decision-making planning horizon for a single flight's forecasts. The flight enters the 30-minute planning horizon at 11:27:00Z when the forecast wheels-off time is 11:56:21Z. Actual wheels-off time is 11:54:00Z; forecasts range from 11:51:00Z (earliest) to 12:07:39Z (latest).

The forecast issuance when the flight entered the planning horizon was used to calculate the predicted wheels-off error and magnitude metrics. For each flight, the predicted wheels-off error was calculated as the difference between the flight's actual wheels-off time and the predicted wheels-off time at the time the flight entered the 30-minute planning horizon. Given the predicted wheels-off issuances provided in the example flight illustrated in Figure 2, the wheels-off error is -00:02:41Z, which represents a late forecast. A highly volatile forecast may be difficult to use in planning, so the wheels-off forecast spread was assessed. The wheels-off forecast spread for a particular flight was defined as the difference between the latest and earliest predicted wheels-off times for that flight over the interval of time from the flight's entrance into the planning horizon until its wheels-off time. Given the predicted wheels-off issuances provided in the example flight illustrated in Figure 2, the wheels-off error spread is 00:16:39Z.

A highly volatile wheels-off forecast may likely manifest itself in the forecasted fix demand counts, so the aggregate hourly fix demand forecast spread was assessed. The hourly fix demand forecasts were selected instead of the more fine-grained 15-minute forecasts to focus on this strategic decision-support aspect. The total predicted hourly demand across twenty-four NY-area departure fixes, equal to the sum of hourly demand forecast from all twenty-four fixes, was calculated for each forecast issuance. The hourly fix demand forecast spread was defined as the difference between the largest and smallest total hourly fix demand for forecast issuances within a 15-minute time period.

RESULTS

The forecast performance metrics were evaluated for a total of twelve days spanning the summer of 2011, two fair weather and ten convective weather days. The scale (widespread or local) and duration (long-lived or short-lived) of convective weather impacts were reported in (DeLaura et al., 2011), which included: four days having short-lived, widespread impacts; one day having long-lived, local impacts; five days having periods of long-lived, widespread impacts.

The IDRP prototype was predominantly used in one facility, the New York TRACON (N90), out of the twelve facilities deployed. The supervisors and traffic management coordinators at N90, responsible for the airspace surrounding the NYC airports, were observed using the tool to monitor trends in fix demands, detect capacity overloads, and identify possible reroutes to avoid congested fixes. The field observer noticed flickering of the fix demand forecasts, but subjects did not explicitly comment on this tool behavior.

Forecast Wheels-Off Analyses

Over 15,000 departure flights were included in the two predicted wheels-off analyses. Wheels-off prediction times were generally constant (or infrequently changing) until aircraft entered into ASDE-X coverage, as shown in the example flight in Figure 2. From this point, wheels-off prediction times changed every minute. Several different prediction behaviors were observed. In some instances, predictions steadily converged toward the actual wheels-off time, and errors decreased as the actual wheels-off time approached. However, forecasts often showed considerable volatility, as the wheels-off time forecasts moved later and earlier, sometimes not approaching the actual wheels-off time until just a few minutes before takeoff.

Predicted wheels-off error measurements were made separately for convective and fair weather days, and are presented as a histogram and a line overlay, respectively, in Figure 3. Median errors are near zero minutes for both datasets. A negative wheels-off forecast error indicates that a flight departed before the predicted wheels-off time (a 'late' forecast). The error distribution falls off more slowly for convective days than fair weather days. Half of the wheels-off prediction errors on convective days fell within the error bound envelope of -10 and plus 12 minutes (except for August 25, when the error envelope reached 20 minutes). The extreme error bound – the ceiling for the highest 10% errors – ranged from 30 to 50 minutes on convective days (with the exception of August 25 and June 22, when the extreme error bounds were 70 minutes and 23 minutes, respectively).

94

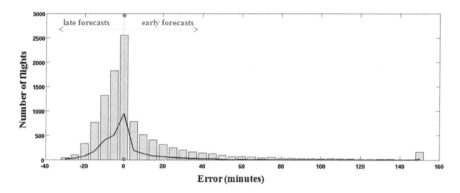

Figure 3. Histogram of wheels-off forecast errors for convective (bars) and fair (line) weather days

Wheels-off forecast spread measurements were made separately for convective and fair weather days. On convective days, the spread was typically 20 minutes or less for many flights, but there was a very long tail to the distribution. Wheels-off forecast spreads on fair weather days was 20 minutes or less for the majority of forecasts. The spread of wheels-off forecasts on convective days was generally around 30 minutes or less for 75% of departures (with the exception of August 25, when the spread was approximately 45 minutes). The extreme spread ranged from 50 to 70 minutes on convective days (with the exception of August 25, when the extreme spread was approximately 90 minutes).

Hourly Fix Demand Forecast Analysis

The hourly fix demand forecast spread statistics (10th, 25th, 50th, 75th, and 90th percentiles) were calculated for each individual day. Predicted hourly fix demand spread was 9 flights or less for 50% of departures (except for September 7, when the spread reached 14 flights). The hourly fix demand forecast spread on convective days was 19 flights or less for 75% of departures (except for two convective weather days, July 29 and September 7, whose spreads were 28 and 34 flights, respectively). The extreme forecast spread ranged from 17 to 55 flights on convective days, and ranged from 8 to 19 flights on fair weather days. Six out of the seven days having the largest spreads of the two most extreme spread bounds (75th and 90th percentiles) occurred in the presence of long-lived weather impacts. Three out of the four days having the smallest spreads of the two most extreme spread bounds incurred widespread weather impacts of short duration. An example of large forecast spreads in the presence of impacted weather is illustrated in Figure 4, which shows increased forecast spread starting 19:45Z after locally impacted weather starts around 18:00Z and continues for 8 hours.

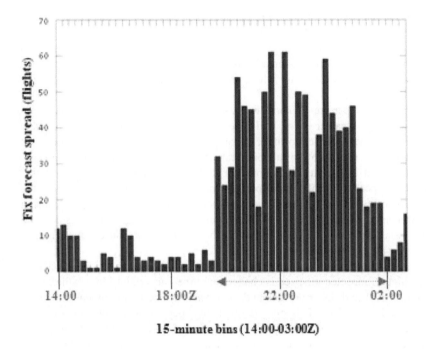

15-minute bins (14:00-03:00Z)

Figure 4. Hourly fix demand forecast spreads, in 15-minute bins, from July 19 at 14:00Z to July 20 at 03:00Z, in 15-minute bins; locally impacted weather from 18:00Z to 02:00Z, increased forecast spread from 19:45Z to 02:00Z.

DISCUSSION AND CONCLUSIONS

Wheels-off forecast accuracy and reliability are important because weather impacts on operations can vary greatly at different departure times and result in different traffic management decisions. Accuracy of the wheels-off prediction can influence the quality of traffic management decisions. Reliability of the wheels-off prediction, for individual flights and their contribution to aggregate hourly fix demand, may also impact the quality of a traffic management decision and may cause users to distrust the decision support.

The wheels-off forecast error metric revealed that, although half the flights had a near zero error across all days, over a quarter of flights had late predictions. The presence of late predictions can give a user the impression that a longer reroute decision time period is available than is actually the case. On convective weather days, errors for 10% of flights were beyond 30 minutes (i.e., 'early' predictions, where actual departure times were more than 30 minutes later than predicted) for all days except one; this exceeds the planning horizon available for users to proactively implement a reroute. The predicted wheels-off error was overall lower (the forecasts were more accurate) on fair weather days. On convective weather days, a quarter of the flights had a wheels-off forecast spread of 30 minutes or more, which increases the uncertainty of departure demand as it may give an inaccurate picture of congestion. The hourly fix demand forecast spread was generally lowest on the two fair weather days and on convective days having short-lived, widespread weather impacts. The forecast spread was highest on convective days characterized by long-lived weather impacts, where most days also had widespread weather impacts.

Overall, the departure demand forecasts were less accurate and reliable on severe convective weather days. Widespread weather impact conditions necessitate the use of impacted airspace to move departures, which itself has a high degree of uncertainty. The uncertainty of airspace capacity in turn makes departure capacity uncertain, which can make predictions about wheels-off times difficult to make. Although the perception of system performance was not explicitly measured, the system instability revealed in this study and noted by a field observer may cause a series of unanticipated consequences in the tool's use. What is not clear is how the system instability affects decision making and whether it causes over-control, paralysis, or poor decisions.

Developing a decision-support tool to enable air traffic managers to effectively manage highly impacted airspace is challenging given uncertainty in weather, pilot behavior, arrival and departure demand, and performance of individual air traffic managers and controllers. This study defined novel performance metrics to evaluate the IDRP tool from a user's point-of-view, exposed areas of system instability, and established specific areas of interest to investigate further. Models that relate errors and reliability in wheels-off, fix demand, and weather impact forecasts to departure throughput should be developed to assess the costs of forecast uncertainty and to determine meaningful forecast requirements. Algorithm improvements that trade dampened forecast response for improved stability should be explored. Finally, a detailed analysis of variations in the underlying flight and route lists may shed light on the usefulness of this tool component in support of reroute implementations.

ACKNOWLEDGEMENTS

The authors wish to acknowledge Richard Ferris, Ngaire Underhill, Darin Meyer, Diana Klingle-Wilson, Hayley Reynolds, and Brad Crowe for their field work in the facilities. This work was sponsored by the Federal Aviation Administration under Air Force Contract No. FA8721-05-C-0002. Opinions, interpretations, conclusions, and recommendations are those of the authors and are not necessarily endorsed by the United States Government.

REFERENCES

Beatty, R., Smith, P.J., McCoy, C.E., Billings, C.E. (2000). Design Recommendations for an Integrated Approach to the Development, Dissemination and Use of Reroute Advisories. *Institute for Ergonomics Technical Report.*

Davison Reynolds, H.J., DeLaura, R.A., Robinson, M. (2010). Field & (Data) Stream: A Method for Functional Evolution of the Air Traffic Management Route Availability Planning Tool (RAPT). *Proceedings of the Human Factors and Ergonomics Society 54th Annual Meeting. 27 September – 1 October. San Francisco, CA, 104-108.*

DeArmon, J., Jackson, C., Bateman, H., Song, L., Brown, P. (2010). Benefits Analysis of a Routing Aid for New York Area Departures. *IEEE/AIAA 29th Digital Avionics Systems Conference (DASC),3-7 October. Salt Lake City, UT, 1.D.3-1-1.D.3-8.*

DeLaura, R.A., Underhill, N., Hall, L.M., & Rodriguez, Y. (2011). Evaluation of the Integrated Departure Route Planning (IDRP) Tool 2011 Prototype. *Project Report (in press), MIT Lincoln Laboratory, Lexington, MA.*

Evans, J.E. (2001). Tactical Weather Decision Support To Compliment "Strategic" Traffic Flow Management for Convective Weather. *4th USA/Europe Air Traffic Management R&D Seminar, 3-7 December. Santa Fe, New Mexico, 1-11.*

Hughes, D. (2011). Application-Oriented NextGen Weather Research. *Journal of Air Traffic Control, 54*(1), 30-33.

Lee, J.D., See, K.A. (2004). Trust in Automation: Designing for Appropriate Reliance. *Human Factors, 46*(1), 50-80.

Masalonis, A., Bateman, H., DeLaura, R., Song, L., Taber, N., & Wanke, C. (2008). Integrated Departure Route Planning. *27th Digital Avionics Systems Conference, 26-30 October. St. Paul, MN.*

Parasuraman, R., Riley, V. (1997). Humans and Automation: Use, Misuse, Disuse, Abuse. *Human Factors, 39*(2), 230-253.

Partnership for New York City (2009). Grounded: The high cost of air traffic congestion. Retrieved February 9, 2012, from http://www.pfnyc.org/reports/2009_0225_airport_congestion.pdf.

Robinson, M., DeLaura, R.A., Evans, J.E., McGettigan, S. (2008). Operational Usage of the Route Availability Planning Tool (RAPT) During the 2007 Convective Weather Season. *Project Report ATC-330, MIT Lincoln Laboratory, Lexington, MA.*

Robinson, M., DeLaura, R., Underhill, N. (2009). The Route Availability Planning Tool (RAPT): Evaluation of Departure Management Decision Support in New York during the 2008 Convective Weather Season. *Eighth USA/Europe Air Traffic Management Research and Development Seminar (ATM2009). Napa, CA.*

Smith, P.J., McCoy, C.E., Layton, C. (1997). Brittleness in the Design of Cooperative Problem-Solving Systems: The Effects on User Performance. *IEEE Transactions on Systems, Man, and Cybernetics-Part A: Systems and Humans, 27*(3), 360-371.

Smith, P.J., Spencer, A.L., Billings, C.E. (2007). Strategies for designing distributed systems: case studies in the design of an air traffic management system. *Cogn Tech Work, 9*, 39-49.

Song, L., Bateman, A., Masalonis, A., Wanke, C. (2009). Integrated Collaborative Departure Traffic Management. *9th AIAA Aviation Technology, Integration, and Operations Conference (ATIO), 21-23 September, Hilton Head, SC, AIAA, Reston, VA.*

CHAPTER 10

Controller-Managed Spacing within Mixed-Equipage Arrival Operations Involving Flight-Deck Interval Management

Christopher Cabrall, Todd Callantine, Michael Kupfer, Lynne Martin, and Joey Mercer

San Jose State University
NASA Ames Research Center
Moffett Field, CA, USA
Christopher.D.Cabrall@nasa.gov

ABSTRACT

New NASA research focuses on integrated arrival operations along efficient descent profiles using advanced scheduling automation, tools to aid air traffic controllers, and airborne precision-spacing automation to enable fuel-efficient arrivals at busy airports during peak traffic periods. This paper describes an initial human-in-the-loop study and presents results that address human factors of controller tools and operational procedures for managing a mix of scheduled arrivals in which some aircraft use Flight-Deck Interval Management (FIM) automation to achieve precise spacing behind their lead aircraft. The results are consistent with prior research and suggest potential enhancements from the ground-side perspective to support mixed-FIM-equipage arrival operations.

Keywords: aircraft arrival management, air traffic controller tools, flight-deck interval management, scheduling

1 INTRODUCTION

A major Next Generation Air Transportation System (NextGen) goal is to enable low-noise, fuel-efficient arrivals with high throughput in congested metroplex areas (JPDO, 2010). Today, aircraft equipped with Flight Management Systems (FMSs) can fly Optimized Profile Descents (OPDs) along Area Navigation (RNAV) routes to provide the required environmental benefits. However, because current air traffic control techniques rely largely on heading adjustments and step-down descents, RNAV OPD operations are only feasible during light traffic conditions. Maintaining OPDs requires controllers to use speed adjustments as the primary means of control, which, without suitable tools, can be difficult (Davison-Reynolds, Reynolds, and Hansman, 2005). Tools for 'Controller-Managed Spacing' (CMS) of scheduled arrivals have been developed in the Airspace Operations Laboratory (AOL) at NASA Ames Research Center, and were found to be useful in enabling OPD operations in simulations with moderately high traffic levels (Kupfer, Callantine, Martin, Mercer, and Palmer, 2011).

A recently inaugurated NASA project called Air Traffic Management Demonstration-1 (ATD-1) seeks to operationally demonstrate the feasibility of fuel-efficient, high-throughput arrival operations using air- and ground-based NASA technologies and Automatic Dependent Surveillance-Broadcast (ADS-B) (Prevot et al., 2012). Under ATD-1, the CMS tools, an advanced arrival scheduler called the Traffic Management Advisor for Terminal Metering (TMA-TM), and advanced avionics for Flight-Deck Interval Management (FIM) will be integrated to form the Interval Management Terminal-Area Precision Scheduling System (IM-TAPSS) (Figure 1). The TMA-TM is an extension to the currently fielded TMA, which is a trajectory-based automation system developed at NASA Ames, that constructs an arrival schedule tailored specifically for high-capacity OPD operations (Swenson et al., 2011). FIM capabilities, as implemented in the Airborne Spacing for Terminal Arrival Routes (ASTAR) algorithm developed at NASA Langley Research Center, enable flight crews to assist air traffic controllers by managing their own speeds to

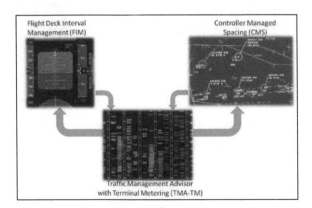

Figure 1. NASA-developed FIM capabilities, CMS tools, and TMA-TM, integrated as IM-TAPSS for ATD-1.

precisely achieve capacity-maximizing arrival spacing (Barmore, Abbott, and Capron, 2005). Following laboratory fine-tuning, verification, and validation, ATD-1 will implement IM-TAPSS in a field prototype for an operational demonstration at a U.S. airport, targeted for 2015.

This paper presents an initial human-in-the-loop simulation conducted, first, to integrate the IM-TAPSS components in the AOL to support follow-on ATD-1 simulations; and, second, to investigate how the CMS tools and operational procedures perform in a mixed-equipage environment where controllers manage the spacing of non-FIM-equipped ('CMS') aircraft, while flight crews manage the spacing of FIM-equipped ('FIM') aircraft. The integration goal was achieved to a sufficient degree to warrant preliminary research on enhancing the CMS tools and operational procedures. The paper first provides background on the IM-TAPSS component technologies and the operational concept IM-TAPSS supports. It then describes the first CMS ATD-1 simulation ('CA-1') in detail. After presenting results derived from the subjective data sets, including participant observation and feedback, post-trial and experiment-summary questionnaires, and a post-simulation debriefing discussion, the paper concludes with recommendations for further research.

2   BACKGROUND

While it principally serves to improve the efficiency of terminal-area air traffic management, IM-TAPSS starts functioning in en-route airspace up to 200 nmi from the terminal-area boundary. Before aircraft begin descending toward the destination airport, TMA-TM performs runway assignments and generates schedules at terminal-area entry fixes, merge points, and runways. The schedules help controllers maximize arrival capacity and strategically coordinate arrival flows from different en-route sectors. TMA-TM freezes the arrival schedule at a preset 'freeze horizon' to provide stable control targets. Schedules are presented as timelines (see Fig. 2) with estimated and scheduled times-of-arrival (ETAs and STAs) with aircraft symbols advancing down the timeline toward the current time at the bottom as they near the scheduling point.

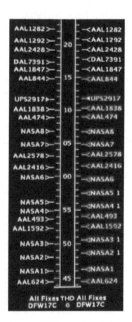

Figure 2. TMA-TM schedule timeline with ETAs on the left (in green), and STAs on the right (frozen STAs in blue; not-yet-frozen STAs in yellow).

Once aircraft are scheduled, en-route controllers can correct schedule errors with lateral maneuvers to absorb delay as they are typically more fuel-efficient at these higher altitudes. Controller tools to support en-route flow conditioning are also

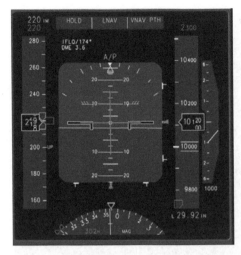

under development at NASA and elsewhere, but are beyond the scope of this paper. For FIM aircraft, controllers issue parameters required by ASTAR (e.g., the planned lead aircraft, the runway spacing interval), so that flight crews can enter them well in advance. Once engaged, ASTAR calculates speed targets for the aircraft to achieve and displays them to the crew. Figure 3 shows a Primary Flight Display (PFD) enhanced with the target speed and FIM speed 'bug' on the speed tape;

Figure 3. PFD with green FIM target speed (upper left), magenta crew-entered target speed below it, and matching green and pink speed bugs on the speed 'tape' (left side).

such information may instead be displayed elsewhere. The ASTAR algorithm also provides speed targets to meet a required time-of-arrival (RTA), while an aircraft is not yet within ADS-B range of its assigned lead aircraft.

Upon entry to the terminal-area FIM aircraft are typically following ASTAR speed commands, while controllers are responsible for issuing speeds to the CMS aircraft. 'Feeder' controllers use the CMS tools, which include schedule timelines, early/late indicators, slot markers, and speed advisories, to monitor and adjust the schedule conformance of arriving aircraft. 'Final' controllers primarily use spacing cones (a pre-existing tool available on some terminal-area controller displays) to space aircraft on final approach. FIM status designators in a FIM aircraft's data block (an '®' for RTA mode or 'S' for paired-spacing mode) have also been introduced as reminders controllers can enter to keep track of FIM operations. Figure 4 illustrates how these tools appear on a controller's Multi-Aircraft Control System (MACS) display in the AOL. Kupfer et al. (2011) describe the core CMS tools in detail; generally they range from simple representations of arrival-schedule information (timeline, early/late indicators), to nominal-trajectory-based translations of schedule information as spatial targets (slot markers), to speed advisories computed using trajectory predictions to place aircraft back on schedule.

3   CA-1 SIMULATION

CA-1 constituted the initial integration step for ATD-1 in which IM-TAPSS components were integrated in the AOL. It served the critical purpose of enabling researchers to begin to assess how the components function together operationally. To provide a preliminary perspective on the CMS tools and key aspects of air traffic

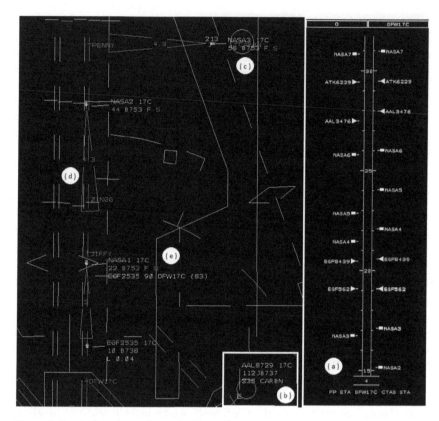

Figure 4. Terminal-area controller tools: (a) timeline, (b) data block with slot marker and speed advisory, (c) data block with slot marker, early/late indicator, and paired-spacing mode designator, (d) spacing cone, (e) RTA-mode designator.

controller procedures and clearance phraseology for managing mixed-FIM-equipage arrival flows, CA-1 was conducted as a fully staffed weeklong simulation in Dallas/Fort Worth (DFW) airspace (Figure 5). Aircraft flew charted OPDs on merging RNAV routes to DFW Runway 17C that were based closely on routes used in prior ASTAR research at NASA Langley.

Nine retired air traffic controllers took part in the simulation. Terminal-area controllers, all of whom had previously participated in CMS research, staffed three Feeder sectors (numbered 258, 259, and 264 in Figure 5) and one Final position (269). Three controllers staffed en-route sectors (24, 25 and 75) and the remaining two served as en-route 'Ghost' and Tower confederates. A mix of general aviation students and pilots flew eight Langley-developed, FIM-equipped Aircraft Simulator for Traffic Operations Research (ASTOR) single-piloted simulators and staffed eight MACS pseudo-aircraft stations used to control CMS aircraft.

The first day of CA-1 included an initial briefing, followed by four one-hour training sessions. Over the next four days, eighteen one-hour experimental trials

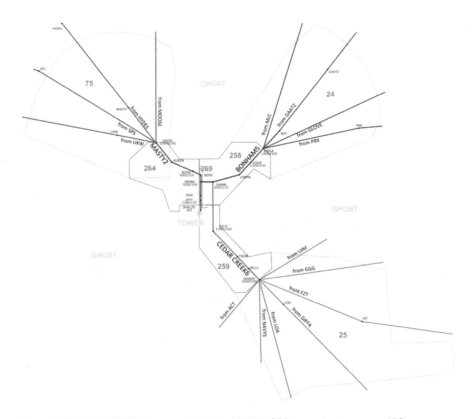

Figure 5. Simulated DFW airspace, with charted RNAV OPDs merging to runway 17C.

were conducted in which three traffic scenarios were presented twice (with different aircraft call signs) under each of three conditions: a 'Full ATD' condition with all the controller tools available, an 'ATD Lite' condition in which terminal controllers did not have timelines or speed advisories, and a baseline condition with no terminal-area controller tools and ASTORs participating but not conducting FIM operations. Digital data, including flight state information, pilot and controller entries, and schedule information, were logged from all MACS and ASTOR stations, as well as from TMA-TM. Controllers completed questionnaires after each experimental trial, as well as a comprehensive post-simulation questionnaire prior to the final debriefing session.

Controllers were asked to maintain charted OPD operations to the extent possible by primarily issuing speeds to separate and space aircraft. In the 'ATD' conditions, operations began with en-route controllers issuing FIM clearances to FIM aircraft while controlling CMS aircraft toward their terminal-area-entry-fix STAs. Outstanding integration issues limited the FIM aircraft to spacing only behind other FIM aircraft; in the traffic scenarios, FIM aircraft were interspersed in various sized clusters with the CMS aircraft. To enable close examination of the behavior of the FIM algorithm in an operational setting, controllers were also requested to allow FIM operations to proceed unimpeded, even when it appeared they should intervene to ensure proper spacing. While these instructions run counter to the IM-TAPSS concept of operations in which controllers are responsible for issuing clearances to FIM aircraft as necessary to ensure separation, they were deemed necessary for supporting the integration objectives of CA-1. The following section presents results from the CA-1 study.

4   RESULTS

A key hypothesis is that CA-1 would affirm performance trends and controller human factors results observed in previous CMS research for IM-TAPSS operations with mixed-FIM-equipage arrival flows. Subjective data from controller questionnaires, researcher observations, and the closing debriefing session show similarities in workload levels and tool preferences, as well as a general acceptance of operational procedures and clearance phraseology.

Post-trial workload ratings fell in a nominal mid-scale range; controllers most often rated workload as 'Manageable' and never rated it as 'Too high.' Real-time Workload Assessment Keypad (WAK) ratings logged every three minutes were generally low throughout CA-1 (M=1.56, SD=0.74; 1:'Very low', 7:'Very high'). FIM operations on average did not increase task complexity (M=4.05, SD=1.01; 1:'Substantially increased complexity', 7:'Made other tasks less complex'), according to combined post-trial responses for both 'ATD' conditions. Issuing and following up on FIM clearances also minimally impacted other controller tasks (M=1.75, SD=1.25; 1:'No interference at all', 7:'Interfered with all other tasks'), and had limited impact on the controllers' confidence in being able to achieve the required inter-arrival spacing at the runway threshold (baseline: M=6.29, SD=0.69; ATD conditions: M=5.96, SD=0.75; 1:'No confidence', 7:'Highest confidence'), according to post-trial questionnaire responses.

CA-1 afforded an opportunity to examine the clearance phraseology required to conduct FIM operations in a voice communications environment. En-route controllers, in particular, issued lengthy initial FIM clearances that included the lead aircraft, required spacing interval, achieve-by point, and RTA (e.g., "NASA5, for interval spacing, scheduled time at runway one seven center is one four, three two, plus fifteen Zulu; cross runway one seven center, four point three nautical miles behind NASA4 on the CEDAR CREEK arrival."). When asked to comment, post-simulation, on 'which clearances were smooth and which clearances were not smooth,' all controllers indicated they had few, if any, issues with the FIM clearances. In the post-simulation debriefing, controllers agreed that splitting the initial clearance into an RTA clearance and an interval-spacing clearance could be beneficial, and that issuing the RTA in two-digit 'chunks' helped read-back accuracy. Controllers also noted that an initial request to "advise when ready to copy" reduced the risk of read-back errors. Post-trial questionnaire reports show that controllers had to repeat FIM clearances 15% of the time; however, on average they found this rate to be 'Very acceptable' (M=5.96, SD=1.65; 1:'Completely unacceptable', 7:'Completely acceptable').

Communications involved with querying FIM pilots about the status of the FIM operations were also investigated. Controllers were observed to make such queries to resolve uncertainty about whether a FIM aircraft was conducting RTA or paired-spacing operations, or whether a pilot was successfully following FIM speed commands. In CA-1 controllers were generally amenable to delayed pilot responses to status requests, reporting that "stand by" responses were either 'always acceptable' or 'usually acceptable.' Confusion sometimes arose concerning the meaning of the term "following" when it was used in status requests, indicating structured phraseology for such requests would be beneficial.

Controllers commented positively about FIM aircraft transitioning from RTA to paired-spacing mode as the assigned lead aircraft entered ADS-B range. The data-block status designators ('®' and 'S') introduced to support FIM operations in CA-1 were generally well received. Controllers rated 'adding/removing/updating' the FIM status indicators favorably in the post-simulation questionnaire (M=5.86, SD=1.68; 1:'Very unreasonable, not at all workable', 7: 'Very reasonable, completely workable'). However, controllers agreed that a toggling scheme, in which the designator would change from '®' to 'S' to blank using a single repeatable command, would improve ease of use.

Other tools received ratings in line with prior research. Feeder controllers preferred the slot markers and used them most often to condition CMS aircraft for merging; the Final controller found the spacing cones most effective for managing spacing on final approach. Figure 6 depicts average Feeder controller ratings for various tools in the ATD conditions in which they were available. On the whole, helpfulness and usability ratings are similar for each tool. They were ambivalent about the spacing cones (used primarily by the Final controller) and the speed advisories (which require further research to be properly integrated with the TMA-TM scheduling scheme). While the helpfulness ratings of slot markers and early/late indicators for managing CMS or FIM aircraft are similar, usability ratings for both

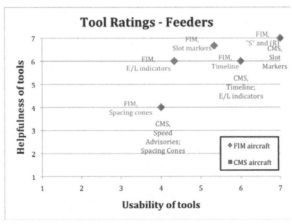

Figure 6. Tool helpfulness vs. usability ratings by the Feeder Controllers

tools are notably higher for CMS aircraft. This is likely due in part to allowing the FIM aircraft to descend unimpeded to support CA-1 integration objectives, so that both the early/late indicators and slot markers often reflected large schedule errors for FIM aircraft. The similar helpfulness ratings suggest controllers would welcome the ability to use these tools to manage FIM aircraft; responses to strategy-related questions in which controllers expressed a desire to have both CMS and FIM aircraft in their slot markers confirm this sentiment.

In the post-trial questionnaires, terminal-area controllers responded that they 'would not have done anything differently with regard to the FIM aircraft' 93% of the time. They also responded that they did not have to change how they worked to accommodate the FIM aircraft 80% of the time. Responses that indicated a need to work differently were generally accompanied by comments about discomfort with FIM aircraft well ahead of or behind their slot markers. Researcher observations suggest that controllers were not entirely sure about the performance of the FIM ASTAR algorithm relative to the TMA-TM arrival schedule reflected in the slot-marker locations. Controllers reported the presence of 'problematic' aircraft in 24% of post-trial questionnaires; the majority of these were FIM aircraft arriving early.

5   DISCUSSION

The CA-1 simulation was the first in a series of integration studies, and allowed an initial investigation of the performance of the IM-TAPSS components working in concert. Removing the limitations imposed to support integration objectives will improve operations and enable further analysis of the IM-TAPSS concept. First, while enabling FIM operations to proceed unimpeded provided useful insights about the operational behavior of the ASTAR algorithm—as well as increased opportunity for pilots to gain experience using it—it was inconsistent with controller separation responsibilities and control strategies. In succeeding simulations controllers will actively manage FIM traffic as necessary, which should improve spacing performance and controller acceptability. Second, en-route controllers did not actively condition the FIM aircraft in CA-1. Applying en-route control to mitigate schedule errors before clearing FIM aircraft to activate ASTAR is expected to allow the FIM aircraft to fly near-nominal speed profiles, and arrive in the terminal-area with small schedule errors. This would have several desirable effects, including enabling ASTAR to command near-nominal speeds, increasing the usability of the slot markers, and reducing the controllers' perceived need to intervene. Third, because the slot markers currently reflect the schedule and nominal speed profile, they do not provide feedback to the controllers on how well FIM aircraft are progressing toward the advised spacing. Further research will address the enhancement of slot markers for FIM aircraft in order to provide such feedback; this is likely to increase slot-marker usefulness and usability, and therefore controller acceptability. Lastly, the simulation software used for CA-1 limited FIM operations to consecutive FIM aircraft, so that FIM aircraft behind CMS aircraft in the arrival sequence had only the RTA mode available. This resulted in controllers having to continuously monitor their spacing. Work is underway to remove this limitation, which should further reduce controller workload and merge-point conflicts, and improve the traffic flow to the Final controllers.

The CA-1 study showed controllers are receptive to the IM-TAPSS mixed-FIM-equipage arrival operations planned for ATD-1. Controllers found the workload acceptable, rated support tools positively, and generally agreed with the proposed FIM clearance phraseology. Research to address the above issues is in progress, and is expected to enable detailed analyses of key performance metrics (e.g., inter-arrival spacing accuracy, throughput), as well as refinements to IM-TAPSS that are necessary to support validation and prototype-development work for the ATD-1 demonstration.

ACKNOWLEDGMENTS

The authors acknowledge the dedicated efforts of Thomas Prevot (ATD-1 Integration Lead), Harry Swenson (ATD-1 TMA-TM Lead), and Brian Baxley (ATD-1 FIM Lead) in support of this research. The authors also thank the members of the Airspace Operations Laboratory who served as experimental observers, participant recruiters, trainers, and analysts.

REFERENCES

Barmore, B. E., T. S. Abbott, and W. R. Capron. 2005. "Evaluation of airborne precision spacing in a human-in-the-loop experiment", AIAA-2005-7781, Proceedings of the AIAA Aviation Technology, Integration, and Operations Conference, Arlington, VA.

Davison-Reynolds, H., T. Reynolds, and R. J. Hansman. 2005. "Human factors implications of continuous descent approach procedures for noise abatement in air traffic control", Proceedings of the Sixth USA/Europe Air Traffic Management Research and Development Seminar, Baltimore.

Joint Planning and Development Office. 2010. "Concept of Operations for the Next Generation Air Transportation System," Version 3.2, Washington, DC.

Kupfer, M., T. J. Callantine, L. Martin, J. Mercer, and E. Palmer. 2011. "Controller Support Tools for Schedule-Based Terminal Operations." Proceedings of the 9th USA/Europe Air Traffic Management Research and Development Seminar, Berlin.

Prevot, T., B. T. Baxley, T. J. Callantine, W. C. Johnson, L. K. Quon, J. E. Robinson, and H. N. Swenson. 2012 (in press) "NASA's ATM Technology Demonstration-1: Transitioning fuel efficient, high throughput arrival operations from simulation to reality", To appear in Proceedings of the Conference on Human-Computer Interaction in Aerospace (HCIAero 2012), Brussels.

Swenson, H., J. Thipphavong, A. Sadovsky, L. Chen, C. Sullivan, and L. Martin. 2011. "Design and Evaluation of the Terminal Area Precision Scheduling and Spacing System," Proceedings of the 9th USA/Europe Air Traffic Management Research and Development Seminar, Berlin.

CHAPTER 11

A Preliminary Investigation of Tower Flight Data Manager Safety Benefits

Amy L. Alexander, Tom G. Reynolds

MIT Lincoln Laboratory
Lexington, MA, USA
amy.alexander@ll.mit.edu, tgr@mit.edu

ABSTRACT

Improvements to current air traffic management technologies and techniques are required to move toward the next generation air transportation system (NextGen). The Tower Flight Data Manager (TFDM) is a prototype air traffic control system consisting of the: (1) Flight Data Manager (FDM) facilitating interaction with electronic flight data, (2) Tower Information Display System (TIDS) providing enhanced surveillance information, and (3) Supervisor Display providing a means for front line managers and traffic management coordinators to interact with strategic and tactical planning and decision support tools. Given that TFDM aims to enable safe and efficient operations under NextGen, it is critical to analyze potential safety impacts and determine what types of real-world safety issues can be prevented or mitigated by TFDM. With this goal in mind, we reviewed 560 National Transportation Safety Board (NTSB) and Aviation Safety Reporting System (ASRS) reports focusing on commercial air carrier operations over a five-year period. Over 100 reports were deemed relevant to TFDM and further analyzed to determine the likelihood that these safety-related events could have been mitigated or prevented by the key TFDM capabilities outlined above. A systematic method for generating probabilistic estimates of benefits for a technology not yet deployed was utilized to produce effectiveness ratings for the various TFDM components.

Keywords: air traffic control, NextGen, aviation safety

1   INTRODUCTION

Improvements to current air traffic management technologies and techniques are required to move toward the next generation air transportation system (NextGen). Over the next several decades, the Federal Aviation Administration (FAA) projects a significant increase in air traffic in the National Airspace System (NAS). Existing air traffic control towers will need to manage this growth while meeting NextGen targets for safe and efficient surface operations. The Tower Flight Data Manager (TFDM) is a prototype air traffic control system designed to help address these needs. TFDM's consolidated display system consists of the:

(1) Flight Data Manager (FDM) facilitating interaction with electronic flight data (Figure 1a). Most towered airports will receive the FDM to support the electronic distribution and tracking of flight data and clearances, supporting situation awareness and reducing workload associated with maintaining an accurate picture of the traffic situation in increasingly complex circumstances.

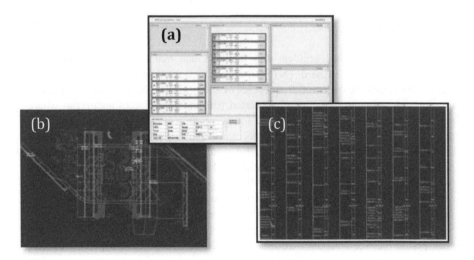

Figure 1 (a) Flight Data Manager (FDM); (b) Tower Information Display System (TIDS); and (c) Supervisor Display.

(2) Tower Information Display System (TIDS) providing enhanced surveillance information (Figure 1b). Airports with surface surveillance capabilities (i.e., Airport Surface Detection Equipment, Model X (ASDE-X)) will also receive the TIDS and therefore benefit from enhanced processing of surveillance data, enabling both intent and state-based conflict detection.

(3) Supervisor Display providing a means for front line managers and traffic management coordinators to interact with strategic and tactical planning and decision support tools (Figure 1c). These tools (e.g., taxi conformance

monitoring) will be introduced at key airports to allow controllers to monitor and manage traffic more effectively and to provide advance notice of hazardous situations.

Given that TFDM aims to enable safe and efficient operations under NextGen, it is critical to analyze potential safety impacts and determine what types of real-world safety issues can be prevented or mitigated by TFDM. With this goal in mind, we conducted a data-driven safety assessment involving a comprehensive review of aviation accident and incident databases to determine the likelihood that these safety-related events could have been mitigated or prevented by the key TFDM capabilities outlined above. Similar assessments have been conducted for other aviation systems such as Runway Status Lights (RWSL; Wilhemsen, 1994). In this paper, we report the number of safety-related events deemed relevant to TFDM by controller position, weather conditions, flight phase and contributing factor (e.g., decision error, adverse mental state) to provide contextual information regarding the types of incidents that can be addressed by TFDM. A systematic method for generating probabilistic estimates of benefits for a technology not yet deployed was then utilized to produce quantitative effectiveness ratings for the phased TFDM components.

2   METHOD

We utilized archived accident and incident data maintained by the National Transportation Safety Board (NTSB) and Aviation Safety Reporting System (ASRS) to determine observed frequencies of safety events. NTSB reports represent thorough investigations of events associated with the operation of an aircraft where any person suffered death or serious injury or any aircraft received substantial damage. ASRS reports are subjective accounts about safety-related aviation events voluntarily submitted by pilots, air traffic controllers, and other aviation industry personnel. Although subject to limitations related to sampling and reporter bias (Chappell, 1994; Degani *et al.,* 1991), reported incidents provide valuable qualitative information regarding the types of hazards, accident precursors, and safety-related issues that could potentially be prevented or mitigated by TFDM.

We specifically reviewed Part 121 accidents/incidents that occurred over a five-year period between January 2005 and December 2009, focusing on Part 121 operations as they represent scheduled commercial air carriers generally operating out of controlled airports. Only completed NTSB investigations and ASRS reports submitted by tower air traffic controllers were utilized to ensure more accurate reporting of causal factors. These selection criteria produced a total event count of 560 (NTSB: 247; ASRS: 313).

A coding spreadsheet was developed to collect relevant data (e.g., airport, tower position) from each of the selected reports. Contributing factors were inferred by the analyst given information provided in the individual reports, and were systematically categorized according to the Department of Defense (DoD) Human Factors Analysis and Classification System (HFACS; Wiegmann and Shappell, 2001, 2003; DoD, 2005), presented hierarchically in Figure 2.

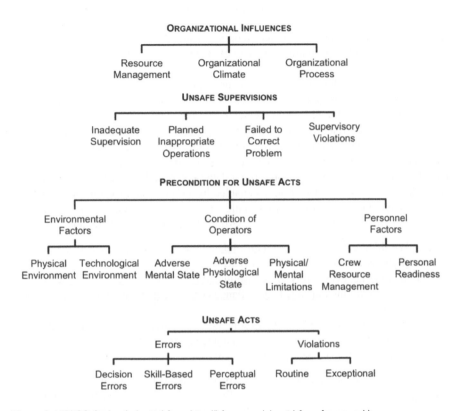

Figure 2 HFACS Codes (adapted from http://hfacs.com/about-hfacs-framework).

It is often the case that accidents and incidents involve multiple failures lining up across various system layers due to failed or absent defenses. Within an air traffic control context, latent failures in the "organizational influences" layer may involve inappropriate processes or a climate conducive to complacency. While TFDM does not directly address organizational influences, this layer is critical in that it can impact performance at all other levels. The next layer refers to "unsafe supervision" and captures strategic issues such as planned inappropriate actions (e.g., maintaining or choosing an airport configuration not aligned with environmental constraints). Decision support tools provided through the TFDM Supervisor Display provide support for this layer. Moving to the next layer in the model, "preconditions for unsafe acts" includes both environmental (e.g., reduced visibility) and operator state (e.g., high workload, low situational awareness) factors. TFDM provides defenses at this layer through improved surveillance and the consolidation of stove-piped systems, allowing easier access to information. The final opportunity for accident prevention is captured by the "unsafe acts" layer where errors or violations may take place. Decision errors (e.g., decision to issue takeoff clearance while another aircraft is landing) and perceptual errors (e.g., misjudging aircraft location) occur at this layer and are targeted by many aspects of

TFDM. Electronic flight data, for example, tracks aircraft state and provides earlier alerting to potentially hazardous conditions (e.g., runway incursions). In addition, decision support tools provide tactical support for monitoring the airport surface and alerting the controller to situations in need of attention (e.g., taxi non-conformance).

A systematic method for generating probabilistic estimates of benefits for a technology not yet deployed (Barnett and Paull, 2004) was then utilized to produce effectiveness ratings for TFDM components. An aviation human factors expert with piloting experience rated the likelihood that individual safety-related events could have been mitigated or prevented by TFDM components. The TFDM components were considered incrementally according to planned implementation phasing; namely, consolidation and electronic flight data (as enabled by the FDM), improved surveillance and conflict detection (as enabled by TIDS), and decision support tools (enabled by the full TFDM tool suite, including the Supervisor Display). Specifically, the rater considered three questions for each analyzed incident:

(1) Would the availability of consolidated/integrated systems and electronic flight data have prevented the event?
(2) Would the availability of consolidated/integrated systems and electronic flight data plus improved surveillance/conflict detection have prevented the event?
(3) Would the availability of consolidated/integrated systems and electronic flight data plus improved surveillance/conflict detection and decision support tools have prevented the event?

Responses to these questions were provided along a five-point scale ranging from "almost definitely no" to "almost definitely yes" with intermediate responses of "probably no," "50/50," and "probably yes." These responses were translated into probabilities as follows:

- Almost Definitely No   0%
- Probably No   25%
- 50/50   50%
- Probably Yes   75%
- Almost Definitely Yes   100%

This method allowed for calculations of incremental effectiveness per TFDM component as well as an aggregate effectiveness rating of the TFDM system as a whole.

3   RESULTS AND DISCUSSION

Following a thorough review of the 560 accident reports meeting the selection criteria defined previously, a subset of the reports were found to be relevant to the TFDM safety analysis. 129 of the reports (25 from the NTSB database and 104 from ASRS) were considered relevant to TFDM. Fifty airports were represented in the reports deemed relevant to TFDM. The median number of analyzed events per airport was one, with a range of one to 14 (five ASRS reports did not indicate airport).

Table 1 presents the number of NTSB and ASRS accident and incident reports associated with various tower positions, weather conditions, and phases of flight. While contributing factors are not presented in this table, we discuss which factors are most prevalent for each category. With respect to tower position, an overwhelming majority of analyzed events included a local controller (note that incidents may have involved more than one controller position). The local controller is responsible for the active runway surfaces, clearing aircraft for takeoff or landing and ensuring that prescribed runway separation exists at all times. Although the time spent during the takeoff and landing phases represents a small portion (~6%) of the total time spent in flight, over half of all accidents occur during these phases (Boeing, 2011). In looking at the breakdown of contributing factors specific to the local control position, the vast majority of cases involved adverse mental states (e.g., high workload; 47) and decision errors (e.g., inappropriate takeoff clearance; 39), reflecting the complexity (e.g., aircraft density, less decision making time, high demand for prompt action; see Koros et al., 2006) involved with operations during the takeoff and landing phases. Other key contributing factors to analyzed incidents involving a local controller include the technological environment (e.g., system failures; 26), skill-based errors (e.g., visual scanning disruptions; 23), and the physical environment (e.g., inclement weather; 22).

Table 1 Tower position, weather condition, and flight phase summary results by database.

	# of Reports		
	NTSB	ASRS	Combined
Tower Position			
Ground	10	17	27
Local	18	92	110
Local Assist	1	1	2
Supervisor	1	11	12
Weather Condition			
VMC	23	83	106
IMC	2	11	13
Marginal	0	4	4
Not Reported	0	6	6
Flight Phase			
Taxi	23	37	60
Takeoff	17	70	87
Climb	0	12	12
Approach	1	8	9
Landing	8	39	47

The majority of analyzed safety events occurred during visual meteorological conditions (VMC). In interpreting these findings, it is important to keep in mind the percentage of time spent in VMC versus instrument meteorological conditions (IMC; implying reduced visibility) at any given airport. For example, the FAA reported that in 2004, Atlanta (ATL) spent 73% of the time in VMC, 16% of the time in marginal VMC, and 11% of the time in IMC (Kang et al., 2007). Assuming similar distributions at other airports, it would be expected that a larger number of safety-related events would occur under VMC than IMC simply due to the amount of time spent under these conditions. With respect to contributing factors, adverse mental states (49) and decision errors (41) account for the vast majority of safety-related events during VMC, while the physical environment (8) was the biggest contributing factor during IMC.

Analyzed incidents represent the full range of tower operations in terms of phase of flight. It is clear from Table 1 that the majority of safety events involved at least one aircraft that was taking off (note that events involving multiple aircraft may represent more than one phase of flight). As mentioned previously, over half of all accidents occur during the takeoff and landing phases. Interestingly, more safety-related events occurred during the taxi phase than during landing across our analyzed incidents. Adverse mental states (44) and decision errors (43) are the leading contributing factors during the takeoff phase, while adverse mental states are most implicated during the taxi (32) and landing (17) phases. There are no clear trends for contributing factors during the climb or approach phases of flight.

As discussed previously, contributing factors to analyzed incidents were classified according to HFACS codes, the results of which are shown in Figure 3. Nine of the 19 HFACS codes were identified as relevant to the TFDM safety assessment. All safety events were associated with at least one HFACS code, with a range from 1 to 4 HFACS codes per incident (mean = 1.8). There appear to be three natural groupings of contributing factors in terms of their frequencies in contributing to analyzed incidents. Decision errors and adverse mental states comprised the majority of coded safety-related events. Skill-based errors and both the physical and technological environments form the second grouping. Perceptual errors, adverse physiological states, crew resource management, and planned inappropriate operations form the final grouping. From a human factors perspective, the breakdowns across these contributing factors could be utilized to drive design requirements in future systems to ensure that proposed solutions (technological or otherwise) actually address existing safety threats. For example, the high frequency of decision errors contributing to safety-related events points to the need for improved decision support systems within the air traffic control tower.

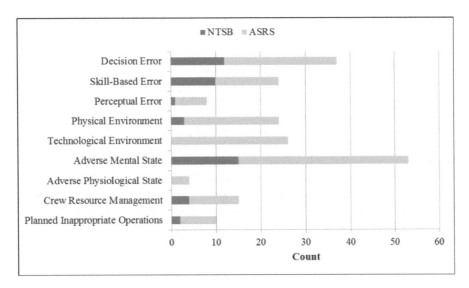

Figure 3 Contributing factors results by database.

Table 2 summarizes the aggregate effectiveness results (averaged over all NTSB and ASRS accidents/incidents considered) according to incremental benefits provided by a phased implementation of TFDM components. TFDM core implementation involving the consolidation of systems as well as the availability of electronic flight data has an estimated effectiveness of 25% in preventing or mitigating analyzed incidents. An example incident that could have been prevented or mitigated by electronic flight data is the Boston 2005 runway incursion (NTSB event ID 20050624X00863) in which two aircraft (EIN132, USA1170) were cleared for takeoff on intersecting runways within five seconds of one another; the FDM could have alerted the controller when the second aircraft (USA1170) was cleared for takeoff and the controller would have been able to immediately cancel the takeoff clearance.

Table 2 TFDM effectiveness ratings

TFDM Component	Estimated Effectiveness Rating
Consolidated/integrated systems and electronic flight data	25%
Plus improved surveillance/ conflict detection	42% increment
Plus decision support tools	15% increment
Total Effectiveness Rating	82%

Adding improved surveillance provides an average incremental effectiveness of 42%. Enhanced conflict detection enabled by this capability could have prevented or mitigated the San Francisco 2007 runway incursion (NTSB event ID 20070610X00701) in which the controller forgot about a landing aircraft (SKW5741) and cleared another aircraft (RPA4912) for takeoff on an intersecting runway. TFDM could have alerted the controller when RPA4912 was cleared for takeoff given surveilled information and conflict detection algorithms indicating that SKW5741 was crossing the landing threshold of the intersecting runway. Finally, adding decision support tools provides an average incremental effectiveness of 15%. An example incident that could have been prevented by a decision support tool is the Denver 2007 runway incursion (NTSB event ID 20070110X00037) in which an aircraft (LYM4216) missed its taxiway turn due to inclement weather and ended up turning onto an active runway on which another aircraft (FFT297) was attempting to land. Taxi conformance monitoring, specifically, could have alerted the controller to this situation when LYM4216 missed its intended taxiway. In total, full implementation of TFDM reveals an effectiveness of 82% in preventing or mitigating safety-related events across analyzed incidents.

4 CONCLUSIONS

This research involved analyzing a subset of NTSB and ASRS reports to determine the likelihood that these safety-related events could have been mitigated or prevented by any of three key TFDM capabilities: (1) consolidation/integration of systems and electronic flight data; (2) improved surveillance; and (3) decision support tools. The findings from this preliminary investigation of TFDM safety benefits indicate that while all three capabilities provide safety benefits in and of themselves, the largest impacts would be realized through the introduction of improved surveillance and enhanced conflict detection. In addition, the findings related to contributing factors, in particular, could be utilized to drive design requirements in future systems to ensure that proposed solutions (technological or otherwise) actually address existing safety threats. Next steps in this research could include expanding the sample size of accident/incident data by examining operational error data (i.e., FAA Operational Error/Deviation System), increasing the time period of interest, and looking at other aviation operations (e.g., Part 129 foreign air carriers, Part 135 air taxi and commuter, and Part 91 general aviation). Importantly, the safety benefits associated with TFDM capabilities are intrinsically linked to the nature of the accidents/incidents contained in the sample of analyzed reports. Expanding the sample size could greatly influence the incremental benefits associated with those capabilities.

The systematic approach used to estimate the effectiveness of TFDM components can be extended to other prospective NextGen systems, as well as to other domains (e.g., medical) in which archived safety-related data are available. To the extent that safety-related data can be represented as actual costs, efforts could be made to monetize safety benefits associated with the prevention of relevant accidents. Furthermore, effectiveness ratings and monetized values could be utilized in conjunction with extrapolation protocols to estimate potential safety benefits in future years. We are currently in the process of monetizing safety benefits associated with the implementation of TFDM over its expected 2015-2035 lifetime.

ACKNOWLEDGMENTS

The authors wish to thank Arnie Barnett, Daniel Howell, James Kuchar, and Jon Holbrook for discussions regarding this work. This work was sponsored by the Federal Aviation Administration under Air Force Contract No. FA8721-05-C-0002. Opinions, interpretations, conclusions, and recommendations are those of the authors and are not necessarily endorsed by the United States Government.

REFERENCES

Barnett, A. and G. Paull. 2004. Effectiveness analysis for aviation-safety measures in the absence of actual data. *Air Traffic Control Quarterly, 12*(3).

Boeing. 2011. Statistical summary of commercial jet airplane accidents. Retrieved from http://www.boeing.com/news/techissues/pdf/statsum.pdf.

Chappell, S. L. 1994. Using voluntary incident reports for human factors evaluation. In N. Johnson, N. McDonald, & R. Fuller (Eds.), *Aviation Psychology in Practice.* Aldershot, England: Ashgate.

Degani, A., S. L. Chappell, and M. S. Hayes. 1991. Who or what saves the day? A comparison of traditional and glass cockpits. *Proceedings of the Sixth International Symposium on Aviation Psychology,* Columbus, OH: The Ohio State University.

Department of Defense (DoD). 2005. Department of Defense Human Factors Analysis and Classification System: A mishap investigation and data analysis tool. Retrieved from http://www.uscg.mil/safety/docs/ergo_hfacs/hfacs.pdf.

Kang, L., M. Berge and M. Carter. 2007. Selecting representative weather days for air traffic analysis. *The Airline Group of the International Federation of Operational Research Societies (AGIFORS) Annual Symposium.*

Koros, A., P. S. Della Rocco, G. Panjwani, V. Ingurgio, and J-F. D'Arcy. 2006. *Complexity in airport traffic control towers: A field study. Part 2. Controller strategies and information requirements* (Technical Report DOT/FAA/TC-06/22). Atlantic City, NJ: Federal Aviation Administration William J. Hughes Technical Center.

Wiegmann, D. A. and S. A. Shappell. 2003. *A human error approach to aviation accident analysis: The Human Factors Analysis and Classification System.* Burlington, VT: Ashgate Publishing Company.

Wiegmann, D. A. and S. A. Shappell. 2001. Human error analysis of commercial aviation accidents: Application of the Human Factors Analysis and Classification System (HFACS). *Aviation, Space, and Environmental Medicine, 72*(11).

Wilhelmsen, H. 1994. Preventing runway conflicts: The role of airport surveillance, tower-cab alerts, and runway status lights. *The Lincoln Laboratory Journal, 7*(2): 149-168.

CHAPTER 12

Initial Evaluation of a Conflict Detection Tool in the Terminal Area

Savita Verma, Huabin Tang & Debbi Ballinger

NASA Ames Research Center
California, USA

Thomas Kozon & Amir Farrahi

NASA Ames Research Center/ UARC
California, USA

ABSTRACT

A conflict detection and resolution tool, Terminal Tactical Separation Assurance Flight Environment (T-TSAFE), is being developed to address the inadequacies of the legacy system in the field, Conflict Alert. Since altitude intent information improves T-TSAFE's conflict detection accuracy, this initial human-in-the-loop test of the tool had three T-TSAFE conditions: altitude clearances entered into the tool by the controllers via keyboard, altitude clearances entered into the tool from the flight deck via Automatic Dependant Surveillance-Broadcast (ADS-B), and no altitude entries. Entering altitude clearances into T-TSAFE was expected to reduce false alerts but did not, possibly due to the short duration of the runs. The test conditions also did not significantly impact the duration of the alert or the controller's response time to the alert. The subjective data showed that controllers favored T-TSAFE over Conflict Alert as used in the field due to the ease of use and perceived reduction in false alerts.

Keywords: conflict detection, terminal area, conflict alert

1   INTRODUCTION

Managing terminal area traffic is challenging due to the density of traffic and complexity of trajectories and separation standards. Conflict Alert is a short time horizon conflict detection tool currently in operational use in both the en route and terminal area, but it is often inhibited or desensitized in the terminal area because it generates a high number of false alerts. Conflict Alert uses only dead reckoning to determine when aircraft are in dangerous proximity to each other. Safety is maintained in the current system but at the expense of capacity, since controller workload is often considered as a limitation to capacity (Majumdar et al., 2002).

The objective of the current work is to develop a reliable and effective conflict alerting system. Tang et al. (2011) augmented the legacy dead reckoning approach with flight trajectory intent information to create a short time-horizon tool called Terminal Tactical Separation Assurance Flight Environment (T-TSAFE). The tool was developed to address the inadequacies of the operational Conflict Alert (CA). A comparison of T-TSAFE against a model of Conflict Alert found that T-TSAFE reduced the false alert rate and provided an average alert lead time of 38 seconds (Tang et al., 2011).

Previous Human-In-The-Loop (HITL) research on the predecessor TSAFE tool was conducted in the en route phase of flight (Homola et al., 2009). The current experiment tests the tool for the first time in the terminal environment with current day operations and technology. The experiment varied levels of altitude intent made available to the algorithm and assessed conflict detection performance as well as controller subjective feedback. Eight recently retired controllers served as participants, each controlling simulated traffic in the Southern California TRACON over approximately eight hours. Objective and subjective data are presented that characterize the performance of the Terminal TSAFE prototype.

2   BACKGROUND

The Federal Aviation Administration (FAA) has forecasted an increase in air traffic demand that may see traffic more than double by the year 2025 (JPDO, 2004) (FAA, 2009). Increases in air traffic will burden the air traffic management system, and higher levels of safety and efficiency will be required. Maintaining current levels of safety will be more difficult in a more constrained and crowded terminal air space. Thus, automation is proposed to aid the terminal area controllers with the task of assuring separation.

Terminal airspace has proven to be difficult for tactical conflict detection automation. The factors that contribute to this difficulty include dense traffic, frequent large turns made by aircraft, imprecise flight plans, a complex set of separation standards and the fact that the aircraft operate close to the minimum separation standards due to compression errors on approaches (Tang et al., 2011). In the current-day environment, Conflict Alert, a legacy system, was shown to be inadequate (Paeilli & Erzberger, 2007). Analysis of Conflict Alert by Friedman-Berg et al. (2008) shows that in the terminal area, controllers respond to CA alerts 56% of the time. Also, they analyzed the duration of the alerts, defined as the time between alert onset and the time when the conflict is resolved. Analyses of Conflict Alert durations showed that 36% of the alerts lasted less than 15 seconds. The short duration of CA alerts does not provide adequate time for the controller to respond and have any effect on the situation. Friedman-Berg et al. (2008) estimated that about 81% of all alerts in the terminal area are false, or nuisance, alerts. If automation can take into account aircraft intent, nuisance alerts are expected to decrease in number, improving the controller's trust in the credibility of the alert. En-route conflict detection automation incorporates some of this intent functionality already, incorporating information about routes and interim altitudes when calculating conflicts, but current terminal area conflict detection automation either lacks this functionality or controllers do not commonly use it.

Most of the research on the parent TSAFE tactical tool has focused on en-route airspace. En-route prototypes have been developed and HITL studies were performed at NASA Ames. These studies compared conflict detection and resolution done manually by the controller to conflict detection and resolution performed by TSAFE. The concept of operations in these studies required new technologies such as Automatic Dependent Surveillance-Broadcast (ADS-B) and a Required Navigation Performance (RNP) of 1.0, making it a mid- to far-term concept. All the aircraft in the test airspace were capable of performing trajectory-based operations via datalink communications. For the flights that maintained their trajectory, TSAFE was responsible for the detection and resolution of strategic conflicts. Trajectory changes to resolve conflicts were uplinked directly to the aircraft without the controller's involvement. Overall results showed that using TSAFE resulted in better resolution of tactical conflicts and fewer separation violations than without TSAFE. These studies investigated TSAFE in the en-route airspace with far-term concepts of operations and technologies. Adapting TSAFE to the terminal environment and using it with a near term or current day concept of operations became the focus of this study.

A new algorithm for tactical conflict detection (T-TSAFE) developed by Tang et al. (2011) aims to address the inadequacies of Conflict Alert and incorporates some of the recommendations made by Friedman-Berg et al. (2008). The T-TSAFE algorithm uses a single analytic trajectory that takes into account both flight intent information and the current state of the aircraft. In addition to the flight plan, it takes into consideration Area Navigation (RNAV) departure routes, segments of nominal TRACON routes, speed restrictions, and altitude clearances inferred from the recorded track data. Tang et al. (2011) compared the T-TSAFE algorithm with a model of the Conflict Alert algorithm using recorded data from Dallas/Fort Worth TRACON that included 70 operational errors. An analysis of fast-time simulation data showed that T-TSAFE yielded a false alert rate of 2 per hour with 38 seconds of lead alert time, giving the controller adequate time to address conflict situations before they became critical. When the algorithm has information about where and which aircraft will level off, fast-time analyses showed further significant reductions in false alerts. The potential benefit from additional altitude intent information was the rationale for asking controllers to enter some commanded altitudes in the current investigation.

Although Terminal-TSAFE has undergone considerable fast-time testing, this study was the first to test terminal TSAFE with humans-in-the-loop, and also to evaluate the algorithm under current-day operational conditions. This study also investigates the feasibility of having controllers enter the level-off altitudes, into the automation, which they also gave the pilots verbally for the purpose of reducing the number of false alerts.

3   METHODS

This study compared T-TSAFE's alerting performance with an emulation of Conflict Alert in a human-in-the-loop test. As noted earlier, TRACON controllers currently do not provide input into Conflict Alert with commanded altitudes, as is current practice for en-route controllers. Prior studies (Tang et. al, 2011) have shown the rate of false alerts drops dramatically if altitude intent information can be provided to conflict detection automation. This is because, with altitude information, the algorithm knows where aircraft will level off and will not predict conflicts based on a presumption of continued descent. This study included one condition that required the controllers to enter the altitudes at which they verbally commanded aircraft to level off, and another condition where the commanded altitude information was provided via short-term intent broadcast over ADS-B from the flight deck.

3.1   Experiment matrix

The three conditions in the experiment described in this paper are shown in Table 1. The experiment included a baseline condition where Conflict Alert was emulated, but those objective results are not described in detail in this paper. The Conflict Alert emulation used in this study was simply represented with only 1000 ft vertical separation and 3 nmi lateral separation requirements between aircraft, and it did not provide meaningful alert data. However, the subjective results comparing Conflict Alert and T-TSAFE still have value because controllers related their experience with the fielded version of Conflict Alert to the T-TSAFE tool. Thus, Conflict Alert data are only discussed in the subjective results.

Table 1. Experiment Matrix

Altitude Intent	Condition
No Altitude Entry	Condition A
Altitude Entry via Keyboard	Condition B
Altitude Entry via ADS-B	Condition C

The three T-TSAFE conditions described in the results were: T-TSAFE with no additional altitude information (Condition A); T-TSAFE with controllers entering verbally commanded level-off altitude, if the altitude was for conflict resolution (Condition B), and T-TSAFE where the controller assigned an altitude for conflict resolution by voice and an ADS-B broadcast returned the altitude entered by the pilot to T-TSAFE (Condition C). In conditions B and C, the assigned altitude was shown in green in the second line of the data tag with 'A' prefixed to the three-digit altitude. Four traffic scenarios were exercised with each of the conditions for a total of 12 runs.

3.2   Air Traffic Control Tools and Procedures

The study simulated five arrival streams and one departure stream into Los Angeles International Airport (LAX) using current airspace and procedures within the TRACON. The scenarios were designed to create many situations that would result in a loss of separation between aircraft unless a controller intervened. Controllers in the study occasionally were able to successfully avoid conflicts for extended periods due to early intervention. It was therefore necessary to add conflicts to the scenarios using observers collaborating with pseudo pilots. Each scenario contained the traffic equivalent of heavy current-day traffic, with all LAX traffic under Instrument Landing System (ILS) simultaneous rules. The controller's goal was to avoid losses of separation. The controllers worked the East Feeder and Zuma feeder sectors, and Downe and Stadium approach sectors in the Southern California TRACON, rotating positions after each run.

Figure 1. T-TSAFE data tags

T-TSAFE used 1000 feet minimum vertical separation, wake turbulence lateral separation standards, and a look-ahead time of 120 seconds to calculate conflicting trajectories. T-TSAFE predicted wake encounters as well as physical losses of separation, and factored in flight plan information and standard procedures, as well as using dead reckoning to predict conflicts. T-TSAFE alerts the controllers to a conflict by placing the number of seconds to predicted Loss of Separation (LoS) on the end of the first line in the data block, and the call sign of the conflicting aircraft in the third line of the data block, both in red. If more than one other aircraft is involved in a conflict, the third line shows the call sign of the aircraft closest to a loss of separation (Figure 1). The controller could also roll the cursor over any aircraft showing a conflict, and the data blocks of all other conflicting aircraft on the display would turn yellow for 5 seconds, e.g., UAL842 and MXA902 as other conflicting aircraft with the conflict shown in Figure 1.

3.3   Experiment Procedures

The study was conducted over a two-week period with two teams of controllers, each participating for a week. Each controller team consisted of four controllers that had retired recently from Southern California TRACON. Both controller teams were briefed on the T-TSAFE concept, the T-TSAFE interface, and the conditions of the study. During each week, the controller team completed twelve runs, four runs each of the three different conditions, rotating through four different traffic scenarios. Controllers rotated between sector positions after each run. Pseudo-pilots flew all the aircraft in the scenarios. All controllers completed questionnaires after every run and took part in a debrief session at the end of the study.

4   RESULTS & DISCUSSION

The metrics discussed in the paper include the following: total number of alerts, total number of false alerts, workload and usability of the T-TSAFE interface.

4.1 Total Number of T-TSAFE Alerts and Total Number of False Alerts

Figure 2 shows that the total number of alerts is similar between the three conditions. Each controller dealt with about 14 conflicts on average every run. There was no statistical difference in the number of alerts per condition. The false alerts were defined as a condition where an alert was provided, but the two aircraft did not lose separation, even though the controller did not intervene from 60 sec before the alert through the predicted loss of separation time. As shown in Figure 2, the number of false alerts was about 1 false alert per run, which were not statistically different between the different conditions. Contrary to the expectation, the altitude entries did not reduce the number of false alerts. This was possibly because controllers were able to enter commanded altitudes for conflict resolution only once or twice each run. It is likely that there was insufficient data to evaluate the impact of altitude entries on the number of false alerts.

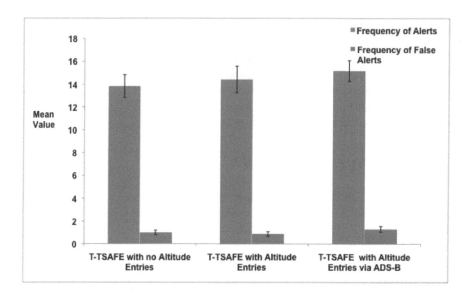

Figure 2. Total Number of T-TSAFE Alerts Compared with False Alerts
(Mean Value Per Controller Per Run)

4.2 Duration of the Alert and Response Time to Alert

The duration of alert was defined as the time between the alert onset and the time when the conflict was resolved, which is dependent on several factors. One factor was the look-ahead time that the T-TSAFE algorithm used to predict conflicts. The average duration of an alert across different conditions was about 110 sec (Figure 3). The mean alert durations were slightly shorter in the two conditions where altitude clearances were provided via keyboard or ADS-B, as compared to the condition with no altitude entries. However, these differences did not reach statistical significance.

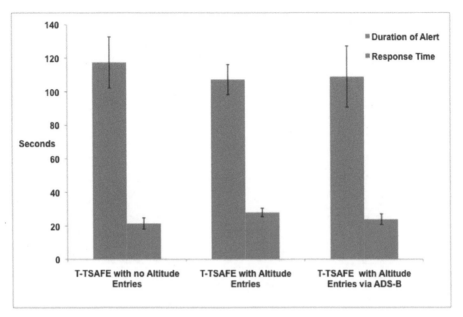

Figure 3. Comparison of Duration of Alert and Response Time to an Alert

The response time to an alert was measured as the difference between the onset of the alert and the time the pseudo pilot responded to the controller's commands. Figure 3 shows that the time to respond to an alert was 22 sec on average. It seems the duration of the alerts was more than adequate for the controller to see the conflict, determine a resolution, inform the pilots, and for the pilots to initiate the commands issued by the controllers. The controllers also mentioned that they used the "time to predicted loss of separation" provided in the data block to prioritize their tasks. The feedback provided by the controllers made it clear that in the TRACON environment they often address other high-priority emerging situations first and act to resolve a conflict in the last 25-30 sec prior to the predicted loss of separation.

4.3 Subjective data - Workload

Participants completed the NASA TLX workload questionnaire (Hart et al., 1988) after every run. As mentioned earlier, the subjective data include a baseline condition where the Conflict Alert model was simulated. Data were collected on each of the six TLX workload measures, and a variable measuring overall workload combining all six of these measures was derived, for a total of seven workload measures. The overall workload variable, also known as the "composite" measure, once derived, was then scaled down to match the 1-to-5 scale for direct comparison with the other six measures. Also, the "performance" measure was analyzed on an inverse scale, so a higher score would actually mean lower performance and a lower score is indicative of better performance. Results on all 7 of these measures, comparing the four experimental conditions, are summarized in Figure 4.

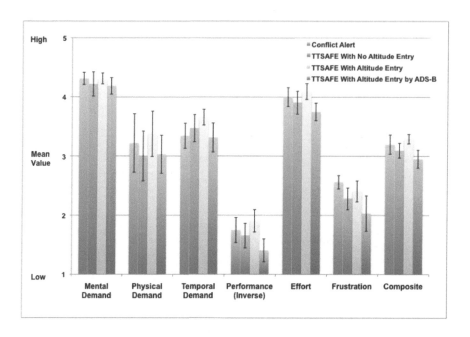

Figure 4. Workload Across the Experimental Conditions

The directionality of the mean values for most of the workload categories shows that controllers reported higher workload when using the Conflict Alert model and T-TSAFE runs where commanded altitude entries were made via keyboard, as compared to the other two conditions. TRACON controllers are not used to making data entries, and they often reported that this task increased their workload. Likewise, the controllers expressed many negative opinions about the Conflict Alert model, so it is not surprising that workload ratings were mostly higher for this condition. Conversely, for most of the workload categories, workload was rated lower during T-TSAFE runs where no commanded altitude entries were required and T-TSAFE runs where commanded altitudes were received via ADS-B. It seems likely that the lower workload ratings were due to the absence of required commanded altitude entries. However, it should be noted that these trends should be viewed with some caution, since the mean differences did not reach statistical significance.

4.4 Subjective data – Procedures

The controllers were asked questions about T-TSAFE and Conflict Alert procedures, which they rated on a scale of 1-to-5. These data were then analyzed using Analysis of Variance. Table 2 shows the mean ratings of the controllers' ability to maintain awareness of potential conflicts. Results indicated that the controllers found it significantly easier to maintain awareness of potential conflicts with T-TSAFE as compared to the Conflict Alert model ($F=7.99$, $df=[1,7]$, $p<0.05$).

These data were validated in discussions between researchers and controllers, where controllers indicated that having the Time-to-Conflict in the first line of the data block provided them with information that helped them prioritize their tasks.

Similarly, T-TSAFE was rated easier to use than Conflict Alert, possibly due to controllers' experience with the high number of false or nuisance alerts that are routine for Conflict Alert as it is currently implemented in the field. There was a significant difference in the acceptability of the T-TSAFE procedures as compared to Conflict Alert (F=5.73, df=[1,7], p<0.05). The explanation of the interface and the logic behind T-TSAFE given to the controllers might explain their levels of higher acceptability of T-TSAFE. The controllers also perceived that T-TSAFE alerts were more timely and useful than Conflict Alerts. This result is consistent with controller feedback, indicating that the duration of T-TSAFE alerts was adequate for the controllers to respond in time to prevent a loss of separation. A statistically significant difference was found for the perceived number of false alerts between T-TSAFE and Conflict Alert (F=8.27, df=[1,7], p<0.05). Again, the controllers were drawing on past experience with Conflict Alert in the field.

Table 2. Mean Subjective Ratings on Procedures (*p<0.05)

	T-TSAFE	SE	Conflict Alert	SE
*Maintain Awareness of Potential Conflict (1=very easy, 5=very difficult)	1.9	0.39	3.0	0.42
Ease of Procedures (1=very easy, 5=very difficult)	1.8	0.31	2.8	0.25
*Acceptability of Procedures (1= low, 5= high)	4.0	0.38	3.1	0.37
Timeliness of Alerts (1=barely enough time, 5=adequate time)	4.1	0.26	3.2	0.38
*Number of False Alerts (1=too many, 5=none)	3.8	0.44	2.5	0.42

5   CONCLUSIONS

Terminal T-SAFE is being developed by NASA as a conflict detection tool to address the inadequacies of Conflict Alert as it is currently used in the field. The selection of the independent variables in this study was guided by previous research on T-TSAFE showing that making altitude intent information available led to a reduction in false alerts. T-TSAFE received commanded altitudes either from controllers entering commanded altitudes issued for conflict resolution or as short-term intent over ADS-B. Results showed that there was no difference in the rate of false alerts across conditions. This could possibly be explained by the fact that controllers do not make too many altitude entries that were expected to impact the rate of false alerts. Also, the duration of the alerts was not affected by the conditions, and neither was the controllers' response time to the alerts.

Subjective data included results comparing a simplified model of Conflict Alert to the three T-TSAFE conditions. The controllers reported that they experienced similar levels of workload, with a slight increase in the physical component of the workload in the condition that required the controllers to make commanded altitude entries via the keyboard. The subjective data on the procedures were favorable towards T-TSAFE over Conflict Alert. The controllers found that they maintained better awareness of potential conflicts and had adequate time to act on the alerts with T-TSAFE. They also reported the T-TSAFE procedures as more acceptable and easier than those with Conflict Alert. Further human in the loop testing of the tool with longer run durations is necessary to better understand the impact of altitude entries on the overall performance of the conflict detection and resolution tool.

5   REFERENCES

Erzberger, H. (2001). The Automated Airspace Concept, Proceedings of the Fourth USA/Europe Air Traffic management R&D Seminar, Santa Fe, New Mexico, USA, December 2001.

Federal Aviation Administration (2009). FAA Aerospace Forecast Fiscal Years 2009-2025. Washington DC, 31.

Friedman-Berg, F., Allendoerfer, K. & Pai, S. (2008). Nuisance Alerts in Operational ATC Environments: Classification and Frequencies. Proceedings of the Human Factors and Ergonomics Society 52nd Annual Meeting.

Hart, S. & Staveland, L. (1988). Development of a Multi-Dimensional Workload Rating Scale: Results of Empirical and Theoretical Research, In P. Hancock & N. Meshkati (Eds.), Human Mental Workload, Amsterdam, The Netherlands, Elsevier, 139-183.

Homola, J., Prevot, T., Mercer, J., Manini, M. & Cabrall, C. (2009). Human/Automation Response Strategies in Tactical Conflict Situations. 28th Digital Avionics System Conference, October, 2009.

Joint Development and Planning Office (2004). The Next Generation Air Transport System Integration National Plan. December 12, 2004.

Kuchar, J. & Yang, L. (2002). A Review of Conflict Detection and Resolution Modeling Methods. *IEEE Transactions on Intelligent Transportation Systems*. Vol. 1, No. 4, December 2000, 179-189.

Majumdar, A., Ochieng, W. & Polak, J. (2002). Estimation of European Airspace Capacity from a Model of Controller Workload. *The Journal of Navigation*. 55. 381-403.

Paeilli, R.A. & Erzberger, H. (2007). Design and Testing of Tactical Conflict Alerting Aid for Air Traffic Controllers.

Prevot, T., Homola, J., Mercer, J., Manini, M. & Cabrall, C. (2009). Initial Evaluation of NextGen Air/Ground Operations with Ground-Based Automated Separation Assurance. Eight USA/Europe Air Traffic management Research and Development Seminar (ATM 2009). June 2009, Napa Valley, CA, USA.

Tang, H., Robinson, J.E. & Denery, D.G. (2011). Tactical Conflict Detection in Terminal Airspace. *Journal of Guidance, Control, and Dynamics*. Vol 34, No. 2, 403-413.

CHAPTER 13

Air Traffic Control Voice Data Analysis to Validate NextGen Procedures

Jonathan Histon[1], Esa Rantanen[2], Cecilia Ovesdotter Alm[2]

[1] University of Waterloo
Waterloo, ON, Canada
jhiston@uwaterloo.ca

[2] Rochester Institute of Technology
Rochester, NY
esa.rantanen@rit.edu, coagla@rit.edu

ABSTRACT

The NextGen ATC modernization effort is introducing new technologies and procedures affecting communication between pilots and controllers. This is creating opportunities for fundamentally new ways of providing air traffic control (ATC) services. In order to take the fullest advantage of these changes it is important to understand the communication needs of pilots and controllers in both current and future environments. This paper reviews previous taxonomies of the content of air-ground communications between pilots and controllers and the most recent findings on the type, frequency, and function of existing communication technologies. Methods of analysis and metrics of relevant variables derived from ATC voice data are presented in order to illustrate the rich and complex communication strategies controllers currently employ. Examples of these analyses and metrics using voice data from multiple sectors at the New York, Indianapolis, and Boston air route traffic control centers are discussed. An initial analysis of the application of these taxonomies and metrics to communications in advanced NextGen environments is presented in order to illustrate the challenges with identifying how to most effectively take advantage of new communication technologies associated with NextGen.

Keywords: air traffic control, NextGen, communications, taxonomies

1   INTRODUCTION

Current and near-future concepts of operations for air traffic control (ATC) rely on effective and efficient communications between the controllers on the ground and the pilots in the air. The limitations of existing voice technologies are well known and have been extensively documented (Van Es, 2004; Zingale, Mcanulty, & Kerns, 2005). They include cross-talk and interference on radio channels, challenges with understanding different languages, and tasks that scale with increasing levels of traffic. A new technology, controller-pilot datalink communications (CPDLC), replaces traditional voice based radio communications between pilots and controllers with a text-based datalink, often described as providing functionality similar to email. This can allow controllers to provide clearances, information items, pass along requests, or otherwise communicate with the pilots of an aircraft. With datalink capabilities, communications are often structured into a series of pre-formatted text-based message types though free text capabilities are also supported. Communication with an aircraft can be passed from one air traffic control sector to another, reducing overhead associated with routine handling of an aircraft. There are several technical standards providing implementation of CPDLC concepts, and its use is becoming increasingly common in oceanic and non-radar environments. Future Air Navigation System (FANS)-1/A provides CPDLC connectivity via the Aircraft Communications Addressing and Reporting System (ACARS) network and is primarily found in use in long-haul oceanic environments (Turner, 2011). Aeronautical Telecommunication Network (ATN)-CPDLC provides connectivity via the Aeronautical Telecommunication Network and is currently in use in the Maastricht Center in Europe (Turner, 2011).

While significant studies and operational deployments have been made of new, datalink enabled, communication technologies (e.g. Willems et al., 2010), there remains a need to develop analytical tools to understand how these new technologies will affect controller decision making and strategies. We review the previous work on both voice and datalink communication systems with a particular focus on the tools, taxonomies, and metrics used to analyze controller-pilot communications and their effect on controller decision making and strategies. Our objective is to document core metrics used for understanding voice based communications and compare and contrast voice-oriented metrics with metrics used for examining datalink communications. We summarize taxonomies used to analyze voice data, provide example data illustrating how these taxonomies can be used in practice, and assess the appropriateness of the taxonomies for assessing datalink based communications.

2   PREVIOUS WORK

Controller-pilot communications have been studied extensively. For example, Prinzo & Britton (1993) provided a summary of common challenges in primarily voice environments; more recently the advent of CPDLC has created significant interest in examining how the technology change will affect controller and pilot performance. Other topics extensively researched through voice communication analyses include controller workload (e.g., Manning et al., 2002), controller situation awareness (e.g., Metzger & Parasuraman, 2006; Gil et al., 2008), communication errors (e.g., Prinzo et al., 2007; Cardosi, 1993), trust and preference for communication modes (e.g., Stedmon et al., 2007; Metzger & Parasuraman, 2006; Sharples et al., 2007), controller strategies (e.g., Histon, 2008; Albuquerque, 2012), and language proficiency (e.g., Prinzo et al., 2008).

2.1 Communications and Workload: An Unintrusive Measure of Critical Aspects of ATC

A predominant theme in the past communication research has been the relationship of communications with controller workload. Because communication is an integral part of controllers' work and because it is so natural, the amount, types, and characteristics of voice communication offer observable manifestations of controllers' task load, and corresponding cognitive load of their current tasks (McGann et al., 1998). Many studies have investigated the correlation of communication loads with controller activity as well as the duration of and frequency of verbal communication events (e.g. Manning et al., 2002). This research has established a convincing connection between workload and communication activity; consequently communication activity is often used as a surrogate, or "on-line", nonintrusive measure of workload.

There are several different measures of communication activities that can be useful for assessing workload. Hurst & Rose (1978) showed that the durations of radio communications were good predictors of behavioral ratings made by expert observer controllers. Porterfield (1997) compared communication times recorded from high-fidelity simulations to concurrently recorded subjective workload estimates (the Air Traffic Workload Input Technique, ATWIT; Stein, 1985) and showed an impressive maximum coefficient of correlation of .88. Also Manning et al. (2002) analyzed traffic samples against ATWIT workload estimates provided by ATC instructors. The results showed significant correlations between ATWIT ratings and total number and duration of communications and individual communication durations, as well as number of instructional clearances. The activity component of task load was also correlated with total number and duration of communications as well as with the number of frequency changes and instructional clearances. Conversely, manipulating communication load has been shown to modify both subjective ratings and objective measures of workload (Casali & Wierwille, 1983).

2.2 Communications, Situation Awareness, and Errors

Another common theme in published research is the important role communication has in supporting controller situation awareness, and the consequences of communication failures for controller errors. Current voice technologies provide transmissions on a shared radio frequency where pilots can hear transmissions to all aircraft. This provides opportunities for multiple pilots to overhear instructions, creating "party-line" information and enhancing pilot situation awareness (Farley, et al., 1998). Datalink communications transmit directly between a flight crew and controller, raising concerns about the ability of flight crews to maintain situation awareness (Signore & Hong, 2000). However, Metzger & Parasaruman (2006) suggests that this loss in awareness partly offset by the ability to retain and review datalink messages, where the text from a command can be stored and recalled later to ease a pilot's working memory and prevent task disruption. A key question raised by Kapp & Celine (2006) is how the use of CPDLC will modify conflict resolution strategies and representations of the traffic situation shared amongst controllers.

The effects of different technologies, traffic situations, and environments on making errors in the communication process have also been a topic of considerable interest. Prinzo et al. (2007) developed complexity metrics of the content of controller-pilot communications in terminal environments and used them to examine the frequency of readback errors. Gil et al. (2008) has argued that the use of text messages in CPDLC will reduce flight crew errors, such as entering erroneous data into the flight computer, by providing a visual, and persistent, reference source. There has also been considerable interest in determining if there are tactical preferences for using voice or datalink communications (Stedmon et al., 2007). Metzger & Parasaruman (2006) and Sharples et al. (2007) have examined differences in the level of trust in each type of technology.

2.3   Variables, Measures, and Taxonomies for Communication Analysis

Throughout the research described above there are a large variety of variables and measures that have been used to capture and analyze communication data. For example, Rantanen and collaborators (Rantanen et al., 2005; Rantanen et al., 2007) analyzed a total of six hours of operational ATC voice data without transcribing the communications but by classifying key elements in them. Table 1 below depicts the variables extracted from these data.

The measures in Table 1 are mostly on a nominal scale (categorical), with the exceptions of time data, which are measured on a ratio scale, and sequence data on an ordinal scale. Despite the limitations of the measurement scales, these data nevertheless lend themselves to many useful analyses, including the correlations with workload, taskload, and situation awareness (described above). The last two rows of Table 1 illustrate a critical component of most previous communication analyses: the use of a taxonomy to cluster, categorize and filter communication activities.

There is a substantial history of development of taxonomies of communication activities in ATC (e.g. Prinzo & Britton, 1993, 1995). Following a taxonomy provides a uniform coding scheme, creating a consistent basis for analysis and interpretation of results. There are several important themes common across the previously developed and presented taxonomies. The first is the importance of identifying a basic unit of analysis; in some cases the analysis unit represents a speech act, identified at the most basic level as "a single utterance used to convey a single action or intention for action." (Prinzo & Britton, 1995) In other cases, the basic unit of analysis is a transmission, or a continuous radio or telephone message by a single sender, from the sender's first utterance to the end of the last utterance in a message (Rantanen et al., 2005). Cardosi (1993) and Histon (2008) used information content as the basis for identifying elemental units of analysis. The variation in choice of unit of analysis reflects differences in the ultimate use and purpose of the analysis: increased granularity can provide raw data needed to examine different types of error mechanisms (e.g. pronunciation slips, substitution errors) at the cost of increased data collection effort (Prinzo & Britton, 1995).

Table 1. Variables extracted from ATC communications analyzed by Rantanen, et al. (2005) and Rantanen et al. (2007).

Variable	Measure
Transmission start and stop: time stamps, duration, response lags	Seconds and milliseconds
Source: controller, pilot, other controllers	Category, tally
Role: initiate or respond to communication	Category, tally
Sequence: number of transmission in a transaction	Ordinal number
Standard phraseology	Likert scale
Speech act of transmission: clearance, advisory, request (information, permission, communication or content clarification), repeat, acknowledgment (full or partial readback, error, callsign, or "roger")	Category, tally
Content of transmission: altitude, heading, route, speed, radio freq., traffic, transponder, weather, other.	Category, tally

Table 2. Synthesis of common categories in taxonomies of controller pilot communications

Generalized Category	Example Elements
Trajectory Modification (Clearances)	Heading, Heading Modification, Altitude, Altitude Restriction, Speed, Approach/Departure, Route/Position
Clarification / Communications about communications	Roger, Say Again, Confirm Message Content
Information Passing / Advising	NOTAM, ATIS, Weather, Sighting, Traffic
Information Requests	Aircraft State, Weather, Sighting, Traffic
Courtesy	Thanks, Greetings, Apology
Other / Non-Codable Remarks	Equipment, Delivery, Other

A second important theme is the discrimination between different types of content in the communication activities. Table 2 provides a synthesis of common categories of events drawn from recent taxonomies used to analyze controller pilot communications (Histon, 2008; Prinzo & Britton, 1995; Rantanen et al., 2005). These categories provide an important tool for distinguishing between the functional purpose of different communication activities. For example, Table 2 shows that common distinctions are made between communication events that provide an instruction or directive to a pilot to take an action (typically to change an aircraft's trajectory) and those that support the exchange of information to support decision making about what those instructions should be.

3 EXAMPLE INSIGHTS GAINED FROM COMMUNICATIONS ANALYSIS: CONTROLLER STRATEGY IDENTIFICATION

The importance of developing appropriate taxonomies for examining NextGen communication technologies is illustrated by Histon's (2008) use of taxonomies of existing voice-based controller-pilot communications to identify important aspects of current controller decision making strategies. Recordings of two way controller-pilot communications were obtained from two internet websites that archive and stream live controller-pilot radio communications using private radio scanners. Observations were collected for a range of six high-altitude sectors and more than 72 hours of data were analyzed (for details see Histon (2008)). The time, aircraft addressed, and content of each transmission from the controller was captured. Based on a preliminary analysis, a taxonomy based on content type was developed resulting in eight content types representing general classes of events. Each content type was further subdivided into individual categories (see Histon 2008, for a complete listing of coded events and descriptions).

Analysis of the coded controller-pilot communication events (Histon 2008) showed that the proportion of transmissions that were commands was approximately 45% and this was consistent across all six ATC sectors. This illustrates the value of a content-based taxonomy analysis for understanding the potential effects of shifts in technology: in the sampled set of enroute sectors, more than half of all transmissions do not change an aircraft's trajectory. This is an indicator of the multiple functions that current voice technology techniques supports that will need to be retained and supported as new technologies, such as CPDLC, are expanded in use.

The analysis was also used to corroborate field observations of the timing of controller planning activities, a key indicator of controller cognitive strategies. Analysis of the timing of controller-pilot commands showed over 25% of the commands occur in the first minute after check-in, consistent with observations collected during field observations. This is key evidence supporting the recognition that events beyond the nominal boundaries of a sector are important factors in cognitive complexity and controller strategies. A controller's "Area of Regard" (see Histon, 2008) conceptualizes this need to consider structural and complexity influences beyond the physical dimensions of the sector.

Finally, by breaking out just events that affect aircraft trajectories ("commands"), unique signatures for different types of sectors can be identified. There are distinct differences in the type of commands used in each sector; this unique functional signature of a sector can be useful for comparing sector equivalencies (e.g. for supporting generic airspace concepts and other advanced atm concepts). The discrete types of commands also supports further analysis of the distinct patterns in the sequences of commands that are observed within individual sectors. Linguistic behaviors often have sequential properties and there are various ways to computationally model such phenomena. Recurring patterns in the sequences of commands can form the basis for standardized 'macros' or predetermined command sequences that could be transmitted in data-link enabled environments. Modeled sequences could also be presented to users in different ways according to sector environment. Albuquerque (2012) conducted a similar analysis over a limited range of sectors.

4   COMMUNICATIONS ANALYSES FOR NEXTGEN VALIDATION

Communications between human operators in aviation systems, pilots communicating with air traffic controllers, pilots communicating with other pilots in the same cockpit, and controllers communicating with other controllers both within and between their sectors, are fundamental to the operation of the National Airspace System (NAS). This fact is unlikely to change with implementation of NextGen although human-human communications are likely to be supplanted to some degree with human-automation and automation-automation communications. Both the ingrained role of communications in aviation and the foreseeable changes to current communication patterns brought about by NextGen make communications analysis a compelling means for validation of NextGen technologies and procedures.

Aviation communications are highly standardized to the minute detail of radiotelephony phraseology worldwide. Standardization will, most likely, only increase with the increased use of CPDLC. Yet, despite the standardization aviation communications analyses have yielded invaluable insights into myriad aspects of human and system performance in a vast range of operations, as we saw in the preceding review. There is no reason to doubt that communications analysis will continue to be a valuable tool in examination of the impact of NextGen on the performance of the system (i.e., NAS) as a whole and in particular on the people working within it.

There are two distinct approaches to communications analyses under NextGen. The first is analysis of communications via the new technologies, most notably CPDLC. However, CPDLC must necessarily be augmented by traditional voice communications for situations that fall outside standard CPDL procedures, including clarifications and emergency communications. These nonstandard voice communications under NextGen will thus become much richer source of much more valuable information than the current, mostly routine, voice communications. Therefore, continual development of methods and metrics of communications analysis is both imperative and urgent.

4.1   CPDLC Research Questions

There appears to be little published work assessing impacts of operational deployments of CPDLC on controller strategies and decision making. As operational use expands into more complex and heavier traffic airspace, there is a need for further analyses of the impact of CPDLC communication technologies on controller strategies. In the previous work, there are relatively few examples of examinations of the relative frequency of different types of transmissions or message content. Gonda et al. (2005) examined the frequency of different message types during a trial CPDLC deployment at Miami Center (ZMA) in the early 2000s and there has been some work as part of technical demonstrations of CPDLC in simulation environments (Willems et al. 2010). However, as yet, there appears to be no formal method for classifying or grouping these transmissions (i.e. by clearance, instruction, etc.) for the purposes of analysis and controller strategy identification. None of the studies attempted to examine the effects of specific CPDLC transmission types on controller performance, but rather generalized all transmissions to measure issues such as response times and workloads.

In assessing and validating NextGen concept of operations involving CPDLC a number of key research questions can be addressed by leveraging the work previously done on communications analysis. One of the most critical questions surrounds examining what communication elements should be delegated and supported by datalink and other NextGen technologies and—more importantly—whether there are communication activities that should remain primarily in the voice channel. This will require careful consideration of appropriate metrics for characterizing the different functions provided by different communication elements. Other questions that CPDLC communications analysis can help address include the value of retaining standardized sequences of commands, the effect of asynchronous nature of datalink and lack of immediate feedback on responses to transmissions, how feasible will CPDLC be in high frequency operational environments, the differences be between military and civilian communications, procedures for restricted airspace and interceptions, and the use free text fields and information content supported through structured pre-formatted message types.

As noted above, it is not only CPDLC communications that will need to be assessed. During the transition to NextGen there will be challenges if controllers operate in an environment of mixed equipage, with some aircraft capable of datalink communications and others not. Metrics and measures of communication performance in mixed environment will be important for assessing the transition path to get to steady state NextGen operations.

4.2   Communication Analyses With Computational Linguistics Methods

Beyond the topics in Table 1, verbal data, spoken or written, can be regarded as unstructured information. Appropriate computational linguistics tools can add new forms of analysis and insights, and make analysis of pilot—controller interactions more efficient. At the moment, such approaches to analyzing and modeling pilot-controller communicative situations are underexplored.

Computational models of language behaviors (based on linguistic clues such as the use and distribution of interlocutors' structural, semantic, and discourse structures) are used in many areas to help identify, predict, or forecast domain-specific phenomena that can be connected with linguistically expressed behaviors. For example, data-driven linguistic analysis could help understand what indicates communicative trouble or the strategies pilot-controller interlocutors use to repair trouble in communication, and potentially which linguistic repair strategies are more effective.

Additionally, computational linguistic analytics can allow analytic comparisons or integration of text-based vs. voice-based communication channels. Speech is rich in information. For example, speech patterns that communicate uncertainty or cognitive load could help further insights about the reasoning processes among air traffic experts, and potentially result in useful practical applications.

5   SUMMARY, CONCLUSIONS AND FUTURE WORK

Despite the technological advances in NextGen, effective communications will remain the backbone of NAS operations. The achievements of decades of aviation communications research offer true and tried methods for systematic validation of many aspects of NextGen as it is implemented. The existing knowledge of the relationships between communications and situation awareness, workload, and airspace complexity also allow for use of communications analysis beyond communication (e.g., CPDLC) aspects of NextGen.

REFERENCES

Albuquerque, E. (2012). Analysis of Airspace Traffic Structure and Air Traffic Control Techniques. Unpublished, Masters Dissertation, Massachusetts Institute of Technology.

Cardosi, K., (1993). An Analysis of En Route Controller-Pilot Voice Communication, *DOT/FAA/RD-93/11*, Washington, DC: FAA.

Casali, J. G., & Wierwille, W. W. (1983). A Comparison of Rating Scale, Secondary-Task, Physiological, and Primary-Task Workload Estimation Techniques in a Simulated Flight Task Emphasizing Communications Load. *Human Factors* 25(6), 623-641.

Farley, T. C., Hansman, R. J., Endsley, M. R., Amonlirdviman, K., & Vigeant-langlois, L. (1998). The Effect of Shared Information on Pilot / Controller Situation Awareness and Re-Route Negotiation. Proceedings of the 2nd FAA/Eurocontrol ATM R&D Seminar. Washington DC: FAA.

Gil, F.O., Vismari, L. F. & Camargo Junior, J.B., (2008) Analysis of the CPDLC Real Time Characteristics and the Mode S Data Link Capacity. *Anais do VII SITRAER Anais... Rio de Janerio: Universidade Federal do Rio de Janeiro*, v.1, p. 92-99.

Gonda, J., Saumsiegle, W., Blackwell, B., & Longo, F. (2005). Miami Controller-Pilot Data Link Communications Summary and Assessment, 6th Eurocontrol/FAA Air Traffic Management Seminar, Barcelona

Histon, J. M. (2008). Mitigating Complexity in Air Traffic Control: The Role of Structure-Based Abstractions. *Doctoral dissertation, Massachusetts Institute of Technology.* Available from *http://hdl.handle.net/1721.1/42006*

Hurst, M. W., & Rose, R. M. (1978). Objective Workload and Behavioural Response in Airport Radar Control Rooms. *Ergonomics* 21(7), 559-565.

Kapp, V., & Celine C. (2006). Designing an ATC CPDLC Environment as a Common Information Space. Proceedings of 2006 IEEE/AIAA 25TH Digital Avionics Systems Conference, Portsmouth, OR.

Manning, C. A., Mills, S. H., Fox, C. M., Pfleiderer, E. M., & Mogilka, H. J. (2002). Using Air Traffic Control Taskload Measures and Communication Events to Predict Subjective Workload. *(DOT/FAA/AM-02/4)*. Washington, DC: FAA Office of Aerospace Medicine.

McGann, A., Morrow, D., Rodvold, M., & Macintosh, M. A. (1998). Mixed-media Communication on the Flight Deck: A Comparison of Voice, Data link, and Mixed ATC Environments. *Int'l Journal of Aviation Psychology 8*, 137–156.

Metzger, U. & Parasuraman, R. (2006). Effects of Automated Conflict Cuing and Traffic Density on Air Traffic Controller Performance and Visual Attention in a Datalink Environment. Int'l Journal of Aviation Psychology 16(4): 343-62.

138

Porterfield, D. H. (1997). Evaluating controller communication time as a measure of workload. *Int'l Journal of Aviation Psychology*, 7(2), 171-182.

Prinzo, O.V. and Britton, T.W. (1993). ATC/pilot voice communications – A Survey of the Literature. *DOT/FAA/AM-03/20.* Washington, DC:FAA

Prinzo, O.V., Britton, T.W., and Hendrix, A.M. (1995). Development of a coding form for approach control/pilot voice communications. *(DOT/FAA/AM-95/15).* Washington, DC: FAA

Prinzo, O. V., Hendrix, A., Hendrix, R., (2007). The Computation of Communication Complexity in Air Traffic Control Messages. Proceedings of the 7th FAA/Eurocontrol ATM R&D Seminar. Barcelona, Spain.

Prinzo, O.V., Hendrix, A.M., & Hendrix, R. (2008). Pilot English language proficiency and the prevalence of communication problems at five U.S. air route traffic control centers. *(DOT/FAA/AM-08/21).* Washington, DC: FAA.

Rantanen, E. M., Maynard, P., & Özhan, D. (2005). The impact of sector characteristics and aircraft count on air traffic control communications. Proc. 13th Int'l Symposium on Aviation Psychology (pp. 491-496). Dayton, OH.

Rantanen, E. M., Naseri, A., & Neogi, N. (2007). Evaluation of airspace complexity and dynamic density metrics derived from operational data. Air Traffic Control Quarterly, 15.1, 65-88.

Sharples, S., A. Stedmon, G. Cox, A. Nicholls, T. Shuttleworth, & J. Wilson. (2007). Flightdeck and Air Traffic Control Collaboration Evaluation (FACE): Evaluating Aviation Communication in the Laboratory and Field." Applied Ergonomics 38.4: 399-407.

Signore, T. L. & Hong, Y. (2000). Party-Line Communications in a Data Link En-vironment. 19th Digital Avionics Systems Conference Vol.1 (pp. 2E4-1-2E4/8).

Stedmon, A., S. Sharples, R. Littlewood, G. Cox, H. Patel, & J. Wilson. (2007) Datalink in Air Traffic Management: Human Factors Issues in Communications. Applied Ergonomics 38.4 : 473-80.

Stein, E.S. (1985). Air traffic controller workload: An examination of workload probe. (DOT/FAA/CT-TN84/24). Atlantic City, NJ: FAA Technical Center

Turner, J. ATS Datalink Communications. (2011) IATA/AEA 50th Joint User Requirement Group Conference, CNS Performance Requirements and Investments. May 30, 2011

Van Es, G. (2004) Air-Ground communication safety study: Analysis of pilot-controller occurrences. Ed.1.0, EUROCONTROL, DAP/SAF, 16.04.2004

Willems, B., S. Hah, & K. Schulz. (2010). En Route Data Communications: Experimental Human Factors Evaluation (DOT/FAA/TC-10/06). Atlantic City International Airport, NJ: FAA William J. Hughes Technical Center

Zingale, C., Mcanulty, M., & Kerns, K. (2005). Human Factors Evaluation of a Digital, Air-Ground Communications System. Proceedings of the 24th AIAA/IEEE Digital Avionics Systems Conference (pp. 5.B.5.1-9).

CHAPTER 14

Comparison of Procedures for Dual and Triple Closely Spaced Parallel Runways

Savita Verma & Debbi Ballinger

NASA Ames Research Center
California, USA
Shobana Subramanian

NASA Ames Research Center/ Dell Perot
California, USA

Thomas Kozon

NASA Ames Research Center/ UARC
California, USA

ABSTRACT

A human-in-the-loop high fidelity flight simulation experiment was conducted, which investigated and compared breakout procedures for Very Closely Spaced Parallel Approaches (VCSPA) with two and three runways. To understand the feasibility, usability and human factors of two and three runway VCSPA, data were collected and analyzed on the dependent variables of breakout cross track error and pilot workload. Independent variables included number of runways, cause of breakout and location of breakout. Results indicated larger cross track error and higher workload using three runways as compared to 2-runway operations. Significant interaction effects involving breakout cause and breakout location were also observed. Across all conditions, cross track error values showed high levels of breakout trajectory accuracy and pilot workload remained manageable. Results suggest possible avenues of future adaptation for adopting these procedures (e.g., pilot training), while also showing potential promise of the concept.

1 INTRODUCTION

The Next Generation Air Transportation System (NextGen) is being designed with the expectation that the volume of the traffic will double by 2025 (Joint Development and Planning Office, 2004). In order to handle the expected traffic demand, airport capacity needs to expand dramatically. To gain such capacity at the airport, runways with centerline distances closer than 2500 ft need to be explored, as they are a potential solution to meeting increased demand. Runways could be built in-between existing dual runways that have greater than 2500 ft separation between them. One of the main challenges of closely spaced runway operations is that capacity is greatly reduced in low visibility conditions. The FAA allows simultaneous instrument approaches on two and three runways spaced at least 4300 ft apart using the Instrument Landing System (ILS), and Precision Runway Monitor (PRM) approaches to runways 3000 ft apart for most domestic airports. Some airports, like San Francisco International (SFO) airport, land flights on parallel runways that are 750 ft apart using the Simultaneous Instrument Offset Approach (SOIA) procedures in lower visibility conditions (Magyratis, 2001). SOIA approaches require the trailing aircraft in the paired approach to obtain a visual sighting of the lead aircraft, which is possible under marginal weather conditions such as a 2100 ft cloud ceiling and 3nmi visibility.

Focusing on closely spaced runways that are 750 ft apart, the current investigation assumes technologies and procedures (described later in this paper) such that arrival capacity is maintained even when weather conditions degrade. While procedures under nominal conditions may not pose as much of a concern, one of the most serious concerns regarding simultaneous landings on runways closer than 2500 ft has been finding off-nominal breakout procedures that are acceptable to the pilots and maintain safe separation. The reduction of runway spacing for independent simultaneous operations increases the likelihood of wake vortex incursion and allows for less maneuvering area if the lead aircraft deviates from its course. Thus, there is a requirement for the calculation of safe and proper escape maneuvers.

The authors conducted a study to investigate off-nominal procedures for dual parallel runways in all-weather conditions (Verma et al., 2008). Further capacity on the airport could be achieved with triple runways 750 ft apart. This led the authors to design and conduct another experiment involving triple runways that were 750 ft apart and included off-nominal conditions (Verma et al., 2009).

This paper provides a comparative analysis of two experiments using runways spaced 750 ft apart when approaches include off-nominal conditions. One study used dual runways (the "2-runway" study) and the other used triple runways (the "3-runway" study). The off-nominal conditions investigated in these two studies included the lead aircraft deviating, or blundering off course, and wake intrusion. This paper will compare the breakout maneuvers of two and three closely-spaced parallel runways and their impact on workload and accuracy of flying the breakout trajectory.

2 BACKGROUND

Airports with parallel runways lose capacity under poor visibility conditions. Hence, there is a need for investigating parallel runway operations that will work under poor weather conditions. For runways 750 ft apart, the safety of simultaneous landings and breakout procedures that might be required due to off-nominal conditions is paramount. Such concepts currently in operation include SOIA and PRM approaches, while others have been developed and investigated in the research (Verma et al., 2008; Verma et al., 2009).

The dual and triple studies analyzed in this paper use a concept developed by NASA in collaboration with the Raytheon Corporation called the Terminal Area Capacity Enhancing Concept (TACEC), which allows paired approaches on runways that are 750 apart in instrument meteorological conditions (Miller et al., 2005). The TACEC concept includes a ground-based processor, which identifies aircraft that could be paired approximately 30 minutes from the terminal airspace boundary. The aircraft are selected for pairing based on several parameters such as relative aircraft performance, arrival direction, and the size of the aircraft's wake. The ground based processor then assigns 4-dimensional (4-D) trajectories to the aircraft in the pair. It is assumed that all aircraft will use differential GPS-enabled, high-precision 4-D flight management system capabilities for the execution of these trajectories. Enhanced cockpit displays provide the trailing pilot with detailed position and some intent information about the lead aircraft, and show a predicted wake for the lead aircraft.

The concept uses breakout trajectories that require a less extreme turn than the maneuvers used in other concepts. This concept that was originally developed for dual runways and then extended to triple runways considers wake prediction data to determine when a breakout is required, and provides for a dynamically generated breakout trajectory that changes as the aircraft flies the approach. Most of the previous concepts did not consider wake data in their concepts or displays.

This paper provides a comparative analysis of the 2-runway and 3-runway experiments by comparing results of the pairs of aircraft in the triple formation (left and center aircraft or center and right aircraft) with the 2-runway pair (left and right aircraft). Procedures for dual runways involving a leading and trailing aircraft pair are compared to procedures from the 3-runway study, when the piloted aircraft was either the center or the right aircraft in the echelon formation (Figure 1). The comparison between the dual and triple runways is meant to evaluate the dual-runway procedures that were implemented to create new procedures for triple runways. The dual and triple runway procedures are compared on level of accuracy for flying breakouts and differences in workload experienced by the pilots. Results on these factors will provide insight into the human factors issues associated with the different positions of the aircraft in the dual and triple formation.

Figure 1: Echelon formation for triples (shaded area below aircraft shows predicted wake turbulence zone)

3 METHODS

3.1 Airport and Airspace Design

Both the 2-runway and 3-runway studies used a common fictitious airport (KSRT) based on the current Dallas/Fort Worth International Airport (DFW) layout and operations, with the exception of the runways which were set 750 feet apart. The west side of the airport was simulated in its south configuration only (18L and 18R for 2-runways, and 18L, 18C and 18R for 3-runways). Equipage to a CAT-IIIB level was assumed.

3.2 Operational Procedures

Flights in the simulation were initiated at about 25 nmi from the airport, with the assumption that they were already placed into aircraft pairs or triples. Approach and departure routes and procedures were similar to those used at DFW. The aircraft flew 4D arrival trajectories and were paired or 'tripled' with aircraft arriving from any of the four meter fixes (NE, NW, SE, SW) located near the edge of the terminal airspace, about 40-60 nmi from the airport. The concept allows for pairing based on aircraft type, performance characteristics and estimated time of arrival. In the study, the pairing was scripted in the traffic scenarios.

The aircraft fly 4D trajectories up to a point in the airspace, referred to as the coupling point, designated at 12 nmi from the runway threshold. From the coupling point onwards, the aircraft fly in a formation such that they were coupled for speed. In the 2-runway study, the trailing aircraft precisely maintained temporal spacing of 15 sec with +/- 10 sec tolerance for error (a window of 5-25sec), behind the lead aircraft to avoid wake of the lead aircraft (Rossow et al., 2005). The path flown by the trailing aircraft in the 2-runway study involved a slew angle of six degrees to the landing runway. The aircraft became parallel at about two nmi from the runway, as shown in Figure 2.

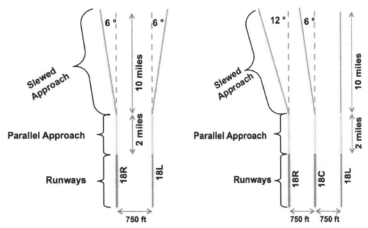

Figure 2: Final approach geometry for operational procedures for dual and triple runways

In the 3-runway study, the center aircraft precisely maintained 12 s spacing behind the lead aircraft, and the right aircraft maintained 24 s behind the lead aircraft beyond the coupling point. As shown in Figure 2, the approach paths of the two trailing aircraft were at designated slew angles from the center of the runway - 6-deg for the center runway aircraft and 12-deg for the right runway aircraft. All three aircraft turned straight-in for the final approach during the last two nmi from the runway.

For both the 2- and 3-runway procedures, onboard automation monitored the paired runway for potential conflicts. Automation also displayed the predicted safe zone from the wake generated by the lead aircraft (and center aircraft for 3-runway procedures). Visual and aural alerts were used to alert pilots to the lead (or center) aircraft blunders or the wake of the lead (or center) aircraft drifting towards the aircraft behind it. The navigation displays (Figure 3) depicted the breakout trajectory as a white line, after the aircraft crossed the coupling point. For the 3-runways study, the breakout trajectory was shown for both the center and the right aircraft. In both studies, the breakout trajectory was dynamically generated considering wake, traffic, structures, and terrain of airport surroundings.

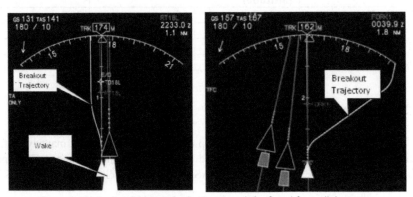

Figure 3: Navigation Displays for final approach for 2 and 3 parallel runways

Breakouts were caused by an intentional lead-aircraft blundering towards the following aircraft, or the wake of the lead aircraft drifting towards the following aircraft. Different locations of the breakout on the arrival path required different breakout maneuvers, which change the angle of the escape trajectory on the navigation displays. When the breakout was required at different altitudes on the arrival path, different bank angles for the breakout maneuvers were used and the curvature of the breakout trajectory changed on the navigation displays. The pilots were required to fly the breakout trajectory manually using the flight director when they received an aural and red visual alert.

For both the 2-runway and 3-runway studies, the breakout performed above 500 ft altitude required an initial bank angle of 30 deg, and the breakout at an altitude between 200-500 ft required an initial bank angle of 10 deg (Tables 1 and 2). The pilots at this stage were instructed to follow the "S" shaped breakout trajectory displayed on the navigation display as accurately as possible (Figure 3). The trajectory was "S" shaped so the final leg of the trajectory became parallel to the runways.

The 3-runway study used similar bank angles to those used in the 2-runway study. In addition, the pilot participants flew different headings based on the position in the echelon. The center aircraft (18C) changed its heading to 20-deg and the right aircraft (18R) changed its heading to 40-deg, giving more space to the center aircraft. The aircraft performing the breakout maneuver also climbed to 3000 ft as part of the breakout trajectory. The final leg of the breakout trajectory parallel to the runways was 1.5 nmi abeam for aircraft flying to 18C and 3.0 nmi for aircraft flying to 18R.

Table 1: Breakout trajectory for dual runways

Runway	Breakout Location	Initial Bank Angle
18 R (2-runway)	> 500 feet	30 °
	200-500 feet	10 °

Table 2: Breakout trajectory for triple runways

Runway	Breakout Location (altitude)	Initial Bank Angle	Initial Heading Change
18 C [3- Runway (Center Ownship)]	> 500 feet	30 °	20 °
	200-500 feet	10 °	20 °
18 R [3- Runway-Right Ownship)]	> 500 feet	30 °	40 °
	200-500 feet	10 °	40 °

3.3 Simulation Platform

For both studies, the human-in-the-loop experiments of breakout maneuvers for paired and triple runways were performed approaches in the Advanced Concepts Flight Simulator (ACFS) located at the NASA Ames Research Center. The ACFS is a motion-based simulator that can be configured to represent current and future cockpits. At the time of this experiment, the simulator had performance characteristics similar to a Boeing 757, but its displays were modified to study advanced flight operational concepts.

3.4 Participants

The study participants were recently retired pilots from commercial airlines. All of them were male and all had experience with glass cockpits. Their average pilot experience was about 38 years, and their average number of years since retirement was less than two.

3.5 Traffic Scenario

For the 2-runway study, the traffic scenario involved two aircraft: (1) The ACFS flight simulator as the trailing aircraft (i.e., the ownship) and (2) A scripted Boeing 747-400 as the leading aircraft.

For the 3-runway study, the traffic scenario involved three aircraft, where the flight simulator (i.e., the ownship) was either the center or right aircraft. The other two aircraft were scripted. When the ownship was in the center position, the aircraft causing the off-nominal situation was the left-most aircraft. When the ownship was in the right-most position, the aircraft causing the off-nominal maneuver was the center aircraft. The off-nominal event was introduced in the scenarios through lead aircraft intentionally deviating off its trajectory or adverse winds causing its wake to drift towards the following aircraft.

4   RESULTS AND DISCUSSION

Statistical results on two dependent variables are reported in the analysis of data generated from the experimental runs: (1) Ownship cross track error, collected digitally during the breakout phase of the simulation flight and (2) The pilots' subjective assessments of workload. Data were analyzed using 3-way Factorial Analysis of Variance with three independent variables, with each independent variable having 2 levels: (1) Number of runways (2 vs. 3), (2) Cause of breakout (aircraft deviation and wake) and (3) Location of breakout (high and low altitude).

4.1 Cross Track Error

Cross track error, collected by the simulator's digital data collection system, is one measure of trajectory accuracy particularly sensitive to breakout maneuvers. Cross track error was measured by the distance between the actual ownship position and the system-generated breakout trajectory position (i.e., the off-course distance), with both positions shown on the Navigation Display. Hence, less cross track error correlates to higher breakout trajectory conformance. For each simulation run, cross track error was averaged across time from the breakout point to the end of the flight.

A statistically significant ANOVA main effect of the number of runways on the ownship's breakout cross track error was found, in comparing the 2-runway and the 3-runway (right ownship) conditions ($F=21.92$, $df=1,15$, $p<0.001$) (Table 3).

Table 3: Breakout Cross Track Error (2-runway vs. 3-runway)

Runway	Mean (ft)	SE	MIN	MAX
2-runway	56.25	3.67	0.65	106.12
3-runway (center ownship)	73.44	11.45	5.83	542.42
3-runway (right ownship)	104.37	11.46	16.26	513.29

The directionality of means for this main effect indicates more cross track error under the 3-runway (right ownship) condition as compared to the 2-runway condition. This could be attributed to having 2 aircraft to the left of the ownship during breakout (3-runway, right ownship), creating an increased sense of urgency on the part of the pilot to escape the cause of the off-nominal situation, i.e., the possible additive effect of wake and/or blunder of both aircraft to the left of the ownship might prompt the pilot to overshoot the breakout trajectory further to the right as a safety measure. Some increased cross track error was also observed under the 3-runway (center ownship) condition as compared to the 2-runway condition, but this difference did not reach statistical significance. This lack of statistical significance might reflect the center position of the ownship, which requires that the pilot maintain safe separation with 2 other aircraft – one to the right, and one to the left of the ownship, thereby posing constraints on aircraft movement to either the right or the left, to maintain adequate separation. The pilot-participants pointed out that this prompted them to exercise a larger degree of vigilance in flying the breakout trajectory, which would explain less cross track error, as compared to the 3-runway (right ownship) condition, even though the right ownship is not much safer than the center aircraft. Mean cross track error values generally indicate reasonable levels of accuracy in flying the breakout trajectory. However, maximum values at the end of the distribution for the triple-runway operations (Table 3) might indicate a need for improved training to prevent the occasional overshoot of the breakout trajectory.

A statistically significant Number of Runways by Breakout Location (altitude) interaction effect on cross track error was also observed. A larger mean cross track error difference between high and low altitude locations was observed under the 3-runway (right ownship) condition, as compared to the 2-runway condition ($F=16.12$, $df=1,15$, $p<0.005$) (Figure 4).

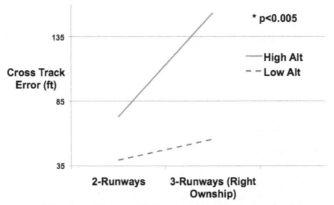

Figure 4: Number of Runways X Breakout Location Interaction Effect:
Cross Track Error (* p<0.005)

This interaction effect is best understood when one considers that breakout procedures for the higher altitudes are more difficult and that procedures are more complex for the 3-runway operations. As postulated above, the 3-runway center-ownship pilot may have exercised special vigilance in flying the breakout trajectory more accurately, due to the aircraft's central location in the triplet echelon, which would account for less cross track error. Since the pilot in the 3-runway (right ownship) condition is mostly concerned about loss of separation with the center aircraft and the possible additive effect of having two aircraft to the left in the breakout formation (wake turbulence and/or track deviation of both aircraft), the pilot may be less concerned about exercising special vigilance in flying the breakout trajectory accurately, but rather, escaping the track deviation or wake turbulence of both aircraft by moving as quickly as possible towards the right, and possibly overshooting the breakout trajectory. Also, since the higher altitude breakout procedures require a more aggressive maneuver (as compared to the lower altitude procedures), the possible tendency for the pilot to overshoot the breakout trajectory further to the right at the higher altitude might reflect a continuation of the already aggressive nature of the required maneuver.

4.2 Workload

Participants completed the NASA TLX workload questionnaire (Hart and Staveland, 1988) after every run. Data were collected on each of the six TLX workload measures, which were combined to derive a composite workload measure, which ranged from 1 (very low workload) through 7 (very high workload).

Table 4 presents statistics on average composite workload, broken down by the number of runways and position of the ownship. Overall, workload can be characterized as moderate. While trends should be viewed with some caution, due to lack of statistical significance, the directionality of means shows increased workload under the 3-runway conditions as compared to the 2-runway condition.

This would make sense, due to the increased geometric complexity of the 3-runway procedures and pilots needing to maintain safe separation with 2 other aircraft (as compared to only 1 other aircraft under the 2-runway condition), thereby increasing pilot workload.

A statistically significant Number of Runways by Breakout Cause interaction effect was observed, in comparing workload for 2-runways and 3-runways (Right Ownship) by Aircraft Deviation and Wake (Figure 5).

Table 4: Average Composite Workload Statistics by Number of Runways

Runway	Mean	SE
2-runway	2.78	0.15
3-runway (Center Ownship)	3.69	0.12
3-runway (Right Ownship)	3.64	0.12

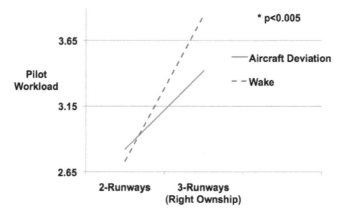

Figure 5: Comparison of Composite Workload for Wake versus Aircraft Deviation Under 2-runway and 3-runway Conditions (Right Ownship) *p<0.005

Figure 5 shows a larger workload difference between Aircraft Deviation and Wake causes under the 3-runway (Right Ownship) condition, as compared to the 2-runway condition (F=12.43, df=1,15, p<0.005). This effect can be explained by the relative complexity of the 3-runway operations and the unstable nature of wake turbulence from possibly two other aircraft to the left of the ownship. Since having two aircraft to the left of the ownship during breakout (3-runway, right ownship) could create an increased sense of urgency on the part of the pilot to escape the cause of the off-nominal situation, i.e., the possible additive effect of wake of both aircraft to the left of the ownship, this increased sense of urgency might cause increased workload. Still, workload remained at manageable levels across all four interaction conditions (i.e., reasonably low, yet high enough to prevent tedium and vigilance decrement), which was further substantiated by pilot-participant feedback during open-ended discussion.

5   SUMMARY AND CONCLUSIONS

A high-fidelity human-in-the-loop flight simulation experiment investigated breakout procedures for Very Closely Spaced Parallel Approaches (VCSPA) with 2 and 3 runways. Results indicate larger cross track error under the 3-runway condition, where the ownship is approaching the rightmost runway, as compared to the 2-runway condition. However, breakout trajectories under both two and three runway conditions were flown with high accuracy, which degraded only slightly in the three runway procedures. Also, pilot workload levels, while higher under the 3-runway condition and the highest for the center-ownship, remained at manageable levels overall. The similarity in off-nominal procedures for dual and triple runways allows for pilots to fly either of the procedures with minimal adaptation. However, the positions of the aircraft in the 3-runway formation will impact procedures. This might suggest the need for further exploration of procedures for switching between 2- and 3-runway operations and future adaptation in pilot training.

REFERENCES

Hart, S., & Staveland, L. (1988). Development of a Multi-Dimensional Workload Rating Scale: Results of Empirical and Theoretical Research, In P. Hancock & N. Meshkati (Eds.), Human Mental Workload, Amsterdam, The Netherlands, Elsevier, 139-183.

Joint Development and Planning Office (2004). The Next Generation Air Transport System Integration National Plan. December 12, 2004.

Magyratis, S., Kopardekar, P., Sacco, N. & Carmen, K. (2001). Simultaneous Offset Instrument Approaches – Newark International Airport: An Airport Feasibility Study. DOT/FAA/CT-TN02/01.

Miller, M.E., Dougherty, S., Stella, J. & Reddy, P. (2005). CNS Requirements for Precision Flight in Advanced Terminal Airspace. Aerospace, 2005 IEEE Conference, 1-10.

Rossow, V. J., Hardy, G. H., and Meyn, L. A. (2005). Models of Wake-Vortex Spreading Mechanisms and Their Estimated Uncertainties, AIAA-2005-7353, ATIO Forum, Arlington, VA, September 26-28, 2005.

Verma, S., Lozito,S., Ballinger,D., Kozon, T., Panda, R., Carpenter, D., Wooten, D., Hardy, G., and Resnick, H. (2009). Evaluation of Triple Closely Spaced Parallel Runway Procedures for Off-nominal Cases, Eighth USA/Europe Air Traffic Management Research and Development Seminar (ATM2009).

Verma, S., Lozito, S., Kozon, T., Ballinger, D., and Resnick, H. (2008) Procedures for Off-Nominal Cases: Very Closely Spaced Parallel Runways Approaches, Digital Avionics System Conference, St. Paul, MN.

CHAPTER 15

Integrated Management of Airport Surface and Airspace Constraints

Philip J. Smith, Amy Spencer, Mark Evans,
Alicia B. Fernandes and Roger Beatty

Ohio State University
Columbus OH
Smith.131@osu.edu

ABSTRACT

This cognitive walkthrough of a departure scenario explores human factors considerations concerned with the integration of several NextGen concepts relevant to Collaborative Air Traffic Management (CATM) and airport surface management, including flexible airspace management, collaborative routing, translation of weather data into traffic management and flight operations decisions and airport surface departure management. These human factors considerations include strategies for distributing work (roles and responsibilities), supporting efficient and effective interactions (coordination vs. collaboration), information exchange requirements, communication requirements (voice vs. digital, potentially including graphics), the design of decision support tools and the incorporation of safety nets.

Keywords: traffic flow management, airport surface management, airline operations control, human factors, flexible airspace management

1   METHODS

The initial cognitive walkthrough conducted for this operational sequence involved a pair of traffic managers, one with significant Air Route Traffic Control Center (ARTCC) experience and one with significant Terminal Radar Approach Control (TRACON) experience. In a separate walkthrough, an airline dispatcher was asked to provide input on those parts of the scenario where airline dispatch expertise was relevant. The interviews were conducted by phone, with displays shared over the internet. The researcher controlled the presentation of the displays, which were embedded in a slideshow. The discussions were recorded for reference in the analysis of the results. Both of the traffic managers retired from the FAA within the past 4 years. The dispatcher was still active at a major airline. One traffic manager had 5 years of experience as a Traffic Management Coordinator (TMC) at the Denver TRACON and 2 years as a TMC at ZLA. The other had 14 years of experience as a TMC at ZMP and ZFW. The dispatcher had 14 years of experience as a dispatcher and 8 years as an airline ATC coordinator.

2   RESULTS

Below, we show a subset of the graphics as presented to the participating traffic managers. We also present the responses of the participants to questions, editing out those statements that do not contribute to an understanding of their inputs. Then, following the description of the walkthrough with the participants, we discuss the important human factors and operational findings at a more abstract level.

2.1   Scenario Walkthrough and Participant Responses

Before walking through a specific scenario, the participants were introduced to the flexible airspace concept and asked some general questions. Below, we present some of the graphics that were shown and the responses to these questions. The flexible airspace management concept described included important features like extending the TRACON boundary out to 90 nm and up to FL250, with a separation minimum of 3 nm and with the potential to open up predefined in arrival airspace and vice versa. Figure 1 shows the SIDs for MJA, the hypothetical airport used for this cognitive walkthrough.

After the description of the flexible airspace management concept, the participants were asked to address 4 questions:
- Are there any potential problems with this expansion of the TRACON airspace?
- Is it necessary to have predefined SIDs for departure routes through arrival airspace and vice versa?
- If arrival airspace is used for departures, who should control the traffic?
- What coordination would be necessary to open arrival airspace up for departures?

Representative responses are provided below.

Potential Problems?
TRACON: "I don't think there are going to be any problems."
ARTCC: "My only concern is that there needs to also be a redesign of the Center sectors. You also have to redesign your holding points in case all of a sudden the TRACON has to shut you off. You'd have to adjust those as well. But that's just a matter of somebody redesigning the airspace."

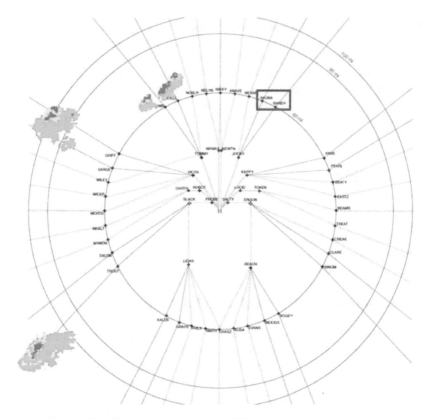

Figure 1. SIDs for MJA with the actual weather at 1930Z.

Pre-Defined vs. Dynamically Designed SIDs and STARs

TRACON: "If SIDs and STARs are defined on the fly, how will these be communicated to the pilots? I can see having a number of predefined SIDs and STARs, and flexibly changing from one SID or STAR to another. Those could be stored in the plane's database."

ARTCC: "If the top of this airspace was greater than FL250, that could be a disadvantage. You could have a big cement block in the middle of your enroute airspace, having to coordinate to get the transcons through. The higher it was, the more conflict with transcons. 250 would probably be the maximum."

Using Arrival Airspace for Departures

TRACON: "To decide whether to use SIDs in the northeast arrival gate, you need better information about the arrival and departure demand. It would be very helpful if it was more accurate than that information is today. And you need to display it in a way that makes it easy to see what the tradeoffs are and to decide when to make the change."

ARTCC: "You need a tool that can integrate arrival demand and departure demand and balance it out. Input from the airlines to inform this decision would also be good. Do they want us to favor certain departures or arrivals?"

TRACON: "I would think you'd give control of the SIDs in the northeast arrival gate to the departure guy. But it might depend on controller workload. If you're just moving departures from the normal SIDs to these other SIDs in arrival airspace, there wouldn't be any problem. But if you're adding a lot more departures, that might be too much for the departure controller. Then you might want to give these additional SIDs to the arrival guy and let him work departures. Today, in my TRACON all controllers worked arrivals and departures, so that shouldn't be a problem with procedures, as long as you give them the right training and experience. That might not be true in all TRACONs, though. ... For the outer fixes, I would definitely split it up so controllers work arrivals and departures in the same area, so they are used to working those arrivals and departures. Downwind and final would be another area for internal arrivals."

Coordination

TRACON: "In terms of the coordination for opening and closing specific fixes, you'd have to configure the computer to make it easy. You'd need coordination between the TMC and the supervisor a few minutes ahead."

TRACON: "There's two different scenarios. One is weather and the other is where the traffic is going [to be volume]. If the weather closes [some Northgate departure fixes], depending on arrivals maybe you can move some to [departure fixes in the northeast arrival airspace]. ... You want to fan them out 15 degrees. It's up to the Tower coordinator and ground control to get them to the right runway in the right order. You need to coordinate with the Tower."

Scenario-Based Walkthrough. At this point, the traffic managers were asked to walk through a specific, detailed scenario, starting at 1500Z and involving the hypothetical airport MJA. In this scenario, at 1500Z, pop-up thunderstorms are predicted to arise starting around 1900 in the expanded terminal airspace.

1930Z: A storm cell develops to the west of the Northgate departure fixes (see Figure 1), and is expected to move east, affecting these fixes until 2300Z. Based on this weather, the traffic managers send out an advisory indicating that they plan to open up 2 departure fixes in the northwest arrival airspace when necessary to expedite departures if and when the storm cell begins to impact departures north. In commenting on this plan to open up departure airspace for arrivals, the traffic managers note:

TRACON: "We would consider the current and forecast weather and the arrival and departure demand. The main consideration is how many arrivals are coming in over the northeast arrival gate. If there are more arrivals than departures, it wouldn't make sense to open [departure fixes in northeast arrival airspace]. You'd just reroute the departures or make them wait. There also might be staffing issues to consider. Who is going to be working that airspace? Training comes in there. A particular area sup might feel comfortable with his staffing, or you might not have the people to staff the sectors properly. When all of a sudden weather blows up, throwing a monkey wrench into the situation, you might not have the right people."

ARTCC: In terms of lead times, it depends on how fast you can bring the right people into the decision making process. You'd want the weather unit involved, for instance."

TRACON: "A lot of times, it comes from the floor, pilots saying we can't do that anymore. It's kind of a talent that varies from Center to Center. Some are proactive so the transition is easy to make; some want to wait and see and thus have to then struggle to get them away from the weather when it pops up quickly."

ARTCC: "If you're going to move the arrivals to another corner post, you need enough lead time to reroute arrivals to another gate. That could take half an hour or 45 minutes to get the traffic away. But if the arrivals are light enough that you can merge the arrivals [to a single northeast arrival fix], then that won't hardly have any impact at all. You'd want to move those arrivals onto their new arrival route outside of 100 miles, so it could take about 10-15 minutes."

2045Z: Continuing the scenario, a dispatcher is planning a flight from MJA (assumed to be located at the site of DFW) to BNA with a planned Out time of 2200Z. The dispatcher in this scenario looks at a display providing information on alternative routes from MJA-BNA (see Figure 2), which incorporates a RAPT forecast into the display as indicated by the colored dots. (Red indicates the route is predicted to be blocked; yellow partially impacted and green open.)

2100Z: In the scenario, the dispatcher files BEATY, but includes EARTZ and BEAMR (all fixes to the east) in the Trajectory Option Set (a set of pre-coordinated alternative routes for the flight) as the added cost in fuel and time is small. He also includes routes through the standard fixes to the north and the departure fixes to the northeast through arrival airspace:

Dispatcher: "In terms of workload, if the advisories were machine-readable like you said, and if the computer automatically generated a list of alternative routes like you've shown me in that display (see Figure 2), it wouldn't be too bad as long as it was easy to indicate which routes to include on the flight release. I do something like this today using CDRs, so this would just be a more refined way of doing that."

Flight: ABC6214 (B757)		Current Time: 2100Z	
Orig: MJA Dest: BNA OFF: 2215Z			
Status	Air Time	Air Fuel Burn	Route
○	137 min	17125 lbs	MJA BEATY TXK J42 MEM GHM5 BNA
○	138 min	17250 lbs	MJA EARTZ TXK J42 MEM GHM5 BNA
○	139 min	17375 lbs	MJA BEAMR TXK J42 MEM GHM5 BNA
○	146 min	18250 lbs	MJA BANDY MLC RZC ARG GHM GHM5 BNA
●	147 min	18375 lbs	MJA AKUNA MLC RZC ARG GHM GHM5 BNA
●	148 min	18500 lbs	MJA NEXAR MLC RZC ARG GHM GHM5 BNA
●	149 min	18625 lbs	MJA KNAVE MLC RZC ARG GHM GHM5 BNA
○	150 min	18750 lbs	MJA NIKKY MLC RZC ARG GHM GHM5 BNA

Figure 2. Dispatcher display.

In response to this segment of the scenario, the interviewed dispatcher noted:
Dispatcher: "If I'm dispatching this flight at 2100, an hour ahead of time, and it's supposed to be Off at 2215, then I need to know what the forecast is as it departs along its departure route. From what you say, RAPT is looking at that time frame (see Figure 2), with red meaning that the route is not likely to be available for a 2215 departure. That's looking out almost 2 hours ahead. Looking at the forecast weather, it seems to me that RAPT is about right."
Dispatcher: "To be really useful, RAPT has to do a better job of predicting than I do. It would certainly be helpful to have a display like this that helps me to quickly focus on which routes to seriously consider. It doesn't have to be perfect, because I'm making a decision about what departure routes to prepare for, how much fuel to put on board. It's not like I'm committing the plane at 2100 to depart at 2215 and fly through whatever weather is present at that time. RAPT has to help me make the right decision about what routes to prepare for more often than I would without it. You said that RAPT looks at echo tops as well. That integration by itself would be helpful since I would have to look at the forecast precipitation and echo tops separately and try to decide whether a route is likely to be available. In terms of specific probabilities, I'm not sure whether that additional detail would be necessary. I would like to know what red, yellow and green mean in terms of likelihoods, though."

2015Z: The flight crew reviews the flight release and concurs.

2030Z: ATC decides to open AKUNA and BANDY (2 departure fixes in northeast arrival airspace in 30 minutes. This decision is based in part on the fact that the demand for departures north will be heavy at that time, while the demand for arrivals from the northeast will be light. FAA and flight operator staff are informed. Controllers start moving traffic from JASON to BLORE (from the more northerly northeast arrival fix to the more southerly northeast arrival fix).

In response to questions about this decision process, the traffic managers stated:
TRACON: "Like we talked about earlier, the minimum lead time depends on how far I need to move the arrivals to JASON. It could be a short as 10 minutes or as long as 45 minutes. Even if it can be done in 10 minutes though, it is often good to be proactive and start earlier instead of taking a wait and see attitude."
TRACON: "To plan this change in the airspace, we'd have the Center and the TRACON talking, with the Tower included. The TRACON TMC would have talked with the area supervisor to make sure there weren't staffing issues. The Command Center would also need to be in the loop to make sure any necessary coordination with other Centers has been done."
TRACON: "We discussed a strategy for dividing up the airspace into 4 areas earlier. I like that idea since you would have the involved departure and arrival sectors in the same area. You'd want to put them near each other to help them communicate more effectively and to develop as a team. It would be very good to have the TMU and the weather unit physically close to the areas as well, so they can talk easily and feel more like they are working together. It's a problem when you put the TMU or the weather unit off in some area away from the controllers."

2130Z: The flight crew receives the PDC showing BEATY as the assigned route.

2200Z: An unexpected pop-up storm has developed that has closed BEATY, EARTZ and BEAMR (see Figure 3). These closures are sent to the other relevant FAA facilities and airline ATC coordinators, dispatchers and ramp supervisors.

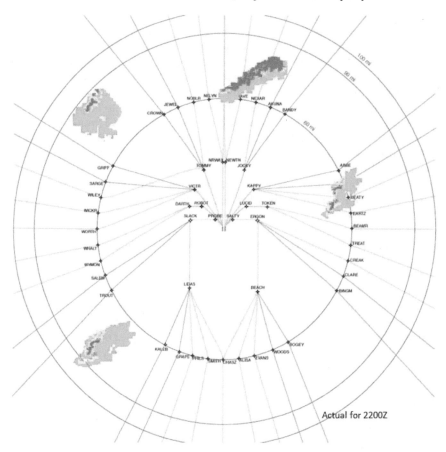

Figure 3. Weather at 2200Z with pop-up storm closing eastern departure fixes.

The evaluation of the traffic managers and dispatcher was as follows.

ARTCC: "Today, this could play out several ways. First, the Center and Tower in conjunction with the TRACON would need to coordinate to determine what reroutes to use. Suppose they had 5 flights over BEATY that needed to be rerouted. They assume they can be rerouted and make the amendment before the flight has pushed back or when it is taxiing out. The flight is informed of the reroute when it talks to clearance delivery. If a flight can't take the reroute, the pilot will tell them."

ARTCC: "It is also possible that a flight was already taxiing out when the storm popped up over BEATY. When the flight contacted clearance delivery, they might ask can we get a reroute over to whatever is open."

ARTCC: "At some airports, the airline's ramp control may also be involved since they have to decide when to push the flight back and how to help ground control to sequence flights. They might check with the Tower to see if a plane has a reroute before pushing him. They could also coordinate with dispatch and the Center to help make sure planes are getting reroutes."

In the scenario, the traffic manager at the expanded TRACON has set up an alert in the surface management tool that highlights flights that are filed though a closed departure fix. This alert is triggered and tells the traffic manager that the currently filed route for the flight to BNA will take it through one of the closed departure fixes. The same is true for flights to IAD, BWI and CVG (see Figure 4). These flights are highlighted in red on the airport surface display, drawing the traffic manager's attention to the affected flights and helping to indicate the order in which these flights should be processed for amendments to still open fixes. One of the tabular displays shows the alternate routes submitted for the flight from MJA-BNA, along with an indication of the weather along those routes (using RAPT data) regarding departure route availability.

The assessments of the display in Figure 4 by the traffic managers were:
ARTCC: "With the expanded TRACON, it seems like the TRACON should have the responsibility for making these amendments. It's possible that in some cases you want the Tower to do so, since they know what the situation is on the ground."
ARTCC: "Regardless of who is responsible for making amendments, some type of alert or indication would be useful. You don't want it to be a nuisance, so it might be more subtle than a blatant alert, like a change in the color or the relevant flight."
TRACON: "From what I see there (Figure 4), that's got pretty much everything I can think of. You'd look at it, pick BANDY and submit it. That would be great."
ARTCC: "Awesome. I don't see anything else. That's all the information you need to make a decision. If you get too much information, it makes the decision take too long. Depending on who is looking at it, though, you'd probably want to be able to expand the map to show the full route for a flight."

Using this display, the traffic manager first amends the route for the BNA flight so that it will depart via BANDY. This amendment is automatically transmitted to the Tower and dispatcher and ramp supervisor. The crew is cleared on this route by the Tower, accepts it and is Off at 2215Z.

3   CONCLUSIONS

The above dialog is very useful in providing an understanding of how an ARTCC TMC, a TRACON TMC and a dispatcher view the integration of four important NextGen focus areas:
- Flexible airspace management.
- Collaborative Routing.

- Translation of weather data into traffic management and flight operations decisions.
- Airport surface departure management.

Below, we summarize some of the relevant human factors and operational requirements that were highlighted relative to such integration.

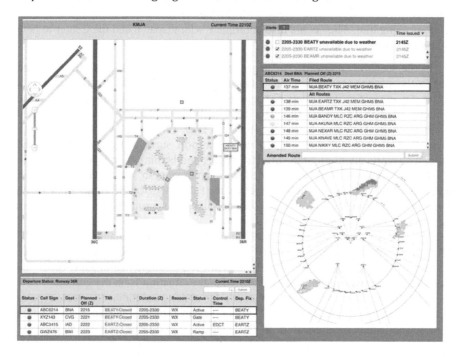

Figure 4. Airport surface and departure fix status displays.

3.1 Planning Tools

Both the ARTCC and TRACON TMC indicated the need for a decision support tool that helps them to make informed decisions about whether and when to use arrival airspace for departures. They posited that:

- This decision support tool needs more accurate information about departure and arrival times in the weather impacted airspace.
- To be useful and usable, this decision support tool needs to support visualization of current arrival and departure demand relative to the current and forecast weather, and visualization of how that demand would change with adjustments in airspace usage for arrivals vs. departures.

3.2 Organizational Design and Physical Layout

The traffic managers strongly emphasized the need to carefully consider the need to easily transition arrival airspace to departures and vice versa. They suggested that the mapping of airspace responsibility to organizational units (to areas within the expanded TRACON) should be carefully considered so that effective teams are likely to develop. Physical proximity of the traffic managers, meteorologist and controllers was also mentioned as a method for enhancing team formation, along with cross-training across both departure and arrival sectors within an area.

3.3 Software to Support the Implementation of a Plan

In addition to developing an effective plan taking advantage of flexible airspace capabilities, tools are required to support its implementation. This includes coordination to make sure everyone is aware of the planned airspace changes, software to actually implement these changes, and software that makes it easy to then make use of the newly opened departure or arrival fixes.

3.4 Use of Surface Management Data to Support Rerouting

To stimulate discussion about the usefulness of surface data in making rerouting decisions, Figure 4 shows a notional display that indicates potentially useful information as well as a notional display concept. In response to this display for the TMC responsible for pre-departure flight amendments, the following responses were generated:

TRACON: "From what I see there (Figure 4), that's got pretty much everything I can think of. You'd look at it, pick BANDY and submit it. That would be great."

ARTCC: "Awesome. I don't see anything else. That's all the information you need to make a decision. If you get too much information, it makes the decision take too long. Depending on who is looking at it, though, you'd probably want to be able to expand the map to show the full route for a flight".

ACKNOWLEDGMENTS

The FAA Human Factors Research & Engineering Group coordinated the research requirements for this work and its principal representative acquired, funded, and technically managed execution of the research service.

Section III

Human Performance in Aviation

Assessing the Impact of NextGen Trajectory Based Operations on Human Performance

Katherine A. Berry, Michael W. Sawyer

Fort Hill Group
Washington, D.C.
Katie.Berry@FortHillGroup.com, Michael.Sawyer@FortHillGroup.com

ABSTRACT

In an effort to modernize the National Airspace System (NAS), the Federal Aviation Administration (FAA) has introduced the Next Generation Air Transportation System. The introduction of new capabilities, decision-support tools, and automation as described in Next Generation Air Transportation System (NextGen) operational improvements offers the potential for changes to the daily activities and tasks of NextGen air traffic controllers. Each change presents the opportunity for both positive and negative effects on the overall safety of the NAS and the role of the air traffic controller. The paper presents a methodology and application for prospectively identifying potential hazards to the performance of air traffic controllers. The Trajectory Based Operation Solution Set was utilized to demonstrate the identification of potential hazards and the most affected sector controller tasks. These identified potential hazards represent rich opportunities for future human factors research.

Keywords: Human Performance, Trajectory Based Operations, NextGen, Safety

1   INTRODUCTION

The FAA is currently executing a considerable transformation of the NAS. NextGen aims to improve the convenience and dependability of air travel while increasing safety and reducing environmental impact. NextGen plans to meet these goals by introducing a variety of new systems and capabilities, such as the introduction of data communications and Automatic Dependent Surveillance-Broadcast (FAA, 2012a). NextGen aims to improve the capacity and efficiency components of the NAS and, perhaps more importantly, NextGen operational improvements aim to enhance the overall safety of the NAS. Many operational improvements aim to provide controllers with decision support tools and automation specifically designed to provide safety enhancements to NAS operations. While NextGen may produce many positive safety improvements, the introduction of each new system and capability also offers the possibility of increasing the human contribution to risk in the NAS (Sawyer, Berry, & Blanding, 2011). This is especially true when considering the system-wide impact and concurrent development of many of the systems (Zemrowski & Sawyer, 2010). From a risk management perspective, research into these effects is needed to address the potential for both positive and negative impacts on the safety of the NAS (FAA, 2011a).

The human factors community of practice has long served as a key player in the enhancement of safety in the Air Traffic Control (ATC) domain. The techniques and tools primarily utilized by human factors practitioners, however, are often retrospective in nature. These techniques and tools serve as aids in the analysis of incident and accident data gathered post-hoc. While these tools and techniques have been and are still valuable in assessing the safety of the NAS, these tools and techniques are limited in regards to analysis of future systems, such as NextGen (GAO, 2011; FAA, 2011a).

The ability to proactively identify potential hazards to human performance associated with a new system before the new system is introduced into the NAS has long been identified as a need (GAO, 2010; GAO, 2011). The integration of proactive human factors safety research into the earliest stages of system design and acquisition could not only reduce industry cost, but improve system design, development, and implementation (EUROCONTROL & FAA, 2010). A standardized approach to proactively identifying and assessing human performance hazards is needed to ensure these hazards are identified, described, and tracked.

1.2   Purpose

The incorporation of proactive human factors safety research in the earliest stages of NextGen development has the ability to both impact system design and improve upon the level of safety of the NAS. This chapter presents an approach to proactively identifying hazards to successful human performance in the NextGen ATC domain and summarizes the key results obtained by applying this methodology to a set of NextGen increments. The results identify the NextGen increments from a set of operational improvements that are most vulnerable to human performance hazards. These identified hazards should serve as the basis for the creation of design and research requirements to minimize the human contribution to system risk.

2   METHODOLOGY

This analysis will focus on the Segment Bravo portion of the NextGen mid-term timeframe. Segment Bravo is comprised of seven solution sets represents changes planned to be implemented beginning in 2016. The solution sets are compellation of related operational improvements grouped by overarching themes (FAA, 2012b). For this analysis, the Trajectory Based Operations (TBO) solution set will be utilized.

2.1   Trajectory Based Operations

With a focus on high-altitude cruise operations, the TBO solution set describes planned changes to the en route environment and the en route sector controller. TBO changes incorporate capabilities, decision-support tools, and automation that may enable the sector controller to manage air traffic by trajectory rather than clearance. The shift to trajectory-based movement offers the potential to improve NAS efficiency by permitting aircraft to fly negotiated flight paths that incorporate both operator preferences and optimal airspace system performance (FAA, 2011b). Five operational improvements comprising 20 incremental changes proposed in the TBO solutions set were considered for this analysis. Operational improvements provide high-level descriptions of the new capabilities, while the increments provide descriptions of the individual capabilities, decision-support tools, and automation necessary to support TBO implementation. Table 2 lists the Segment Bravo TBO increments.

2.2   Human Performance Hazard Identification Process

The Human Error and Safety Risk Analysis (HESRA) tool was utilized to perform the assessment of the Segment Bravo TBO solution set (FAA, 2009). The HESRA tool is based on Failure Modes and Effects Analysis in which potential failure modes and their effects and relative risk are identified at each step of the process being analyzed. For each individual scenario step, the potential hazard conditions and human performance hazards were identified through panel discussions with human factors and ATC experts. The worst credible effect or outcome of each potential human performance hazard was described. The severity, likelihood, and recovery were also qualitatively estimated for each human performance hazard and its associated effect. ATC and human factors experts were utilized to identify hazards and then to estimate their impact.

This analysis utilized the FAA's (2008) Safety Management System (SMS) likelihood and severity scales. The use of FAA likelihood and severity scales was essential to ensure that identified hazards could be managed in the existing FAA safety process. The analysis employed the HESRA scale for recovery as the SMS manual lacks a recovery or detection classification scheme. For each hazard a Risk Priority Number (RPN) was calculated by multiplying the likelihood, severity, and recovery rating. There are five RPN risk categories ranging from extremely low risk to extremely high risk as shown below in Table 1 (FAA, 2009). The identified hazards were then prioritized based on the RPN in order to identify the NextGen operations most vulnerable to human performance hazards.

Table 1 HESRA Risk Priority Number Categories

Category	Definition/Action
Extremely Low Risk	• No system or safety implications • No further design or evaluation efforts required
Low Risk	• No significant system or safety implications • Unlikely that significant design, training, or procedural changes will be required
Moderate Risk	• Potentially significant system or safety implications • Possible that significant design, training, or procedural changes will be required
High Risk	• Significant system or safety implications • Likely that significant design, training, or procedural element will be required
Extremely High Risk	• Critical system or safety implications • Significant design, training, or procedural changes are required before the system is deployed

While the focus of this analysis is on the en route sector controller, hazards relating to the actions or inactions of other NAS actors were considered if the sector controller would be the actor responsible for recovering from that hazard. The NextGen actor list and descriptions were utilized for this process (FAA, 2012b). Additionally, for those hazards that identified the sector controller as the actor, the NextGen job task was also identified for the hazard. The job descriptions for the NextGen mid-term en route controller were utilized (AIR, 2011).

3   RESULTS AND DISCUSSION

The HESRA hazard identification process identified many potential NextGen induced human performance hazards. The qualitative rating of severity, likelihood, and recovery from the analysis was then used to prioritize the operational scenarios based on the opportunities for introducing human performance hazards. Table 2 presents a prioritized list of TBO increments based on the relative risk of potential human performance hazards. The increments noted with a "—" represent instances where no human performance hazards were identifying through the analysis of the increment.

Table 2 TBO Human Performance Hazards Relative Risk

Increment Description	Relative Risk
Aircraft-to-Aircraft and Aircraft-to-Airspace Problem Resolution	1
Aircraft-to-Flow Management Time Resolution	2
Automated Handoff	3
Aircraft-to-Weather Area Problem Resolution	4
Airspace Configuration Selection	5
Conformance Monitor	6
Interval Management – Delegated Separation	7
Planned SAA Release Areas	8
Airspace Configuration Evaluation and Formulation	9
Dynamic SAAs	10
Interval Management during Descent using Flight Data Capability	11
Aircraft to Severe Weather Notifications	12
Determine Meter Points and Identify Flights to Meter Points	13
Airspace Configuration Execution	14
Meet TBFM Constraints Use Require Time of Arrival Capability	15
Wake Separation	—
Airspace Design and Collaboration Tool	—
Airspace Configuration Model	—
Oceanic 4D Trajectory Coordination Capability	—
Interval Management during Cruise using Ground Automation	—

The analysis of increment "Aircraft-to-Aircraft and Aircraft-to-Airspace Problem Resolution" resulted in hazards with the highest relative risk. This increment describes the capability of providing the sector controller with ranked resolutions for both aircraft-to-aircraft and aircraft-to-airspace problems. Like most operational improvements, the benefit of the capability is clear. This capability will allow a problem to be resolved in a quicker and more efficient manner. However, potential human performance hazards are also associated with the increment. For example, over time the sector controller could potentially become complacent due to the addition of the ranked resolution tool. Automation could present the controller with the ranked resolutions, and the controller—due to the over-trust in the automation—could select the first ranked resolution without reviewing the resolution in detail. In most cases, the resolution provided by the automation would be accurate and no problems would arise. However, in the worst credible outcome the resolution generated by the automation would be inadequate to resolve the problem. In this instance, design requirements are needed to ensure the automation is capable of presenting accurate and reliable resolutions to the controller.

Furthermore, the controller must be made aware of the types of information or situations that are not considered in the automation's resolution generation algorithm. Automation design guidelines suggest that the preferred practice is to have the automation monitor the human, rather than the human monitor the automation (Billings, 1997). Finally, research requirements are necessary to determine the appropriate level of trust in automation for sector controllers.

The analysis of increment "Automated Handoff" resulted in the third highest relative risk. This increment describes Segment Bravo capability of automated handoffs. This capability differs from other automated handoff capabilities in that if automation finds no potential conflicts or issues with a handoff then the handoff process and transfer of communications is completed without controller input. While this increment will potentially help to lessen the controller's workload and permit the controller to focus on higher priority issues, it also creates the potential for new human performance hazards.

Table 3   Example NextGen Human Performance Hazards

Operational Improvement	102114	102118
Increment Description	Conformance Monitor	Interval Management - Delegated Separation
Hazard Condition	Monitor aircraft conformance and provide an alert on the Radar Console when an aircraft's track or position information indicates that the aircraft is laterally deviating from its assigned route.	Increment enables a controller to evaluate situations suitable for delegating separation— checking if either aircraft is predicted to have an aircraft-to-aircraft problem
Human Performance Hazard	Weather or excess nuisance alerts causes controller to suppress all conformance monitor alerts. Controller is unaware of nonconformance.	Controller fails to identify weather condition requiring cancellation of delegated separation.
Worst Credible Outcome	Controller remains unaware of nonconformance due to suppression of alerts. Aircraft continue on present trajectories. Potential for airspace violation or loss of separation.	Aircraft encounter weather. Pilots request deviations. Controller cancels delegated separation and issues corrective vectors to aircrafts.

For illustrative purposes, one human performance hazard will be explained. The first hazard in Table 3 originated from the Conformance Monitor increment. For this capability, automation may monitor aircraft conformance to the assigned route.

If the aircraft laterally deviates from the route, the automation may alert the sector controller on the radar console. It is possible that given the presence of adverse weather or other situations the automation could result in a disproportionate amount of nuisance alerts causing the controller to suppress all conformance monitor alerts. During the time of alert suppression, the controller could fail to notice an aircraft laterally deviating from its route, resulting in a potential airspace violation or loss of separation. While this hazard might not occur frequently, the potential for a significant adverse outcome dictates further research related to differentiating between true and nuisance alerts.

3.1   NextGen Actor Analysis

The actor associated with each potential hazard was identified and used to determine the relative risk introduced by each actor. Table 4 lists the relative risk results from the NextGen hazard actor analysis. The NextGen sector controller position has the highest relative risk based on the TBO Segment Bravo solution set. This can be attributed to the characteristics of the TBO solution set, which focuses primarily on the tools and capabilities being provided to the sector controller. The traffic management coordinator, automation, and front line manager were also identified in the analysis as hazard actors. These non-controller induced hazards are of interest as the sector controller will be responsible for ensuring a safe recovery from each hazard.

Table 4   Human Performance Hazard Actor Analyses

Actor	Relative Risk
Sector Controller	1
Traffic Management Coordinator	2
Automation	3
Front Line Manager	4

3.2   NextGen Job Tasks Analysis

The NextGen mid-term job task was identified for each hazard associated with the sector controller. For this analysis, only one job task was associated with each hazard. Table 5 lists the top five job tasks resulting from the sector controller human performance hazards. The identification of the NextGen sector controller job task affected by each hazard provides insight into the impact of the TBO capabilities on the controller's job. Two NextGen job tasks tied for the job task with the highest relative risk. Task 152: "Determine appropriate action to resolve conflict situation" and Task 174: "Determine appropriate action to resolve airspace violation" resulted in the most associated human performance hazards. Many of the TBO increments described decision-support tools and automation to aid the controller in identifying and remedying various types of aircraft and airspace problems.

Table 5   Top Five Impacted NextGen Job Tasks

Task	Relative Risk
T152: Determine appropriate action to resolve conflict situation	1
T174: Determine appropriate action to resolve airspace violation	1
T205: Re-evaluate traffic sequence	2
T268: Determine altitude or route change to bypass severe weather	3
T206: Issue revised control instructions if required	4
T077: Monitor aircraft progress through radar coverage area	5

3.3   Future NextGen Research Areas

This analysis and other similar analyses should be utilized to guide future human performance research for the NextGen Segment Bravo environment. Rather than broadly researching an ATC NextGen topic, researchers can focus their efforts on one of the many human performance issues identified in this analysis. First, the hazards from the TBO increments resulting in the top relative risks should be examined in further detail, and research efforts should be made to determine mitigation strategies for those increments and their associated human performance hazards. Many of the top ranked increments are similar in that decision support tools will be provided to the controller to aid in identifying and resolving a variety of problems. For these increments, it is possible to examine the capability of problem resolution as a whole and then develop detailed research plans and mitigations for the specific increments and associated human performance hazards.

While this analysis can aid in identifying future NextGen research areas, this analysis should also aid in the development of NextGen design requirements. The NextGen research areas should focus on developing design requirements that mitigate the human performance hazards and their worst credible outcome. Additionally, it would be beneficial to identify—and when necessary, mitigate—the more likely outcome of the various human performance hazards. For example, a human performance hazard could result in the worst credible outcome of a loss of separation or the more likely outcome of a nuisance alarm. While the occurrence of a single nuisance alarm does not result in a high risk situation, the impact of nuisance alarms over time should be taken into consideration when developing mitigating design requirements.

Furthermore, TBO research efforts should be focused on the sector controller as evident in Table 4, but should also extend to other NAS actors, such as the traffic management coordinator and the front line manager. The actions of the traffic manager, in particular, directly impact the actions of the sector controller. If the traffic manager fails to identify or accurately assess the impact of weather on a sector, the result could be severe weather impacting many aircraft in a sector or even sector overload.

Future human performance research should also focus on the activities and tasks of the controller most impacted by the TBO operational improvements and increments as listed in Table 5. Many of the top tasks identified involved the decision making process of developing a plan for resolving a problem situation. While the TBO increments describe decision-support tools expected to aid in this process, research should be conducted in the presentation and development of these tools. The controller should be able to understand the reasoning of the various algorithms behind the decision-support tools in order to obtain controller support of the tools.

It is important to note that this analysis should serve as an initial foundation for future NextGen human performance research. NextGen human performance research should not focus only on testing a hypothesis, but also on identifying mitigation strategies for the human performance hazards identified in this analysis. These mitigation strategies should include the development of NextGen design requirements. In addition, future research should continue in conducting similar assessments of other NextGen Segment Bravo solution sets, such as the flexible terminals and airports solution set. This methodology should also be extended to identify human performance hazards associated with off-nominal situations and scenarios and to the flight deck perspective, as well.

4   CONCLUSION

In this assessment, hazards to successful ATC human performance in the midterm Segment Bravo NextGen environment were identified for the TBO solution set. The HESRA methodology was utilized to prioritize the hazards based on the RPN, which encompasses likelihood, severity, and recovery. The resulting hazard assessment produced a list of human performance hazards. The hazards were classified based on TBO increment, NextGen actor, and en route job task. Since this assessment is in the pre-mitigation stages, future work should include developing risk mitigations strategies and research plans addressing these hazards and their associated job tasks. By implementing effective risk mitigation strategies, the human factors community has the ability to be involved in the design phase of many NextGen operational improvements. Additionally, these findings can also be utilized in creating a research plan for mid-term NextGen operations that addresses these key hazards and issues and in developing strategic design requirements to aid in mitigating the hazards.

5   ACKNOWLEDGEMENTS

We would like to acknowledge the FAA's Human Factors Research and Engineering Group (ANG-C1) for funding this project and similar work. Additionally, we would like to acknowledge the ATC and human factors subject matter experts who helped to assess the operational scenarios.

REFERENCES

American Institutes for Research. (2011). Job Description for the NextGen Mid-Term ARTCC Controller. Retrieved 2011, from https://www2.hf.faa.gov/HFPortalNew/.

Billings, Charles E. (1997). *Aviation Automation: The Search for a Human-Centered Approach*. Mahwah, New Jersey: Erlbaum.

EUROCONTROL & Federal Aviation Administration. (2010). Human Performance in Air Traffic Management Safety. EUROCONTROL/FAA Action Plan 15 Safety

Federal Aviation Administration. (2008). Air Traffic Organization: Safety Management System Manual. Retrieved 2012, from http://www.faa.gov/about/safety_efficiency/sms/resources.

Federal Aviation Administration. (2009). Human Error and Safety Risk Analysis for Federal Aviation Administration Air Traffic Control Maintenance and Operations. Retrieved 2012, from http://www.hf.faa.gov

Federal Aviation Administration (2011a). Destination 2025. Retrieve 2011, from www.faa.gov/about/plans_reports/media/Destination2025.pdf.

Federal Aviation Administration. (2011b). Trajectory Based Operations. Retrieved 2012, from http://www.faa.gov/nextgen/implementation/portfolio/sol_sets/tbo/index.cfm.

Federal Aviation Administration. (2012a). NextGen Implementation Plan. Retrieved 2012, from http://www.faa.gov/nextgen

Federal Aviation Administration. (2012b). NAS Enterprise Architecture. Retrieved 2012, from https://nasea.faa.gov/

Government Accountability Office. (2010). Aviation Safety: Improved Data Quality and Analysis Capabilities Are Needed as FAA Plans a Risk-Based Approach to Safety Oversight (GAO-10-414). Retrieved from http://www.gao.gov/products/GAO-10-414.

Government Accountability Office. (2011). Aviation Safety: Enhanced Oversight and Improved Availability of Risk-Based Data Could Further Improve Safety (GAO-12-24). Retrieved from http://www.gao.gov/products/GAO-12-24.

Sawyer, M., Berry, K, & Blanding, R. (2011). Assessing the Human Contribution to Risk in NextGen. In the Proceedings of the Human Factors and Ergonomics Society Annual Meeting, 2011, Las Vegas, NV.

Zemrowski, K, & Sawyer, M. (2010). Impacts of Increasing Reliance on Automation in Air Traffic Control Systems. Journal of Air Traffic Control, 52 (4), 49-55.

CHAPTER 17

Improving Multi-tasking Ability in a Piloting Task through Action Videogames

Dan Chiappe[12], Mark Conger[3], Janet Liao[12], Kim-Phuong L. Vu[12], Lynn Caldwell[4]

[1]California State University, Long Beach, [2]Center for Usability in Design and Accessibility, [3]Northrop Grumman Corporation, [4]Air Force Research Lab

Dept. of Psychology
California State University, Long Beach
1250 Bellflower Blvd.
Long Beach, CA 90840
dan.chiappe@csulb.edu

ABSTRACT

We examined whether action videogames can improve multi-tasking in high workload environments. Two groups with no action videogame experience were pre-tested using the Multi-Attribute Task Battery (MATB). It has two primary tasks; tracking and fuel management, and two secondary tasks; systems monitoring and communication. In addition, participants were given subjective measures of workload and situation awareness (SA) – the NASA TLX and Situation Awareness Rating Technique (SART). One group served as a control, while a second played action videogames for 5 hours a week for 10 weeks. We found the videogame treatment enhanced performance on the secondary tasks without interfering with the primary tasks. However, measures of workload and SA did not reveal an effect of videogame treatment. Our results demonstrate action videogames can increase the ability to take on additional tasks. However, this may not be revealed by subjective judgments because these are most likely determined by primary task performance.

Key words: Multi-tasking, videogame training, situation awareness

1   INTRODUCTION

Technology has greatly expanded the information available to system operators. This increase in system complexity places considerable demands on their attentional capacity. Whether the domain is nuclear power plant operation, aircraft piloting, fire-fighting, military command and control, or air traffic control, operators increasingly need to coordinate the performance of multiple tasks, but have limited attentional resources to do so (Endsley, Bolte & Jones, 2003). A central challenge in such contexts is maintaining situation awareness – the understanding needed to operate a complex system in a rapidly changing task environment (Chiappe, Strybel & Vu, in press; Endsley, 1995). In this study, we examine whether playing action videogames can increase people's ability to multitask and to maintain situation awareness in a high workload task environment.

The potential usefulness of action videogame training to enhance multi-tasking is supported by experiments finding that such a regimen can yield improvements in spatial cognition and attention among non-gamers (Spence & Feng, 2010). In particular, action videogames, in comparison to other genres, have been found to broaden the attentional visual field. This is likely due to the extensive attentional demands action videogames place on individuals. Green and Bavelier (2003), for example, found that participants that played action videogames for as little as 10 hours improved their ability to identify peripherally presented targets compared to those that played non-action videogames. Importantly, a follow up study by Green and Bavelier (2006) showed that the increased ability to process peripheral information did not come at the expense of carrying out a primary task on information presented at the center of fixation.

The current study investigated the ability of action videogames to improve multi-tasking using the Multi-Attribute Task Battery (MATB), as well as subjective measures of situation awareness (SART) and workload (NASA TLX). The MATB consists of two primary tasks that require constant monitoring and action (tracking and fuel management), as well as two secondary tasks that have to be performed intermittently and on peripherally presented information (communications and systems monitoring). Participants were selected based on their lack of experience playing action videogames. One group of participants played action videogames over a period of 10 weeks for approximately five hours per week, while a second group did not.

We predicted that those participants receiving the videogame treatment would perform better on the post-assessment session of the MATB. In particular, we predicted that videogame participants would do better on the communications and systems monitoring tasks (i.e., the secondary tasks) than control participants. Following Green and Bavelier (2006), we also predicted that this improvement in the ability to carry out additional secondary tasks would not come at the expense of their performance on the primary tasks of tracking and fuel management. Finally, we predicted that those participants playing action videogames would have lower subjective workload ratings and higher subjective situation awareness ratings at the end of the study compared to control participants.

2   METHODS

2.1   Participants

Participants were 49 undergraduates (Mean age 22 years, range 18-36) from California State University, Long Beach. In the videogame group, there were 12 males and 12 females, and in the control group, 12 males and 13 females. Participants were screened using a selection questionnaire that asked about health history and experience playing action videogames. Over 600 individuals filled it out, and we selected those that reported having no prior or current experience playing action videogames and no major health problems.

2.2   Materials

Multi-tasking ability was assessed using the MATB (Comstock & Arnegard, 1992). It requires the simultaneous performance of four tasks under high workload conditions – tracking, systems monitoring, fuel management, and communications tasks (See Figure 1).

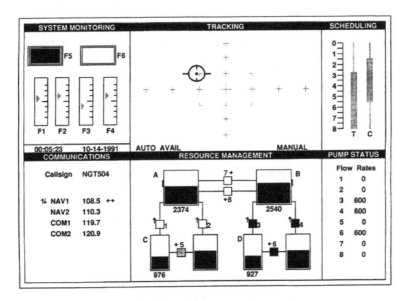

Figure 1 The Multi-Attribute Task Battery (MATB) task from Comstock and Arnegard (1992).

The *tracking* task requires the manual operation of a joystick to keep a target in the center of a tracking window. The *systems monitoring* task includes monitoring lights and dials. For the lights, when one goes out, participants are required to press a button to turn it back on. For the other designated light, it requires them to press a button when it is lit to turn the light back off. For the dials, if they drift outside of a specified range, participants have to reset them by pressing a button. The *fuel management* task requires the activation of virtual pumps to fill two tanks, maintaining them as closely as possible at an ideal level. The *communications* task requires individuals to respond to auditory communication instructions to a specific call-sign. The MATB was presented to participants on a Dell Desktop computer using a 20" × 20" computer screen.

Participants also filled out measures of situation awareness (SART, i.e., the Situational Awareness Rating Technique) and workload (NASA Task Load Index). The SART (Taylor, 1990) is a subjective assessment of situation awareness that includes 10 questions, three surveying the demand on attentional resources imposed by the task, four surveying participants' supply of attentional resources, and three surveying their understanding of the situation. Each question of SART required an answer on a 7-point scale. Ratings were averaged for each dimension, and combined using the formula of SART-Combined = Mean Understanding Rating – (Mean Demand Rating – Mean Supply Rating). The NASA-TLX (Hart & Staveland, 1988) is a subjective assessment of workload that asks operators to rate their workload along six dimensions. These include mental demand, physical demand, temporal demand, performance, effort and frustration. Each of these is rated on a scale, and the unweighted ratings were combined into a total workload score out of 100.

Videogame participants were given a Sony PlayStation 3™ console to use at home during the 10 weeks of treatment. In addition, these participants were given two games to start: Ghost Recon Advanced War fighter 2™, and Unreal Tournament 3™. Once they finished these games, they were given additional games that included Medal of Honor™, Vanquish™, Bioshock 2™, and Resistance 2™. These games are all first person shooter games. Players have the same point of view as the main character in the game, and their task is to shoot at enemy targets, while monitoring the status of their ammunition and weapons systems, as well as other team members. The games were chosen for the high mental, visual, and attentional resources that they demand. Participants' progress through these games was recorded using weekly game-playing diaries that asked them to record the amount of time played, as well as rating the enjoyment of each session on a 1 to 5 scale.

2.3   Procedure

Participants in the control and videogame playing groups were brought into a Psychology lab for a pre-assessment using the MATB task. The task was administered in four high-workload sessions: One 10 minute (familiarization), one 20 minute (warm-up), and two 30 minute (practice and test) sessions. The last of these was used in the analyses reported below. Participants were given 5 minute breaks after the 20 minute session, and between the two 30 minute sessions. After completing the MATB, participants were given the NASA TLX and SART. They were told to rate their workload and situation awareness based on the last 30-minute test session on the MATB.

Once they completed the pre-assessment with the MATB, participants in the videogame condition were brought into an adjacent room to be trained on the

PlayStation 3™. Experimenters taught participants how to setup the console with the laboratory TV. They were then assisted by the experimenter on the training modes of the games Ghost Recon 2™ and Unreal Tournament 3™. This was to ensure that they understood how to orient their characters, the goals of the games, and basic maneuvers such as how to use weapons and positioning. The participants were then instructed on how to complete their weekly game-playing diaries.

Videogame participants were asked to play a minimum of 5 hours per week for 10 weeks, though they were encouraged to play as much as possible during each week. They were also given leeway in terms of how that game playing was spread out during the week. If participants did not comply, they were asked to make up the hours of game-playing the following week. Two participants were dropped from the study for not playing the required amount of hours. The remaining averaged a minimum of 5 hours of videogame playing across the 10 weeks of treatment.

All participants in the videogame and control conditions were brought back for a post-assessment on the MATB after 10 weeks. They completed a 10 minute (re-familiarization), a 20 minute (warm-up), and two 30 minute (practice and test) sessions. They also completed the NASA TLX and the SART. In addition, eight of the participants in the control group, and eight of the videogame participants were brought in for a midterm assessment on the MATB 5 weeks after the pre-assessment. This was done to determine whether effects of the videogame treatment are evident half way through the treatment.

3   RESULTS

3.1   Post-treatment Effects on MATB Performance

To examine whether playing action videogames for 10 weeks affected the ability to multitask on the MATB, we conducted a series of $2\times2\times3$ mixed factors ANOVAs. Condition (videogame vs. control) is the between-subject factor, and session (pre-assessment vs. post-assessment) and interval (first 10 min, second 10 min, vs. third 10 min) are the repeated measures factors. Interval was included because it allowed us to examine whether there are any changes in performance throughout the session either because of fatigue, or practice effects. Separate ANOVAs were conducted for variables collected for each of the four subtasks of the MATB. Detailed analyses are reported in Chiappe, Conger, Liao, Caldwell, and Vu (submitted). We only summarize the most relevant findings here.

3.1.1 Communications task

The analysis examining correct communications RT (COMCRT) revealed a significant interaction between session and condition, $F(1, 45) = 5.02$, $p < .05$. As shown in Table 1, the control condition did not show a decrease in RT between pre- and post-assessments, but the videogame condition did.

An analysis of the standard deviation of the RT to correctly respond to communications (COMCSD) revealed a marginally significant interaction between session and condition, $F(1, 45) = 3.32$, $p = .075$. For the control group, the SD did not decrease between the pre- and post-assessments, but for the videogame condition there was a decrease in the SD between the two assessments.

Table 1 Performance on the communication task of the MATB

Task	Videogame	Control
COMCRT		
Pre-assessment	3.11	3.07
Post-assessment	2.54	2.99
Difference	0.57	0.08
COMCSD		
Pre-assessment	1.40	1.41
Post-assessment	1.07	1.49
Difference	0.33	-0.08

3.1.2 Systems Monitoring task

There was an interaction between interval, session, and condition with respect to time-out errors to the dials, $F(2, 90) = 5.39$, $p = .006$. For the control condition, the interaction between session and interval was not significant, $F < 1$. For the videogame condition, however, the interaction was significant, $F(2, 44) = 3.07$, $p = .056$. Specifically, in the pre-assessment, videogame participants displayed an inverted U function across the three intervals. They had fewer time out errors in the first interval ($M = .29$), but these increased during the higher workload, second interval ($M = .92$), and then decreased again in the final interval ($M = .21$). However, in the post-assessment session, time out errors did not reveal a similar increase due to increasing workload. Instead, the errors decreased from the first ($M = .25$) and second interval ($M = .25$) to the third interval ($M = .08$). In comparison, for the control group their post-assessment revealed an increase in errors from the first interval ($M = .49$) to the second interval ($M = .85$), and then dropped slightly in the third interval ($M = .69$). Thus, in the post-assessment session, the videogame participants displayed a greater ability to manage an increased workload, while the control group did not.

3.1.3 Tracking and Fuel management tasks

For the tracking and fuel management tasks, main effects of condition were obtained, with videogame participants having a lower error on the tracking task, $F(1, 45) = 9.10$, $p = .004$, and a smaller deviation from the ideal on the fuel management task, $F(1, 42) = 4.84$, $p = .033$ (three participants were dropped because their data was more than two standard deviations from the mean.)

3.2 Post-treatment Effects on Situation Awareness and Workload

3.2.1 Situation Awareness Rating Technique (SART)

We conducted separate 2×2×2 mixed factors ANOVAs for the combined SART score and each sub-scale, with session (pre-assessment vs. post-assessment) as the repeated-measures factor, and condition (videogame vs. control) and gender (male vs. female) as the between-subjects factors (See Table 2).

The results revealed no interactions between session and condition for any of the SART variables, with all $Fs < 1$. There were also no main effects of condition for the Demand and Understanding variables, $Fs < 1$, for the Supply variable, $F(1, 45) = 1.29$, $p = .26$, or for the Combined score, $F(1, 45) = 1.19$, $p = .28$. In short, the videogame treatment did not lead to significant improvements in subjective ratings of situation awareness.

Table 2 SART scores for pre-assessment and post-assessment sessions

Scale		Videogame	Control
Demand			
	Pre-assessment	4.18	4.23
	Post-assessment	4.06	3.95
	Difference	0.12	0.28
Supply			
	Pre-assessment	4.93	4.76
	Post-assessment	4.72	4.35
	Difference	0.21	0.41
Understanding			
	Pre-assessment	4.03	3.72
	Post-assessment	4.53	4.31
	Difference	-0.50	-0.59
Combined			
	Pre-assessment	4.78	4.27
	Post-assessment	5.19	4.71
	Difference	-0.41	-0.44

3.2.2 NASA-TLX

We conducted a 2×2×2 mixed ANOVA, with session as the repeated-measures factor, and condition and gender as between-subjects factors (see Table 3). There were no significant main effects and no significant interactions with condition (all $Fs < 1$).

Table 3 NASA TLX scores for pre-assessment and post-assessment sessions

Session	Videogame	Control
Pre-assessment	57.95	62.19
Post-assessment	56.08	59.83
Difference	1.87	2.36

3.3 Midterm Effects on MATB Performance

As noted, eight videogame and eight control participants were brought in for a midterm assessment on the MATB, on the sixth week after the pre-assessment session. Descriptive statistics are listed in Table 4.

Table 4 Midterm performance on the MATB secondary tasks for 16 participants

Task	Videogame	Control
Communications task		
COMCRT		
Pre-assessment	3.32	3.18
Midterm assessment	2.95	3.26
Difference	0.37	-0.08
COMCSD		
Pre-assessment	1.57	1.37
Midterm assessment	1.60	1.52
Difference	-0.03	-0.15
Systems monitoring task		
DLSTO		
Pre-assessment	0.68	0.71
Midterm assessment	0.58	0.96
Difference	0.10	-0.25

None of the interactions between session and condition were significant for the communications task ($Fs < 1$) or for the systems monitoring task ($Fs < 1$). There was a marginal effect of condition on the tracking task with videogame participants performing better, $F(1, 12) = 4.03$, $p = .068$, though the interaction between session and condition was not significant, $F(1, 12) = 1.50$, $p = .28$. There was also a marginal effect of condition on the fuel management task, $F(1, 12) = 3.81$, $p = .075$, though the interaction between session and condition was not significant, $F(1, 12) = 1.49$, $p = .246$. The midterm results, though not significant, thus yield a similar pattern to the post-test results. The failure to find significant effects is likely due to the small number of participants that completed the midterm test.

4   DISCUSSION

Our study supports the claim that action videogame play for 10 weeks can enhance the ability of operators to multi-task. In particular, videogame participants improved their performance on the secondary tasks of the MATB; monitoring systems and communications, without sacrificing their performance on the primary tasks of tracking and fuel management. The midterm results with a limited number of participants demonstrate that these enhancements of performance may be evident after just 5 weeks of game play, though more participants are required to convincingly demonstrate this claim.

The present findings are consistent with previous research that demonstrates a widening of the attentional visual field as a result of action videogame play. Our findings are noteworthy, however, because they demonstrate that the broadening of attention can span domains other than the visual domain. They are also important because they show an enhancement of attention in a task that is more complex than the ones typically used to explore the effectiveness of videogame training.

Despite finding improvements in the ability to take on additional tasks among videogame participants, the treatment did not lead to improvements in subjective judgments of situation awareness or workload. We argue that this is due to the fact that these subjective ratings are likely to be most influenced by performance on the tasks that demand constant attention – the primary tasks, which did not reveal improvement as a result of training.

To conclude, videogame training appears to be an effective, enjoyable, and cost-effective tool for enhancing multi-tasking, which offers a potential skills enhancement tool for workload-intensive occupations. Future research will help to identify the amount of videogame play that is required to yield specific improvements in operator performance.

REFERENCES

Chiappe, D., T. Strybel, and K.-P. L. Vu. In press. Mechanisms for the acquisition of Situation Awareness in situated agents. *Theoretical Issues in Ergonomics Science.*

Chiappe, D., M. Conger, J. Liao, L. Caldwell, and K-P. L. Vu. Submitted. Improving multi-tasking ability through action videogames.

Comstock, J., and R. Arnegard. 1992. The Multi-attribute Task Battery for human operator workload and strategic behavior research. Technical Report 104174. Hampton, VA: NASA Langley Research Center.

Endsley, M. R. 1995. Toward a theory of situation awareness in dynamic systems. *Human Factors* 37: 32-64.

Endsley, M. R., B. Bolte, and D. Jones. 2003. *Designing for situation awareness: An approach to human-centered design.* London: Taylor & Francis.

Green, C. and D. Bavelier. 2003. Action video game modifies visual selective attention. *Nature* 423: 534-537.

Green, C. and D. Bavelier. 2006. Action video game experience alters the spatial resolution of attention. *Psychological Science* 18: 88-94.

Hart, S., and L. Staveland. 1988. Development of NASA-TLX (Task Load Index): Results of empirical and theoretical research. In. *Human mental workload* , eds. P. Hancock, and N. Meshkati. Amsterdam: North Holland, 139-183.

Spence, I., and J. Feng. 2010. Video Games and Spatial Cognition. *Review of General Psychology* 14: 92-104.

Taylor, R. 1990. Situation awareness rating technique (SART): The development of a tool for aircrew systems design. In. *Situation Awareness in Aerospace Operations* (AGARD-CP-478). NATO-AGARD, Neuilly Sur Seine, France: 3/1-3/17.

CHAPTER 18

Dimensionality Effects on Weather Avoidance

Veranika Lim

Leiden University
Leiden, The Netherlands
Veranika.Lim@gmail.com

Walter Johnson

NASA Ames Research Center
Moffett Field, USA
Walter.Johnson@nasa.gov

ABSTRACT

The effects of different display perspectives on pilot performance in weather and traffic avoidance tasks were evaluated using a 3 (display format: 2D toggle display, 3D toggle display and a 3D flexible display with a continuously manipulable viewpoint) x 2 (weather phenomena: convection and turbulence potential) within subject experiment. Results showed efficiency and situation awareness advantages when using a 2D toggle display. However, the 3D display with a continuously manipulable viewpoint performed similarly, and was superior to the 2D toggle display for safety. Additionally, different weather seemed to be best supported by different display formats and pilots were most conservative in convective weather conditions. This study provides guidance on ways to display weather and traffic information in order to support pilots' safe navigation around weather hazards.

Keywords: human-computer interaction, dimensionality, human performance.

1   INTRODUCTION

The current air traffic control system does not meet the scalability or flexibility demands of the future (www.jpdo.com). In response to this, the Federal Aviation

Administration (FAA) introduced The Next Generation Air Transportation System (NextGen), a proposed evolution from a ground-centric system to a satellite-based air traffic management system that allows more information to be shared between the air and ground, as well as between flight decks. This change is expected to result in an increase in pilots' situation awareness and understanding of the populated airspace, thus supporting problem-solving tasks such as the prediction, detection, and avoidance of conflicts with other aircraft. The improvement of information sharing across all operators in the air traffic management system will support NextGen concepts of operations like Free Flight, defined as *"a safe and efficient flight operation capability under instrument flight rules in which the operators have the freedom to select their path and speed in real time"* (Wickens, Mavor, Parasuraman & McGee, 1998). To incorporate Free Flight, the flight deck would need to be equipped with appropriate information and tools, giving pilots the possibility to select their flight path and speed while enhancing safety and efficiency. In addition to conflict avoidance with other aircraft, equipped flight decks could use these tools to negotiate weather. As weather is one of the most common hazards for pilots (http://www.transtats.bts.gov), providing better weather information to the flight deck has been an extremely important effort in aviation research.

In a study of actual thunderstorm, penetrations and deviations (Rhoda and Pawlak, 1999) clearly demonstrated the flight crew's need for better weather information. In addition to questions of safety, the current weather information system is limited (Comerford et al., 2009). Because of these limitations, the FAA is working on providing more up-to-date and easy to understand weather information. As the technology needed for weather sensing and real time weather information collection improves, and begins to deliver ever greater amounts of data to the cockpit, the next question is how this information should be displayed in order to enhance pilot's situation awareness, which is defined as *"the perception of the elements in the environment within a volume of time and space, the comprehension of their meaning and the projection of their status in the near future"* (Endsley, 1999). A tool that has been tested extensively in simulation-based studies is the 3D (three-dimensional) volumetric Cockpit Situation Display (CSD) developed by the Flight Deck Display Research Lab (Granada, Dao, Wong, Johnson, & Battiste, 2005). The CSD integrates traffic, weather and terrain information into one display supporting both traditional 2D (two-dimensional) and advanced 3D visualization modes. The integration of traffic and weather is beneficial because different types of information, (e.g., weather and traffic) supporting common tasks or mental operations (e.g., simultaneously avoiding traffic and weather) have a close task or mental proximity, and therefore, should be rendered proximate in perceptual space/display space (Wickens & Carswell, 1995; Boyer & Wickens, 1994). There is, however, no consensus on the best way to display integrated weather and traffic information. The CSD, which continues to serve as a testbed for evaluating display concepts, is a good candidate examining this integration. The main goal of the present study was to design an experiment that will help to find answers on how to integrate weather and traffic information using the CSD.

A main design feature of the CSD is the use of perspective for representing information to pilots. Numerous studies have compared the use of a 2D coplanar display (a display with a top down view and a vertical view of the airspace visualized simultaneously side by side) with a 3D display (one display with three dimensions of location along the lateral, vertical, and longitudinal axes visualized on a flat 2D screen). Some of these studies found that the 2D coplanar display provides overall performance advantages, relative to 3D displays, for avoiding pre-defined protected zone conflicts (O' Brien & Wickens, 1997; Boyer & Wickens, 1994). Performance benefits and costs associated with these display formats have also been found to be, in part, a function of the tasks they support. A 2D coplanar display, for example, seems to support better performance on focused attention tasks while a 3D display supports better performance on integration tasks (Wickens, Merwin & Lin, 1994). The disadvantage of the 2D coplanar display, however, is the cost of visually scanning between two panels creating cognitive demands to reconstruct a 3D environment (Wickens, 2000, as described in Alexander & Wickens, 2001). A 3D display does not create cognitive demands to reconstruct a 3D environment but has been found to be ambiguous in representing precise location and trajectory of aircraft. If spatial ambiguity is the main cause of poorer resolution performance in 3D displays, then including a continuously manipulable viewpoint (providing pilots with the possibility of viewing the 3D information from many different angles) should disambiguate 3D displays and therefore improve performance. Thomas and Wickens (2008) studied the effect of a continuously manipulable viewpoint using the FDDRL' s Cockpit Display Traffic Information (an older version of the CSD) and found that it does eliminate ambiguity costs, resulting in pilot performance comparable to when a 2D coplanar display is used. The present study compares three displays (a display with a continuously manipulable viewpoint, a 2D toggle, and a 3D toggle display with two fixed perspective views) in an integrated traffic and weather avoidance task. The goal of the study is to determine which display is superior, and if any such superiority depends on type of weather being depicted (convection and turbulence potential).

2   METHOD

2.2   Participants

Twenty-one commercial airline pilots (11 Captains and 10 First Officers) with experience in using weather radar, TCAS (Traffic and Collision Alert System) and TMS (Thermal Management System) participated in this study. Pilots were recruited from local airports and through various channels at NASA Ames Research Center and all had flown at least 1000 hours in glass cockpit aircraft.

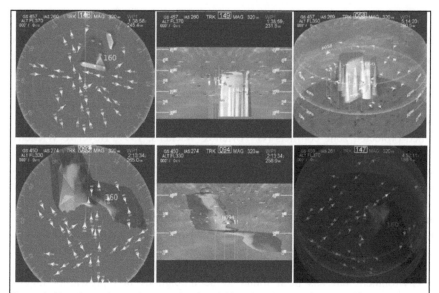

Figure 1. Top down views of convection (top left) and turbulence (bottom left); rear views of convection (top middle) and turbulence (bottom middle); perspective views of convection (top right) and turbulence (bottom right).

2.3   Procedures and Apparatus

A training session was designed to ensure that pilots possessed the skills needed to successfully participate in the experiment. The experimental interface used a 19 inch monitor and a single mouse as input. Pilots interacted with the CSD, a flight deck tool designed to present an aircrew with the location/state/flight plans of most surrounding air traffic, along with the weather. The integrated Route Assessment Tool (RAT) provided the ability to create and visualize in-flight route modifications. Pilots were observed from a remote station and a questionnaire was administered after the experiment to collect subjective evaluations regarding clutter and usefulness of the displays as well as pilots' preferences.

2.4   Experimental Design

A 3 (display format: 2D toggle display, 3D toggle display and the 3D flexible display with a continuously manipulable viewpoint) x 2 (weather phenomena: convection and turbulence potential) within subject design was used. See Figure 1 for exemplars of the displays. The first display had a toggle option between a top down view and a rear view of the airspace. The second display had a toggle option between two different perspective views, one view 30 ° from above and 60 ° to the right of ownship and one view 60 ° from above and 30 ° to the left of ownship. The third display had a continuously manipulable viewpoint that could be moved by pressing the right mouse button and dragging the cursor around the screen. This display could be set anywhere in the vertical range of 0 ° (top down) to 90 ° (rear vertical view) or laterally from 0 ° to 360 °. Convective weather was created by the FDDRL with a weather editor and consisted of 3 intensity layers. The turbulence potential was derived using the Flight Path Tool (http://aviationweather.gov) for the 7th of December 2009, 7.30 pm UTC time and consisted of 1 intensity layer.

There were four replications of each condition for each participant, with the participant's aircraft (ownship) starting from different angles relative to the weather. To control for practice effects, the order of display format presentation was counterbalanced across participants. Pilots were randomly assigned to three groups, and then to one of three display condition orders. Pilots were told that ownship was a commercial category type aircraft flying under Instrument Meteorological Conditions (IMC), (ceiling are less than 1000 feet and/or the visibility is less than three miles). The pilot's main task was to make a route modification around weather and avoid traffic as efficiently and safely as possible using the RAT. This information integration task required divided attention between weather and traffic; both were critical for maneuver selection during all trials. The standard operational procedure was to consider any level of weather as a hazard that needed to be avoided. Relative speed and altitude of ownship and other traffic was kept constant during the experiment. Pilots could change the display to their own preference before making any route modifications.

Different dependent measures were used to assess efficiency, safety and situation awareness (SA). Every trial started with a default path through weather that needed to be modified into a conflict free route. Path length and number of waypoints reflected efficiency of route modifications. Path length was measured as the lateral distance in nautical miles of the new route taken by the participant from the starting position to a fixed final destination. The number of waypoints reflected the number of ownship heading changes. Closest approach to weather, resolution time and weather penetration were gathered to reflect safety. The closest approach to weather is the closest lateral distance in nautical miles between the executed route and the outer weather layer. The resolution time to make a route modification was recorded in seconds from the start of the trial until the execution of the proposed route modification. Weather penetration reflected an executed route that penetrated any weather formation. SA (Endsley, 1999) was assessed at the end of each trial using multiple choice questions presented on the screen, according to the Situation Present Awareness Method (SPAM). SPAM is based on the premise that much of SA involves simply knowing where in the environment to find a particular piece of information, as opposed to remembering what that piece of information is (Durso et al., 1995 as described in Jeannot, Kelly and Thompson, 2003).

3   RESULTS

Although some differences were found between Captains and First Officers (Captains being more conservative), the overall effects were in the same direction, and these differences will, therefore, not be reported in this paper. Three x two repeated measures within subjects ANOVAs are presented for each different dependent measure.

3.2   Efficiency

Significant main effects on path length were found for display format (F (2, 166) = 10.15, $p < .01$) and weather phenomenon (F (1, 83) = 5.00, $p < .05$). There was also a significant interaction between display format and weather phenomenon (F (2, 166) = 173.60, $p < .01$). As seen in the left panel of Figure 2, for the display of convective weather, performance was best with the 2D toggle display, worst with the 3D toggle display, and intermediate with the 3D flexible display. Much the reverse was true when turbulence potential was displayed, with performance being best for the 3D toggle display, worst for the 2D toggle display, and intermediate for the 3D flexible display. It is worth noting that while performance using the two toggle displays was highly dependent on the type of weather they were showing, performance with the 3D flexible display did not change much as a function of weather phenomenon.

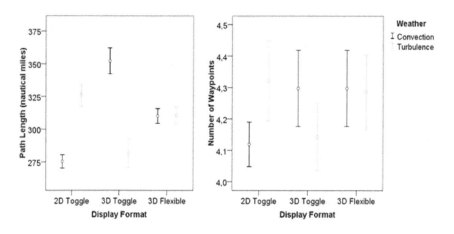

Figure 2. Performance efficiency.

Only an interaction effect (F (2, 166) = 6.64, $p < .01$) was found in the analysis of the number of waypoints. The right panel of Figure 2 shows that, for the convective weather displays, pilots made fewer heading changes with the 2D toggle display than with either of the 3D displays. These two 3D convective weather displays in turn, had about the same number of heading changes. For the turbulence potential displays pilots made fewest heading changes when using the 3D toggle display than when using the 2D toggle and 3D flexible displays, for which in turn they made approximately the same number of heading changes.

3.3   Safety

Significant main effects were found for display format (F (2, 166) = 18.54, $p <$ .01) and weather phenomenon (F (1, 83) = 152.78, $p < .01$) on the closest approach distance (Figure 3, left panel). For both types of weather phenomena pilots' routes came closer to the weather in the 2D toggle and the 3D flexible condition in comparison to the 3D toggle condition. For all three types of displays they also flew closer to weather in the turbulent weather condition than in the convective weather condition.

For resolution time the only significant effect was associated with weather phenomenon (F (1, 83) = 23.35, $p < .01$). For all three display types the pilots were faster in making a route modification in the convective weather condition than in the turbulent weather condition (Figure 3, right panel).

Significant main effects on weather formation penetrations were found for display format (F (2, 166) = 4.63, $p < .05$) and weather phenomenon (F (1, 83) = 20.55, $p < .01$), along with a significant interaction (F (2, 166) = 8.11, $p < .01$). As seen in Figure 4, for the display of convective weather with the 2D toggle display, no weather penetrations were found, while when turbulence potential was displayed, pilots penetrated weather 19.0 % of the cases. It seems that the 2D toggle display supports the two types of weather significantly different. Hence, the type of weather seems to be a critical determinant in what type of display to use.

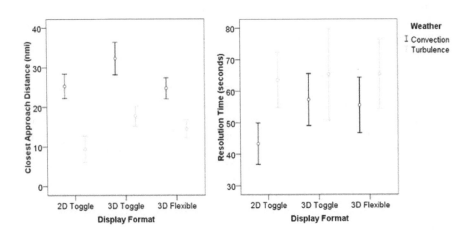

Figure 3. Safety performance.

3.4   Situation Awareness

A significant effect was found for display format (F (2, 164) = 5.29, $p < .01$) on accuracy; pilots had an accuracy of 81% in answering SA questions when using the 2D toggle display while accuracy was 65.5% when using the 3D toggle display.

Their reaction time was affected by display format (F (2, 166) = 8.38, p < .01) and weather phenomenon (F (1, 83) = 9.34, p < .01) as pilots were faster in answering SA questions in the 2D toggle condition than in the 3D toggle or the 3D flexible condition. Pilots were also faster in the convective weather condition than in the turbulent weather condition. Data showed no speed accuracy trade-offs.

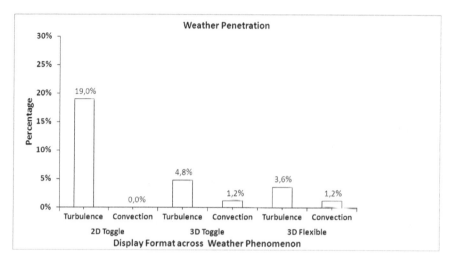

Figure 4. Percentages of weather formation penetrations.

3.5   Subjective Evaluations

Evaluations of display clutter and weather visualizations for each display format were administered with a post questionnaire. Clutter develops when the number of objects and the density of information around a given point increases, causing difficulties in distinguishing relevant from irrelevant information (Kroft and Wickens, 2001). The 2D toggle display was seen as cluttered by 42.9 % of all pilots resulting in a difficulty to assess the scope of the weather. The 3D toggle display was seen as cluttered by 71.4 % of all pilots, because of the inability to determine its height and whether the new route would be in conflict with weather. The 3D flexible display, allowed pilots to constantly change the viewpoint and gave them control of display set-up and the ability to make better sense of the spatial information. Therefore, only 28.6 % of all pilots rated this display as cluttered.

4   DISCUSSION AND CONCLUSION

The main goal of this study was to evaluate the effects of different display perspectives on pilot performance. The findings found here support previous studies regarding the advantages of using a 2D display, but it also support (O' Brien &

Wickens, 1997; Boyer & Wickens, 1994; Thomas & Wickens, 2008) findings on the elimination of ambiguity costs when using a continuously manipulable viewpoint; small differences were found in pilot performance in the 2D toggle versus the 3D flexible conditions. Although the 2D toggle display seemed to give better SA, this advantage could be explained by the fact that pilot's are currently using 2D displays. The most interesting finding in this study, however, was that the 3D display with a continuously manipulable viewpoint outperformed the 2D toggle display for safety for the most critical measure, as it resulted in fewer weather penetrations. Pilot performance not only depended on the display being used but also on the type of weather that was displayed; convective weather conditions resulted in more efficient and conservative routes (farther from weather) with better SA. A conservative behaviour towards convection might have been due to experience with current radar systems convection displays, while there are few if any turbulence potential displays on the flight deck (turbulence warnings are communicated verbally by the ATC). On the other hand it may have nothing to do with the type of weather per se, but might be due to the strikingly different 3D convective and turbulence hazard zones. As pilots were given the task to avoid up to the outer weather layer, the way it was visualized was not expected to affect performance. Future research should include other weather phenomena and the dynamic characteristic of weather.

Although Wickens, Merwin and Lin's (1994) found results concluding that a 3D display supports better performance on integration tasks, in this study, we found that this depends on what weather is visualized. As noted above, different weather phenomena and/or different 3D hazard zones, may be supported by different display perspectives; the 2D toggle display was more suitable for displaying convective weather while the 3D toggle display was more suitable for turbulent weather with regards to efficiency and safety. Pilot performance when using the 3D flexible display did not differ between the two weather conditions. A question remains whether or not this display might therefore be considered as a 'better' display.

There are a number of limitations in this study. First, the study was computer based and trials were short which might have been too repetitive for pilots. Therefore, work in a more complex cockpit environment needs to be done before it is safe to generalize to that environment. A second limitation is the use of only two static weather visuals. Having a different and unique weather visual for every trial would prevent participants from recognizing the weather from former trials and allowing dynamic weather changes would be more realistic. Also, the unique structures of the two hazard zones in this study may have been very important. Further research on all of the above issues is important before firm conclusions can be drawn about the pros and cons of 2D vs 3D displays.

ACKNOWLEDGMENTS

We would like to thank Dominic Wong for software programming support, Patrick Cravalho for assistance in recruiting participants and Doreen Comerford and

Robert W. Koteskey for their support on the development of weather scenarios. We also like to thank Arik-Quang V. Dao, Summer L. Brandt and Joel Lachter for their comments and reviews along the study, Guido Band for his supervision and especially Kevin Jordan for his overall support in different aspects during the study.

REFERENCES

Alexander, A. L. and C. D. Wickens. 2001. Cockpit display of traffic information: The effects of traffic load, dimensionality, and vertical profile orientation (ARL–01–17/NASA–01–8). Savoy: University of Illinois, Aviation Research Lab.

Boyer, S. B. and C. D. Wickens. 1994. 3D weather displays for aircraft cockpits. (Tech. Report ARL-94-11/NASA-94-4).

Comerford, D., S. Granada, W. W. Johnson, V. Battiste, A-Q. Dao, and S. Brandt. 2009. Pilots' weather related decision making: Results from a workshop and a survey. NASA/TM-2009-215384.

Endsley, M. R. 1999. Situation awareness in aviation systems. In: D. Garaland, J. A. Wise & V. D. Hopkins (Eds.), Handbook of Aviation Human Factors. London: Lawrence Erlbaum Associations Publishers.

Granada, S., A. Dao, D. Wong, W. W. Johnson, and V. Battiste. 2005. Development and integration of a human-centered volumetric cockpit situational display for distributed air-ground operations. In *Proceedings of the 13th International Symposium on Aviation Psychology*. Columbus, OH: Department of Aerospace Engineering, Applied Mechanics, and Aviation, The Ohio State University.

Jeannot, E., C. Kelly, and D. Thompson. 2003. The Development of Situation Awareness Measures in ATM Systems. Report Eurocontrol HRS/HSP-005-REP-01.

Kroft, P. D. and C. D. Wickens. 2001. Integrating aviation databases: effects of scanning, clutter, resolution and interactivity. *Presented at the 11th international symposium on Aviation Psychology*, Columbus, OH: The Ohio State University.

O'brien, J.V. and C. D. Wickens, 1997. Cockpit displays of traffic and weather information: effect of dimension and data base integration. (Tech. Report ARL-97-3/NASA-97-1).

Rhoda, D. A. and M. L. Pawlak. 1999. An Assessment of Thunderstorm Penetrations and Deviations by Commercial Aircraft in the Terminal Area. Report prepared for NASA Terminal Area Productivity Program, Center-TRACON Automation System (CTAS) activity.

Thomas, L. C. and C.D. Wickens. 2008. Display dimensionality and conflict geometry effects on maneuver preferences for resolving in-flight conflicts. *Human factors,* 50(2), 576-588.

Wickens, C. D. and C.M. Carswell. 1995. The proximity compatibility principle: Its psychological foundation and its relevance to display design. *Human factors*, 37(3), 473-494.

Wickens, C. D., A. S. Mavor, R. Parasuraman, and P. McGee. 1998. Airspace system integration: The concept of free flight. *The future of air traffic control*, Washington DC: National academy press.

Wickens, C. D., D. H. Merwin, and E. Lin. 1994. Implications of graphics enhancements for the visualization of scientific data: Dimensional integrality, stereopsis, motion, and mesh. *Human Factors, 36,* 44–61.

CHAPTER 19

Quantifying Pilot Visual Attention in Low Visibility Terminal Operations

Ellis, Kyle K.E., Bailey, Randall, E., Arthur III, J.J. (Trey), Latorella, Kara A., Kramer, Lynda J., Shelton, Kevin J., Norman, Robert M., and Prinzel, Lance J.

National Aeronautics and Space Administration
Langley Research Center
Hampton, VA USA
Kyle.Ellis@NASA.GOV

ABSTRACT

Quantifying pilot visual behavior allows researchers to determine not only where a pilot is looking and when, but holds implications for specific behavioral tracking when these data are coupled with flight technical performance. Remote eye tracking systems have been integrated into simulators at NASA Langley with effectively no impact on the pilot environment. This paper discusses the installation and use of a remote eye tracking system. The data collection techniques from a complex human-in-the-loop (HITL) research experiment are discussed; especially, the data reduction algorithms and logic to transform raw eye tracking data into quantified visual behavior metrics, and analysis methods to interpret visual behavior. The findings suggest superior performance for Head-Up Display (HUD) and improved attentional behavior for Head-Down Display (HDD) implementations of Synthetic Vision System (SVS) technologies for low visibility terminal area operations.

Keywords: eye tracking, flight deck, NextGen, human machine interface, aviation

1   INTRODUCTION

Since its inception, NASA Langley Research Center (NASA LaRC) has conducted research in the field of Aeronautics, with an ever increasing interest in the human factors associated with aviation safety and the pilot/ flight deck interface. With this research scope, LaRC conducts research in flight deck interface design, employing various approaches to quantify pilot visual behavior (Comstock, Coates, & Kirby, 1985); (Spady & Waller, 1973).

Today, NASA LaRC is working with state-of-the-art technology including the integration of remote eye tracking systems inside its commercial aviation simulators with minimal impact on the flight deck environment (Latorella and Ellis, 2010). An important aspect of eye tracking is to quantify visual behavior and to associate this behavior with quantitative metrics related to a pilot's visual attention, cognitive processing, workload, and awareness or saliency of visual information (e.g., traffic awareness, prominence and importance of outside visual cues and references, effectiveness of visual alerts). Differences in pilot behavior across display variations can lead to significant findings in regard to the utility of various display types, location of display, complexity in usability, and operational comparisons in various flight environments.

The development of quantitative behavioral metrics using eye tracking is explored by comparing eye tracking indicators to the behavioral indicators of display use and flight technical data. A review of the metrics developed for NASA's NextSafe-2 experiment is presented to highlight the capabilities of NASA LaRC's eye tracking installation and associated analytical methodologies.

The NextSafe-2 experiment was a fixed-base simulation experiment investigating whether a lower decision height or reduced visibility minima is supported by the use of Synthetic Vision Systems (SVS). A SVS display was compared to a baseline blue-over-brown primary flight display in approach and landing operations (Ellis, Kramer, Shelton, Arthur III, & Prinzel, 2011). Along with flight technical analysis, eye tracking results were utilized, contrasting these display technologies with flight technical performance and human behavioral differences. The approach to oculometer hardware integration, methodologies for data reduction, and the impact of results are discussed. Further, this work points out future research needs to fulfill the development of quantitative behavioral metrics.

1.2   Background

Remote eye tracking systems first determine head position in all six degrees of freedom. This is done by two dimensional image recognition using several key facial characteristics. Points such as the eye corners, nostrils, corners of the mouth, ears, etc are identified and measured in relative pixel distance. Combining the located image points using two cameras of known position allows for 3D image processing, producing 6 degree of freedom head position values. Eye tracking is then measured by determining the center of the pupil through contrast image processing, relative to a glint reflection, provided by infra-red light sources of known location on the iris that indicates the center of the eye itself. By calculating the known distance between these two points, trigonometry is used to calculate a vector between the two points. A three dimensional eye gaze vector can be calculated in reference to a world coordinate system, such as a flight deck. A minimum of two cameras are required to perform 3 dimensional calculations (Ellis & Schnell, 2009).

Reducing the data has historically been done with some scientific variance. Lookpoints from the gaze vector calculation are determined by the intersection point of the gaze vector with a predefined area of interest (AOI) where an AOI is within the subject's visual environment, such as the primary flight display, or navigation display. Sub-AOIs may also be defined (e.g., the attitude indicator or airspeed read-out). Lookpoints may also be used in the determination of fixations and saccadic movement. There are no standard definitions of fixations and saccades, however, most research loosely defines a fixation as "a relatively stable eye-in-head position within some threshold of dispersion (~2 deg) over some minimum duration (200ms), and with a velocity threshold of 15-100 degrees per second" (Jacob & Karn, 2003). For our research, fixations are defined as spatial and time dependent, characterized by occupying a spatial radius of 2 degrees for a minimum of 200ms. Saccade movements are defined as the movement from one fixation to the next, measured in both angular distance and velocity.

2   METHOD

The Smart Eye remote eye tracking oculometers have been integrated into two of the simulation cabs in NASA Langley's Cockpit Motion Facility. (Latorella, et al., 2010)describes the initial Integrated Flight Deck installation, which collected both crew members' synchronized eye movement data. The following describes a similar installation in the Research Flight Deck (RFD) simulator for the left pilot seat. Each system is capable of using anywhere from 2 to 8 cameras, outputting real-time eye gaze data across a 100$^+$ degree visual tracking range. The Smart Eye system was selected due to its superior performance, coupled with the flexibility in the placement of cameras and illuminators. This flexibility was crucial for nearly seamless integration into the flight deck, minimizing intrusion to the flight deck environment, thus, maintaining a high level of simulator fidelity.

2.2   Flight Deck Integration

The cameras provided the desired level of eye gaze accuracy (approximately 2-3 degrees), without compromising simulation fidelity during the NextSafe-2 experiment. Figure 1 shows the locations of the cameras and illuminators installed in the RFD simulator. Data was collected for the left seat pilot in the NextSafe-2 experiment; however, synchronized collection of a dual crew flight deck has been installed and successfully utilized in NASA Langley's Integrated Flight Deck (IFD) for another HITL experiment investigating Data Comm interaction on the flight deck (Norman, et al., 2010).

Validation and verification of system output was performed, assessing data collection quality and accuracy commensurate with previous experiments (Latorella, et al., 2010). The functional head box - the tracking region of the subject, centered on the designed eye point of the flight deck - was approximately one cubic foot. Trade-offs in eye tracking camera placement and lens size were made to create a head box capable of capturing all pilot head movements (as dictated by the experiment tasks) yet maintain sufficient resolution for the image processing.

Figure 1. Camera and Illuminator Locations in RFD

2.3   Data Collection

The raw data stream output of the two Smart Eye computers was recorded when the simulator was operating and was combined with simulator state data. The data were synchronized and time stamped for each trial.

The collected eye tracking data for the NextSafe-2 experiment is shown in Table 1. AOIs were explicitly outlined by defining a world coordinate model of the interior flight deck and out-the-window (OTW) scene. The eye gaze vector information, especially when coupled with AOI information, provides explicit insight into which displays the pilots focused their attention, as well as visitation frequency and when the attention occurred.

Table 1. Eye Tracker Output Parameters for NextSafe-2

Frame Rate	Eye Position X
Frame Number	Eye Position Y
Head Position X	Eye Position Z
Head Position Y	Gaze Direction A
Head Position Z	Gaze Direction B
Head Position Quality	Gaze Direction C
Head Heading	Gaze Heading
Head Pitch	Gaze Pitch
Head Roll	Gaze Direction Quality
Time Stamp	Pupil Diameter
User Time Stamp	Pupil Diameter Quality
Closest World Intersection	

2.4   Data Reduction

The data were reduced to support specific research objectives. Aviation research often uses specific windows of time or "snapshot data" (i.e., data collected at operationally significant points in time or space) to parse out the data and answer meaningful experimental questions and hypotheses. Specific to eye tracking applications, it is important to also define the data reduction parameters based upon the expected human behavior as demanded by the experiment tasks to elicit the best results.

For the NextSafe-2 experiment, specified flight segments during the low visibility instrument approach and landing operations were defined, shown in Figure 2, and associated analyses were conducted based on assumptions of pilot visual behavior within these segments. Eye tracking data were used to support or refute the use of SVS to improve the safety and performance of instrument approach to landings.

Within the "instrument segment," the use of SVS was assumed to not significantly change the pilot's visual behavior from that of conducting the instrument approach using the "blue-over-brown" baseline display condition. The only difference in the displays concepts was the presence of terrain and landing runway information on the PFD for the SVS concept.

Around the 150 ft Height above threshold (HATh) point – the Decision Height (DH) for the instrument approach - the instrument-to-visual transition flight segment eye tracking data was analyzed to identify if SVS improved the pilot's ability to visually transition from instrument to visual conditions. The contention was that SVS, because of its intuitive depiction of the runway, would induce a more efficient visual scan outside the airplane to find the runway and align the aircraft for landing.

Within the Visual Segment, it was assumed that SVS would not affect visual behavior or attention. The pilot's visual attention was assumed to be ~100% directed out-the-window, attending to the pre-flare (line-up laterally and vertically) and preparing for the visual landing, including 'clearing the runway' or visually searching for the presence of other aircraft, vehicles, objects, or animals on the runway.

Similarly, within the Flare Segment (flare to touchdown), it was assumed that SVS would not affect visual behavior or attention. The pilot's visual attention was assumed to be 100% OTW, attending to the visual landing.

Lastly, within the Landing Segment (touchdown to a safe taxi speed), it was assumed that SVS would not affect visual behavior or attention. The pilot's visual attention was assumed to be 100% out-the-window, attending to the roll-out, tracking the runway centerline, and decelerating to a safe taxi speed.

Based on these assumptions, percent of eye gaze (i.e., attention) OTW and number of transitions between OTW and head down was calculated for each analysis segment as a comparator across display conditions.

Another inquiry to pilot behavior was the determination when pilots transitioned from instrument-flight to OTW (i.e., visual-flight) to complete the approach to land task. Determining the specific altitude, distance from threshold, and direction of the glance required new methods to analyze the data structured upon eye movement science.

In development of the transition of attention algorithm, the AOI for which the lookpoint resided simplified the analysis. Time spent in an AOI by the lookpoint denoted attention, either head-up OTW or head-down on the instrument panel.

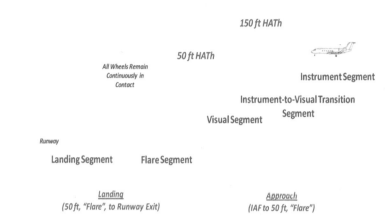

Figure 2. Eye Tracking Analysis Segments

Eye tracking data was analyzed by creating a binary value to specify if the pilot attention was OTW or not. If the lookpoint was tagged in the OTW AOI, a value of 1 was assigned for that data point, if not, a zero was designated. If there was no calculated lookpoint due to loss of eye tracking, head tracking was utilized by referencing a running average of head tracking pitch. If the head pitch was greater than the head-up reference pitch, that data point was tagged as OTW, and a value of 1 was assigned.

A moving time window of two (2) seconds, calculating an average of the past 2 seconds of the binary OTW value, was used to determine when the first occurrence of both a glance and a full transition OTW occurred. This 'gated average' value was then compared to a set-point threshold of 10% (200ms) to signify the occurrence of the "initial glance", and 100% identifying a full transition to OTW, shown in Figure 3. The algorithm reports the reference times when the transition set points of 10% and 100% are reached, and simulator and eye tracking data such as altitude above field-level (AFL), distance from threshold, or gaze direction at that instant in time are recorded. The 200ms time duration was chosen as the initial glance threshold based upon 200 ms being the experimental definition of a single fixation.

The 2-second time window was chosen due to the eye movement behavior of the pilots. Several size windows of time were reviewed to process the eye movement signal, including 10s, 8s, 6s, 4s, 2s, and 1s, and evaluated over each pilot's visual behavior to estimate when the defined transitions occurred. The time window size specifically impacts the full transition metric. An exceedingly large window neither quickly nor accurately captures a full transition point. In a approach and landing environment, small differences in time, (as small as 1 second of elapsed time) can result in large lateral and vertical distances. A window that is too small may result in an inaccurate depiction of a full transition, in which pilot attention is falsely classified as fully transitioned OTW when it may only be an extended glance between transitions.

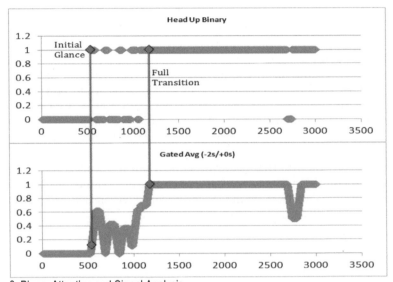

Figure 3. Binary Attention and Signal Analysis

2.5   Data Analysis

The data were analyzed across the pilot subjects to determine if statistical significance existed in three operational comparisons; SVS display location (HUD vs. HDD), SVS HDD Equipage (SVS vs No SVS), and effects of single versus crewed pilot operations. The results presented herein specifically focus on eye tracking and its analysis. The operational results from the NextSafe-2 experiment are discussed elsewhere (Kramer, et al., 2011).

3   RESULTS

The NextSafe-2 experiment observed a data collection rate of 87% across all experiment scenarios. Eye tracking was robust across all segments of the approach, with consistent and operationally rational results, providing strong operationally comparable human performance metrics. Spatial accuracy was observed in calibration of the eye tracking system for each subject to the order of approximately 2 to 3 degrees. The spatial accuracy proved sufficient due to the decisive pilot eye behavior in transitioning across gross AOIs on the HDDs and OTW. The spatial accuracy could have been improved but was not necessary to meet the research objectives.

3.2   Flight Simulation Study Results

The analysis of the eye tracking data was grouped into SVS operational comparisons, yielding several significant findings in SVS location and crewed versus single pilot operations.

In Figure 4, differences in head-up percentage across eye tracking segments revealed that pilots retain a significant level of attention head down, even in "head-up" flight segments (i.e., ~30% during the visual segment of flight, and as much as 10% in the flare segment). This information reveals that pilots allocate a significant attention to guidance and instrumentation during these "visual" segments, suggesting this information is still utilized until the beginning of the flare. The HUD, with or without SVS, allows for this information to be observed while maintaining an OTW view, an impossible behavior to achieve with only HDD presentation of guidance and instrumentation. Transition count data between HDD and OTW, shown in Figure 4, provides insight into how pilots without a HUD transition their attention during the visual segment, indicating, on average, that two transitions occur (OTW to HDD, HDD to OTW). However, combining information from both transition count and head up percentage, the transition to head down accounts for 30% of the head down time, suggesting pilots are still flying guidance in the visual segment and not merely referencing it.

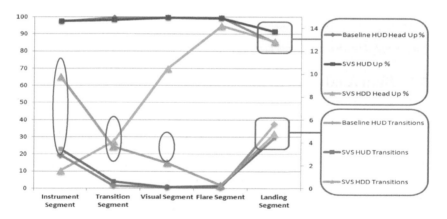

Figure 4. SVS Operational Comparison SVS HUD vs. SVS HDD

The specific altitude at which visual transition occurred was examined. The data showed that pilots using a HDD with SVS maintained their visual attention on the HDD until closer to the DH than without SVS. The data also showed that the time between a pilot's initial and full transition to OTW. Table 2 shows the operational comparisons between baseline HDD and SVS HDD indicate the initial glance and full transition of attention, as defined previously. Transition of attention data also shows the difference in altitude for SVS operations are less than that without SVS. This finding, coupled with the findings of initial glance being closer to DH with SVS, suggest pilots operating with SVS are making a more decisive transition from HDD to OTW, thereby suggesting greater confidence in their trajectory and position and a more efficient visual behavior.

Table 2. SVS Operational Comparison Visual Transition Altitude

		Baseline HDD	*SVS HDD*
Initial Glance AGL (ft)	**Mean**	367	276
Full Transition AGL (ft)	**Mean**	197	178

A comparison of utilization of SVS with crewed vs. single pilot operations was also conducted. This analysis was made to evaluate the influence of crew assistance on visual behavior and attention. Surprisingly, no significant differences in visual attention were found during the approach and flare segments. Significant findings were, however, observed during the landing/roll-out segment, shown in Figure 6, indicating single pilots made several more transitions between the HDD and OTW. This is explained by the difference in task loading between the two comparisons. Crewed operations allow for the pilot flying to maintain attention OTW with the other crew member providing speed, runway remaining, and turn-off information callouts especially using the NextSafe-2 advanced airport moving map display. These tasks are critical in the rollout phase, made particularly more difficult in low visibility operations, and are all tasked to the individual pilot in single crew operations. This information is only available head-down, requiring the single pilot to transition with increased frequency to retain critical attention OTW while at the same time collecting the necessary information from the HDDs.

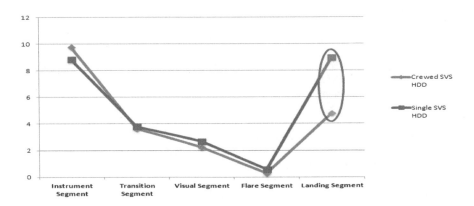

Figure 5. Crew of 2 vs. Single Pilot SVS Operational Comparison OTW/HDD Transition Count

4 CONCLUDING REMARKS

The eye tracking capabilities integrated into the flight simulators at NASA Langley provide a critical capability to HITL research. The measures of attention and transition of attention algorithm used in the NextSafe-2 experiment demonstrate the significance and efficacy of characterizing pilot visual behavior. The data provide strong support to the efficacy of both HUD and SVS operations to improve the efficiency of pilot visual behavior and furthermore, provide a rationale to describe variability observed in flight technical performance. Without eye tracking these results would not have been discovered, and flight technical performance would be the sole comparator to define the operational advantages or disadvantages to SVS and are not typically significant. The methodologies produce quantifiable results across operational comparisons revealing visual behavior data affording researchers improved understanding of pilot performance.

ACKNOWLEDGMENTS

This work was jointly sponsored by the NASA Aviation Safety Program, Vehicle Systems Safety Technologies project, led by Dr. Steve Young (Project Scientist), and the FAA Human Factors R&D Project for NextGen, led by Dr. Tom McCloy, Mr. Dan Herschler, and Mr. Stephan Plishka. Mrs. Terry (Stubblefield) King of the FAA Flight Standards was instrumental in the motivation, planning, and conduct of this work.

REFERENCES

Comstock, J. R., Coates, G. D., & Kirby, R. H. (1985). Eye-Scan Behavior in a Flight Simulation Task as a Function of Level of Training. *Human Factors Society, 29th* (p. 5). Baltimore, MD: Human Factors Society.

Ellis, K. K., & Schnell, T. (2009). *Eye Tracking Metrics for Workload Estimation in Flight Deck Operations.* Iowa City: University of Iowa.

Ellis, K. K., Kramer, L. J., Shelton, K. J., Arthur III, J. (., & Prinzel, L. J. (2011). Transition of Attention in Terminal Area NextGen Operations Using Synthetic Vision Systems. *Human Factors and Ergonomics Society 55th Annual Meeting.* Las Vegas, NV: Human Factors and Ergonomics Society.

Jacob, R. J., & Karn, K. S. (2003). Eye tracking in human-computer interaction and usability research: Ready to deliver the promises. *The Mind's eye: Cognitive The Mind's Eye: Cognitive and Applied Aspects of Eye Movement Research*, 573 - 603.

Kramer, L. J., Bailey, R. E., Ellis, K. K., Norman, R. M., Williams, S. P., Arthur III, J. J., . . . Prinzel III, L. J. (2011). Enhanced and Synthetic Vision for Terminal Manuevrering Area NextGen Operations. *SPIE Defense, Security, and Sensing.* Orlando, FL: SPIE.

Latorella, K., Ellis, K. K., Lynn, W., Frasca, D., Burdette, D., Feigh, C. T., & Douglas, A. (2010). *Dual Oculometer System for Aviation Crew Assessment.* San Francisco: Human Factors and Ergonomics Society Annual Meeting.

Norman, R. M., Baxley, B. T., Ellis, K. K., Adams, C. A., Latorella, K. A., & Comstock, J. R. (2010). *NASA/FAA Data Comm Airside Human-In-The-Loop Simulation.* Washington DC: Federal Aviation Administration.

Spady, A. A., & Waller, M. C. (1973). The oculometer - A new approach to flight management research. *AIAA Visual and Motion Simulation Conference.* Palo Alto, CA: AIAA.

CHAPTER 20

Awareness of Need for Immediate Communication Predicts Situation-Awareness and Performance in an ATC Training Course

Zach Roberts, Hector Silva, Thuy Vo, Ryan Boles, Nora Gomes, Matt Iwan,
Lindsay Sturre, Dan Chiappe, Thomas Z. Strybel

California State University, Long Beach, Center for Human Factors in Advanced
Aeronautics Technologies (CHAAT)

Dept. of Psychology
California State University, Long Beach
Long Beach, CA
dan.chiappe@csulb.edu

ABSTRACT

NextGen promises to alter the way that ATCos manage traffic for the purpose of increasing the capacity of the NAS. This will largely be accomplished through the use of various automation tools. However, because the tools are likely to be introduced gradually, ATCos will have to learn both, manual and NextGen tools. The present study examines the order in which these tools ought to be trained (manual tools first vs. NextGen tools first), and whether effective training procedures need to take into account individual differences of students. Specifically, we examined differences in awareness of need for immediate communication (ANIC) as an indirect measure of how proficient student controllers are at acquiring and maintaining Situation Awareness (SA). We found that student controllers that were higher in ANIC performed better on the SPAM measure of SA. They also had fewer LOS and a faster time through sector on an ATC simulation. Importantly, they were also unaffected by training order, while those lower in ANIC had greater SA when they were trained with manual tools first. We thus recommend introducing manual tools first, as doing so is less likely to lead to complacency with respect to SA among the less proficient students.

Key words: Situation awareness, air traffic control, training

1   INTRODUCTION

The goal of the NextGen air traffic management system is to accommodate increases in air traffic density without compromising the safety of the NAS (JDPO, 2011). This increase, however, cannot be accomplished with current day tools based on pilot and air traffic controller (ATCo) radio communications. This is because verbal communications are a main source of ATCo workload, the latter being a major bottleneck currently limiting airspace capacity (Durso & Manning, 2008). To deal with the limits posed by the current day infrastructure, NextGen will introduce new automation tools and technologies that will significantly alter the way that ATCos manage traffic (FAA, 2011). Potential tools include conflict resolution probes, improved displays of weather information, and Data Communications (Data Comm). Although text-based Data Comm is currently being used in some centers, in NextGen Data Comm will be integrated with conflict probes to allow conflicts to be resolved graphically and uplinked without text input. The tools comprising the NextGen system will likely be introduced in a gradual manner because they depend on the willingness of air service providers to upgrade existing aircraft, meaning that the airspace in the near term will contain a mix of equipped and unequipped AC. ATCos will therefore need to be proficient at managing traffic using both manual and NextGen tools. The present study examined how future ATCos should be trained to use current day and NextGen tools in the same airspace, and whether individual differences among students need to be taken into account when designing training programs for this purpose.

The NextGen tools used in the present study include a conflict trial planner that allows for route modifications and conflict resolutions, as well as integrated Data Comm. The conflict-alerting program automatically alerts controllers 8 minutes in advance of a Loss of Separation (LOS). This reduces the need for controllers to scan their radar displays for potential conflicts between equipped AC. The trial planner with automatic conflict probing allows controllers to determine whether a potential route change will produce unintended conflicts between AC. This tool reduces the need for projection of the future state of AC, the number of voice communications, and the working memory requirements associated with keeping a message active until a command has been executed (Hinton & Lohr, 1988).

A critical factor that will determine the success of NextGen is how voice-based and NextGen tools are taught to controllers. These should be taught in a way that ensures optimum levels of situation awareness (SA), workload and performance when both sets of tools are used simultaneously. Should a whole-task training strategy be used wherein both tools are learned together or should each be trained individually using part-task training? Possibly, NextGen tools will have to be learned using a part-whole task strategy, in which controllers learn to manage traffic with current day and NextGen tools separately before practicing them together (Sohn, Douglass, Chen & Anderson, 2005). If tools are first learned sequentially, is there an optimal order that they should be introduced? It may be beneficial to teach voice-based tools first, as these are more difficult to learn, and must be relied on in cases of automation failure. Alternatively, it may be beneficial to teach NextGen tools first, as these are easier to learn than the attention-demanding manual procedures. This would free up cognitive resources for controllers to learn other aspects of air traffic control, such as scanning, sector characteristics and traffic patterns. Training with NextGen tools first, however, can have unintended consequences. For instance, it could lead to greater degrees of complacency and over-reliance on automation on the part of controllers—common side effects of high levels of automation (Parasuraman, Sheridan, & Wickens, 2000).

In one of the few investigations of training strategies for the NextGen airspace, Billinghurst et al. (2011) compared two training orders using a part-whole training strategy. Students were trained in either manual procedures first then NextGen tools, or vice-versa. Training effectiveness was measured in terms of performance, workload and SA when managing mixed-equipage traffic. Student performance, measured by number of LOS, was affected by the order of training in that fewer LOS were made immediately after manual training. Importantly, training order did not affect workload, SA, or measures of controller efficiency.

Another factor that needs to be considered, however, is whether one method of training will be equally effective for all individuals (Paas & van Gog, 2009). Some students may be less proficient initially at certain skills that determine the optimal order for training tools because of differences in previous experience, cognitive ability or personality factors (e.g., Salden, Paas, & van Merrienboer, 2006). The present study reports on additional data from Billinghurst et al. (2011) to determine whether individual differences in the proficiency at acquiring and maintaining SA may have mediated their findings. Specifically, we examined student controllers' awareness of the need for immediate communication (ANIC). This was determined by to the extent to which they took steps to correct errors in communication; either their own or those of pilots, and the frequency with which they issued expedited commands, thereby demonstrating awareness of the need for immediate action. Previous studies suggest that there are likely individual differences in this regard, as the communications of pilots and controllers often include errors and these are not always corrected. For example, Morrow and Rodvold (1992) found procedural deviations in 15% of controller communications, and in 19% of pilot communications. Cardosi (1993) showed that controllers do not always correct pilot errors in communication. This is important because many air traffic incidents can be linked to failures in communication (Roske-Hofstrand & Murphy, 1998).

Assessing proficiency using ANIC is reasonable given the central role that communication plays in the acquisition and maintenance of SA by controllers

(Roske-Hofstrand & Murphy, 1998). Moreover, verbal communication is a known source of ATCo workload and previous work has shown that one attribute of experienced controllers their use of workload management strategies (Redding et al., 1991; D'Arcy & Rocco, 2001). Specifically, Redding et al., (1991) showed that experts have strategies for identifying aircraft that can be expedited through the sector, thus reducing the number of aircraft that need attention.

This study sought to validate the ANIC measure as a tool for measuring SA by examining whether it is related to performance on another SA measurement tool – the Situation Present Assessment Method (SPAM; Durso & Dattel, 2004). Moreover, we examined whether ANIC interacted with order of training (manual first vs. NextGen tools first) in predicting the SA and performance of controllers. For example, students more proficient at acquiring SA may be less affected by training order than those that are less proficient. To address these questions, we conducted secondary analyses of the dataset used by Billinghurst et al. (2011).

2   Methods

Participants

Eleven students from Mt. San Antonio College, an FAA Collegiate Teaching Initiative (CTI) school, participated. Students were enrolled in a 16-week internship at the Center for Human Factors in Advanced Aeronautics Technologies (CHAAT) at CSULB. They took part in a midterm test on an ATCo simulation after 9 weeks of the internship and in a final exam taken a week after the 16-week internship.

Materials

Simulations were presented using the Multi Aircraft Control System (MACS) software (Prevot, 2002). The program simulates manual (verbal) and Data Comm (digital) communications between ATCos and pseudo-pilots. The software was used to simulate en route sector ZID 91. Participants were presented with three counterbalanced scenarios in which the percentage of AC equipped with NextGen tools (i.e., conflict alerting, trial planner with conflict probes, and Data Comm) varied by scenario half. The first scenario consisted of 25% NextGen equipped AC for the first half and 75% for the second half. The second consisted of 50% equipped AC for both halves. The last consisted of 75% equipped AC for the first half and 25% for the second. Scenarios also varied in traffic density with the second half of each scenario featuring approximately twice as many AC as the first half.

Procedure

Students were trained on both manual (verbal) commands and NextGen ATCo tools in different orders. One group had manual training during the first 8 weeks of the course, while a second group received NextGen tools training first during this same period. At week 9, training was reversed, and the group first trained with manual tools received NextGen training, while those trained with NextGen tools first received manual training. The students then took their midterm exam, where they had to manage both NextGen equipped and unequipped aircraft in the simulation environment. Students continued training with the new tools from weeks 10 through 16 so that by the end of the course they had received equal amounts of training on each tool type. At week 17, all participants took the final exam that once again required them to manage traffic in a simulation environment.

Design

A 2(training order: manual training first vs. NextGen tools training first) X 2(Test: Midterm vs. final) X 2(Traffic density: Low vs. High) X 2(Awareness of Communication Errors: Low vs. high) X 3(Percent equipped aircraft per scenario half: 25%-75%, 50%-50%, 75%-25%) mixed factorial design was used. The between-subjects variables were training order and Awareness of Need for Immediate Communications (ANIC). ANIC was defined by the number of errors that the student ATCos corrected in terms of their own and pilot clearances, the number of repeats issued, and the number of expedited commands issued. The variable was calculated based only on the verbal communications, not the Data Comm exchanges. These were added up across the three scenarios for both the midterm and final tests. Participants were sorted into two levels based on a median split performed on the totals. The total number of communication errors corrected and expedited commands issued for the low-ANIC group was 17.17 ($SE = 3.48$) and for the high-ANIC group it was 38.40 ($SE = 4.56$), $t(9) = -3.77$, $p = .004$. The dependent variables for the study were: (1) SA measured by the percent correct and reaction time to answer probe questions correctly using SPAM, (2) Performance safety measured by the number of Losses of separation (LOS), and (3) Performance efficiency measured by the average AC time through sector.

3   Results

3.1 Situation Awareness probe accuracy and RT

SA data were analyzed using a 2x3x2x2x2 mixed factors ANOVA with test, traffic density, and equipage as within-subject factors, and training order and awareness of need for immediate communications (ANIC) as between-subjects factors. Separate ANOVAs were computed for the dependent variables of probe accuracy (i.e., percent correct) and RT to respond to probe questions correctly.

Beginning with probe accuracy, ANOVA revealed a main effect of ANIC, $F(1,6) = 11.87$, $p = .014$. Those in high-ANIC group had greater accuracy in answering probe questions ($M = 79\%$, $SE = 2\%$) than those in low-ANIC group ($M = 70\%$, $SE = 2\%$). There was also an interaction between ANIC and traffic density, $F(1,6) = 19.08$, $p = .005$. Subsequent analyses revealed that there was no effect of traffic density for the high-ANIC group, $F(1,3) = 4.19$, $p = .13$, but there was a significant effect for the low-ANIC group, $F(1,3) = 15.45$, $p = .03$. Specifically, those in the low-ANIC group were more likely to answer probe questions correctly when traffic density was high than when traffic density was low as shown in Figure 1. Although non-significant, students in high-ANIC group showed the opposite pattern.

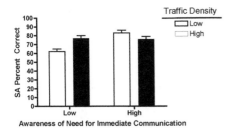

Figure 1 SA accuracy for high and low-ANIC as a function of traffic density

The analysis also revealed an interaction between training order and ANIC, $F(1,6) = 8.72$, $p = .026$ (see Figure 2). Tests of simple effects revealed that for those in high-ANIC group, there was no difference in probe accuracy between participants trained in NextGen tools first and those trained with manual tools first, $F(1,3) = 1.31$, $p = .33$. However, for participants in low-ANIC group, those trained in manual tools first were more likely to answer probe questions correctly than those trained in NextGen tools first, $F(1,3) = 7.98$, $p = .066$.

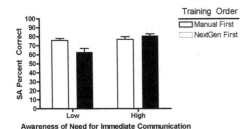

Figure 2 SA accuracy for high and low-ANIC as a function of training order

Finally, there was an interaction between training order and traffic density, $F(1,6) = 11.1$, $p = .016$. For participants trained with manual tools first, probe accuracy was the same when traffic density was low ($M = 79\%$, $SE = 3\%$) and high ($M = 75\%$, $SE = 2\%$), $F(1,4) = 1.23$, $p = .33$. However, participants trained with NextGen tools first had lower probe accuracy in the low traffic density condition ($M = 66\%$, $SE = 3\%$) than in the high traffic density condition ($M = 79\%$, $SE = 3\%$), $F(1,2) = 24.72$, $p = .038$.

Turning now to the RT to respond to probe questions correctly, ANOVA revealed a main effect of traffic density, $F(1,6) = 12.48$, $p = .012$. Participants were faster when density was low ($M = 10.30$s, $SE = 1.31$) than when it was high ($M = 11.99$s, $SE = .99$). There was no main effect of ANIC, $F<1$. However, ANIC did interact with traffic density, $F(1,6) = 6.25$, $p = .046$. As shown in Figure 3, for participants in the low-ANIC group, there was an effect of traffic density, with RT being greater in the high density condition than in the low density condition, $F(1,3)$ = 47.15, p = .006. For the high-ANIC group, there was no effect of density, $F<1$. Based on probe latency, traffic density affected the SA of the low-ANIC group, but not the high-ANIC group.

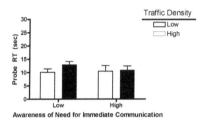

Figure 3 SA probe RT for high and low-ANIC as a function of traffic density

3.2 Workload

Workload was measured by the RT to the ready prompt for probe questions, workload ratings throughout the scenario and the number of probe questions answered. For ready RT, a significant effect of ANIC was found, $F(1,6)$ = 16.2; p = .007. The mean RT for the low-ANIC group (M =7.0s; SE = .6s) was almost twice that of the high-ANIC group (M = 3.7s; SE = .5s). A significant interaction between ANIC, training sequence and equipage, $F(2,12)$ = 6.96, p = .01 was also found, as shown in Figure 4 . Students with high-ANIC responded faster across all equipage levels. Although those receiving manual training first responded faster at 25 and 50% equipage levels, the interaction was non-significant (p = .21). The interaction of scenarios and training order was significant for low-ANIC students, $F(2,6)$ = 6.85, p = .028. Low-ANIC students receiving NextGen training first responded much slower at 50% and 75% equipage levels, compared to all other groups.

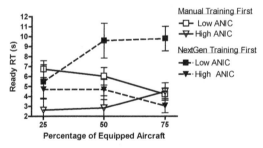

Figure 4 Ready RT as a Function of Equipage and Training Order for High and Low-ANIC Students.

Students answered more questions under low density traffic conditions (M = 5.4; SE = .2) compared with high density traffic (M = 4.7; SE =.3), $F(1.6)$=23.05; p=.003. The main effect of ANIC and its interaction with traffic density were not significant, yet the differences between high and low density in the number of questions answered was greater for low-ANIC students. Finding that fewer questions were answered by low-ANIC participants in the high density condition may be responsible for the finding that SA probe accuracy for low-ANIC students increased with traffic density.

3.3 Performance of ATCo tasks

Additional analyses were carried out to examine the effect of ANIC on the safety and efficiency with which participants performed ATCo tasks in a simulation environment, and to determine whether ANIC is related to measures of performance. Specifically, we compared high and low-ANIC groups in terms of the number of Losses of Separation (LOS) and average time of AC through the sector. A t-test revealed that participants in the high-ANIC group had fewer LOS ($M =$ 3.40, $SE = .81$) than participants in the low-ANIC group ($M = 6.33$, $SE = 1.02$), $t(9)$ = 2.178, $p = .057$. Furthermore, there was an effect of ANIC for time through sector, $t(9) = 2.175$, $p = .058$. The high-ANIC participants had shorter time through sector ($M = 582.80s$, $SE = 3.61s$) compared to those in the low-ANIC group ($M =$ 598.64s, $SE = 5.89s$). In short, those in the high-ANIC group were safer and more efficient at performing ATCo tasks.

4   Discussion

The results of this study support the claim that individual differences in awareness of the need for immediate communications (ANIC) is an important variable—one that to our knowledge has not previously been examined in the aviation literature. We found that students that were more likely to correct communication errors (theirs and those of pilots), and that were more likely to expedite commands, had greater SA and performed ATCo tasks more safely and efficiently. Specifically, high-ANIC participants were less likely to have LOS and moved traffic more quickly through their sector compared to low-ANIC participants. High-ANIC participants were also more likely to answer SA probe questions correctly, and managed workload more efficiently, a result that helps to validate ANIC as a measure of SA. This is desirable because it is likely that errors in communication lead to performance failures by producing inaccurate SA. In contrast, taking steps to ensure proper communication has taken place is likely to lead to better overall SA, lower workload and consequently better performance.

Consistent with the analysis of Billinghurst et al. (2011), we found no overall effects of training order on SA. However, we did find that training order interacted with ANIC, with low-ANIC participants being more affected by training order than high-ANIC participants are. Those that were in the high-ANIC group had just as good SA in both training orders, but those that were in the low-ANIC group had better SA if they were trained with the manual tools first. Moreover, those trained with NextGen tools first had higher workload for some equipage mixtures. This suggests that the best training strategy may be to introduce manual tools first. Although this would not affect participants proficient at acquiring SA, it would benefit those that are less proficient.

The benefit of training with manual tools first is also supported by the interaction between training order and traffic density. Those trained with NextGen tools first had lower SA under low traffic density conditions. However, those trained with manual tools first did not feature a similar decrement as a result of low traffic density. It is possible that training with manual tools first produces scanning patterns that become habitual, and less dependent on particular traffic conditions.

The results of the present study reveal that the participants in high-ANIC group were more robust and less affected by factors such as training order and traffic density. With respect to the latter, high-ANIC participants maintained equally high levels of SA irrespective of traffic density, while low-ANIC participants had lower SA under low traffic density conditions.

A key question that needs to be addressed has to do with the basis of individual differences in the ability to acquire and maintain SA. Why are some participants more adept at this? Billinghurst et al. (2011) and the present investigation found no change in SA over the course of the semester, suggesting that individual differences may be important. It is possible, for example, that the difference between individuals in correcting communication errors, expediting commands, and maintaining SA irrespective of traffic conditions reflects differences in conscientiousness. Conscientious individuals are more self-disciplined, careful, thorough, are less likely to procrastinate, and are more productive (Costa & McCrae, 1992). Durso, Bleckley and Dattel (2006) found conscientiousness predicted ATCo SA and performance, and we are presently investigating its importance in an ATCo training study. It is important to note, however, that other cognitive factors may also be important, including differences in working memory capacity and executive functioning.

To conclude, the present study found that awareness of the need to engage in immediate communications, either to correct errors or to expedite commands, is an important variable that needs to be taken into consideration when designing training programs. It reflects differences in the ability to acquire and maintain SA, and contributes to safe and efficient performance of ATCo duties. The personality and cognitive factors that underlie this ability, however, require further investigation.

REFERENCES

Billinghurst, et al., 2011. Should students learn general air traffic management skills before NextGen tools? Paper presented at the 2010 HFES annual convention.

Cardosi, K. M. 1993. An analysis of en route controller-pilot voice communications. (Report No. DOT/FAA/RD-93-11). Washington: DC. U.S. Department of Transportation/Federal Aviation Administration, Research and Development Service.

Costa, P. T. & McCrae, R. R. (1992). *NEO personality Inventory professional manual.* Odessa, FL: Psychological Assessment Resources.

Durso, F. T., & Dattel, A. (2004). SPAM: The real-time assessment of SA. In S. Banbury & S. Trembley (Eds.) *A Cognitive Approach to Situation Awareness: Theory, Measures and Application* (pp. 137–154). New York: Aldershot.

D,Arcy, J-F. & Della Rocco, P.S. (2001). Air Traffic Control Specialist Decision Making and Strategic Planning – *A Field Survey. Federal Aviation Administration Technical Note DOT/FAA/CT-TN01/0.*

Durso, F., & Manning, C. (2008). Air traffic control. In C. M Carswell (Ed.), *Reviews of human factors and ergonomics* (Vol. 4, pp. 195–244). Santa Monica, CA: Human Factors and Ergonomics Society.

Durso, F., Bleckley, K. & Dattel, A. (2006). Does situation awareness add to the validity of cognitive tests? *Human Factors, 48,* 721-733.

Federal Aviation Administration (FAA) (2011). NextGen Implementation Plan. Retrieved on July 17, 2011 from http://www.faa.gov/nextgen/media/ng2011_implementation_plan.pdf

Hinton, D. A. and G. W. Lohr. 1988. Simulator investigation of digital data link ATC communications in a single-pilot operation (NASA Technical Paper 2837). Hampton, VA: NASA Langley Research Center.

Joint Planning and Development Office (2011). Concept of operations for the next generation air transportation system version 1.2. Retrieved October 2011, from http://www.jpdo.gov/library/NextGenConOpsv12.pdf.

Morrow, D. & Rodvold, M. (1992). Analysis of problems in routine controller-pilot communication. Paper presented at the Annual meeting of the American Psychological Association, Washington, DC.

Paas, F. & T. vanGog. 2009. Principles for designing effective and efficient training of complex cognitive skills. Reviews of Human Factors and Ergonomics, 5, 166-194.

Parasuraman, R., Sheridan, T. B., & Wickens, C. D. (2000). A model for types and levels of human interaction with automation. *IEEE Transactions on Systems, Man and Cybernetics – Part A: Systems and Humans, 3,* 286-297.

Prevot, T., J. Homola, and J. Mercer. 1988. Human-in-the-loop evaluation of ground-based automated separation assurance for NextGen. The 26th Congress of International Council of the Aeronautical Sciences (ICAS), Anchorage, Alaska, September 2008.

Redding, R.E., Ryder, J.M., Seamster, T.L., Purcell, J.A., & Cannon, J.R. (1991). Cognitive task analysis of en route air traffic control: Model extension and validation (Report to the Federal Aviation Administration). McLean, VA: Human Technology, Inc.

Roske-Hofstrand, R. J. and E. Murphy. 1998. Human information processing in air traffic control. In: M. Smolensky and E. Stein (Eds.), *Human Factors in Air Traffic Control* (pp. 65-113). San Diego, CA: Academic Press.

Salden, R. J. C. M., F. Paas, and J. J. G. Van Merrienboer. 2006. Personalized adaptive selection in air traffic control: Effects on training efficiency and transfer. Learning and Instruction, 16, 350-362.

Sohn, M., S. Douglass, M. Chen, and J. R. Anderson. 2005. Characteristics of fluent skills in a complex, dynamic problem-solving task. Human Factors, 47, 742-752.

CHAPTER 21

Examining Pilot Compliance to Collision Avoidance Advisories

Amy R. Pritchett, Elizabeth S. Fleming, Jonathan J. Zoetrum, William P. Cleveland, Vlad M. Popescu, Dhruv A. Thakkar

Georgia Institute of Technology
Atlanta GA
amy.pritchett@ae.gatech.edu

ABSTRACT

This paper describes an experiment with airline pilots in an integrated air traffic control and flight simulator facility, examining a range of factors impacting pilot interaction with TCAS. Through arrivals and approach-intercept, pilots experienced a range of factors experienced in realistic operations, including situations where the controller does and does not call out the traffic, where the traffic flow is and is not being executed properly by the traffic, and a range of relative trajectories to the target. Intruder trajectories were purposefully designed to establish a range of resolution advisories at varying points throughout the flights.

This paper focuses on measures of pilot compliance to TCAS RAs. Pilots were urged at all stages of the experiment to 'use their best judgment' and to 'act as they would in real operations.' Thus, the measures of compliance here are not indicators of raw pilot ability to comply with TCAS RAs, but instead are intended as predictors of what pilots feel they actually do in the context of air traffic operations.

Keywords: TCAS, collision avoidance, human factors, air traffic control.

1 INTRODUCTION

NextGen concepts of operation will likely reduce aircraft separation and implement a shift towards flight deck separation assurance responsibilities. The Traffic alert and Collision Avoidance System (TCAS) has been in operational use for approximately 20 years. While this system has been extremely successful in reducing the risk of midair collisions, human factors concerns remain when considering pilot interactions with TCAS, as reflected in lower rates of pilot compliance to TCAS advisories than assumed in safety studies, and other impacts on interaction with air traffic control. [1,2,3,4,5]

Transition to NextGen often focuses on changes to nominal air traffic control. However, while TCAS is often framed as separate to -- or opposing -- air traffic control, i.e. as a backup, redundant system that invokes resolution advisories (RAs) when something within the system has failed, it cannot be viewed separately from air traffic control. Instead, TCAS may also be viewed as introducing another dynamic component into the flight deck. This additional dynamic includes the information about traffic provided by TCAS, including its persistent Traffic Situation Display (TSD) as well as the occasional traffic advisory (TA) and resolution advisory (RA). [6] Further, TCAS adds the dynamic of its TAs and RAs requiring the pilot to search for information and make an assessment, both being decision making and control actions. This is coupled with the pilot's interaction with air traffic control, including their ability to overhear party-line communications. [7,8,9] Thus, NextGen will require changes to collision avoidance and separation assurance roles, responsibilities, displays and alerts, demanding additional human factors insights supporting flight deck and air traffic procedures, and the design of algorithms, information displays, command displays and automated systems for collision avoidance and separation assurance.

This paper describes an experiment with airline pilots in an integrated air traffic control and flight simulator facility, examining a range of factors impacting pilot interaction with TCAS. Through arrivals and approach-intercept, pilots experienced a range of factors experienced in the current day, including situations where the controller does and does not call out the traffic, where the traffic flow is and is not being executed properly by the traffic, and a range of relative trajectories to the target. Intruder trajectories were purposefully designed to establish a range of resolution advisories at varying points throughout the flights.

The results presented here focus on measures of pilot compliance to TCAS RAs. Pilots were urged at all stages of the experiment to 'use their best judgment' and to 'act as they would in real operations.' Thus, the measures of compliance here are not indicators of raw pilot ability to comply with TCAS RAs, but instead are intended as predictors of what pilots feel they actually do in the context of air traffic operations.

2 DESCRIPTION OF FLIGHT SIMULATOR STUDY

2.1 Participants

Sixteen airline pilots participated. All of the participants were male, ranging in age from their mid-20's up to 59 years old. Eight held the rank of Captain in their airline, seven were ranked as first officers, and one did not respond to the question.

Eight of the sixteen of the pilots reported having received some form of military training in their aviation career. Eight of the pilots reported being very familiar with the airport used for this study (Dallas- Fort Worth), seven of the pilots had some familiarity with the airport, and one pilot reported having no familiarity.

2.2 Experimental Task

The pilot's task was to fly a Standard Arrival Route (STAR) and to perform all required checklists, beginning with the approach briefing and checklist at the start of the flight. The participant acted as the Captain, sitting in the left seat of a fixed-based simulator. A researcher who was familiar with the controls of the aircraft simulator posed as the First Officer (FO). The FO provided the duties of the 'Pilot Not Flying' in airline operations, which focus on managing the aircraft systems and interacting with air traffic control. The pilot was briefed on the specific simulator's FMS and Mode Control Panel (MCP) in initial training runs, and was also informed that the FO was proficient in the autopilot's operation and could be asked to provide any commands to the autoflight system.

Typically, the flights began around an altitude of 10,000 to 20,000 feet and lasted 15 minutes. The flights ended during the approach intercept, i.,e. when the aircraft was within 'one dot' of the localizer beam indicating the approach course. The weather was calm with no wind. However Instrument Meteorological Conditions (IMC) applied for the duration of the flight, as there were no out-the-window visuals provided by the simulator; thus, the pilot could only reference a traffic situation display and air traffic communications for information about the traffic situation, and could not visually acquire a target.

In each flight, pilots encountered two traffic events. Some of events resulted only in 'traffic advisories' from TCAS and required no maneuvering; the more severe events resulted in a 'resolution advisory' from TCAS which displayed a vertical maneuver to the pilot. An air traffic controller was controlling the pilot's aircraft as well as other aircraft in the vicinity of the airport, and the pilot could hear the party-line communications to the other aircraft on the shared voice frequency. Before some events the pilot was given an explicit 'call-out' by the air traffic controller (e.g. "GT123, traffic your 3 o'clock, 1000 feet below") and before other events the air traffic controller communicated with the aircraft causing the traffic event, providing 'party-line information' which could be useful in predicting the development of the traffic event. Thus, the pilot's tasks included monitoring disparate sources of traffic information and evaluating potential maneuvers as displayed by TCAS or instructed by air traffic control.

2.3 Apparatus

As shown in the schematic in Figure 1, the simulator apparatus included several components. Most notable to the participant, the 'Reconfigurable Flight Simulator' software provided an emulation of a two-crew flight deck in which the participant sat in the left seat to act as Captain and Pilot Flying. The flight displays and underlying models of the aircraft dynamics and aircraft systems were based on a B747-400, and had been used in prior studies with airline pilots. [10,11,12] A basic physical mockup of the flightdeck mounted computer monitors in approximately the correct location for each, an provided the captain with a side-stick at his left hand and mockup of a throttle quadrant's structure on his right. A large touch screen in front of the throttle quadrant provided the Control Display Unit (CDU), Mode Control Panel (MCP), control's for the captain's ND, Engine Indication and Crew Alerting Display (EICAS), and levers for flaps, gear and speedbrakes; the pilot could also us a mouse or trackball. The pilot was also provided with an approach checklist and all appropriate charts.

An experimenter station for air traffic control used the air traffic display and traffic emulation capabilities provided by the FAA's Traffic Generation Facility (TGF), populated with traffic flows recorded by the FAA from real operations. Dynamic models of intruder aircraft steered the intruder. Through a priori analysis, their relative trajectories were pre-determined such that traffic events would create specific TCAS advisories. An experimenter acting as air traffic controller interacted with the pilot according to proper air traffic procedures via a voice channel established using an aviation intercom. Another experimenter acted as 'party-line,' i.e. providing the voice communications of all other aircraft in the area.

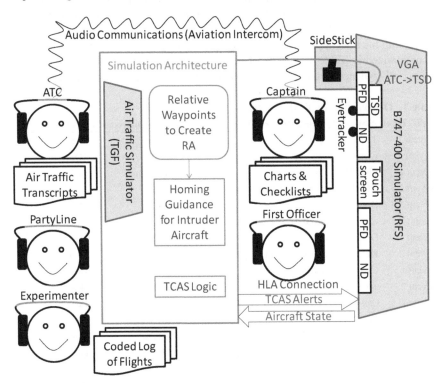

Figure 1. Schematic of the simulator setup, including both flight simulator and air traffic simulator

TCAS was emulated according to the Minimum Operational Performance Standards (MOPS) for TCAS [13]. The PFD provided the RA maneuver guidance both in terms of a red-arc on the vertical speed tape and a 'trapezoid' on the attitude indication indicating the pitch attitude corresponding to the vertical speed red arc.

The TCAS TSD was provided by a 7 inch display mounted in the flight simulator above the ND, with its range slaved to the ND range selected by the participant. However, it was driven by the computer providing the air traffic simulator and intruder generation, which had knowledge of the location of the other aircraft. The TSD also followed all conventions required by regulatory material [13,14,15].

2.4 Independent Variables and Experiment Design

The independent variables described include whether pilots received air traffic callouts from the controller or could overhear party-line information; traffic density; and the intruder trajectory relative to the aircraft, which included the type of advisory displayed to the pilot by TCAS (an 'active' RA displaying a vertical maneuver away from the intruder, a 'crossing' RA displaying a vertical maneuver through the intruder's altitude when the closure rate prevented a maneuver away, or a TA alone without an RA). One special case involved an aircraft under Visual Flight Rules (VFR) crossing 500' below (legal, but can trigger TCAS advisories) the pilot's aircraft. A second special case created an RA while the pilot was intercepting his final approach course. Altogether the pilots flew eight 'data' flights as summarized in Table 1. The experiment design used an 8th order Latin-Squares design, fully balancing run order within each of two sets of eight participants.

Table 1. Description of Traffic Events

Flight / Event	Advisory Type and Intruder Trajectory	Traffic Information	Traffic Density
A-Event 1	TA	Partyline	Heavy
A-Event 2	Climb RA	Callout	Heavy
B-Event 1	Climb RA	Conflicting	Light
B-Event 2	TA- While on intercept approach	None	Light
C-Event 1	Descend RA	Partyline	Heavy
C-Event 2	Climb RA, caused by VFR traffic	Callout	Heavy
D-Event 1	Descend RA	Partyline	Light
D-Event 2	Preventive RA while on intercept approach	None	Light
E-Event 1	Crossing Descend RA	Callout	Heavy
E-Event 2	TA	Partyline	Heavy
F-Event 1	Crossing Descend RA	Callout	Heavy
F-Event 2	TA	Partyline	Heavy
G-Event 1	TA- Caused by VFR traffic	Partyline	Light
G-Event 2	TA	Callout	Light
H-Event 1	TA	Partyline	Heavy
H-Event 2	Descend RA	Callout	Heavy

2.5 Dependent Variables

Objective measures included the simulator's record of aircraft location, attitude, and pilot's control actions. During a traffic event, the time of TA and RA were also recorded, as was the maneuver displayed by a TCAS RA through time (which could include 'strengthening' or 'weakening' maneuver rates). The air traffic simulator logged the location of neighboring aircraft.

In post-hoc categorization, the pilot's response was analyzed over the duration of each RA and for 30 seconds after "Clear of Conflict." As shown in Figure 2, compliance at any instant in time to a TCAS RA was defined as, after allowance for a five second reaction time, the pilot meeting (or exceeding) the vertical rate displayed by TCAS and the pilot not performing any horizontal maneuver. Horizontal maneuvers were defined as the aircraft's roll angle exceeding 5 degrees.

The measure 'compliance percentage' compares durations of time in which the pilot was in compliance to the total duration of the RA. For example, if an RA lasted for 30 seconds of which the pilot was in compliance for 15 seconds, their compliance percentage for the total duration of the RA would be 50%. The measure 'total change in altitude' over the duration of the RA was also calculated by comparing the altitude at the start of the RA to that at "Clear of Conflict."

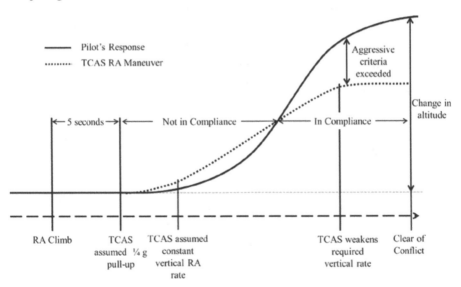

Figure 2. Vertical profile of TCAS RA and measures of pilot response

3 RESULTS

Analysis used SPSS statistical software package. ANOVA used a general linear model treating pilots and run order as random effects and ATC interaction (callout, party-line information or conflicting air traffic instructions), traffic density, and RA type as fixed effects. Specifically, three RA types were examined: "(Corrective) Climb," "(Corrective) Descend," "Crossing Descend," and "Preventive" RA's. Preventive RAs, while tested in this experiment, did not require maneuvering and thus were not considered appropriate to these measures of compliance.

Examining compliance percentage, as shown in Table 2, ANOVA found that neither run order, nor ATC interaction, nor traffic density had significant effects. Between-pilot variance was significant. As shown in Figure 3, significant differences were found between the RA types

Table 2. ANOVA of compliance percentage

Source		Type III Sum of Squares	df	Mean Square	F	Sig.
Intercept	Hypothesis	590218.833	1	590218.833	1838.548	.000
	Error	5470.365	17.040	321.024		
Traffic Density	Hypothesis	52.004	1	52.004	.254	.616
	Error	19880.882	97	204.958		
RA Type	Hypothesis	4503.710	2	2251.855	10.987	.000
	Error	19880.882	97	204.958		
ATC Interaction	Hypothesis	383.936	2	191.968	.937	.395
	Error	19880.882	97	204.958		
Pilot Number	Hypothesis	6248.334	15	416.556	2.032	.020
	Error	19880.882	97	204.958		
RunNumber	Hypothesis	1310.648	7	187.235	.914	.499
	Error	19880.882	97	204.958		

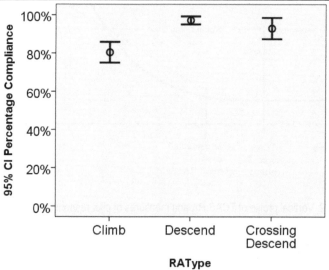

Figure 3. Percentage compliance with each type of RA (Mean and 95% confidence interval)

Examining the measure 'change in altitude, ANOVA (Table 3) found that ATC interaction did not have significant effects. Unlike the prior measures, however, traffic density (heavy or light) and run order were found to have a significant effect. Like the previous measures, type of RA also had a significant effect, with greater change in altitude with the Descend and Crossing Descend RAs compared to the Climb RA type, as shown in Figure 4.

Table 3. ANOVA of change in altitude

Source		Type III Sum of Squares	df	Mean Square	F	Sig.
Intercept	Hypothesis	72491673.39	1	72491673.39	250.122	.000
	Error	5646822.03	19.484	289825.31		
Traffic Density	Hypothesis	441592.40	1	441592.40	5.543	.021
	Error	7727144.29	97	79661.28		
RA Type	Hypothesis	3104015.53	2	1552007.77	19.483	.000
	Error	7727144.29	97	79661.28		
ATC Interaction	Hypothesis	457587.37	2	228793.69	2.872	.061
	Error	7727144.29	97	79661.28		
Pilot Number	Hypothesis	4902604.19	15	326840.28	4.103	.000
	Error	7727144.29	97	79661.28		
Run Number	Hypothesis	1267400.35	7	181057.19	2.273	.035
	Error	7727144.29	97	79661.28		

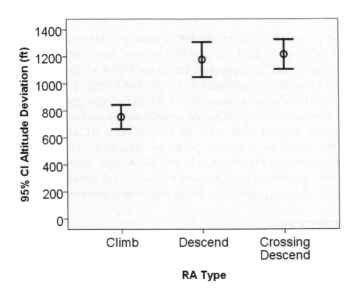

Figure 4. Change in altitude with each type of RA (Mean and 95% confidence interval)

4 DISCUSSION AND CONCLUSIONS

A consistent effect with the type of RA was found in all measures of compliance, including percentage compliance within each RA, rate of full compliance, change in heading and change in altitude. In this experiment, lower compliance (percentage within the duration of an RA, and full compliance rates) was found with 'climb' RAs compared to both 'descend' and 'crossing descend' RAs, and smaller altitude changes were made.

These results should be evaluated in the context of the air traffic operations within which the pilots were operating: standard terminal arrivals (STARs) and intercept of their final approach course, throughout which the pilots' aircraft was either level or descending at the start of the traffic event. Further, in these air traffic operations the pilots were either cleared to, or expecting to be cleared to, a lower altitude; thus, a 'descend' RA often did not contravene any air traffic instructions (or mirrored pilot expectations for their next air traffic instructions).

On the other hand, no significant differences were found in compliance between 'descend' and 'crossing descend' RAs, despite the differences between them. The 'crossing descend' RAs are, by their nature, generally time-critical and require fairly high vertical rates; they may appear counter-intuitive to pilots as they required the pilot to descend through the altitude of traffic below them. That no differences were found suggests these attributes – required vertical rate and relative trajectory to traffic visible to the pilot on the TSD – may not be the most important factors in describing pilot compliance.

Regulatory material (with some limited exceptions) requires pilots to comply to TCAS, and describes the assumed pilot response [14]. However, complete pilot compliance was not manifested in this experiment consistently. Instead, the same measures of pilot compliance to TCAS RAs resulted in significant variation between pilots.

This experiment purposefully varied the traffic density (light and heavy) and "ATC interactions" (whether the pilot was provided relevant party-line information, a callout, or conflicting air traffic instructions before the TCAS advisories). These independent variables examined hypotheses that these factors may influence pilot trust in air traffic control or other assessments of the traffic situation. However, they were found to have no significant effect on compliance measures.

Throughout, results suggest that pilot interaction with TCAS cannot be examined outside the context of air traffic operations. The pilots' phase of flight, assigned route and immediately-cleared altitude may all strongly interact with pilot responses to RAs: here, descend-sense RAs were followed more closely, but climb-sense RAs may receive the same preference during departure/climb phases of flight.

ACKNOWLEDGMENTS

The authors are grateful for the time of the pilots who participated, specifically the 2 pilots who helped vet the simulator, traffic events and experiment design, and the 16 pilots who acted as participants. Additionally, the authors would like to acknowledge Henry Tran for his assistance in running the experiment and acting as the 'party-line' voice.

This material is based upon work supported by a Cooperative Agreement (DTFAWA-10-C-00084) with the Federal Aviation Administration (FAA) Human Factors Research and Engineering Group, with Tom McCloy acting as technical manager. The authors also gratefully acknowledge the time and expertise provided by Wes Olson of MIT Lincoln Labs as a technical consultant.

REFERENCES

Olson, W. & Olszta, J. 2010. TCAS monitoring in the US national airspace. Presentation at the Aerospace Control and Guidance Systems Committee Meeting.

Olson, W. 2011. TCAS Operational Performance Assessment. Presentation at the Aviation InfoShare Meeting, Memphis TN.

Coso, A.E., Fleming, E.S., & Pritchett, A.R. 2011. Characterizing pilots' interactions with the aircraft collision avoidance system. International Symposium of Aviation Psychology, Dayton, OH.

Vallauri, E., 1997. Suivi de la mise en oeuvre du TCAS II en France en 1996, (Survey of the 1996 TCAS II implementation in France.) Centre D'Etudes de la Navigation Aerienne: Toulouse.

Law, J. 2005. Incorrect use of the TCAS traffic display, ACAS II Bulletin. Eurocontrol.

Olson, W. 2009. Impact of traffic sysmbol directional cues on pilot performance during TCAS events, Proceedings of the 28th IEEE/AIAA Digital Avionics Systems Conference. Orlando, FL

Pritchett, A.R. 2012 (submitted) The pilot as an expert fallible machine during traffic events, HCI-Aero 2012.

Midkiff, A.H. & Hansman, R.J. 1993. Identification of important 'party line' information elements and implications for situational awareness in the datalink environment. Air Traffic Control Quarterly, 1(1), 5-30.

Pritchett, A.R. & Hansman, R.J. 1997. Variations among pilots from different flight operations in party line information requirements for situation awareness, Air Traffic Control Quarterly, 4(1), 29-50.

Ippolito, C.A. & Pritchett, A.R. 2000. Software architecture for a reconfigurable flight simulator, Proceedings of the AIAA Modeling and Simulation Technologies Conference, Denver, CO.

Chen, T.L. & Pritchett, A.R. 2001. Cockpit decision aids for emergency flight planning, Journal of Aircraft, 38(5), 935-943.

Pritchett, A.R. & Yankosky, L.J. 2003. Pilot-performed guidance in the national airspace system, Journal of Guidance, Control and Dynamics, 26(1), 143-150.

RTCA, 1997. Minimum Operational Performance Standards for Traffic Alert and Collision Avoidance System II Airborne Equipment., Washington DC.

FAA, Advisory Circular No. 120-55B: Air Carrier Operational Approval and Use of TCAS II. 2001: Washington, D.C.

FAA, 2005. Advisory Circular No. 120-86: Aircraft Surveillance Systems and Applications. Washington, D.C.

CHAPTER 22

Nighttime Offshore Helicopter Operations – Identification of Contextual Factors Relevant to Pilot Performance

Felipe A. C. Nascimento, Arnab Majumdar, Washington Y. Ochieng

Imperial College London, U.K.
f.a.c.nascimento@imperial.ac.uk

ABSTRACT

When visibility is poor offshore helicopter operations are prone to accidents. Given that a significant increase in nighttime activities is predicted in the near future, this chapter proposes a systemic approach to identify the factors underlying such accidents. It presents the results of accident analysis, interviews of pilots from five different scenarios, and a review of the literature on performance shaping factors in human reliability analysis (HRA) techniques. The results are used to compile a comprehensive list of the factors that shape the ability of pilots to fly at night in the offshore helicopter domain, thus guiding future data collection and safety interventions.

Keywords: Offshore helicopters; Oil and gas aviation; Accident analysis; Task analysis; Qualitative methods; Human Reliability Analysis

1   INTRODUCTION

Helicopter operations are vital for oil and gas exploration and production activities though they are the biggest contributor to the overall risk of fatal accidents in the offshore environment (OGP, 2010). To balance that, stakeholders in the oil and gas industry have developed their own aviation policies and standards that often extrapolate safety regulations in many countries and benchmark safety requirements for the wider helicopter community. However, despite such commendable leadership and many investments, offshore helicopter operations at night are still a major concern to both regulators and operators.

Ross and Gibb (2008) analysed offshore helicopter accidents that happened between 1990 and 2007, and concluded that the nighttime accident rate was more than 5 times greater than that in daylight. Since their study, other accidents (e.g., OGP, 2010) and incidents have indicated that flying helicopters at night in the offshore environment remains a risky business in need of thorough investigations.

This need is enhanced given an expected increase in nighttime offshore helicopter activities in the upcoming years. Increasingly, the search for energy sources is moving towards the Poles, presenting new challenges to helicopter operations as daylight hours are considerably shorter during several months of the year. In other regions, growing energy production has required greater number of night flights, for example in the Middle East and Brazil (Nascimento et al., 2012b). Additionally, the evacuation needs of the growing offshore population during medical emergencies or natural catastrophes (e.g., hurricanes), including at night, have increased commensurately. Details of the drivers to more intense nighttime offshore helicopter operations can be found in Nascimento et al. (2011, 2012c).

Although many investments have occurred to develop the infrastructure needed for safer offshore helicopter operations in degraded visual environments (DVE), Nascimento et al. (2011, 2012a, b, c) have pointed out that the scant scientific knowledge of the risk factors at stake, coupled with the lack of a holistic approach to safety (i.e., a look at the problem in integral systemic manners) might still impede further reductions in accident numbers. A review of the human factors literature in visual perception, spatial orientation, situation awareness and decision-making clarifies the paramount need for a systemic approach to safety in this domain.

The human ability to move in space depends on the information captured by many senses, on the robustness of the internal models used to interpret the incoming stimuli (Cheung, 2004) and on the availability of cognitive resources for this matching process (Cheung et al., 1995). Therefore, orientation results from both 'top-down' (interpretative) and 'bottom-up' (neural sensing) perceptual mechanisms (Gibb et al., 2010), of which vision accounts for more than 80% of the inputs needed for spatial 3-dimensional (3-D) locomotion (Newman, 2007).

Situation awareness (SA), as a more thoughtful process which precedes decision-making, is also heavily reliant on both incoming stimuli and internal schemas. Endsley (1995) for example, places information processing as pre-requisite to all stages of SA, underpinned by the perception of the elements in the current situation (Fig.1). This view connects the perceptual array (dominated by vision in aviation) to the knowledge base of the individual in similar manners to Gibb et al.'s (2010) concepts mentioned above.

226

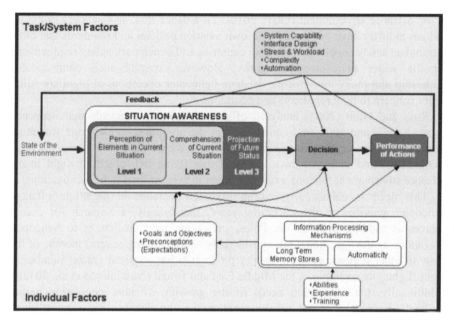

Figure 1 – Situation Awareness (Endsley, 1995)

Models of collective SA in multi-crew operations (e.g., Greenberg et al., 2005) have also acknowledged the importance of internal schemas to SA and orientation, which Gibb et al. (2010) found increasingly important as the availability of raw stimuli decreases, e.g. in degraded visual environments (DVE). Whereas various human visual mechanisms work effortlessly and efficiently to de-bias abundant visual cues, e.g. in nominal daylight conditions, they all fail to provide unambiguous images as external conspicuity decreases, leading to the prevalence of interpretation in processing incomplete visual pictures (Nascimento et al., 2011, 2012c, Gibb et al., 2010). This process, coupled with a persistent over-estimation of visual capabilities forged by the living experience in daylight (Andrade, 2011) has caused far too many accidents in aviation, often over-simplistically attributed to flawed pilot decision-making and actions (e.g., EHEST, 2010; U.S.JHSAT, 2010).

Because the cognitive strategies (e.g., mental models) that enable the perception needed for orientation, situation awareness and decision-making are a by-product of the operator's experience with the system which hosts the tasks (e.g., Fig.2), there is a need to understand how systemic interactions actually frame operators' perception (thus performance) and contribute during normal operations to suboptimal behaviours in higher workload or complex conditions (Carvalho et al., 2009). Therefore, a systemic approach is required for the identification of the hazard factors in nighttime offshore helicopter operations and specification of informed interventions.

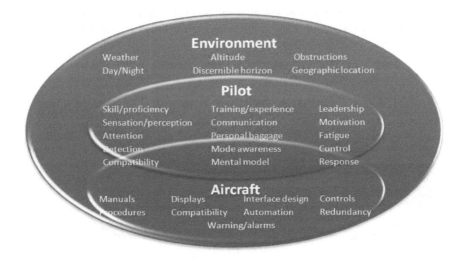

Figure 2 - The aviation system, adapted from Gibb et al. (2010)

The aim of this chapter is to identify which systemic factors (hereby labeled 'Contextual Factors') are relevant to crew performance in nighttime offshore helicopter transportation, with a view to guide future safety initiative.

2 METHODOLOGY

This investigation was undertaken in three main steps: accident analysis, expert knowledge elicitation and a review of HRA techniques.

2.1 Accident analysis

This was aimed at developing a high-level understanding of the problem at hand and at identifying the most critical accident types and flight phases during nighttime offshore helicopter operations. Details of the dataset, variables investigated and statistical tests applied can be found in Nascimento et al. (2012c).

2.2 Expert knowledge elicitation

Given the results and limitations of the previous step, a purposive sampling strategy was designed to enable data collection at representative offshore helicopter operating scenarios. Additionally, seventeen carefully-chosen subject matter experts (SMEs) were consulted before the laying out of a task description and interview questions which generated comparable data across the scenarios. Using talk-through cognitive task-analysis in individual semi-structured interviews, the hazards faced by pilots during the most critical phase of flight were discovered and grouped in a novel, unbiased and bespoken hazards template tailored to the needs of the offshore helicopter industry. The scenario and pilot sampling strategies, as well the specific methods applied to 4 out of 5 scenarios have been detailed in Nascimento et al. (2011, 2012c), and summarized in Figures 3 and 4.

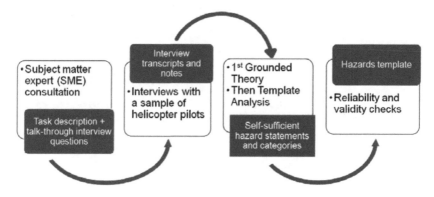

Figure 3 – Methodology for expert knowledge elicitation, with sub-step results in the blue boxes (adapted from Nascimento et al., 2012c)

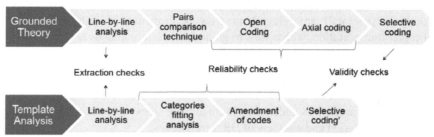

Figure 4 - Methodology for analysis of textual data from the interviews, with checks undertaken (adapted from Nascimento et al., 2012c)

2.3   Review of critically chosen HRA techniques

This step was aimed at ensuring that all determinants to helicopter pilot performance in nighttime offshore helicopter operations had been captured by the previous steps. In order to achieve this, the hazard categories developed were compared against previous taxonomies of performance shaping factors in HRA techniques in various domains (e.g., Kirwan, 1992; Kirwan and Ainsworth, 1992; Hollnagel, 1997; Hollnagel, 2006; Kirwan and Gibson, 2007; Subotic et al., 2007). This process also helped validating the developing results and enabled the harmonisation of categories to the terms of well-established techniques in the field. Amendments to and groupings of hazard categories resulted, which were then re-checked against a randomly selected sample of narrative data from the interviews to reassure the reliability and validity of categories.

3   RESULTS

3.1   Accident analysis

Details of the accident analysis undertaken have been reported in Nascimento et al. (2012c). This section provides a summary of the most relevant findings in support of the choices made for the remainder of the research (all refer to worldwide offshore helicopter operations undertaken between 1997 and 2007).

- Fatal accidents were 15 times more frequent at night than in daytime, with the fatal accident rates significantly differing between the daytime and nighttime on an annual basis.
- The numbers of fatally injured people in each nighttime accident was significantly greater than in daylight.
- Accident causes were significantly associated with lighting conditions, with more pilot-related accidents occurring at night.
- Controlled flight into terrain or water (CFIT/W) was the most common type of accident, accounting for 18.2% of all and 84.6% of the nighttime accidents, with a significant association to the nighttime.
- The approach-and-landing phases concentrated the majority of accidents, accounting for over 32% of the total and 42% of the nighttime accident.
- Across all lighting conditions, there was a significant association between accident causes and phases of flight, with a relatively high number of pilot-related accidents in the approach-and-landing phase.

Given the results, a worst case scenario strategy was chosen for expert knowledge elicitation, thus focusing on the hazards related to shortcomings in pilot performance during the approach-and-landing phase of nighttime flights.

3.2   Expert knowledge elicitation

Given the sampling strategies, five offshore helicopter operating scenarios were investigated, using varying numbers of interviewees as follows:

- British Northern North Sea (Aberdeen), 33 pilots.
- British Southern North Sea (Blackpool), 3 pilots.
- Brazilian Campos basin (Macaé, Cabo Frio, Rio de Janeiro), 28 pilots.
- Norwegian North Sea (Stavanger), 3 pilots.
- Spain (Reus, Amposta, Bilbao), 6 pilots

From the narratives, 2480 self-sufficient hazard statements were extracted, representing pilots' experiences with factors affecting their ability to fly night visual approach segments offshore. The template developed was formed by exhaustive 103 hazard categories, grouped in 13 codes, split into two sections ('Contextual Factors', with 8 codes and 78 categories; and 'Impacts on Crews'), as partially reported in Nascimento et al. (2012c).

3.3   Review of critically chosen HRA techniques

Comparison of the template against various HRA techniques led to the fusion of some hazard categories according to common mechanisms of human performance breakdown. The final list of 'Contextual Factors' which shape pilot performance in nighttime offshore helicopter flights is show in table 1 below.

Table 1 – 'Contextual factors' relevant to pilot performance at night offshore

		Pilots	% of pilots
Training related factors		60	82%
1	Currency of training	53	73%
2	Experience	30	41%
3	Quality of mission-specific training	16	22%
4	Simulator quality	15	21%
5	Usage of simulators	8	11%
6	Aircraft-specific training	6	8%
Procedural factors		55	75%
7	Communication standards	30	41%
8	Guidance for choice of instrument flight profile	30	41%
9	Standard flight profile for the visual segments	27	37%
10	Standards for automation usage	15	21%
11	Guidance for use of available aircraft resources	3	4%
Aircraft related factors		52	71%
12	Automation capabilities	39	53%
13	Aircraft resources	24	33%
14	Inherent handling qualities	15	21%
15	Cockpit design	13	18%
16	Field of view	8	11%
17	Engine power available	5	7%
Platform and surroundings related factors		47	64%
18	Awareness of obstacles at the destination	29	40%
19	Sufficiency of visual cues for approach	16	22%
20	Turbulence	13	18%
21	Quality of weather reports	13	18%
22	Stability of visual cues	10	14%
23	Awareness of obstacles to flight path jutting from the sea	8	11%
24	Sufficiency of visual cues for landing	5	7%
25	Conspicuity of visual aids	2	3%
Internal factors		38	52%
26	Comfort in doing the task	25	34%
27	Self-induced pressure for mission accomplishment	23	32%
28	Complacency or self confidence	10	14%
29	Inherent aptitude to fly at night	4	5%
30	Regularity of feeding patterns	1	1%
Environmental factors		37	51%
31	Illusiveness of weather conditions	27	37%
32	Celestial illumination	21	29%
33	Maintenance of visual references through to landing	6	8%
34	Crosswinds	5	7%
35	Speed of weather changes	3	4%

Regulator factors	37	51%
36 Meteorological minima to be abided by	20	27%
37 Aeronautical infrastructure for nighttime flying	9	12%
38 Legislation stringency	8	11%
39 Ease for nighttime flying practice allowance	7	10%
40 Airspace congestion	4	5%
Organisational factors	29	40%
41 Commercial pressure	19	26%
42 Flight programme	7	10%
43 Reporting / learning culture	5	7%
44 Sponsorship for training-only sorties	4	5%
45 Hiring standards	2	3%
46 Representativeness of pilots in contractual decisions	1	1%
Teamwork / communications	21	29%
47 Teamwork attitude/seniority gradient	17	23%
48 Rosters	7	10%

4.0   DISCUSSION

This discussion focuses on the findings across all interviews undertaken instead of the regional differences. This is important to avoid drawing conclusions based on the low sample sizes of the British Southern and Norwegian North Seas and Spain.

The list developed revealed direct agreement with relevant HRA taxonomies in important aspects, which further indicated its validity. For example, Hollnagel's (1997) dichotomous error phenotypes and genotypes correspond to the 'Impacts on Crews' and 'Contextual Factors' sections of the template, respectively. In both cases, the former lists the ways in which errors manifest themselves in observable manners, whereas the latter refers to the mechanisms, embedded in the context of operations, which contribute to such errors. This section discusses deals with the 'Contextual Factors' (Table 1) rather than the 'Impacts on Crews'.

Using the numbers of participants who commented on the codes and categories as an indicative of the importance of such codes and categories to the problem at hand has been practiced by many authors (e.g., Flanagan, 1954; Butterfield et al., 2005; Bertolini, 2007; Nascimento et al., 2011, 2012a, 2012b). In this respect, both the 'training-related factors' code and the 'currency of training' category unveiled an important agreement among pilots from all scenarios in that the frequency with which they practice offshore nighttime operations is too poor to ensure that the skills needed are properly developed and maintained. This has fundamental implications to operators based in high latitude regions, where night operations are barely feasible during the summer months. Further development of simulation technology for example, could ameliorate this problem.

Regarding the code 'procedural factors' that followed, a more spread distribution of commentators per categories indicated the need for standardisation in different aspects of the flights, which requires commitment from the operators and has yet been stressed in the industry (e.g., Kontogiannis and Malakis, 2011).

Amidst the 'aircraft-related factors', participants highlighted that not having high levels of automation contribute to impair their performance. Therefore, more advanced auto-pilots, potentially offering auto-hover or auto-land capabilities, should be pursued.

The 'platform and surroundings-related factors' code has in turn pointed out that unawareness of the position of obstacles at the destination are still a factor to compromise pilot performance, as attentional resources are devoted to scanning the surroundings for obstacles at the expense of flying the aircraft. Overcoming this issue would require tighter crane stowage procedures by the platform/ship operators or publishing more credible platform/ship charts, for example.

The remaining codes and categories were mentioned by comparatively fewer participants, suggesting that, based on the experiences of the sample investigated, these issues are less relevant to pilot performance or happen in less alarming frequencies. Other hypothesis for the variations in the frequencies of comments can be found in Nascimento et al. (2012c).

7.   CONCLUSIONS

Flying offshore helicopters at night is a complex task, as shown by the large number of contextual factors found to influence pilot performance. This endorses current scientific knowledge whereby only concerted efforts by the different stakeholders in the offshore helicopter industry (i.e., a systemic approach to safety) can redress the problems of unacceptably high accident rates.

ACKNOWLEDGEMENTS

The authors are grateful to the Lloyd's Register Educational Trust and Santander Universities for sponsoring most of the investigation reported in this paper. Additionally, all pilots who faithfully shared their experiences are also mentioned.

ANDRADE, E. B. 2011. Excessive confidence in visually-based estimates. *Organizational Behavior and Human Decision Processes,* 116, 252-261.

BERTOLINI, M. 2007. Assessment of human reliability factors: A fuzzy cognitive maps approach. *International Journal of Industrial Ergonomics,* 37, 405-413.

BUTTERFIELD, L. D., BORGEN, W. A., AMUNDSON, N. E. & MAGLIO, A. T. 2005. Fifty Years of the Critical Incident Technique: 1954-2004 and Beyond. *Qualitative Research,* 5, 475-497.

CARVALHO, P. V. R. D., GOMES, J. O., HUBER, G. J. & VIDAL, M. C. 2009. Normal people working in normal organizations with normal equipment: System safety and cognition in a mid-air collision. *Applied Ergonomics,* 40, 325-340.

CHEUNG, B. 2004. Nonvisual Spatial Orientation Mechanisms. *In:* PREVIC, F. H. & ERCOLINE, W. R. (eds.) *Spatial Disorientation in Aviation.* Reston, VA: AIAA.

CHEUNG, B., MONEY, K., WRIGHT, H. & BATEMAN, W. 1995. Spatial Disorientation-Implicated Accidents in Canadian Forces, 1982-92. *Aviation, Space, and Environmental Medicine,* 66, 579-585.

EHEST 2010. EHEST Analysis of 2000 – 2005 european helicopter accidents. Cologne, Germany: EASA.

ENDSLEY, M. R. 1995. Toward a theory of situation awareness in dynamic systems. *Human Factors,* 37**,** 32–64.

FLANAGAN, J. C. 1954. The Critical Incident Technique. *Psychological Bulletin,* 51**,** 327-358.

GIBB, R., GRAY, R. & SCHARFF, L. 2010. *Aviation visual perception: research, misperception and mishaps,* Ashgate.

GREENBERG, R., COOK, S. C. & HARRIS, D. 2005. A civil aviation safety assessment model using a Bayesian belief network (BBN). *The Aeronautical Journal,* 109**,** 557-568.

HOLLNAGEL, E. 1997. *Cognitive Reliability and Error Analysis Method CREAM,* Halden, Norway, Elsevier.

HOLLNAGEL, E. 2006. *Capturing an Uncertain Future: The Functional Resonance Accident Model* [Online]. Paris: Mines de Paris. Available: http://www.skybrary.aero/bookshelf/books/402.pdf [Accessed 24/03 2010].

KIRWAN, B. 1992. Human error identification in human reliability assessment. Part 1: Overview of approaches. *Applied Ergonomics,* 23**,** 299-318.

KIRWAN, B. & AINSWORTH, L. K. 1992. *A Guide to Task Analysis,* London, Taylor and Francis.

KIRWAN, B. & GIBSON, H. 2007. CARA: A Human Reliability Assesment Tool for Air Traffic Safety Management - Technical Basis and Preliminary Architecture. *Fifteenth Safety-Critical Systems Symposium.* Bristol, UK.

KONTOGIANNIS, T. & MALAKIS, S. 2011. A systemic analysis of patterns of organizational breakdowns in accidents: A case from Helicopter Emergency Medical Service (HEMS) operations. *Reliability Engineering & System Safety.*

NASCIMENTO, F. A. C., MAJUMDAR, A., JARVIS, S. & OCHIENG, W. Y. 2011. Safety Hazards in Nighttime Offshore Helicopter Operations. *37th European Rotorcraft Forum 2011.* Ticino Park, Italy.

NASCIMENTO, F. A. C., JARVIS, S. & MAJUMDAR, A. 2012a. Factors Affecting Safety During Night Visual Approach Segments for Offshore Helicopters. *The Aeronautical Journal,* 116.

NASCIMENTO, F. A. C., MAJUMDAR, A. & JARVIS, S. 2012b. Nighttime approaches to offshore installations in Brazil: Safety shortcomings experienced by helicopter pilots. *Accident Analysis & Prevention,* 47**,** 64-74.

NASCIMENTO, F. A. C., MAJUMDAR, A., OCHIENG, W. Y. & JARVIS, S. R. 2012c. Assessing the hazards of nighttime offshore helicopter operations. *Transportation Research Board 91st Annual Meeting.* Washington, D.C.: TRB.

NEWMAN, D. G. 2007. An overview of spatial disorientation as a factor in aviation accidents and incidents. Canberra: ATSB.

OGP 2010. Safety performance indicators – 2009 data. International Association of Oil & Gas Producers.

ROSS, C. & GIBB, G. 2008. *A Risk Management Approach to Helicopter Night Offshore Operations* [Online]. Available: http://asasi.org/papers/2008/ Risk%20Approach%20to%20Night%20Offshore%20Operations%20Presented%20by %20Gerry%20Gibb%20&%20Cameron%20Ross.pdf [Accessed 15/10 2010].

SUBOTIC, B., OCHIENG, W. Y. & STRAETER, O. 2007. Recovery from equipment failures in ATC: Determination of contextual factors. *Reliability Engineering & System Safety,* 92**,** 858-870.

U.S.JHSAT 2010. Calendar Year 2006 Report to the International Helicopter Safety Team. USA: U.S. Joint Helicopter Safety Analysis Team.

CHAPTER 23

Assessing Air Traffic Control Performance in Loss of Separation Incidents: A Global Review

Marie-Dominique Dupuy, Arnab Majumdar

Imperial College London
United Kingdom
md.dupuy@imperial.ac.uk, a.majumdar@imperial.ac.uk

ABSTRACT

Air Traffic Control (ATC)-related incidents such as those related to separation, e.g. Loss Of Separation (LOS), are currently analyzed for the purpose of detecting failures or errors from or within the Air Traffic Management system, i.e. negative performance of the system. However, it is possible to assess the capability of the system and air traffic controllers in managing successfully such incidents using a different approach for their analysis based on an incident model that considers a characterization of the intrinsic functions of ATC. This paper presents the main attributes of the LOS analysis framework that enable *(i)* the enhancement of the way incidents are currently analyzed individually and *(ii)* the assessment of ATC safety performance. The initial results of the analysis of a database of 191 SRIs showed that in 72% of cases, ATCO(s) in interaction with the whole system successfully recovered the incidents and 19% were not fully recovered by ATC alone and necessitated the support of airborne recovery. These findings help identifying the 9% incidents that failed to be managed, thus enabling more focus on them.

1   INTRODUCTION

The Air Traffic Management (ATM) system is complex and characterised by the continuous interaction of human operators, technology and procedures to ensure the safe and efficient conduct of aircraft from departure to destination points. Air Traffic Controllers (ATCOs), are the crucial operators of Air Traffic Control (ATC), the dynamic component of ATM, and perform in ensuring that aircraft are safe from conflicts with others, airspace, terrain or obstacles during all phases of flight. The examination of precursors to accidents in the form of specific incidents related to the loss of required separation, i.e. Loss of Separation Incidents (LOS) helps in understanding how the risks of failure of the system or the non-expected performance that may result in incidents and in worst case scenario, accidents. Incidents are usually considered to indicate a weakness in safety performance of ATM. However, by means of an appropriate investigation methodology, it is possible to extract more meaningful information than typically collected, to understand and analyse how ATC manages an incident, i.e. how controllers performs in detecting and successfully resolving any emerging conflicts or recovering critical situations.

This paper outlines a method to assess ATC safety performance in a two-step process, the first of which is to consider the progressive development of LOS (sequential model). The second is to provide an appropriate model of the functions of ATC. Such a model of ATC can be developed based on that of ATM processes proposed for the Single European Sky ATM Research (SESAR) program (EPISODE 3, 2007) and refined for this research. It is characterized by nine intrinsic functions: traffic sequencing, traffic adjustment, traffic separation, collision avoidance, communication, navigation, surveillance, workload management and safety nets. This functional model characterizes the various activities of ATC in tactically managing the traffic, the controller's cognitive performance generic attributes and the supporting tools that help them achieved their tasks.

The literature review in Section 2 reviews the current approach to identify positive human performance during ATC operations and identify the lack of use of incident data to demonstrate this. Section 3 presents the model and data captured based on the LOS analysis framework, which considers the ATC functions-model useful for the assessment of incident management by ATCOs. Section 4 presents the initial results of the ATC safety performance assessment based on the descriptive analysis of 191 LOS examined through the proposed framework and Section 5 concludes the paper with a discussion of the results and the use of the new framework for the assessment of ATC, including controllers' performance, in managing LOS.

2   LITERATURE REVIEW

The monitoring of safety performance levels in the ATM system as required by ICAO SMS (ICAO, 2009) is currently achieved in different ways depending on the organization. The organization is required to monitor its operations, including changes in the operational environment, degradation in operational processes, facilities, equipment conditions, and human performance that could reduce the effectiveness of its existing safety risk controls (ICAO, 2009). With respect to human performance monitoring, the EUROCONTROL/FAA Action Plan 15 (EUROCONTROL, 2010) recommends specific human factors techniques that can be categorized into *(i)* human performance observations and *(ii)* identification of human performance issues from safety databases.

Examples of human performance observations include observational safety surveys based on Normal Operations Safety Survey (NOSS) and Day 2 Day Safety Surveys (D2D) methodologies that focus on how controllers perform daily and require a large amount of data (EUROCONTROL, 2011). NOSS is a methodology for the collection of safety data during normal air traffic service operations, which exclude accidents and incidents (EUROCONTROL, 2011). It is based on the Threat and Error Management (TEM) framework, which assesses how controllers, as part of their normal operational duties , employ countermeasures to keep threats, errors and undesired states from affecting the safety margins in ATC daily operations (ICAO, 2005). Threats are defined as events or errors that occur beyond the influence of controllers due to various contextual complexities. Errors include actions or inactions by controllers that can lead to deviations from intentions or expectations. And undesired states are unintended traffic situations that results in a reduction in safety margins. Therefore, the focus of TEM is on "positive" safety, i.e. what controllers do right.

D2D are a variant of observational surveys during which trained operational staff (other controllers) observe objectively their peers in their normal working situations to record how apparent techniques and tasks are employed. The observations focus on techniques and positive behaviors that benefit safety (EUROCONTROL, 2011). D2D have been found useful in revealing situations which controllers do not seem to recognize as being risky, such as distraction, time pressure, on-the-job-training and handover (Isaac, 2009). Similarly to NOSS, D2D focuses on what controllers do right and to publicize these good practices.

Besides observational surveys to monitor the positive human performance of controllers, the investigation and analysis of safety-related occurrences such as accidents and incidents generally tend to focus on the failures of the system or non-optimal performance, including human performance issues, that led to such outcome (ICAO, 2009). Since accidents are rare, the tradition in safety has been to examine precursors to collision accidents in particular, in the form of specific incidents related to the loss of required separation, i.e. LOS. Numerous studies on the analysis of such incidents have been published and focused on the identification of human factors explaining the causes of incidents. A review of the main studies have classified the findings into *(i)* the factors relating to the various professional tasks ATCOs are meant to perform to fulfill the ATC functions, i.e. when ATCOs are in charge and have control through their actions and decisions, and *(ii)* factors relating to the interaction of the ATCO with the rest of the ATM system or external to it (e.g. organizational factors, airspace and traffic context, and environmental conditions), i.e. when ATCOs have no direct control on them (Dupuy, 2012). For example, communication issues between controllers and pilot or between controllers represent a redundant issue.

Thus, LOS incidents are currently not seen as a way of assessing the controller's performance in a "positive" manner. Their analysis mainly highlights the weaknesses of the system and/or its human element. However, it can be considered that incidents do happen and the controller and its interface with the wider ATM system, performs successfully at resolving these most of the time. This thinking goes in line with the concept of resilience engineering that undesired events, such as incidents, may arise anytime as a result of the normal variability of the system performance (EUROCONTROL, 2009). It is possible to analyze incidents to demonstrate the resilience of the system and its human operator, in absorbing all normal and abnormal events or perturbations. For example, in a study based on geometrical attributes of traffic, ATCOs have been found to be efficient at both resolving LOS incidents and mitigating potential conflicts before they lead to a LOS (Surabattula et al., 2010). There are also various ways to categorize LOS based on the level of intervention of the system required to manage them. In the European severity and risk classification scheme, the use of the controllability criterion is a reasonable approach to assess the degree of intervention of ATC in dealing with any particular incident considered (EUROCONTROL, 2005). The level of controllability over the incident, i.e. how much the system contributed to resolve the incident, moderates the severity classification of the incident. Consider the example of an incident where the controller has an adequate awareness of the position and trajectory of the traffic involved in a LOS. Such an incident should be assessed as less severe than an incident with identical geometrical configuration, but where the controller did not notice the potential conflict. Likewise observational surveys, it is possible to assess positively the safety performance of ATC and ATCO's performance.

When the controller's performance is assessed in particular, his/her attributes can be classified into three categories of cognitive performance: *(i)* situation assessment, *(ii)* planning/decision making, and *(iii)* plan/decision implementation (Hadley et al., 1999). This decomposition of performance is essential and is currently taken into account in the model of ATC presented below and necessary to assess the effectiveness of ATCOs in managing LOS.

3   METHODOLOGY & DATA COLLECTION

To assess the performance of the ATC system in safely managing LOS, there is a need for two generic models: one that assesses the generic development of an incident and the other that characterizes ATC into its functions. These are described below in turn. These are followed by the presentation of the LOS analysis framework and database derived from the analysis of LOS data.

3.1   Incident Development Model

The basic assumption of the LOS analysis framework is that all emerging conflicts have a number of steps in common. This concept of common steps or causes is based on Heinrich's hypothesis that common causes are at work between accidents and incidents (Heinrich, 1931). A review of a sample of official LOS investigation reports from a number of civil aviation organizations, including regulators and air navigation service providers (ANSPs), has demonstrated that it is possible to identify the key common steps in the emergence of a conflict. This identification criterion relies on the distinction between a typical, innocuous conflict (i.e. any situation between two aircraft that has to be dealt with as part of ATC daily operations) and an imminent incident, where the conflict actually becomes a LOS.

This distinction is supported by the activity of ATC from the moment that an air traffic controller deals with two aircraft flight plans and has to provide efficient routings that ensure sufficient separation, until the process fails and the controller needs to resolve the conflict by any means. This is the case even if it implies the provision of inefficient routes for the aircraft. When this point is reached, safety prevails over efficiency and the trajectories of the aircraft may be readjusted later, only after the incident is fully resolved. Therefore, there is a change in the type of instructions ATCO(s) should provide, as shown in Figure 1. The step at which the emerging conflict becomes a LOS is when there is a real safety separation problem for the controller.

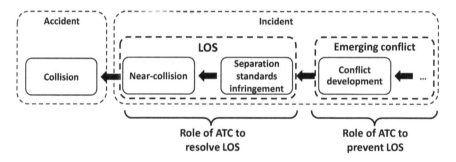

Figure 1 LOS profile and ATC role

This incident profiling is closely related to a geometrical problem, where conflicting paths can be characterized in four dimensions and the distances between the conflicting parties can be measured. But all LOS are not the same and separation standards vary according to the phase of flight or airspace regulation. The consequence of this is that LOS tend to be analyzed individually, thereby making common factors more difficult to identify and prevent the development of effective mitigation measures. Identifying the incident profile as proposed enables a consistent and standard mapping of the two common incident steps, independently from separation standards and regulation.

3.2   ATC Functional Model

ATC is usually described through aerodrome, approach and area services that relate to the appropriate phases of flight. However, to analyze LOS, ATC can be considered from a processes perspective, as presented in the SESAR ATM Process Model (EPISODE 3, 2007). Nine intrinsic functions of ATC, independent of the phase of flight and specific ATC operational unit or centre, can be identified. The dissection of ATM processes described in the SESAR model provides a basis for the identification of these specific functions, as these produce the best summary of the attributes that enables ATC to perform its stated duties. Four primary and five secondary functions have been described to highlight the major principles that underpin ATC operations. The four primary functions describe the essential tasks that any controller is trained to perform: ATCOs sequence the traffic on certain routes (**traffic sequencing**, TSeq), tactically implement routing or sectors adjustments for unplanned changes (**traffic adjustment**, TA), ensure that separation standards are maintained at all time (**traffic separation**, TS) and if this fails, recover any LOS when it occurs (**collision avoidance**, CA). The distinction between the latter two functions corresponds to the incident profile explained above: the former is to ensure efficient separation and the latter is for essential separation.

The five secondary functions (or support functions) describe procedural duties or supporting technologies necessary to ensure the four primary functions are performed adequately. The **communication** function considers all methods of sharing data, including flight plan info notification and coordination between ATC centers, and not just pilot-controller communication (whether analogue or digital). The **navigation** function considers the means from which an ATCO proposes routings to the aircraft including trajectory adjustment and instruction (e.g. speed control or heading instructions) and how the ATC/ATM navigational system (both ground and airborne equipment and facilities) supports the controller's job. The **surveillance** function encompasses all aspects that enable a controller and more generally the ATC system to have the most precise and accurate situation awareness (traffic presence, position and trajectory, conflict awareness and applicable rules awareness). The **safety nets** function is related to any aspect that emphasizes the detection of the two main steps of an incident, i.e. emerging conflict and imminent LOS, while the **workload monitoring** function is characterized mainly by means of detecting any excessive workload that is useful for example, for implementing staffing readjustment.

Such a concise and standard characterization of ATC operations into nine generic functions presents the advantage of providing a comprehensive set of ATC activities whose performance can be assessed individually and in relation to others. When used as part of a checklist during the incident investigation process, each primary function can be assessed. Similarly, the secondary functions can be assessed simultaneously to provide an explanation of any failure of the primary functions. It allows for the consistent identification of interactions (or their absence).

To take into account the cognitive performance of controllers, each ATC primary function is characterised by up to three task attributes: *(i)* detect, *(ii)* plan and *(iii)* implement the plan (i.e. sequencing plan, adjustment plan, conflict resolution plan or recovery plan) as shown in Table 1. Only the CA function is characterised by tasks *(i)* and *(iii)* as there is much less time to elaborate different plans compared to the TS function, and once the LOS is detected, the recovery process (planning and execution of the plan) is immediate. It is important to note that the fulfilment of each task can be affected by both the ground-based ATC and the airborne system. For example, the ground-based ATC effect includes cases where a traffic sequencing situation may not have been detected because the ATCO did not read the flight strip adequately, or the execution of a plan may have failed because the communication equipment did not function properly and prevented the timely delivery of the instruction to the pilot. Examples of the airborne-side effect include cases where a traffic separation went undetected because an unknown aircraft infringed into a controlled airspace without prior authorization, or cases where a plan is not executed properly because an aircraft did not comply with the ATC instruction.

Table 1 Primary ATC functions attributes

Functions	Duties and attributes
Traffic sequencing (TSeq)[a]	(i) Detection of a sequence need (ii) Preparation/Agreement with the sequence plan (iii) Execution of the sequence plan
Traffic adjustment (TA)[a]	(i) Identification of the adjustment need (ii) Preparation/Approval of the adjustment (iii) Execution of the adjustment
Traffic separation (TS)	(i) Detection of emerging conflict (ii) Preparation of a conflict resolution plan (iii) Execution of the conflict resolution plan
Collision avoidance (CA)	(i) Detection of LOS (iii) Execution of the LOS recovery plan

a. Can be simultaneously achieved with TS

3.3   LOS Analysis Framework

The investigation process for a LOS follows a logical and consistent path that identifies the sequential causes in the incident profile, as shown in Figure 2. The progressive change from an emerging conflict to a LOS and thence to a potential collision corresponds to the LOS profile. Information on the initial conditions is collected to recreate the emerging conflict situation and, based on this, a comprehensive description of the initial traffic flow configuration is provided for the emerging conflict considered in the incident. This description includes the: traffic conditions (e.g. geometrical conflict configuration, aircraft performance); airspace and sector conditions (e.g. sector characteristics); and external and environmental conditions (e.g. weather).

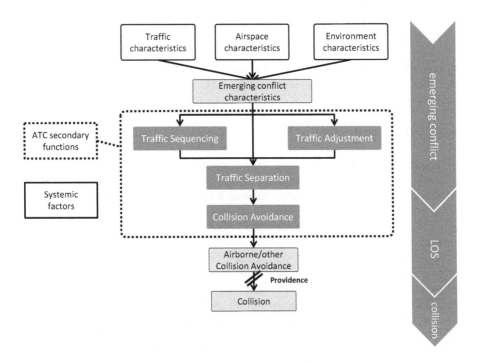

Figure 2 LOS model, path from an incident to an accident

From this initial configuration, an ATCO has to ensure that the traffic is well separated from other traffic or obstacles (i.e. as part of the TS function), and emerging conflict characteristics can still be modified if the TSeq and/or TA functions are activated. For example, a controller can identify well in advance that two aircraft should be sequenced to transit to another sector in a certain order, and provides at the same time instructions that ensure the aircraft are well separated and the flow is efficient. In this case, the TSeq and TS functions work in parallel. There are situations where the TSeq and TA functions are not activated, e.g. routine flight plans put two aircraft in a conflicting situation, so only the TS function is concerned. If the TS function fails (e.g. the emerging conflict was not detected, or it was detected but not resolved), the ATCO has to correct the situation and provide resolution manoeuvres to re-establish adequate separation distances as part of the CA function. Finally, if the CA function fails, only airborne collision avoidance or providence (i.e. the geometrical config-uration of the conflict) can impact the course of the LOS to prevent it from becoming a collision.

Using this LOS model, it is possible to assess whether the performance of ATCO(s) was adequate in such situation, and if not, whether anything hindered its adequacy. The performance of each supporting function is evaluated individually to identify any failure to support the ATCO in performing each primary function in a nominal way. This allows the impact of specific causes, such as a communication problem during the TSeq function compared to the same problem occurring during the CA function, to be assessed.

Using this LOS analysis framework, incident are analysed in an integrated manner where all aspects from the initial conditions to the progressive deterioration into a LOS and finally its management by ATC through the functions are fully considered.

3.4   Data collection

To analyse incidents following this framework, sufficient data need to be available in full investigation reports. A sample of 191 LOS incident reports have been analysed using the analytical framework presented above and this led to a consistent database. These investigations reports were obtained from various sources including publicly available reports obtained from specialised accident and incident investigation bureaux, e.g. UK Airprox Board (UKAB), Swiss Aircraft Accident Investigation Bureau (AAIB), and from regulatory authorities or ANSPs from several countries. The sample covers the following European countries: Ireland, UK, Sweden, Norway and Switzerland; for the Americas, the USA; and for the Asia-Pacific region: New Zealand and Malaysia. Accessibility to such reports is often restricted given the sensitivity of the information, in particular the human operators' details. Therefore, the non-publicly published reports have been provided under a confidentiality agreement with the relevant organizations solely for this research. The diversity of regions of the world covered by the sample ensures the diversity of incident scenarios and to test the transferability and adequacy of the framework with various incident investigation report templates. It is also important to ensure that the source of information comes from countries with reliable safety data. ICAO Member States, including those cited above, are audited under ICAO's Universal Safety Oversight Audit Program (USOAP), based on their Lack of Effective Implementation (LEI) indicator used to measure audit results against traffic volume (measured in number of departures) in 2009. The most favorable USOAP results are those having the lowest LEI percentage, under 30% (ICAO, 2011), and all countries selected for this research fall into this category with their LEI below 20%.

4   INITIAL RESULTS

Figure 3 shows the percentages of failed tasks for each ATC primary function. In all cases, the *inadequate execution of plan* is the most frequent failure before *inadequate planning* and *inadequate detection*. The overall outcome for each function provides a general assessment of the performance of that function. The TSeq and TA functions have been performed inadequately in only 9% and 3% of cases respectively. While the TS function performance is rated as inadequate in 98% of cases. This is typical for a database that contains LOS where separation minima have been infringed. In 2% of the cases, the results of the investigation have confirmed that no infringement of separation minima actually occurred but the LOS was still reported due to safety level that seemed affected. For all cases where the TS failed, 70% have been adequately recovered through the CA function. This indicates the good performance of ATC in recovering LOS. However, 19% were not fully recovered by ATC alone and necessitated the support of airborne recovery. The rest have not been recovered at all, although an accident did not occur due to the conflict geometrical configuration and aircraft speeds (e.g. the aircraft were diverging).

This distinction between LOS where there is no awareness or attempt of resolution by ATCO(s) and those where the ATCO(s) is (are) actively involved in the resolution and recovery process allows for the: *(i)* assessment of ATC performance and, *(ii)* identification of the weaknesses in the system that allow some incidents to be poorly managed. The results above show that ATCOs in interaction with the whole system, perform well at managing LOS, thus providing an indicator of ATC safety performance based on the rate of incidents well managed out of those poorly or not managed. More focus needs to be given to the 9% incidents that failed to be managed, to identify any significant differences with those well managed. A methodology to achieve this is being developed and proposed for future research.

Besides the assessment of each primary function, the secondary functions are assessed in parallel to describe any issue that explains the failure of certain primary functions' tasks or their more systemic impact on the incident. Communication issues have been cited in 55% of all LOS and as summarised in Figure 3, for 16%, 40% and 19% of LOS, these occurred during their TSeq, TS and CA functions respectively. These findings support the findings of previous research reported in (Dupuy, 2012), which indicates a non-negligible recurrence of communication problems (e.g. Rodgers et al, 1998; Cardosi and Yost, 2001; Majumdar and Ochieng, 2003). The database allows more detailed analysis and identification of the type of communication issue (e.g. whether it is an issue with the format of the information shared or the means) and the precise task it has impacted. Issues related to surveillance have also been frequently cited and were identified mostly alongside the inadequate detection task of the ATC primary functions. For example, with respect to the TS function, the following issues have been identified: lack of awareness of the traffic (trajectory, position) or the conflict, lack of monitoring of separation distance between identified traffic; and lack of awareness of certain external conditions. The reporting of these results in depth and the other functions assessment is beyond the scope of this paper and for details see (Dupuy, 2012).

For each incident there may have been more than one issue with secondary functions that cannot be shown on Figure 3. However, this can be filtered from the database, and thus can provide a clear and precise mapping of the effect of these problems during the course of an incident. This can be transcribed into a one page format that can accompany long investigation reports and can be easily shown to the operational staff involved in the incident and its reporting for feedback.

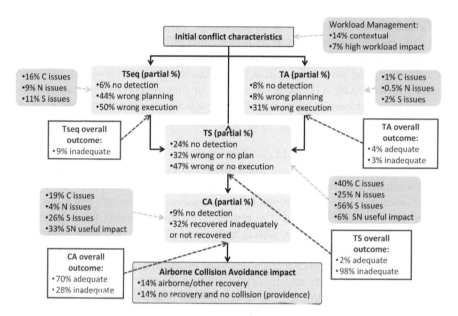

Figure 3 ATC primary and secondary function analysis (% of LOS unless stated)
(C:communication, N:navigation, S:surveillance, SN: safety nets)

5   CONCLUSION

A novel incident analysis framework has been developed to allow the assessment of ATC performance, including ATCO(s), in successfully managing LOS. The framework considers the sequential profiling of incidents and the characterisation of ATC into nine functions to provide a mean to model the various LOS scenarios. Each ATC function can be analysed in a systematic way to identify their effect on the incident development, including positive and negative effects. Such framework provides an efficient summary of the critical elements involved in the incident unfolding and can support the structured identification of causal and contributory factors.

The analysis of 191 LOS based on this methodology enhances safety analysis in three major ways while taking into account the ATCO's cognitive performance: *(i)* the detailed analysis of an individual incident, *(ii)* the aggregate analysis of LOS databases, and *(iii)* the assessment of ATC safety performance.

The framework has the potential to provide indicators of the safety performance of the ATM system and can lead to recommendations targeted at the technology, procedures or users, including controllers and pilots, in relation to the fulfilment of each ATC function. It improves the understanding of incidents where inefficient management occurred in order to target those shortcomings. With this methodology, incidents and their management are considered as part of the routine tasks of the system's performance, rather than simply as an evidence of negative performance.

REFERENCES

Cardosi, K. and Yost, A., 2001. Controller and pilot error in airport operations: A review of previous research and analysis of safety data. Technical Report DOT/FAA/AR-00/51, FAA Civil Aerospace Medical Institute.

Dupuy, M.D., 2012. Framework for the analysis of separation-related incidents in aviation. Unpublished doctoral dissertation, Imperial College London, UK.

EUROCONTROL, 2005. Harmonisation of Safety Occurrence Severity and Risk Assessment,

EUROCONTROL, 2009. A White Paper on Resilience Engineering for ATM,

EUROCONTROL, 2010. Human Performance in Air Traffic Management Safety, A White Paper, EUROCONTROL/FAA Action Plan 15 Safety, September 2010

EUROCONTROL, 2011. Ensuring Safe Performance in ATC Operations: Observational Safety Survey Approaches - A white paper, EUROCONTROL/FAA Action Plan 15 Safety.

EPISODE 3, 2007. ATM Process Model - Guide,

Hadley, Gerald A., Jerry A. Guttman, et Paul G. Stringer. 1999. Air Traffic Control Specialist Performance Measurement Database.

Heinrich, H.W., 1931. Industrial accident prevention a scientific approach. McGraw-Hill, New York.

ICAO, 2005. Threat and Error Management (TEM) in Air Traffic Control,

ICAO, 2009. Doc 9859 - Safety Management Manual,

ICAO, 2011. Flight Safety Information Exchange, From http://www.icao.int/fsix/.

Isaac, Anne, Victoria Brooks, Nicola Jordan, et Magnus McCabe, 2009. « Preventing the drift into failure: How do we know when we get it right ». HINDsight Edition 8 Winter '09

Rodgers, M. D., Mogford, R. H., and Mogford, L. S., 1998. The relationship of sector characteristics to operational errors. Technical report, FAA Civil Aerospace Medical Institute.

Surabattula, D., Kaplan, M. & Landry, S.J., 2010. Controller Interventions to Mitigate Potential Air Traffic Conflicts. In 2010 Human factors and Ergonomics Society (HFES) conference.

Section IV

Automation and Function Allocation

Towards a New Methodology for Designing Dynamic Distributed Cooperative Human-Machine Systems

Fabio Tango[1], Rainer Heers[2], Martin Baumann[3],
Catherine Ronflé-Nadaud[4], Jan-Patrick Osterloh[5], Andreas Lüdtke[5]

[1] Centro Ricerche di Fiat, Strada Torino 50, 10043 Orbassano, Italy
[2] Visteon Innovation & Technology GmbH, 50170 Kerpen, Germany
[2] Deutsches Zentrum für Luft-und Raumfahrt, Lilienthalplatz 7, 38108 Braunschweig, Germany
[4] Ecole Nationale de l'Aviation Civile 7, avenue Edouard Belin, 31055 Toulouse Cedex 4, France
[5] OFFIS Institute for Information Technology, Escherweg 2, 26127 Oldenburg, Germany,
fabio.tango@crf.it, rheers@visteon.com, martin.baumann@dlr.de, catherine.ronfle-nadaud@enac.fr, {osterloh, luedtke}@offis.de

ABSTRACT

New challenges for Human Machine System arise from the ever increasing complexity in each embedded system component as well as from the distributed nature of future systems. Therefore, a new methodology for development and evaluation of Dynamic and Distributed Human Machine Systems (DCoS) is required. This methodology will need to be supported as well with efficient Methods, Techniques and Tools for system development. The proposed methodology will be used in the project D3CoS (www.d3cos.eu) for the

250

development and evaluation of DCoS in four different domains. Described here is a an initial version of this methodology and some evaluation results.

Keywords: human machine system, embedded system, system development methodology

1   INTRODUCTION

The objective of the European project D3CoS (Designing Dynamic Distributed Cooperative Human-Machine Systems) is to develop new and affordable methods, techniques and tools (MTTs) that will support different steps in the industrial development process of Dynamic Distributed Cooperative Human-Machine Systems (DCoS) in the Transportation domain. Each DCoS consists of human and machine agents that share the performance of common tasks by using and sharing common resources. An example of a DCoS is a swarm of Unmanned Aerial Vehicles (UAV), which represent (machine) agents, performing a search and rescue task. In this case the resources can be the capabilities of the single UAV, the portion of flight space to be shared with all the agents involved, etc. These UAVs may be supervised and controlled by several human ground controllers (human agents). Another example from the automotive domain is a distributed cockpit electronics systems comprising several embedded system machine agents and taking the specific driver situation in a traffic environment into account. The results of this project shall allow for an improved development process for DCoS and in turn reduce costs and time to market. In the following, we will describe the theoretical background, which defines the common ground for our DCoS method. Starting from this, an initial version of the DCoS method has been defined, and is currently under evaluation.

2   THEORETICAL BACKGROUND

As an underlying theoretical background we defined a generic framework for cooperative human-machine systems to abstract the commonalities and peculiarities of any given cooperative transportation system. The framework defines important concepts such as cooperation modes and control within a multi-agent system. Within the D3CoS project, a DCoS (as depicted in Figure 1) is defined as "*a dynamic cooperative distributed human-machine or machine-machine system that can change its configuration (tasks, agents, resources and the links (allocations) between them) over time*" (Tango et al, 2011, p. 28).

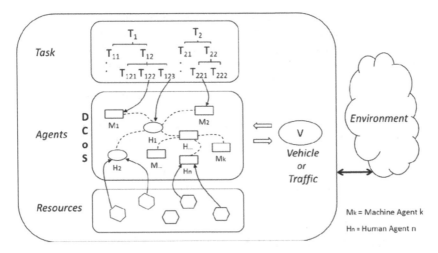

Figure 1: Overview on needed Parts of a DCoS

The General DCoS Framework is based on theoretical work around Multi-Agent-Systems (MAS) in general, the role of human agents in the MAS, and challenges of agent-to-agent cooperation in MAS. Hence, the following working definition for a Multi-Agent-System is used in the D3CoS project: *"MAS are those systems including multiple autonomous entities with either diverging information, or diverging interests, or even both."* (Tango et al 2011, p. 10). Following this definition collaboration and/or cooperation between agents are critical concepts to derive basic principles for interactions between agents. According to Biggers & Ioerger (2001), agents within a system can interact in: (a) sharing or filtering information, (b) computing and combining partial results, (c) negotiating contracts, and (d) in bidding on items. These interaction and cooperation principles are required to understand the complexity of agent-to-agent interaction and cooperation in general terms.

In addition to pure, machine-only MAS, in D3CoS especially MAS consisting of one or more machine agents and one or more human agents, and such MAS are becoming the focus of embedded system development. In order to derive a DCoS Methodology, the specific capabilities and limitations of human operator have to be taken into account. From a human operator's perspective, "the critical thing about sharing tasks [or performing tasks together with other agents, machine or human] is to keep everyone informed about the complete state of things" (Norman, 1993, p. 142). In terms of Hoc's (2001) framework of human-machine cooperation, this means to establish and keep a common frame of reference between all agents, including both machines and humans. Intelligent machine agents need to have a model of the human agent current state to provide appropriate support and the human agents need to have a mental model about the system's current state to understand the machine agents actions, as well as to select their own next actions appropriately. Especially the design of interfaces for human agents in such complex technical environments is a great challenge as the designer has to find the solution for providing the relevant information at the appropriate time in an efficient and easy to use way. Following the Adaptive Automation approach based on a dynamic function allocation principle (Endsley and Kaber, 2008; Hoc, 2010), as well as the "cooperation with automation" principle, leads towards a strong need to analyze and manage cooperation and competition between agents.

This General DCoS Framework was used to derive implications for a DCoS Methodology, which shall:

- address specific needs for Human-Machine cooperation and system development based on the definition of the three entities (agents, tasks and resources) defined above;
- follow an iterative development process including appropriate iterative cooperation, system and performance evaluations;

The next section will detail such a methodology.

3   DCOS METHODOLOGY

Based on this generic framework, an initial version of a DCoS development methodology (integrating some MTTs) has been developed. Currently this methodology is still under preparation and it is used in the first project cycle during the development of demonstrators in four different domains (manned aircraft, unmanned aircraft, maritime, and automotive). Within the D3CoS methodology, three general system design phases are outlined.

First, during a DCoS composition phase, the requirements are defined for human and machine agents, as well as for work division between human and machine agents. Thereafter, the devices, the machine agents themselves, the network and wiring details are specified. Then, all hardware devices, embedded software and mechanical components, as well as network and wiring components are developed. Finally, hardware and software components, DCoS network and agent-agent cooperation is assessed.

In a second step, the DCoS interaction at first includes the requirements definition for interactions between machine agents and human agents. Then, interaction modes between human and machine agents within the DCoS network are specified. Also, all interactions between machine agents, human agents and between both are developed. And finally, system performance, effectiveness of agent-to-agent interaction and usability for human-machine agent interaction is measured.

In the third step, the DCoS interface, the requirements are defined for interfaces and protocols between machine agents and human agents. Thereafter, physical, software and communication interfaces for machine agents and user interfaces for human agents are specified. Then, all physical, software and user interfaces for each device are developed. Finally, interface performance for machine agents and usability performance measures for human-machine agents interfaces are tested.

In addition to the three methodological steps outlined above, a parallel DCoS evaluation process is added to assess system and human operator performance, system robustness and efficiency, as well as network and communication efficiency. Each design phase is again divided into design steps. Table 1 provides an overview on the steps that have to be performed in the D3CoS methodology. See also Tango et al (2011) for further details.

DCoS Composition		8a. Evaluation & Test for DCoS Composition
1. DCoS Composition	- study the tasks of the cooperative system - study the agents and resources needed to perform the tasks - define the composition of the DCoS: select agents to compose the cooperative system - make DCoS network resources available	- establish and define performance measures for DCoS system, human operator and system robustness - refine measures for scenario/use case-specific evaluation - define measures for TARF efficiency evaluation
2. Scenario Identification	- identify different types of situations/scenarios/use cases in which the cooperative system will perform the tasks, with agents and resources	
3. TARF Definition	- for each situation/scenario/use case, define the appropriate Task Allocation and Resource Function (TARF) - likely, a superposition of TARFs for a DCoS is required - for some situations/scenarios, simple algorithms and cooperation mechanisms and resources may be sufficient / for more complex situations more and more sophisticated algorithms are required	
DCoS Interaction		8b. Evaluation & Test for DCoS Interaction
4. Human-Machine Interaction Design	- define the interaction between the agents - define, which human agent interacts with any machine agent/s - define protocol and communication channels	- define performance measures for DCoS interaction - include human and machine agents - include protocol and communication efficiency
5. Machine-Machine Interaction Design	- define the interaction between the machine agents - define, which machine agent interacts with any other machine agent/s - define protocol and communication channels	
DCoS Interface		8c. Evaluation & Test for DCoS Interface
6. Human-Machine Interface Design	- implement the devices and physical interfaces that support the cooperative interaction between each human and 1-n machine agents, for each TARF	- define performance measures for DCoS interfaces - include human and machine agents - include communication and network efficiency
7. Machine-Machine Interface Design	- implement the devices and physical interfaces that support the cooperative interaction between machine agents, for each TARF	

Table 1: Cross-Domain Design and Evaluation Steps in a DCoS Methodology

Overall, the following implications and conclusions can be derived from the General DCoS Framework for a DCoS Methodology. There seems to be a very strong need and very high importance of establishing effective and efficient cooperation and interaction between all kinds of human and machine agents in dynamic situations by applying effective means of communication. In addition, to allow a strong coordination of activities, the overall MAS may benefit from efficient human behavior detection and analysis to allow the technical partial-MAS to act accordingly, taking human cognitive capabilities and limitations as well as social interaction habits and preferences into account. This means that, already in the first stage of the DCoS Methodology (the DCoS composition level), the requirements for cooperation, interaction and communication have to be taken into account, including human agent's needs and, in a similar way, the capabilities and the technical needs for machine agents and for DCoS network.

In addition to the DCoS point of view on the methodology, one should not forget that there are already existing system development processes. Thus, we also applied a system development perspective, which comprises four phases.

In the first phase, DCoS Requirements Capturing, the key DCoS requirements for the definition of human and machine agents, as well as the work division between human and machine agents is prepared. In addition, requirements for interactions between each agent involved – covering all machine agents and all human agents – are set. Finally, requirements for interfaces and protocols between machine agents and human agents are defined.

In the second phase, System Specification, all devices and machine agents, as well as network and wiring details, are specified. Thereafter, all interaction modes between human and machine agents within the DCoS network are defined. The specification of physical, software and communication interfaces for machine agents and of user interfaces for human agents provide a full system specification.

In phase three, System Development, all hardware devices, embedded software and mechanical components, as well as the connecting networks and wiring components are developed and prepared. Also, all interactions between machine agents, human agents and both are developed.

Finally, in the fourth phase, Evaluation & Test, the DCoS network, the hardware and software components, are tested to assess system performance, the effectiveness of agent-to-agent interaction and machine agents interface performance. In addition, usability performance is measured for human-machine agent interaction.

Table 2 shows, how the tasks in the DCoS Methodology match to the tasks of a System Development process.

DCoS Methodology System Development Process	DCoS composition	DCoS interaction	DCoS interface
Requirements Capturing	• requirements definition for human and machine agents • requirements definition	• requirements definition for interactions between machine agents and	• requirements definition for interfaces and protocols between machine agents and

		human agents
	for work division between human and machine agents	• requirements definition for the network as established by machine agents	
System Specification	• specification of devices and machine agents • specification of network and wiring details	• specification of interaction modes between human and machine agents within the DCoS machine and human agent network	• specification of physical, software and communication interfaces and network for machine agents • specification of user interfaces for human agents
System Development	• development of all hardware devices, embedded software and mechanical components • development of network and wiring components	• development of all interactions between machine agents, for human agents and between both	• development of all physical, software and user interfaces for each device
Evaluation & Test	• testing of hardware and software components • testing of the DCoS network itself	• testing of system performance • testing the effectiveness of agent-to-agent interaction • testing for usability performance measures for human-machine agents interaction	• testing of machine agents interface performance • testing of usability performance measures for human-machine agents interfaces

Table 2: Match between tasks of the DCoS Methodology and a System Development Process

4 METHODS, TECHNIQUES AND TOOLS

In order to achieve the overall project objective, the improvement of a DCoS development, the DCoS Methodology should be supported by different Methods, Techniques and Tools, which are developed within D3CoS. The general idea of the project is that partners building DCoS demonstrators use and assess the benefit of MTTs dedicated to DCoS development. Overall, demonstrators are intended to be build in four industrial domains: Manned Aircraft, Unmanned Aircraft, Automotive and Maritime. This development and the evaluation of the demonstrators will be performed in three development cycles. Each cycle should then improve the MTTs as well as the demonstrators. Therefore, at the beginning of the D3CoS project, requirements for the demonstrators in the four domains were collected, while the MTTs have been matched to these requirements. In Cycle 1, the methodology and some MTTs have been applied for the first time in the different domains.

Thanks to the theoretical approach, it was possible to propose a guideline to the industrial partners in order to help them in the design of Dynamic Distributed Cooperative Human-Machine Systems. We proposed a eight step strategy as described above. In correspondence with these steps, MTT development partners supplied the demonstrator building partners with the available tools. Some of them were under development duing Cycle 1, but, for Cycle 2, tools for each step of the Methodology will be provided. Eight clusters have been identified, summarizing in Figure 3, which define the "MTTs Landscape":

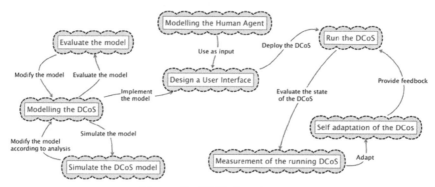

Figure 2 : MTTs Landscape

In general, the DCoS composition includes Agents, Tasks, Resources and Links, as depicted in Figure 1. In order to facilitate this process, an XML Schema has been developed based on very simple scenarios, given by industrial partners for the four domains. The aim of this schema is to describe which agents need to interact with which agents, with which protocols and thanks to which communication modalities. Different cooperation modes as described above were considered. For Cycle 1, the schema was written by hand, but for the next cycles a tool, which allows graphical specification of a DCoS with export to XML code according to the schema, shall be developed. In addition, the schema shall be used to feed simulation platforms in order to perform a "high level" simulation. A preliminary version of a platform for the DCoS modeling and simulation has been developed and tested in the first cycle. In addition, a tool named PED, for *Procedure EDi*tor, for the specifications of procedures or for hierarchical task trees, has been provided and used in the Manned Aircraft domain, in order to specify the standard operation procedures, which are needed in the demonstrator scenarios for the first cycle. In the automotive domain, a specific software development framework dedicated to distributed cockpit electronics systems was applied and refined for an on-board DCoS demonstrator.

In parallel, a general methodology for the development of so-called Design Patterns for DCoS has been proposed. Design patterns shall be used in a similar way as known from the software architecture design. Thus, they shall provide either abstract or concrete solutions to frequently occurring problems. In the field of DCoS development such design patterns will, for example, provide a solution for the assignment of tasks among the available agents based on the requirements of DCoS, the task and agent characteristics, as well as the available resources. A data base of such design patterns will be developed during the project. The XML Schema described above will be used for the description of the patterns.

5   EVALUATION

Evaluation of the MTTs is currently ongoing. Here, some preliminary results for two MTTs are presented: PED and the simulation platform (both tools have been developed by the partner OFFIS).

PED allows specifying procedures, i.e. the tasks of an agent, in a graphical way. PED has been used in the manned aircraft domain by a domain expert, to model the tasks a Pilot has to perform during landing. After a short initial learning phase, the pilot was able to easily model the needed tasks. Some minor issues problems with the editor have been solved by providing patches. A further evaluation of PED and the derivation of new requirements is planned.

The development of the simulation platform has started at the beginning of the project. The simulation platform is based on the open source implementation of the IEEE 1516 High Level Architecture standard by ONERA, called CERTI (https://savannah.nongnu.org/projects/certi). In order to facilitate rapid connection of simulators to a simulation, OFFIS developed a simplified interface. For example, with this interface, a partner was able to connect a Delphi GUI to the simulation within three days. In addition, also several Matlab Simulink models have been connected to the simulation. One result of the evaluation is that, setting up and maintaining the different configuration files needed by a HLA simulation, is error prone and time consuming. Therefore, the aforementioned tool for graphic specification of a DCoS should also be able to create the needed configuration files.

6   CONCLUSIONS AND NEXT STEPS

A new methodology for the design of dynamic, distributed and cooperative human-machine systems (DCoS) is under development in the D3CoS project (www.d3cos.eu) to meet the requirements coming from an always increasing complexity in embedded systems. The initial version of this methodology as presented here is based on a combination of well-established industrial system development processes with methodological steps which are foreseen as necessary for future DCoS development. In addition, a general theoretical framework has been defined for DCoS, based on theoretical approaches related mainly to multi-agent system theories. Key elements of this framework include agent, task and resource function definitions. Overall, strong interactions between human agents and machine agents and between the machine-agents themselves as well as dynamic changes in task and resource allocations were taken into account in this framework. For an efficient support of DCoS development specific MTTs are developed and used for domain-specific demonstrators. During the overall course of the project, the presented methodology as well as the above mentioned MTTs and the underlying theoretical assumptions will be investigated in further details and will be refined. The final objective is, to develop a DCoS Methodology that integrates and extends existing theoretical approaches as well as some insight, experience and refinements coming from the application of this methodology in MTT usage and DCoS demonstrator development.

ACKNOWLEDGEMENTS

The authors of this paper would like to thank Cristine Kalina, Frank Rister, and Matthias Vyshnevskyy for evaluating several MTTs, as well as all project partners who supported the in development of the DCoS methodology.

This research has been performed with support from the EU ARTEMIS JU project D3CoS (http://www.d3cos.eu/) SP-8 - GA No.: 269336. Any opinions, recommendations, findings, or conclusions expressed herein are those of the authors and do not necessarily reflect the views of the ARTEMIS JU and/or the EC.

REFERENCES

Biggers, K.E. and Ioerger, T.R. (2001). Automatic Generation of Communication and Teamwork within Multi-Agent Systems. Applied Artificial Intelligence, 15(10), 875-916.

Endsley, M. and Kaber, D. B. (1999), Level of automation effects on performance, situation awareness and workload in a dynamic control task, Ergonomics, 42, 462-492

Hoc, J.-M. (2010). Human and Automation: a matter of cooperation. Published in "HUMAN 07", Timimoun: Algeria (2007)

Martin, J. (2002). Organizational culture: Mapping the terrain. Thousand Oaks, CA: SageMartin

Norman, D. (1993). Things that make us smart. Cambridge, Massachusetts: Perseus Books.

Tango, F., Baumann, M., and Heers, R. (2011), D3-01 Generic DCoS Framework - Preliminary Version. Research Report D3-01, D3CoS Project No. 269336, www.d3cos.eu

CHAPTER 25

DataComm in Flight Deck Surface Trajectory-Based Operations

Deborah L. Bakowski[1], David C. Foyle[2], Becky L. Hooey[1], Glenn R. Meyer[3], Cynthia A. Wolter[1]

[1]San Jose State University at NASA Ames Research Center
[2]NASA Ames Research Center
[3]Dell Services, Federal Government
Moffett Field, CA
debi.bakowski@nasa.gov

ABSTRACT

The purpose of this pilot-in-the-loop aircraft taxi simulation was to evaluate a NextGen concept for surface trajectory-based operations (STBO) in which air traffic control (ATC) issued taxi clearances with a required time of arrival (RTA) by Data Communications (DataComm). Flight deck avionics, driven by an error-nulling algorithm, displayed the speed needed to meet the RTA. To ensure robustness of the algorithm, the ability of 10 two-pilot crews to meet the RTA was tested in nine experimental trials representing a range of realistic conditions including a taxi route change, an RTA change, a departure clearance change, and a crossing traffic hold scenario. In some trials, these DataComm taxi clearances or clearance modifications were accompanied by 'preview' information, in which the airport map display showed a preview of the proposed route changes, including the necessary speed to meet the RTA. Overall, the results of this study show that with the aid of the RTA speed algorithm, pilots were able to meet their RTAs with very little time error in all of the robustness-testing scenarios. Results indicated that when taxi clearance changes were issued by DataComm only, pilots required longer notification distances than with voice communication. However, when the DataComm was accompanied by graphical preview, the notification distance required by pilots was equivalent to that for voice.

Keywords: NextGen, STBO, surface operations, taxi, displays, DataComm

1 SURFACE TRAJECTORY-BASED OPERATIONS

Surface operations are one aspect of the design of the next generation (NextGen; JPDO, 2009) of the National Airspace System. Research efforts in this phase of flight have focused on the development of surface traffic management (STM) systems for air traffic control (ATC) to provide taxi clearances enabling efficient airport operations and improved throughput. One such example of an STM system is the Spot and Runway Departure Advisor (SARDA), which estimates spot-to-runway taxi times based on the current average (Jung, 2010). Future STM systems will have associated aircraft arrival times (i.e., required time of arrival, RTA) at active runway thresholds so that aircraft can cross with minimal or no delay, and at the departure runway, enabling aircraft departure queue sequencing. Such future "full capability" STM systems will require aircraft to reach specified locations on the airport surface with relatively precise timing. For the flight deck, in order to achieve an RTA, pilots should be provided the aircraft speed required to reach the specified location at the specified time, since speed (actually, thrust) is what the pilots control with the throttles. These NextGen taxi operations with "taxi clearances and RTAs" are termed surface trajectory-based operations (STBO).

The use of Data Communications (DataComm) in surface operations has several potential benefits (e.g., reduced radio congestion, elimination of misunderstandings, see Wargo and D'Arcy, 2011). However, one concern is that DataComm may not be appropriate for use in time-critical situations. Specifically, it has been suggested that taxi route changes during taxi and time-critical events may require that ATC revert to voice communication (Jakobi, 2007; Wargo and D'Arcy, 2011).

1.1 Previous STBO Research

Several experiments have assessed speed-based taxi clearances and their effect on pilots' ability to meet an RTA from an initial ramp spot to the departing runway (Foyle, Hooey, Bakowski, Williams, and Kunkle, 2011). Foyle et al. (2011, Expt. 2) explored pilots' ability to meet an RTA using a commanded speed, with no additional flight deck equipage. Pilots were instructed to comply with the commanded speed on straight segments and accelerate/decelerate "aggressively". The primary measure of pilot performance on the taxi task was RTA error, calculated by subtracting the RTA from the observed arrival time. Results revealed that requiring pilots to comply with a commanded speed, without additional flight deck equipage, produced unacceptably large RTA errors.

Foyle et al. (2011, Expt. 3) explored whether requiring pilots to follow specific acceleration/deceleration speed profiles and follow the speed command more precisely would decrease RTA error. By requiring pilots to taxi within +/- 1.5 kts of a commanded speed and with a specific speed profile, RTA error was quite accurate. However, these requirements required excessive visual attention to the head-down speed display. Fourteen out of eighteen pilots reported that the demand of maintaining the required speed conformance range in actual operations would compromise safety.

The results of these two studies demonstrated that: 1) Simply incorporating a commanded speed requirement into the taxi clearance alone was not sufficient for good RTA performance; and, 2) When pilots were given clearances that required them to control their aircraft according to precise acceleration/deceleration speed profiles in addition to a commanded speed, RTA conformance was quite good -- but, that added precision caused pilots to spend an unacceptable amount of time viewing/tracking the speed display. This suggests the need for a flight deck technology solution to aid pilots in safely meeting the speed/time requirements.

The technology solution investigated by Foyle et al. (2011, Expt. 4) was a speed indicator on the primary flight display (PFD) driven by an "error-nulling" algorithm. The speed algorithm dynamically compensated for speed-maintenance deviations by adjusting the current advised speed according to the remaining time and current distance to the RTA location. In addition to improved RTA error, pilots reported that the use of the algorithm/display did not compromise their out-the-window attention.

1.2   Present Study

The goals of the present study were three-fold:

1) Test an improved version of the error-nulling algorithm/display that more precisely calculated and presented the speed required to reach the departure runway at an RTA. Specifically, advised speed was computed based on remaining time, distance, number of turns, turn speed, and specific rate of acceleration/deceleration.

2) Evaluate that algorithm/display for robustness under multiple realistic conditions. The impact of the improved error-nulling algorithm on pilots' ability to meet an RTA to the departing runway was tested under robust traffic and surface operation conditions, including taxi route changes, departure clearance changes, RTA changes, and a 2-min hold scenario.

3) Assess the use and integration of flight deck DataComm in STBO. The present study also explored DataComm integration with flight deck displays. This integration allowed DataComm information to be displayed graphically, which may improve situation awareness and extend the use of DataComm communication in time-critical situations.

2   METHOD

Twenty commercial pilots, with a mean age of 52 years (range of 36 – 63 years), participated in the study. Nine of the pilots were currently Captains and 11 were First Officers. The mean flight hours logged was 9,885 hours (range of 600 – 17,000 hours). Pilots were paired by company affiliation to form 10 two-person crews.

2.1   Flight Simulation

The study was conducted in the Airport and Terminal Area Simulator (ATAS), in the Human-Centered Systems Laboratory at the NASA Ames Research Center. The airport environment was the Dallas-Fort Worth International Airport (DFW) with high visibility and distant fog/haze conditions. The forward, out-the-window scene was depicted on four LCD displays, with a total viewing angle of 140 deg. The modified-B737NG cockpit included a Primary Flight Display (PFD), Navigation Display (ND), and Flight Management System (FMS) Control Display Unit (CDU) on both crew members' sides, and a shared Taxi Navigation Display (TND) and DataComm display with a touchscreen interface. Aircraft controls included a tiller on the Captain's side, toe brakes, throttles, and emergency brake. The physical and handling characteristics of the aircraft were modeled after a B757.

2.2   RTA Speed Algorithm

Each departure taxi clearance included a required time of arrival (RTA) at the queue area of the departure runway. To aid the pilots in arriving on time, the flight deck was equipped with an algorithm that computed the straightaway speed required to precisely meet the RTA. The RTA algorithm dynamically computed the advised speed by accounting for remaining distance, remaining time to RTA, number of turns, an assumed acceleration/deceleration rate of 1 kt/sec, and a turn speed of 10 kts (per standard operating procedures, SOPs). Taxi clearance RTAs were calculated such that the initial advised straightaway speed was 15 kts. The algorithm was dynamic and compensated for the pilot slowing down or speeding up by appropriately increasing or decreasing the advised straightaway speed.

2.3   STBO Experiment Displays

The PFD was modified for taxi operations by expanding (doubling) the speed scale from 0-60 kts. Advised straightaway speed, as calculated by the RTA algorithm, was displayed as a magenta analog pointer ("speed bug") on the speed tape and digitally in magenta directly above the speed tape (see Figure 1, panel 1). As shown in Figure 1 (panel 2), upon entering a turn, the magenta advised speed bug dropped to 10 kts (per taxi SOPs), while the white inner speed bug continued to dynamically indicate the straightaway speed required to meet the RTA. The PFD also included the current speed, shown as a sliding indicator with digital value inside (18 kts in Figure 1, panel 1), RTA time (Zulu) in magenta, and time remaining to the RTA (min:sec) in the white box.

A Taxi Navigation Display (TND) depicted the airport layout to aid the pilots in airport navigation. The ownship aircraft's position, shown as a white chevron, and other aircraft traffic within the ownship's 1,250 ft declutter circle, shown as yellow aircraft icons, were updated in real time (see Figure 1, panel 3). When the initial taxi clearance, sent via the DataComm, was accompanied by preview information, the TND populated with details of route: a graphical route on an overview map, clearance text, RTA, time remaining to RTA (T Rem), route distance, and advised straightaway speed in cyan (see Figure 1, panel 3). Once accepted by the flight deck, clearance information was loaded into the PFD and the TND updated with the taxi clearance displayed as a magenta route, in perspective, track-up, and with text of the accepted clearance. When a mid-route taxi route change or RTA change was accompanied by preview information, the RTA, time remaining, distance remaining, and advised speed of the pending route were displayed in cyan. Turn graphics and route text were also displayed for taxi route changes. RTA location in the departure runway queue area was displayed as a white, dashed line across the ownship route on the TND, visible on the TND as the pilot neared it.

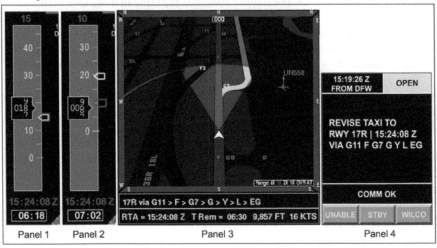

Figure 1 PFD: on straightaway (panel 1), and in turn (panel 2). TND: pending taxi route change with preview information in cyan (panel 3). DataComm: pending taxi route change (panel 4).

The DataComm touchscreen interface was located aft of the throttles between the two pilots. At the start of the trial, the flight deck received a DataComm with an expected taxi and departure clearance, followed by an initial taxi clearance. Taxi route changes (Figure 1, panel 4), RTA changes, and departure clearance changes were delivered via the DataComm during taxi. The DataComm followed a format similar to the European Airport Movement Management by Advanced Surface Movement Guidance and Control System, Part 2 project (EMMA2; Airbus, Thales, DSNA, 2009). When a clearance was delivered via DataComm, three touchscreen response buttons were available to the pilot to respond to ATC: Unable, Standby, and Wilco. The DataComm display included the message sent time in the upper left corner, an indicator of message status in the upper right hand corner (i.e., "OPEN" while ATC was awaiting a response from the flight deck, or "WILCO", "STBY", or "UNABLE" after a response was selected and sent to ATC), and status of the connection (i.e., "COMM OK" or "RECEIVED BY ATC" when ATC received the crew's response). After the crew responded "WILCO" to a clearance, the DataComm text turned magenta, as an indication of acceptance.

2.4 Experimental Design

Each crew experienced nine experimental trials (see Table 1 in Results, Section 3). Taxi routes were constructed with 11,883 ft average distance and 8 min 2 sec average duration. The presentation of trials was randomly ordered and counterbalanced for each of the 10 crews. Each trial started at a 'ramp spot', had a unique route, and ended at the departure queue area.

Four taxi route change trials were included with pilots receiving notification either 500 ft or 2,000 ft before the required turn, and with either Preview or No Preview information on the TND. Route distance was held approximately constant (within 950 ft), and the RTA and runway remained the same in these four trials.

Two RTA change trials consisted of a revised RTA that was changed to approximately one minute earlier. These were given with either No Preview (57 sec earlier) or with Preview information (50 sec earlier) shown on the TND. In order to meet the amended RTA, an 18-kt straightaway speed was required in both trials.

Two departure clearance change trials required the First Officer to reprogram the FMS by selecting a different departure and transition. The change occurred either early (at 25% into the route), or late (at 75% into the route). These clearance changes were sent and responded to via the DataComm.

In the crossing traffic hold scenario, ATC gave instructions via voice to stop for two 777 aircraft -- the RTA and route remained in effect. The crew was delayed for 127 sec before being cleared by voice to continue taxi. Required straightaway speed increased to 21 kts during the 127 sec that passed while being held.

2.5 Procedure

Pilots were told that their primary goal in each trial was to minimize RTA error to the greatest extent possible so as to arrive at the departing queue area at the designated time. They were explicitly instructed to taxi as they would in a B757 aircraft, and never to taxi faster than would be safe in the real world. Pilots were informed that the advised straightaway speed was provided as an aid to help them reach the RTA point on time, but that it was not a requirement to constantly follow it. They were also told that the algorithm was dynamic, and that it assumed a 10-kt turn speed and 1 kt/sec acceleration/deceleration rate. Pilots were instructed to not "track" the advised speed indicator, but rather to use it as a strategic indicator.

Four initial taxi trials were presented to allow each crew to become familiar with the simulator, FMS, and experimental procedures. These trials also provided an opportunity for the pilots to experience the types of scenarios to be tested: a taxi route change, an RTA change, and a departure clearance change. Pilots were not provided with information about the crossing traffic hold scenario prior to its occurrence (or during familiarization trials).

Each trial began with an expected taxi clearance sent to the flight deck via DataComm. The clearance included the expected taxi route, departure runway, RTA, and departure clearance. Conceptually, pilots were told that an expected taxi clearance would be received at the gate, 30 min prior to pushback. In addition to paper airport charts, the TND showed an overview map of DFW airport with the current ownship location. The expect-clearance DataComm allowed pilots to orient themselves to their initial location, taxi route, and departure runway. Crews were instructed to use this time to thoroughly review and discuss the taxi clearance and carefully plan their taxi route. The First Officer was responsible for managing the DataComm and programming the Flight Management System (FMS) for the initial departure clearance and departure clearance change while taxiing.

After completing taxi route planning and FMS entry at the gate, the simulation advanced to a ramp departure spot. Upon receiving the crew's 'ready to taxi' call, the experimenter initiated an ATC-sent DataComm with the initial taxi clearance to the flight deck. In three No-Preview trials (see Table 1), the initial taxi clearance was not accompanied by preview information on the TND. The six remaining trials included preview information on the TND. In particular, the preview of advised speed allowed pilots to quickly understand whether, given the route distance and number of turns, meeting the RTA would be reasonable. The graphical depiction of the taxi route on the overview map facilitated understanding of the clearance. In all trials, pilots were to review and accept or reject the initial taxi clearance with RTA as soon as possible after receiving it, because the time to the RTA was counting down. Time remaining ("T Rem" in cyan on the TND in Figure 1, panel 3), prior to acceptance of the clearance, was only visible to the crew in trials with Preview.

If the crew accepted the clearance, the First Officer responded by selecting "WILCO" on the DataComm, and the Captain released the emergency brake and began taxi. When the clearance was accepted, the RTA, time remaining, and advised speed were loaded into the aircraft avionics and displayed on the PFD (see Figure 1, panel 1), while the magenta route was displayed on the TND (panel 3).

During each of the nine experimental trials, the crew received a taxi route change, an RTA change, a departure clearance change, or a verbal instruction from ATC to stop for crossing traffic. When a taxi route change was accompanied by preview information (see Figure 1, panel 3), pilots were able to use the cyan graphical route depicted on the TND to note their position and locate the required turn. When an RTA change was accompanied by Preview, pilots were able to use the advised speed to determine whether it was reasonable/acceptable to meet the revised RTA.

In the event that the crew decided they could not comply with the initial taxi clearance or any of the three change clearances, pilots were to notify ATC by responding "UNABLE" via DataComm. If pilots decided they were unable to meet the RTA at any time during taxi, they were instructed to contact ATC via voice. When pilots reported they were unable to meet the RTA, by either DataComm or voice, ATC instructed them to "continue taxiing and minimize RTA delay."

Conceptually, pilots were told that they would receive their takeoff clearance after crossing the RTA location in the queue area. For this reason, they were responsible for completing any FMS programming changes prior to reaching the RTA location. Pilots were instructed to continue taxiing after crossing the RTA location; with the trial ending shortly after that point.

Following completion of the trials, pilots estimated the minimum required notification distance using voice, DataComm text only, and DataComm with Preview for: a taxi route change, an RTA change, a runway change, and a departure clearance change. Pilots estimated these minimum notification distances on a post-study airport chart showing a taxi route with a distance scale.

3   RESULTS

The primary measure of pilot performance on the taxi task was RTA error, calculated by subtracting the RTA from the observed arrival time. (Negative RTA errors indicate early arrival; positive RTA errors indicate late arrival.) As shown in Table 1, RTA error is very good (near zero) across all conditions.

Table 1 Mean RTA Error in Seconds (S.E.)

Taxi Route Change (Same RTA)	Speed and Turn Graphics Preview	No Preview
500 ft turn notification	-0.43 (1.11) [1]	4.14 (1.81) [5]
2,000 ft turn notification	-0.99 (1.19)	1.39 (1.02)*
RTA Change (15 kts to 18 kts)	Speed Preview	No Preview
57 sec earlier	--	3.71 (1.01) [1]
50 sec earlier	0.08 (0.29)	--
Departure Clearance Change (Early)	-2.30 (1.09)	
Departure Clearance Change (Late)	-1.57 (1.21)	
127-sec Crossing Traffic Hold (via voice)	4.96 (3.76) [2]	

Notes: Superscript indicates number of crews (of 10) that responded "UNABLE"
* One crew stopped because they misunderstood the clearance, arriving 161.34 sec late. Only this one outlier was removed from the RTA error analysis.

DataComm responses other than "WILCO" are noted as superscripts in Table 1. Note that even though pilots responded "UNABLE" in some cases, they were still able to meet the RTAs with relatively small RTA error (with one exception noted).

Statistical tests did show some differences among means: With a Taxi Route Change (Table 1 top two rows), RTA error was greater with No Preview than with Preview, $F(1,9)=4.89$, $p=.05$; and, for RTA Change, RTA error was greater with No Preview than with Preview, $t(9)=3.44$, $p<.01$. There was no significant effect of location of departure clearance change (early or late in route) on RTA error.

It should be noted, however, that despite these significant findings, the important point is that these are all very small RTA errors, especially when put in the context that the taxi routes were, on average, more than 8 minutes in duration. (All RTA errors were also equivalent to zero, except for the No Preview 500-ft taxi route change, $t(9)=2.29$, $p<.05$, and the No Preview RTA Change, $t(9)=3.67$, $p<.01$.) The most notable "robustness finding" is that despite being held for crossing traffic for more than 2 minutes (specifically, 127-sec), RTA error was under 5 seconds, and was statistically equal to (i.e., not different from) zero, $t(9)=1.32$, $p>.05$.

As in Foyle et al. (2011), to assess visual attention to the error-nulling speed display, pilots (only the last n=12 pilots tested) were asked, using a 5-point scale, where 1=Rarely, 2=Seldom, 3=Sometimes, 4=Frequently, and 5=Most of the Time, "During this trial, how often did you find yourself focusing on the speed and/or time displays when you should have been paying attention to the external taxiway environment?" Results (*M*=2.13, *SD*=0.62) indicated that pilots responded that the error-nulling algorithm/display did not negatively impact their visual attention to the out-the-window airport view, as was found in Foyle et al. (2011, Expt. 3).

3.1   DataComm Notification Distance

On a post-study questionnaire, pilots estimated the minimum distance that they were willing to accept a taxi route change, an RTA that was 1 min earlier, a change to a parallel runway, and a departure clearance change. Pilots provided distances for three communication types: DataComm with Preview information, DataComm text only (No Preview), and voice only. Participants indicated their responses on a taxiway map overlaid with a distance scale. Mean (and S.E.) minimum notification distances (ft) are shown in Table 2.

Table 2 Mean Minimum Notification Distance in Feet (S.E.)

Minimum Notification Distance (ft) Required before Taxi Route Change			
Type of Change	DataComm with Preview	DataComm No Preview	Voice Only
Taxi Route Change	895 (178)	1,378 (119)	928 (149)

Minimum Notification Distance (ft) Required before Runway Takeoff Clearance			
Type of Change	DataComm with Preview	DataComm No Preview	Voice Only
RTA Change (1 min earlier)	4,856 (489)	5,825 (488)	5,675 (536)
Change to Parallel Runway	2,763 (395)	3,025 (386)	2,785 (417)

Type of Change	DataComm (No Preview)	Current Day Voice Environment
Departure Clearance Change	2,600 (307)	2,986 (387)

For a 1-min RTA change, pilots required a shorter notification distance for DataComm with Preview, $F(2,38)=8.14$, $p<01$. For a taxi route change, pilots required a longer notification distance for DataComm No Preview, $F(2,38)=11.22$, $p<.01$. That is, pilots indicated that DataComm with Preview (i.e., preview of advised speed or route change graphic) could be used in place of voice in these scenarios. There was no significant effect of communication type on minimum distance notification to a parallel runway change or a departure clearance change.

4   DISCUSSION

The present study tested a version of an error-nulling algorithm that more precisely calculated advised speed by accounting for remaining time, distance, number of turns, turn speed, and acceleration/deceleration rate. The improved RTA algorithm was tested under robust traffic and surface operation conditions, including taxi route changes, departure clearance changes, RTA changes, and a 2-min traffic hold scenario. RTA error in the present study was near zero across all trials, indicating that pilots were able to meet the RTA with very little time error, and that meeting this RTA did not interfere with visual attention out-the-window.

Typically, it has been noted (e.g., Jakobi, 2007; Wargo and D'Arcy, 2011) that DataComm may not be appropriate for time-critical situations. In this study, pilots reported that for taxi route modifications with a graphical preview of the required turn and for RTA changes with a preview of the advised speed, DataComm requires the same, or less, notification distance as voice communication. This potentially important finding suggested by these results is that DataComm *with graphical preview information* may be usable in more situations than previously considered (presumably, of course, when non-compliance would not result in an unsafe condition). Clearly, more research is needed, but these results suggest that with DataComm information integrated and displayed on a flight deck display, reverting to voice communication may not be required for some time-critical communications as was previously considered.

ACKNOWLEDGMENTS

NASA Airspace Systems/NextGen Concepts and Technology Development/Safe and Efficient Surface Operations funded this work. The authors thank Rob Koteskey and Christina Kunkle of San Jose State University for their help with this study.

REFERENCES

Airbus, Thales Air Systems, DSNA. 2009. "EMMA2 CPDLC Trials in Toulouse." Accessed February 21, 2012, http://www.dlr.de/emma2/meetdoc/DemoDayMalpensa/9_Demo_Day-CPDLC_Toulouse-Public.pdf

Foyle, D. C., B. L. Hooey, D. L. Bakowski, J. L. Williams, and C. L. Kunkle. 2011. "Flight deck surface trajectory-based operations (STBO): Simulation results and ConOps implications." Ninth USA/Europe Air Traffic Management Research and Development Seminar (Paper 132), EUROCONTROL/FAA. Berlin, Germany.

Jakobi, J. "Operational Concept for TAXI-CPDLC." Third AP21 Workshop, NASA Ames, October 31, 2007. Accessed February 21, 2012, http://www.dlr.de/a-smgcs/AP21/2007/10_Jakobi/TAXI-CPDLC_Jakobi_006.ppt

Joint Planning and Development Office. 2009. Concept of Operations for the Next Generation Air Transport System, v3.0. Accessed February 21, 2012, www.jpdo.gov/library/NextGen_ConOps_v3%200.pdf

Jung, Y. C., T. Hoang, J. Montoya, G. Gupta, W. Malik, and L. Tobias. 2010. "A concept and implementation of optimized operations of airport surface traffic," 10th AIAA Aviation Technology, Integration, and Operations (ATIO) Conference, Ft. Worth, TX.

Wargo, C. A. and J. F. D'Arcy. 2011. "Performance of Data Link Communications in Surface Management Operations." Aerospace Conference 2011 IEEE, pp 1- 10.

CHAPTER 26

An Initial Investigation of the Impact of Operator-Automation Goal Divergence in the Tower

Laura D. Strater[1], Jon Holbrook[2], Tina Beard[3], Haydee Cuevas[1] and Mica Endsley[1]

[1]SA Technologies, Inc.
Marietta, GA 30066
laura@satechnologies.com
[2]San Jose State University
[3]NASA Ames Research Center

ABSTRACT

The overall goal of the Next Generation (NextGen) Air Transportation System is to significantly increase the safety, security, capacity, efficiency, and environmental compatibility of air transportation operations. To achieve this goal, we must consider how the introduction of increased automation under NextGen will impact the human operators working with these systems. Toward this end, we report on a study that evaluated the utility of two airport surface operations decision support systems, one that provides recommendations in close compliance with current operator goals, and a second that addresses critical NextGen goals, but provides recommendations based on factors outside the controller's current decision model. Findings offer initial insights on the impact of operator-automation goal divergence on operator situation awareness.

Keywords: goal divergence, automation, situation awareness, NextGen

1   INTRODUCTION

A total transformation of the U.S. National Airspace System is underway in order to meet the anticipated demand for air travel while continuing to maintain the high standards of safety and efficiency that air travelers expect. This effort, the development of the Next Generation (NextGen) Air Transportation System, aims to achieve increases in capacity, throughput, and efficiency, while maintaining the highest possible standards of safety and minimizing the environmental impact of air travel (FAA, 2009). NextGen is not an incremental change to the current air transportation system, but is a more significant transformation, altering the physical systems, the processes and procedures, and even, to some extent, the day-to-day goals and objectives of the operators who will implement the changes.

1.1 Implementation of Automation in NextGen

NextGen is predicated on the implementation of high levels of automation, and automation functioning at higher levels of autonomy than generally found in current systems. Thus, we must consider how the introduction of increased automation under NextGen will impact the human operators working with these systems. The successful implementation of NextGen will require that careful attention be paid to decisions such as levels of autonomy, allocation of functions between humans and automation, or even between diverse roles, such as air traffic controller and pilot, and the thoughtful application of lessons learned from automation implementation in both aviation and other domains. Unfortunately, the introduction of automation has not always yielded the expected gains in efficiency or throughput. Generally, even when very reliable automation is implemented, human decision makers hold the final responsibility for system performance. Therefore, to ensure the successful integration of more advanced automation into current operations, it is important to fully understand the factors that may influence automation usage decisions, in general, and operators' trust in automation, in particular.

Automation usage decisions are influenced by a wide range of psychological, cognitive, social, affective, situational, and system-design factors including level of automation, automation reliability and utility (both actual and perceived), automation consistency, feedback, operator biases (e.g., automation bias, self-serving bias), and operator attitudes, (e.g., trust, self-efficacy) (Lee and See, 2004; Parasuraman and Riley, 1997). Human operators' beliefs about and trust in automation mediate their subsequent reliance on automated systems, ranging from the extremes of over-reliance and complacency to under-reliance and mistrust (Lee and See, 2004). This can lead to at least two potentially problematic situations, misuse and disuse (Parasuraman and Riley, 1997). In *misuse* (over-reliance), operators blindly follow the judgments made by the automation, thereby abdicating their role of system supervisor. With *disuse* (under-reliance), operators either ignore the automation's recommendations or delay action until system judgments can be verified, increasing decision time (Grounds and Ensing, 2000). While the source for these errors differs, either type of error can have critical consequences. Therefore, properly calibrating operators' trust in automation and their automation usage decisions can help to reduce both misuse and disuse of automation.

1.2   Goal Divergence and Trust in Automation

User trust in automation consists of trust in the competence of the system as well as trust in the intentions of the system (goal alignment) (Cramer et al., 2008). *Competence* refers to the perceived skill of the system, that is, the extent to which it is able to perform its functions effectively. With regard to *intentions*, users must also perceive an alignment between the goals of the system and their own goals. Actually, this goal alignment is critical for understanding between any two decision entities, whether they are human, automated, or mixed. If two decision entities make decisions using the same goals, even if the goals are prioritized differently, each decision maker can understand the decisions made by the other. However, when the two decision entities have different goals, or only partially aligned goals, the decisions of one entity may seem irrational, illogical, or foolish to the other. Thus, operators' trust in automation and their automation usage decisions may be influenced by the degree to which a match exists between the operator's goals (what the operator is seeking to accomplish) and the automation's goals (what the automation was designed to do) (see Figure 1). When operator and automation goals' are not aligned, or are only partially aligned, operators may become increasingly confused or frustrated by the system's behavior, and, consequently, may not trust or use the automation, even when necessary for successful task performance. Such less than ideal automation usage decisions can potentially lead to inefficient operations, and, in extreme cases, irrecoverable errors and disastrous accidents.

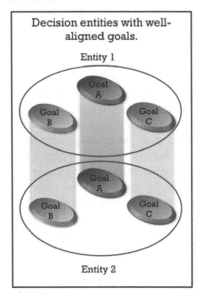 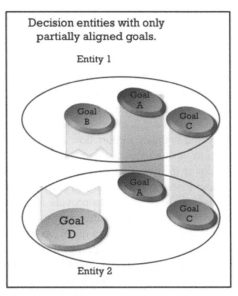

Figure 1 When two decision entities share common goals, their decisions are generally comprehensible and rational to each decision-maker; however, when decision entities utilize different goals as the basis for decisions, decisions outcomes may seem irrational to one another.

Support for the relevance of goal alignment in human-automation interaction can be found in the automation literature. An FAA report by Ahlstrom, Longo, and Truitt (2002) cites the importance of ensuring that the relationship between user tasks and functions and automation functions (e.g., display, control, decision aid, and information structure) is clear to the user; that is, users need to be able to clearly see how the automation facilitates the completion of their assigned tasks. Similarly, automated systems and interfaces should be consistent with user expectations. The report also states that the design of automation should be based on human-centered goals and functions; that is, the functions that the system will perform should be based on the human-centered criteria (goals) of the system.

Drawing from the case-based reasoning literature within the context of expert systems (Roth-Berghofer and Cassens, 2005), one could argue that the more complex automated systems become, the more explanation capabilities operators will expect when interacting with such systems. Moreover, such explanation capabilities (the automation's ability to make its reasoning process and resulting actions transparent to users) may have a significant effect on operators' trust in and acceptance of the guidance provided by the system. Other research has also highlighted the effect of system transparency (i.e., how easily users can understand how the system functions and behaves) on operator trust and acceptance. Specifically, a transparent system allows users to assess the degree to which a system's goals and reasoning processes are in line with their own goals and reasoning in relation to the task they are trying to perform (Cramer et al., 2008; see also, Miller, Pelican, and Goldman, 1999).

The concept of goal alignment has significant implications for ensuring successful human-system integration in NextGen operations. For humans and automation to coordinate their actions effectively, each should be working toward the same goal, or at a minimum understand the goal that each is seeking to accomplish. Yet, as new advanced automation is being developed to support the requirements of NextGen, it is likely that the design objectives and proposed functions for these systems will not always correspond to the goals for operators using the current system. For example, in current operations, the primary goal of the Ground Controller is to monitor and orchestrate ground traffic to maximize aircraft safety and maintain airport flow rate. To achieve this goal, the Ground Controller works to push aircraft to the queuing area as quickly as possible, without regard to backups at the queue (unless queuing area fills up). Similarly, the goal of the Spot Release Planner (SRP) tool, evaluated in this study, is to generate an optimal schedule for spot release to provide maximum runway throughput (Malik, 2009). However, the SRP is also designed to achieve better fuel management by holding aircraft at spot rather than in queue, as there is a significant difference in the power consumption of aircraft at spot, generally operating on a single engine, and those in queue, with two engines powered up. Controllers would be expected to be more receptive to the SRP, and work more effectively with this tool, once they understand that the SRP has a somewhat different goal as well as understood the reasons behind that goal (e.g., fuel savings).

1.3   Situation Awareness in Aviation Automation

In the aviation domain, maintaining a high level of situation awareness (SA) is one of the most critical challenges for personnel. A vast portion of a pilot's, controller's, dispatcher's or other Air Traffic Control personnel's job involves developing SA and keeping it up-to-date in a rapidly changing environment. SA can be defined as the *perception* of elements in the environment within a volume of time and space (Level 1 SA), the *comprehension* of their meaning (Level 2 SA), and the *projection* of their status in the near future (Level 3 SA) (Endsley, 1995). In a simplified sense, SA can be considered an internalized mental model of the current state of the operational environment. This integrated picture forms the central organizing feature from which all decision making and action takes place.

Poorly implemented automation can have significant implications for situation awareness, and ultimately, the performance of the human-automation team. Of particular interest to this study is the problem of operators misunderstanding what the system is doing and why it is doing it that is necessary for accurate comprehension and projection (Level 2 and 3 SA).   With comprehension and projection compromised, significant errors can occur as operators struggle to insure compliance between what the automation is doing and what they want it to do. Wiener and Curry (1980) first noted these problems in studying pilots working with automated system. The goal divergence issue is a specific instance of this operator understanding problems. The loss of SA that results from this misunderstanding is a significant concern in a safety critical domain such as aviation.

SA is often investigated using a unique form of cognitive task analysis called goal-directed task analysis (GDTA) (Endsley and Jones, 2012). The product of the GDTA is a goal hierarchy that shows the operator's goals, with the related decisions and information requirements across the three levels of SA (perception, comprehension, and projection). The decisions are presented as questions that must be addressed in order to meet the operator's goals. The SA requirements focus not only what data the operator needs, but also on how the data are integrated or combined to address each decision. The higher levels of SA— comprehension (Level 2 SA) and projection (Level 3 SA) — are emphasized in this analysis, each a function of the decision requirements associated with the goals. In the analysis, each goal is explored separately and all of the decisions that support goal attainment are defined separately. Thus, because of its goal-driven focus, the GDTA methodology is uniquely beneficial for understanding and evaluating the effects of various function allocation decisions and implementations on goal divergence between human operators and the automated system used to perform their tasks.

1.4   Study Objectives

A simulation exercise was conducted at the NASA Ames Research Center's Future Flight Central in San Jose, CA, to assess the algorithms of two Spot and Runway Departure Advisory (SARDA) decision support system (DSS) tools, one designed to support the Local Controller and the other designed to support the

Ground Controller. The overall objective of the study was to investigate the effectiveness of the decision aids in sequencing and scheduling aircraft, however, the design of the algorithms also allowed us to investigate the impact of goal alignment on operator situation awareness. Our hypothesis was that operator SA would be higher when making independent decisions than when working with a decision aid under conditions of poor goal alignment; however, there would be no decrement in SA when working with a decision aid that shares operator goals.

In this study, the Runway Scheduler (RS) tool provides recommendations in close compliance with current Local Controller operator goals, while the Spot Release Planner (SRP) addresses critical NextGen goals, but provides recommendations based on factors outside the Ground Controller's current decision model.

2   METHOD

2.1   Participants

Two experienced Tower controllers participated in this study. Both participants had more than 20 years experience as Tower Controllers, including the Ground and Local Controller positions evaluated in the study. The simulation exercise occurred across ten days and participants switched roles each day, so each participant was the Local Controller for 5 days, and the Ground Controller for 5 days.

2.2   Interface Display Design and DSS Tools

Two SARDA DSS tools were evaluated during the study: Spot Release Planner for the Ground Controller and Runway Scheduler for the Local Controller. The SRP has a significant goal to hold aircraft longer at the spot that borders the ramp area in order to reduce fuel consumption, unlike current operational goals which move the aircraft into the departure queue quickly for departure sequencing. The Runway Scheduler is a runway management tool that was used by the Local Controller to sequence and schedule departures, arrivals and runway crossing, much like current day controllers. Each tool provided aircraft sequencing and release timing information to the controller, and was implemented in two interface conditions, one utilizing the aircraft datatag, in physical proximity to the aircraft icon on the map, and another using a timeline view, with a separate window showing timing and sequencing information. This information was also available in the aircraft datatags, but the timeline view provided another way to visualize the data. A baseline condition allowed controllers to sequence the aircraft without support from the decision aids. Traffic was also manipulated, with a moderate and high traffic condition.

2.3   Situation Awareness Assessment

SA queries were developed as an objective state-of-knowledge assessment of SA for each role, with a total of 12 SA queries for the Local Controller and 10 for the Ground Controller. Due to the complexity of the simulation environment (e.g., the use of pseudopilots "flying" the aircraft in the high-fidelity simulation), it was not possible to interrupt the simulation, so the battery of SA queries was administered only once per trial, immediately following trial completion.

2.4   Design

A 3 x 2 between-observations design was adopted, with interface display design (Baseline, Datatag, and Timeline) and traffic levels (normal and high) as the independent variables, and overall performance on the SA queries as the dependent variable. Eight trials were completed in each condition, for a total of 48 trials.

3   RESULTS

A one-way analysis of variance, with interface display design (Baseline, Datatag, and Timeline) as the independent variable, was performed on overall SA performance for the Local Controller and Ground Controller. However, since cell size and, by extension, the overall power of this analysis was relatively low, and due to the exploratory nature of these analyses, post-hoc analyses were performed to evaluate trends in the data.

For the Local Controller (goal alignment condition), results for overall SA performance showed no significant main effects or interaction effect across conditions. However, paired-samples comparisons using Fisher's LSD test indicated a slight trend with observations under the Baseline condition ($M = .78$, $SD = .10$) yielding higher overall SA scores when compared to observations under the Timeline condition ($M = .70$, $SD = .13$), with $p = .16$ (Figure 2).

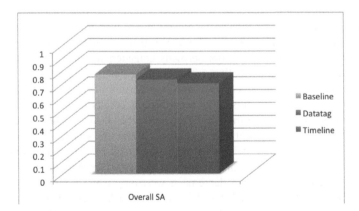

Figure 2 Overall SA results for the Local controller, in the goal alignment condition, show no significant differences in SA between the baseline condition and the two decision aid conditions, though there is a slight downward trend.

For the Ground Controller, no significant main effect of condition on overall SA performance was found, F (3, 27) = 1.69, p = .19. However, post-hoc paired-samples comparisons using the Fisher LSD test revealed that the Baseline condition (M = .72, SD = .13) yielded significantly higher SA scores when compared to the Datatag condition (M = .52, SD = .22, p = .06), and the Timeline condition (M = .53, SD = .13, p = .05).

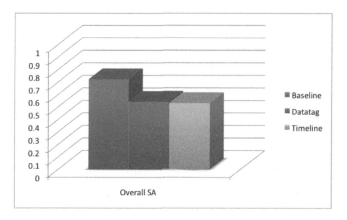

Figure 3 Overall SA results for the Ground controller, in the goal divergence condition, show decreases in SA in post hoc analysis of the two decision aid conditions.

4   DISCUSSION

Results indicated that Ground Controllers showed a pattern of decreased SA when using the SRP advisories. This pattern manifested in an objective measure of SA (SA queries). No significant decrease in objective SA was found for Local Controllers. A possible explanation for this finding is that the SRP advisory, by factoring in data that the operator would not normally consider, was more difficult for operators to understand, and thus challenged the development of their SA.

Discussions during the debrief interviews with participants provide additional possible insights into the study's results. With regard to the results for the Ground Controller, participants remarked that they preferred doing the "heavy pushes" with the timeline advisor because it required less "thinking." While this strategy is good for managing one's own workload, it is often to the detriment of SA, and this is borne out in the SA query scores. This preference was expressed for both the Local and Ground Controller, however, and SA decrements were only seen for the Ground Controller. Another possible reason for the decrements in SA with the timeline condition is the fact that the operator has less need to scan the environment, as the timeline view provides easy access to the defined sequence, without the need for scanning. With the datatag condition, controllers must still scan to find the aircraft that are sequenced for next release. With that in mind, it was surprising that there was no significant difference in SA between the datatag and timeline conditions. Perhaps the additional time afforded by the ease of use of the timeline view allowed the operator to gain SA in the timeline condition.

A possible reason for the lack of variance in the Local Controller SA query scores is that the position is more rigid, and the timeline advisor was confined (as the controller was) in the sorts of changes that can be made to departure clearance schemes. The timeline advisor was, therefore, less disruptive to the Local Controllers, and seemed to generally follow courses of action that they would have implemented in the absence of the advisor. The participants did sometimes express surprise at when the advisor chose to cross traffic, but our SA queries pertaining to this subject focused on the location of the crossing traffic, rather than on the timing of the cross. Future research is needed to determine whether the slight trend indicating decreased SA with the decision aid is a significant difference for the Local Controller position.

5   CONCLUSIONS

This study's findings offer initial insights on the impact of operator-automation goal divergence on operator situation awareness. The findings supported the hypothesis that there are no decreases, or lower decreases, in situation awareness with automation implementation when automation goals are more closely aligned with those of the operators. In the goal alignment condition, the Local Controller did no show reduced SA when using the decision aid in either of the decision-aiding conditions. Additional research is needed to determine whether the slight downward trend in SA with the decision aid in this condition is significant with a larger pool of participants. In the goal divergence condition, the Ground Controller showed decreases in SA for both decision-aiding conditions.

Other possible explanations for this result are that the Ground Controller decisions are more complex, that is, there are a larger number of possible answers within the decision space, and this complexity is the basis for the SA decrement seen in this condition. However, the decrease in SA with the datatag condition, which requires scanning the Ground Control area to determine which aircraft is next (and should therefore be expected to increase SA), may indicate that the decrease in SA is less about the decision space, and more related to the nature of the decision.

This study also highlights the need for further research to investigate the impact of the addition of new operational goals in NextGen, and of conducting research to assess the best way to introduce automation with goals that diverge from those of the operator. These findings support the hypothesis that SA will be lower in cases of goal divergence, however, it provides only preliminary support, and further research is needed to better understand the linkages between goal alignment, situation awareness, automation usage decisions, and human-automation team performance. In sum, although operator and automation goals will not always overlap completely (i.e., perfect goal alignment), assisting operators in gaining an accurate understanding of the automation's goals and intentions will lead to better calibration of trust and appropriate automation usage decisions. This can be accomplished through training on automation functioning (e.g., Ahlstrom et al., 2002), in general, as well as specific training on the alignment between operator and automation goals. Designing for optimal levels of system transparency can also address this issue (see Endsley and Jones, 2012).

ACKNOWLEDGMENTS

Work on the research reported in this paper was partially supported by a contract from the NASA Ames Research Center. The opinions, views, and conclusions contained herein, however, are those of the authors and should not be interpreted as representing the official policies, either expressed or implied, of the National Aeronautics and Space Administration (NASA), the U.S. Government, or the organization with which the authors are affiliated.

REFERENCES

Ahlstrom, V. K. Longo, & T. Truitt. (2002). *Human factors design guide update* (Report Number DOT/FAA/CT-96/01): A revision to chapter 5 automation guidelines. Technical Report Number DOT/FAA/CT02/11. Washington, DC: Federal Aviation Administration.

Cramer, H., V. Evers, S. Ramlal, M. van Someren, L. Rutledge, N. Stash, L. Aroyo, & B. Wielingan. (2008). The effects of transparency on trust in and acceptance of a content-based art recommender. *User Modeling and User-Adapted Interaction, 18 (5)*, 455-496.

Endsley, M. R. (1995). Toward a theory of situation awareness in dynamic systems. *Human Factors, 37 (1)*, 32-64.

Endsley, M. R. & D. G. Jones (2012). *Designing for situation awareness: An approach to human-centered design* (2nd ed.). Boca Raton, FL: Taylor & Francis Group.

Federal Aviation Administration (FAA) (2009b). *NextGen Implementation Plan.* Washington, D.C: Author. www.faa.gov/nextgen.

Grounds, C. B. & A. R. Ensing (2000). Automation distrust impacts on command and control decision time. *Proceedings of the XIVth Triennial Congress of the International Ergonomics Association and 44th Annual Meeting of the Human Factors and Ergonomics Society, 3*, 855-858. Santa Monica, CA: Human Factors and Ergonomics Society.

Lee, J. D. & K. A. See (2004). Trust in automation: Designing for appropriate reliance. *Human Factors, 46(1)*, 50-80.

Miller, C. A., M. Pelican, & R. Goldman (1999). "Tasking" interfaces for flexible interaction with automation: Keeping the operator in control. *Proceedings of the International Conference on Intelligent User Interfaces*, Redondo Beach, CA, 5-8 January 1999.

Parasuraman, R. & V. Riley (1997). Humans and automation: Use, misuse, disuse, abuse. *Human Factors, 39(2)*, 230-253.

Wiener, E. L. & Curry, R. E. (1980). Flight deck automation: Promises and problems. Ergonomics, 23(10), 995-1011.

CHAPTER 27

Varying Levels of Automation on UAS Operator Responses to Traffic Resolution Advisories in Civil Airspace

Caitlin Kenny, Lisa Fern

San Jose State University
NASA Ames Research Center, USA
Caitlin.A.Kenny@ NASA.Gov

ABSTRACT

Continuing demand for the use of Unmanned Aircraft Systems (UAS) has put increasing pressure on operations in civil airspace. The need to fly UAS in the National Airspace System (NAS) in order to perform missions vital to national security and defense, emergency management, and science is increasing at a rapid pace. In order to ensure safe operations in the NAS, operators of unmanned aircraft, like those of manned aircraft, may be required to maintain separation assurance and avoid loss of separation with other aircraft while performing their mission tasks. This experiment investigated the effects of varying levels of automation on UAS operator performance and workload while responding to conflict resolution instructions provided by the Tactical Collision Avoidance System II (TCAS II) during a UAS mission in high-density airspace. The purpose of this study was not to investigate the safety of using TCAS II on UAS, but rather to examine the effect of automation on the ability of operators to respond to traffic collision alerts. Six licensed pilots were recruited to act as UAS operators for this study. Operators were instructed to follow a specified mission flight path, while maintaining radio contact with Air Traffic Control and responding to TCAS II resolution advisories. Operators flew four, 45 minute, experimental missions with four different levels of automation: Manual, Knobs, Management by Exception, and Fully Automated. All missions included TCAS II Resolution Advisories (RAs) that required operator attention and rerouting. Operator compliance and reaction time to RAs was measured, and post-run NASA-TLX ratings were collected to measure workload. Results showed significantly higher compliance rates, faster responses to TCAS II alerts, as well as less preemptive operator actions when higher levels of automation are implemented. Physical and Temporal ratings of workload were significantly higher in the Manual condition than in the Management by Exception and Fully Automated conditions.

Keywords: unmanned systems, collision avoidance, automation

1   INTRODUCTION

Continuous demand for the use of Unmanned Aircraft Systems (UAS) has put increasing pressure on airspace operations in civil airspace. This demand is driven by two main advantages that UAS have over manned aircraft, perceived cost efficiency and the minimization of risk to pilots' lives (Gertler, 2012). The need to fly UAS in the National Airspace System (NAS) in order to perform missions vital to national security and defense, emergency management, and science is increasing at a rapid pace due to the foreseen advantages of their use. In addition to limiting UAS usage for civilian applications, current Federal Aviation Administration (FAA) restrictions on UAS access to the NAS constrain the U.S. military's ability to fulfill regular training requirements to prepare UAS operators for combat (DoD UAS ExCom NAS Access Working Group, 2010).

In order to ensure safe operations in the NAS, operators of unmanned aircraft, like those of manned aircraft, may be required to maintain separation assurance and avoid loss of separation and conflicts with other aircraft while performing their mission tasks. A commonly used conflict avoidance measure for manned aircraft is TCAS II, or the Tactical Collision Avoidance System. TCAS II is a transponder based system that provides Traffic Advisories (TAs) to alert the pilot of incoming traffic, and Resolution Advisories (RAs) to provide pilots with instructions for avoiding conflicts within a five second time limit (FAA, 2001). Resolution Advisories provided by TCAS II are limited to vertical maneuvers only; examples include "Climb, Climb" and "Descend, Descend". While the FAA has ruled TCAS II alone to be unacceptable for UAS flight due to latencies that are inherent to UAS operation and the differing flight characteristics from manned aircraft (FAA, 2011), it is reasonable to expect that a TCAS II – like system could be part of a layered solution involving an integrated traffic display and conflict alerting system.

Those latencies present in UAS operations may create the need for automation to assist operators in responding successfully to conflict alerts, regardless of the conflict avoidance system, or suite of systems, used to provide them. The wide variability of automation capabilities in current unmanned systems raises the question about the effect of human-automation interaction on the ability of operators to respond safely and timely to conflict alerts. Some present day UAS (e.g. MQ-1 Predator) require manual control of the flight control systems, and operator tasks closely resemble those traditionally associated with manual flying. Other UAS (e.g. RQ-4 Global Hawk) are highly automated, with operators flying in pre-programmed waypoint-to-waypoint navigation mode under a supervisory capacity. Most current systems, however, fall somewhere in between fully manual and fully automated with a wide variety of partially-automated control and navigation interfaces. When latencies are present in the control loop, it becomes important to know whether operators can respond quickly enough to potential conflicts given the type of control input and automation capability present.

This experiment investigated the effects of varying levels of automation on UAS operator performance and workload while responding to conflict resolution instructions provided by a Tactical Collision Avoidance System II (TCAS II) during a UAS mission in high-density airspace. The purpose of this study was not to investigate the safety of using TCAS II on UAS, but rather to examine the effect of automation on the ability of operators to respond to traffic collision alerts provided by a commonly used system.

2   METHOD

2.1   Participants

Six pilots were recruited to participate in this study. All were males (averaged 29.5 years) with an average of 2727 flight hours. Total flight hours ranged from 250 to 5300 hours. No pilots reported military or UAS flight experience. Eligibility was limited to participants who had normal or corrected to normal vision and were under 40 years old. Participants were required for approximately seven hours each and were compensated for their participation in the study.

2.2   Displays Setup

Participants were given two computer monitors to observe and manipulate: the Multiple-UAS Simulator (MUSIM) on the right, and the Ames 3D Cockpit Situation Display (CSD) on the left (Figure 1).

Multiple-UAS Simulator (MUSIM). This experiment used the Multiple-UAS Simulator (MUSIM), a full description can be found in Fern & Shively (2009). The current simulation configuration of MUSIM differed only slightly in that it utilized a 1:1 operator to vehicle interface with a generic fixed wing flight control model input with generic Mid-Altitude Long Endurance (MALE) UAS parameters. Ownship airspeed was fixed at 90 kts for the entire experiment. MUSIM was separated into four Graphical User Interfaces (GUIs): a map display indicated the position and flight path of the UAS in purple waypoints, a multi-function display (MFD) indicated UAS status and behavior, a TCAS II (Tactical Collision Avoidance System II) alert box provided textual and auditory Traffic Advisories (TAs) and Resolution Advisories (RAs), and a timer.

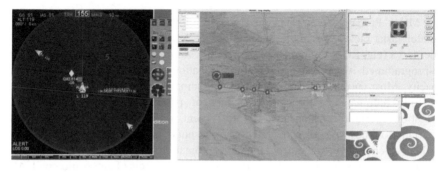

Figure 1 Displays set up with Multiple-UAS Simulator (MUSIM) (right) and Ames 3D Cockpit Situation Display with traffic information and TCAS II alerts (left). MUSIM interfaces include the map display (left), MFD (top right), TCAS alert box (lower right) and timer (far right).

Ames 3D Cockpit Situation Display (CSD). The Ames 3D Cockpit Situation Display was used to display TCAS II information in its basic 2D planar view (for a full description see Granada, Dao, Wong, Johnson & Battiste, 2005). The CSD had an ownship-centric view of surrounding airspace and utilized TCAS II symbology to alert operators of potential collisions. Participants were able to adjust the horizontal viewing distance from 10-640 nm, though no other manipulations were allowed on the CSD during this experiment.

2.3   Experimental Design and Mission Details

A within-subjects design was used to study operator performance and workload measures while flying a signal intelligence mission using a MALE UAS in high density Southern California TRACON airspace (LAX terminal area). Four different levels of automation for responding to TCAS II RAs were counterbalanced across four different mission flight paths.

Levels of Automation. Operators were given four different levels of automation to assist them in responding to TCAS II RAs: Manual, Knobs, Management by Exception and Fully Automated. In the Manual condition, operators flew the UAS in a waypoint-to-waypoint control mode. This was the baseline control mode for all conditions. Flight paths were edited by clicking each waypoint individually to activate the editing function. Once in editing mode, new altitudes were input manually and waypoints were clicked and dragged to new locations. In order to respond to an RA in the Manual condition, the operator would click on the next waypoint, manually input the new altitude, and commit the change. In Knobs, operators utilized the MFD Control and Status Page to quickly edit the altitude of the aircraft when an RA was received. Use of the Knobs input automatically applied the altitude change to the next waypoint on the aircraft's path, and was introduced as a "quick response" control option similar to dialing an analog knob in a cockpit. To change additional waypoints on the flight path, operators had to use the Manual control mode. In Management by Exception, participants either accepted or rejected an automated altitude edit in the MUSIM TCAS II alert box when RAs were received. In the Fully Automated condition, altitude changes in response to RAs were automatically applied, with feedback provided to the operator of the altitude change in the MUSIM TCAS II alert box. All automated responses were in compliance with the RA instructions. In both the Management by Exception and Fully Automated conditions, all other route edits not in response to RAs were done through the Manual control mode.

Missions. Four training and four experimental missions were developed for this experiment. Training scenarios were 10 minutes long, and provided rerouting practice for the operators in the four different levels of automation. Experimental missions were 45 minutes long and differed only in the assigned waypoints and altitudes in the flight path that the operator was instructed to follow. All missions included TCAS II alerts that required operator attention and rerouting for RAs, though timing of conflicts and severity differed between missions to reduce predictability.

Mission Objectives. Participants were instructed to fly a signal intelligence mission with a MALE UAS in Southern California TRACON airspace with three mission objectives: 1) to fly the assigned mission flight path as closely as possible, 2) to respond to TCAS II alerts for collision avoidance by either climbing or descending in accordance with the Resolution Advisory, and 3) to communicate with Air Traffic Control (ATC). The first objective required operators to fly through predetermined mission flight paths while maintaining set altitudes at each waypoint and remaining as close to the original flight path as possible. The second objective required operators to monitor the TCAS II alerts on both the MUSIM and CSD displays and reroute the UAS's path to avoid conflicts while staying as close as possible to the original route. The third objective required operators to maintain radio contact with ATC while flying their route; tasks included calling in altitude requests, flight path edits, waypoint check-ins and responding to any received ATC communications.

2.4   Procedure

Participants were required to fill out an informed consent form and a demographic survey. Training was given to the operators before each experimental condition, with workload and situation awareness probes administered throughout.

Training Sessions. Operators were given a short briefing introducing MUSIM, the CSD and mission objectives after completing paperwork. Self-paced PowerPoint slides were provided before each experimental session detailing how to edit waypoints and altitudes appropriately for each condition. Training scenarios were then completed and lasted 10 minutes.

Experimental Sessions. The experimental sessions were blocked by level of automation and flight path using a Latin Square. Participants completed four experimental missions during the simulation. In each mission, operators received workload probes after each RA, administered verbally by the experimenter. Before each scenario, participants were given a practice mission to familiarize themselves with the level of automation present. After each scenario, participants completed a NASA-TLX (Hart & Staveland, 1988) to measure workload and a 10-D SART (Situation Awareness Rating Technique; Selcon, Taylor, & Koritas, 1991) to measure situation awareness. At the end of the day, operators completed a Post-Simulation Questionnaire asking more in depth questions on workload and SA.

3   MEASURES

3.1   Objective Performance

Response Time. Response Time (RT) to RAs was measured in seconds from when the TCAS II alert was given until the operator committed a flight path change in the Manual and Knobs conditions. In the Management by Exception condition, RT was measured in seconds from when the operator was alerted of the RA to when they responded by clicking either "Accept," or "Reject," on the MUSIM TCAS alert box. If the operators did not respond within 5 seconds, the alert box timed out and automatically adjusted the altitude of the UAS. Reaction Time was not measured in the Fully Automated condition as operators did not need to respond to the RAs.

Operator Response Rate. Operator Response Rate measured the percentage of RAs that operators responded to (correctly or incorrectly) out of the total that occurred.

Compliance Rate. Compliance Rate was measured as the percentage of RAs that operators correctly complied with out of the total number that occurred.

Pre-Emptive Response Rate. Pre-Emptive Response Rate was measured as the percentage of times that the operator began a route edit in anticipation of an RA out of the total number of RAs that he responded to.

3.2   Subjective Ratings

NASA TLX. Workload was measured post-scenario with a 10-point NASA TLX. Participants rated their workload on six dimensions: Mental Demands, Physical Demands, Temporal Demands, Performance, Effort and Frustration.

Additional Measures. Additional Subjective Ratings were collected, though not discussed in this paper. Additional measures include the 10-D SART, workload probes administered during trials, a post-trial questionnaire and a post-simulation questionnaire that were given to further measure workload and situation awareness.

4   RESULTS

The data were analyzed using a one-way repeated measures analysis of variance (ANOVA) with Levels of Automation as the independent variable. Post hoc analyses utilized Bonferroni pair-wise comparisons. The results are organized by type of measure.

4.1 Objective Performance

Response Time. Response times to TCAS II alerts were found to be significantly faster in the Management by Exception condition (M = 3.45; SE = .35) compared to both Knobs (M = 12.20; SE = 2.56) and Manual (M = 10.75; SE = 1.44), $F(2, 12)$ = 8.408, $p < .05$ (Figure 2).

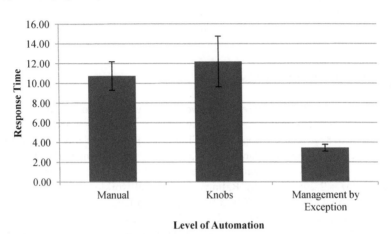

Figure 2 Response Time to TCAS II alerts by Level of Automation

Operator Response Rate. There was not a significant difference in operator response rates across the different conditions, $F(2, 12)$ = 2.602, p = .115. However, there appeared to be a trend toward higher response rates in Management by Exception (M = 95.92; SE = 2.64) than in both Manual (M = 85.86; SE = 4.09) and Knobs (M = 74.40; SE = 10.25) (Figure 3).

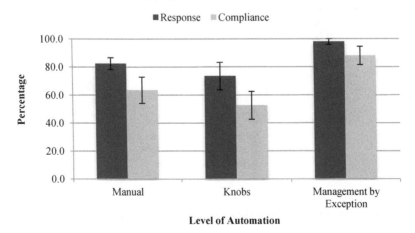

Figure 3 Response and compliance rates by level of automation.

Compliance Rate. Compliance rates for Management by Exception ($M = 89.71$; $SE = 6.04$) were significantly higher than both Manual ($M = 74.40$; $SE = 10.58$) and Knobs ($M = 63.78$; $SE = 12.27$), $F(2, 12) = 7.233$, $p < .01$ (Figure 3). Several reasons were noted for noncompliance; operators did not feel a collision would occur, horizontal route edits were performed instead of vertical, and accidental inputs of wrong altitudes were made.

Pre-Emptive Response Rate. Operators made significantly more pre-emptive responses in the Manual condition ($M = 6.21$; $SE = 2.96$) and the Knobs condition ($M = 42.09$; $SE = 8.20$) than in Management by Exception ($M = 6.20$; $SE = 2.96$), $F(2, 12) = 8.705$, $p < .01$.

4.2   Subjective Ratings

NASA TLX. Significant differences between the levels of automation were found in the physical and temporal dimensions of workload, $F(3, 18) = 3.358$, $p < .05$, and $F(3, 18) = 4.078$, $p < .05$, respectively (Figure 4). Ratings of physical workload were significantly higher for Manual ($M = 3.43$; $SE = 1.02$) compared to Knobs ($M = 2.57$; $SE = 1.04$), Management by Exception ($M = 2.21$; $SE = .69$), and Fully Automated ($M = 2.29$; $SE = .68$). Ratings on temporal workload were significantly higher for Manual ($M = 6.00$; $SE = .59$), compared to Knobs ($M = 3.50$; $SE = .29$) and Fully Automated ($M = 3.93$; $SE = .93$). Ratings of temporal workload in Management by Exception ($M = 4.43$; $SE = 1.09$) were not found to be significantly different than any of the other levels of automation.

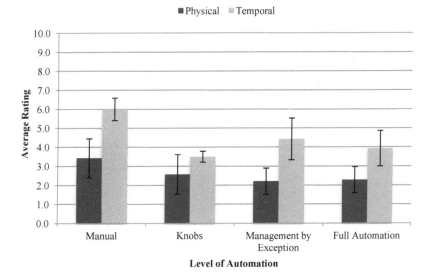

Figure 4 NASA-TLX ratings on the Physical and Temporal dimensions of workload by level of automation.

5   DISCUSSION

In order for UAS to safely operate in the NAS, they must be able to successfully avoid conflicts and respond appropriately when collisions are imminent. This study examined the effects of varying levels of automation on the ability of UAS operators to respond to traffic collision alerts in high-density airspace. The results indicate that response times and compliance rates for unmanned aircraft operating with lower levels of automation could be unacceptable in the NAS environment. The required five-second or less response time was exceeded in both the Manual and Knobs levels of automation by more than five seconds on average. In the event that a UAS is relying on satellite communication and data links, it will have additional operational delays of up to two seconds. This could create an aircraft response time to resolution advisories of 12 seconds or more; over twice the current five-second standard for manned aircraft.

Current regulations allow for noncompliance with TCAS II alerts only when pilots have the conflicting aircraft in visual sight and can ensure separation (FAA, 2001). Although difficult to find accurate reports for actual TCAS II compliance rates for manned aircraft, compliance with RAs has been estimated between 50 – 60%. This is supported by Olson & Olszta (2010), who found a compliance range of 41 – 59%, and Pritchett & Hansman (1997) who reported approximately 61% of pilots complying within the five second limit, and a compliance range of 33 – 71% when including pilots who began to edit for RAs preemptively or late.

Compliance rates across levels of automation in this experiment often exceeded these previously reported rates, with averages for Manual, Knobs and Management by Exception being 74%, 64% and 90%, respectively. Possible reasons for this difference include among others: a lack of an out-the-window view to be used in visual separation (in which application of the FAA regulation to UAS operation should result in 100% compliance); unfamiliarity with the flight characteristics of a MALE UAS causing a reliance on automated, opposed to personal, collision avoidance judgments; taking longer than five seconds to decide on a course of action in the Management by Exception condition (forcing the automation to comply with the resolution advisory); trust in automation; and ease of use for conflict avoidance in the higher automated conditions. Further research is required to better understand the benefits of automation for increased safety and collision avoidance while operating in the NAS.

Interestingly, the Manual condition was found to perform better than the Knobs condition overall with a lower average response time, and a higher average response and compliance rate, despite Knobs being introduced as a "quick response" control input. The lower performance for Knobs goes with a trend of more preemptive, or anticipatory, responses made in that condition. Operators remarked on how the Manual and Knobs conditions took "too long," to perform the edits required for collision avoidance, and would begin to preemptively edit before an RA was given in order to avoid the impending collision.

Overall, slow response times for Manual and Knobs, combined with 1.5 times greater compliance rates for Management by Exception illustrate the need for some level of automation to assist UAS operators in responding quickly and appropriately to collision avoidance alerting while operating in the NAS. However, the lower workload scores for Management by Exception compared to the Fully Automated condition indicate a need for the operator to remain on the loop and capable of overriding the automation when necessary. Continued research on the effects of human-automation interaction on the safe operation of UAS in the NAS is needed if the demands for civil and public UAS operations are to be met.

REFERENCES

Department of Defense, UAS ExCom NAS Access Working Group. (2011) Department of defense final report to congress on access to national airspace for unmanned aircraft systems. Retrieved from
http://www.acq.osd.mil/psa/docs/report-to-congress-ana-for- uas.pdf

Federal Aviation Administration. (2001) Air Carrier Operational Approval and Use of TCAS II. Advisory Circular 120 -55B, Washington, DC, FAA, para 11.b. (2).

Federal Aviation Administration. (2011). Evaluation of candidate functions for traffic alert and collision avoidance system ii (TCAS II) on unmanned aircraft system (UAS) (FAA/AFS-407). Washington, DC: FAA Unmanned Aircraft Program Office.

Fern, L., & Shively, R. J. (2009). A comparison of varying levels of automation on the supervisory control of multiple UASs. In Proceedings of AUVSI's Unmanned Systems North America 2009, Washington, D.C..

Gertler, J. (2012). U.S. Unmanned aerial systems. Congressional Research Service. Retrieved from: http://fpc.state.gov/documents/organization/180677.pdf

Granada, S., Dao, A. Q., Wong, D., Johnson, W. W., & Battiste, V. (2005) Development and integration of a human-centered volumetric cockpit display for distributed air-ground operations. In Proceedings of the 12th International Symposium on Aviation Psychology, Oklahoma City, OK.

Hart, S., & Staveland, L. (1988). Development of NASA TLX (task load index): Results of empirical and theoretical research," In P. Hancock and N. Meshkati (Eds.), Human Mental Workload (pp. 139-183). Elsevier, Amsterdam, 1988.

Pritchett, A., & Hansman, R.J. (1997). Pilot non-conformance to alerting system commands during closely spaced parallel approaches. Proceedings of the Digital avionics systems conference (pp. 9.1-1 to 9.1-8). Irvine, CA: AIAA/IEEE Explore. 10.1109/DASC.1997.637290

Selcon, S.J., Taylor, R.M., and Koritas, E. (1991). Workload or situational awareness?: TLX vs SART for aerospace systems design evaluation. Human Factors, 35, 62-66.

CHAPTER 28

Automation in ATM: Future Directions

Francisco Sáez Nieto, Rosa Arnaldo Valdés

Polytechnic University of Madrid
Madrid, Spain
franciscojavier.saez@upm.es
rosamaria.arnaldo@upm.es

Eduardo García González

CRIDA
Madrid, Spain
egarcia@crida.es

ABSTRACT

The current Air Traffic Management (ATM) Systems need to develop their businesses and strengthen their value chain as a whole to be able to cope with the expected growth in air traffic. Today's ATM processes are not sufficiently geared, or flexible, to maintain the schedules of the airspace users.

It has been broadly recognized that the future ATM capacity and safety objectives can only be achieved by an intense enhancement of integrated automation support. To ensure overall performance of ATM system, the high-level automation principles described in this study are to serve as a guide for its development.

The main objective pursued by the "research" into "Higher Automation Levels in ATM" is to explore unconventional and high risk areas, involving new technologies and concepts around the theme "Toward higher levels of automation" in future Air Traffic Management Systems.

This study aims to provide basic and applied research that will be required to bring the ATM System into the high degrees of automation required by NextGen and SESAR and beyond, up to and including full automation.

At the same time, this research also seeks to apply existing methodologies and techniques in new ways, potentially applying scientific disciplines that have not previously been brought to bear in the air traffic domains

Keywords: automation, air traffic management, system performance

1   HERITAGE IN AIR TRAFFIC MANAGEMENT

Air Traffic Management (ATM) encompasses all airborne and ground-based functions and services required to ensure the safe and efficient movement of aircraft during all of the planning and execution phases of the flights. Within current approach ATM is based on two major functions:

- Air Traffic Services (ATS): aimed to provide and ensure safety and efficiency of air traffic during all phases of flight, including Air Traffic Control (ATC) service, flight information service, and alerting service.
- Air Traffic Flow Management (ATFM): a function to ensure an optimum flow of air traffic through airspace, respecting all Air Traffic Flow constraints such as airspace sector capacity, ATC, and airport capacity constraints.

Although the ATFM is already a European integrated function, Europe ATM does not have a single sky managed at a European level. European airspace is still largely organized and managed along national borders. Current status leads to around up to 600 ATC sectors over Europe.

Airspace is divided into ATC sectors typically handled by one executive air traffic controller, able to control only a limited number of aircraft, denoted as "sector capacity". The existing method to attend any increase of traffic demand, in a given high density airspace, is to proportionally increase the number of available ATC sectors. By doing so in today's congested areas, the size of the resulting sectors becomes too small and then it is impracticable to handle air traffic, mainly due to the unfeasibility of execution of separation instructions inside the proper sector.

At the same time, European airspace is amongst the busiest in the world with over 33,000 flights on busy days, and the number of flights per day is expected to double by 2030. To accommodate this traffic density and growth airports will need to make better use of existing resources in a more cost-efficient way.

Mitigation of sectors and airports capacity limits is currently the responsibility of the ATFM. Any capacity constraints within the European airspace and airports are managed by CFMU coordinating with ANSP and airspace users. ATFM measures result in on ground delays and/or rerouting and are taken each time air traffic demand exceeds capacity limits in any of the sectors or airports.

Although in the past decade ANSPs, Airline Operators, Airports and the CFMU have managed to cope with a significant traffic growth in an acceptably safe and expeditious manner (with delays being historically low at 1.9mins./flight); the current ATM system shows clear signs of saturation.

1.1   The European Answer To ATM Bottlenecks

After several intentions to integrate the European airspace leaded by EUROCONTROL and ECAC, the European Commission launched in 2000 the ambitious Single European Sky (SES) initiative to overcome the fore mentioned bottlenecks by reformulate the entire architecture of European air traffic management. It proposes a legislative approach to coordinate actions to meet future capacity and safety needs at a European rather than a local level. The key objectives are to:

- Restructure European airspace as a function of air traffic flows;
- Create additional capacity; and
- Increase the overall efficiency of the air traffic management system.

As relevant part of the Single European Sky initiative, SESAR (Single European Sky ATM Research) represents its technological dimension. It will help create a "paradigm shift", supported by state-of-the-art and innovative technology. The SESAR programme investigates how to accommodate the growth in air traffic, to optimize the use of airspace, to reduce delays, and to improve the overall safety performance of the European ATM system. The five key features of the SESAR Master Plan to achieve the ATM paradigm shift are (SESAR D5):

- Moving from airspace towards 4D trajectory-based operations, such that each aircraft follows its preferred route and arrives at its desired time of arrival; the so-called Reference Business Trajectory (RBT);
- Moving towards a network centric approach, underpinned by a System Wide Information System (SWIM), such that all parties involved have access to relevant and most up-to-date flight information;
- Dynamic airspace management, facilitated by a central network, to enhance coordination between aviation authorities;
- New and innovative technologies for more precise navigation and surveillance in order to optimize airspace and airport capacity;
- Allocate a central role for the human, but supported by a high degree of automation to reduce workload, optimize airspace capacity, and maintain a sufficient level of safety in complex, high-traffic, and time-critical situations.

By developing these five key features SESAR aims to accomplish a:

- 10-fold increase in safety,
- 3-fold increase in capacity, and
- 50 percent reduction in ATM costs per flight.

In order to fully meet the safety and other performance targets of the future ATM System several parading shifts are required. SESAR will help to create these paradigm shifts.

- Shift from Airspace – Based operation towards a Trajectory – Based operation concept.
- Shift from Tactical Management towards a more Proactive system,
- Shift from a controller-based system towards a more distributed system

1.2   Heritage in ATM Automation

Automation in the new medium term ATM concepts, like those of SESAR and NextGen, is mainly focused and limited to facilitate a substantial reduction of controller task load per flight operation, while also meeting pre-set safety, environmental and economic goals. To address the controller task load issue, three lines of action are included in the future ATM concept:

- Automation for the routine controller task load supported by better methods of data input and data management;
- Automation support to conflict / interaction detection, situation monitoring and conflict resolution;
- A significant reduction in the need for controller tactical intervention, by reducing the number of potential conflicts using a range of de-confliction methods acting on a larger time frame, and allocating more tactical interventions to the pilots.

The three lines of action when included in the concept will require a significant increase in integrated automation support. Improved interaction and visualization techniques shall support the operators to execute their functions and allow them to be aware of the systems status at all times. The role of the human in the ATM system will consequently shift from mainly controlling tasks to rather monitoring and, if necessary, intervening tasks following the common perspective of human centred automation principles. This changing role will invoke new support concepts at different levels of automation. In parallel, new tasks for the human within the ATM system will drive the evolution for enhanced automation techniques.

The automation support of the human roles must be developed and implemented in a way that fosters trust and confidence by the human in the automation functions. Experience gained during the successful implementation of automation to the cockpit or automation in the Unmanned Aircraft world, formerly known as UAS (Unmanned Automated System), can be used in designing automation in other areas, especially ground based functions of ATM.

Current SESAR Concept of Operations has been developed under the understanding that "humans will constitute the core of the future ATM System's operations". The way it is conceived today, this human-centered philosophy places significant constraints on the system design. Automation is not envisaged as a team process but rather a human vs. machine function allocation problem. The ATM automation has evolved at a remarkably slow pace and today is still based on paradigms and conceptions that have not fundamentally changed for decades. Nevertheless, the technical developments in computer hardware and software now make it possible to introduce automation into virtually all aspects of human-machine systems. The analysis of the state of the art in automation and its application in other safety critical industries could bring new ideas on how higher and efficient levels of automation can be implemented in ATM.

Given these technical capabilities the question to be answered with the next level ATM system is: which functions of the system should be automated, to what level and where should these functions logically be implemented? Appropriate selection is important because automation does not merely supplant but changes human activity and can impose new coordination demands on the human operator.

Within this contextual framework, the challenge to the future ATM system is to achieve the best global, so system wide trade-off between new roles derived from a centric or autonomous system and the right level of automation of each of the four steps in any ATM function (acquisition, analysis, decision and implementation).

2   THE NEW PARADIGM SHIFT IN AUTOMATION IN ATM

To cope with the predicted air traffic demand in Europe, in an efficient and competitive manner, ATM will have to adopt its capacities based on SES-SESAR objectives. SESAR approach, as NextGen, contains important components based or taken directly from ATM heritage that in fact will facilitate the implementation of change processes.

But in order to go further than SESAR and NextGen a paradigm shift is required in ATM Automation. This shift will be focused on ATM invariant processes and will allow for new role assignments based on three interdependent criteria, having overall system performance as main driver for ATM automation.

Beyond this implementation strategy and paying interest in the pure ATM context, it can be identified that ATM is essentially composed, as any other management activity, by invariants determined by its goals and limitations.

The goals of ATM are in broad sense two fold; to provide the required separation between aircraft that permits to maintain the safety standards established for the air transport and, at the same time, to maintain air transport as a competitive transport mode by providing required efficiency to users, being environmentally friendly and socially valuable. These goals will only require service providers when autonomy cannot cope with them under any situation.

Limitations of ATM, considering that aircraft are supported by the air, contained in atmosphere, and while the air transport payload is taken and delivered at airports, will be the behavior of atmosphere and airport capacity constraints.

Facing these ATM invariants, automation role should be based on overall system performance rather than in heritage inertia or competitiveness human vs. machine. Operational changes will be supported by new roles for managing aircraft trajectories, separation provision and trajectory de-confliction, leading to adopted procedures for both the ATM function and the aircraft operator.

The new roles for the different actors in the long term future air traffic management and the future degree of automation is today a matter of conceptual proposals, based on conceptual or theoretical considerations and the general assumption that establishes: "humans will still be central corner stones as managers and decision-makers". Nevertheless, the only principle about the assignments of new roles should be that the "decision about aircraft trajectories could be taken by the actor that is ideally placed at each moment and for the given scenario to take such a decision" and the impact on overall system performances should be its main driver.

Evolution from a purely tactical intervention model towards a more strategic trajectory management concept and progressive introduction of more autonomous and decentralized operation are key concepts in the paradigm shift.

Implementation of these future ATM concepts will change the human role in the future European ATM system, and his relationship with the automated processes. Who is the best player against overall system performance should be the choice criterion.

The choice of this best player will be derived from the consideration of the three decision criteria:

- **Optimal decision time** for all and any relevant event in ATM, which involves strategic vs. tactical planning layer;
- **Optimal decision place,** which involves centric vs. autonomous consideration to decide, which place is suited best to take the required decision;
- **Optimal decision player**, which involves human vs. automated player considerations.

2.1 The Three Interdependent Dimensions for the Paradigm Change.

Automation as partner of humans will play an important role in the allocation of these new functions. The selected level of automation will be closely related to the level of strategic/tactical and centralized/autonomous implementation of each required function. This goes in line with the strategy to allow the best logically placed actor to take action to implement the globally best decision in every system context.

It is important to recognize that this strategy may offer multiple solutions for a given problem context as the combination of both, the actor and the function, that may lead to equally efficient solutions. However, with changing the problem context, dedicated solutions may show up weaknesses and vulnerability to the ATM system.

Consequently, a comprehensive problem use case model for the dedicated ATM system component is needed beforehand. It is aimed at defining transfer functions between the different dimensions to identify the complete set of potential solutions, allowing adjusting solution strategies when aggregating from the system component level to the overall system, including interface requirements between the components. Once again, a crucial interface is seen between the airborne and the ground ATM component, where interfaces do not clearly exist yet.

New roles for different actors in the long term future ATM are, even today, a matter of conceptual proposals; the overall system performance as central metric should be the main driver. The roles assignment will conceptually follow the presented three dimensional model comprising:

- Dimension one: searching for the "best time" to take a decision. This deals with finding the right time to act along the strategic to the pre-tactical and tactical planning layer;
- Dimension two: searching for the "best place" to take a decision. This deals with deciding on whether giving pace to a centralized function versus place the function to a (set of) actors and let them decide autonomously;
- Dimension three: searching for the "best player" to take a decision: This deals with deciding to what extent human or an automated function should participate in the decision while taking into consideration that this decision may invoke different logical strategies as the human rather acts on an aggregated system monitoring level whereas the automated function will run at dedicated function control levels.

The following section elaborates potential dependencies between these dimensions.

2.2 "Best Decision Time"

It is well understood that increasing look-ahead time applied to planning and support functions regardless whether automated of manual, will decrease reliability and validity of the function outcome, e.g. resolution advisories for aircraft or (work) load conflicts. The nature of this accuracy degradation relies on the quality of the expected system behavior, so the data intent quality. The quality in turn relies on the sensors available to generate the data. All these contributors are known to be non-deterministic and context dependent. A generalized description is not available. If the ATM system suffers disruptions in servicing due to e.g. complex weather or simply system time-outs, heuristic behavior could be observed for these parameters.

The ATM concept will consequently analyze the root-cause chains and elaborate adopted descriptions, which are context dependent and allow handling of typical disruptions patterns. Based on these findings, the concept shall deliver advisories on what functions act on detected system deviations and when.

As stated, it is assumed that this dimension is hardly coupled with the other dimensions "best decision place" and "best player". The proof-of-concept will have to confirm that thesis.

The "best time" approach as such consists then in determining the optimal moment to initiate an event that generates or modifies any flight trajectory. This should be applied to all flights, on an "ad hoc" and context basis.

It is obvious that during the strategic planning phase "what if" simulations will have to be run to assess safety and efficiency management strategies.

As more and more information become available, these have to be integrated and used to minimize uncertainties. This is especially important for weather information that has to be included in decision support to assess impacts on capacity and individual trajectories.

Considering that the overall system performance is taken as the main driver to decide whether the decision or event should be initiated, the following questions still need to be answered.

- Will the overall system maintain the required stability under the decided "best time"? This is important, because in an evolving ATM system during the various stages of the planning, buffers and feedback loops are necessary to react on unexpected changes.
- What is the impact of uncertainties in a system when most decisions are taken a long time in advance? E.g. how can this system react on a late passenger? Would it become mandatory to act like those airlines which "won't wait, if you're late"?
- How can an adaptive system be designed where the degree of strategic decisions can be chosen, e.g. depending on uncertainty and/or others factors?

- Do the greater number of functions allocated at strategic planning layers imply more complex and rigid operational scenarios?
- Can ATM system deliver required safety and efficient when most of decisions are allocated at tactical level?

2.3 Best Decision Place

The main questions regarding the second dimension "Decision Place" relates to whether a centric or an autonomous decision approach shall be used in a future ATM system and what correlation shall exists between them. A key aspect to be considered about where to locate the required ATM function will be the impact on the resulting level of complexity.

The implementation will also consider the question on when and where autonomy is a useful extension to the current ATM network and how it can be integrated into the ATM network. As an example, with the planned transition to 4D-trajectory-based operations, as set out in SESAR, a new segregated airspace structure, containing both unmanaged and managed airspace (UMAS/MAS) is being implemented. For these airspaces, while the separation of aircraft is planned to remain under an air navigation service provider as centralized ground function, it is also foreseen to delegate to the airspace user in unmanaged airspace as decentralized and autonomous function.

With regard to the future ATM system, it remains open, whether this segregation concept is able to cope with the growing capacity requirements. It is assumed, that airspace boundaries between UAS and MAS shall be taken flexible like today's Conditional Route and Airspace concept but extending them to a larger set of different airspace user groups with different target functions. As such, either a more complex segregation concept or the return to non-segregated procedures are potential path paved and will be subject to investigation in that second dimension.

Considering new technologies and strategies within research projects, e.g. NextGen and SESAR, a shift to autonomous decision (support) tools at tactical level are candidates for implementation. On the other hand the strategic planning is nowadays centralized, represented by the CFMU, showing good performance until now.

The integration of all ATM components, split into the airborne and ground component on an highly aggregated level, is a key task for a successful automation in the overall ATM system and consequently calibrating the "best place" strategy. It will further be investigated, in which scenario (centric or autonomous) automation will provide a higher overall system performance and if the automation may be limited by complexity, due to high traffic density, solving time restriction or deficit information condition.

It is foreseen that, in a controlled scenario, a certain threshold will become apparent beyond which a human will suffer from information overload if not supported by supporting functions running automatically. Consequently, this dimensional analysis complies with the strategic concept of increasing automation as central means to improve ATM system.

Furthermore an analysis must be performed regarding whether tactical decisions imply autonomous and fully automated processes or, if a strategic decision making process implies centric controlled scenarios.

Considering that the overall system performance is taken as the main driver to decide where the decision or event should be initiated, the following questions still need to be answered.

- What is the level of correlation between complexity and centric processes?
- What is the level of correlation between autonomy and centric process?
- Up to what extend are segregated airspaces structures the solution to the questions of where and when implementing autonomy?
- Do tactical decisions imply autonomous and fully automated processes?
- Is high traffic density/complexity a key factor limiting autonomy?
- In which scenario (centric or autonomous) will automation provide higher overall system performance?
- Does strategic decision making imply centric scenarios?

2.4   Best Decision Player

This third dimension reveals the question on what should be the role of automation and whether and to what extent humans will remain within the trajectory management processes. As the future ATM concept calls for more strategic planning functions to allow for enhanced system efficiency, despite unavoidable interventions such as those generated by de-confliction functions reached through higher automation, this is a crucial answer.

It is the nature of strategic systems to provide increasingly complex solution outcomes (such as e. g. multi-dimensional re-routing advisories for conflicting aircraft several minutes ahead of reaching their closest approach point); the decision obviously requires "management qualities" to handle the situation. These management capabilities are typically not fully addressed by automated functions. As such, a dilemma between two of the main ATM concept targets may occur and priority must be granted to either: Act for more strategic functions where human skills still plays a relevant role although supported by automation, or increase the level of automation where only humans monitoring skills are required.

The question of "best player" is attributed to the area of function allocation, and has been central to the area of automation for a long time. One ongoing initiative in aviation is Collaborative Decision Making which is based on the principle that the actor with the best information should provide that piece of information to be available to all concerned stakeholders. In the context of automation, "best player" means that the actor that is best suited based on some continuum of parameters should perform a function. The basis for selection and grading of such parameters is at the heart of the issue and has been subject to much investigation over the years.

Considering that the overall system performance is taken as the main driver to decide where the decision or event should be initiated, the following questions still need to be answered:

- Should trajectory management (e.g. Trajectory deconfliction, even tactical decisions) be fully automated?
- To what extent do strategic decisions require human intervention?
- How can uncertainty be managed in automated systems?
- Are the current frameworks for automation, cognition and human factors enough to capture ATM singularities?
- Is a fully automated air transport system socially/psychologically acceptable?
- Can the ATM system be decomplexified through automation?
- How to deal with transition issues when implementing higher levels of automation?
- How can resilience be taken into account in automated systems design?
- Does uncertainty require human centered decision making?

The dimension "best player" is interrelated to the other dimensions "best time" and "decision place", where "best time" has been mentioned in previous paragraphs. The dimension "decision place" (central or distributed) is related to "best player" in that the ATM system will inhabit not only the two categories humans and automation, but each category may be divided into sub-categories with their own function allocation, the most apparent ongoing effort being the current re-definition of tasks for air traffic controllers and pilots.

ACKNOWLEDGMENTS

The work presented in this paper is the result of a joint effort by all the members of the HALA! SESAR Research Network.

REFERENCES

SESAR Master Plan D5, April 2008, document no. DLM-0710-001-02-00

Wickens, D. C., Mavor, A., Parasuraman, R., & McGee, J. (1998). *The Future of Air Traffic Control: Human Operators and Automation.* Washington, DC: National Academy Press.

Parasuraman, R., Mason, G., Wiener, E. L. (2008). Humans: Still Vital After All These Years of Automation. Human Factors, 50 (3), 511–520.

Hollnagel, E (2002). Cognition as control: a pragmatic approach to the modelling of joint cognitive systems.

Dave Nakamura Aircraft Working Group Co-Chair (2009). JPDO NextGen. Trajectory Based Operations (TBO), Conference Aircraft Considerations.

Ian A B Wilson (2007)., 4-Dimensional Trajectories and Automation Connotations and Lessons learned from past research. ICNS Conference.

Kaber, D. B., & Endsley, M. (2004). The effects of level of automation and adaptive automation on human performance, situation awareness and workload in a dynamic control task. Theoretical Issues in Ergonomics Science, 5, 113–153.

Parasuraman, R., Barnes, M., & Cosenzo, K. (2007). Adaptive automation for human-robot teaming in future command and control systems. International Journal of Command and Control, 1(2), 43–68.

Section V

The Next Generation Air Transportation System (NextGen)

CHAPTER 29

Innovative Flight Deck Function Allocation Concepts for NextGen

Emmanuel Letsu-Dake, William Rogers, Michael Dorneich, Robert De Mers
Honeywell Aerospace
Golden Valley MN, USA
Emmanuel.letsu-dake@honeywell.com

ABSTRACT

This paper presents two competing flight deck design concepts developed based on different directions that flight deck designs may take in the future because of unpredictable forces and factors. Future operations envisioned in the Next Generation Air Transportation System (NextGen) involve more complex and precise operations, more available information for the flight crew and air traffic control, more highly automated and complex systems, and increased flight crew tasks and responsibilities. To mitigate these potential NextGen issues, it is necessary to identify the functions of a critical element of future aviation systems—aircraft flight decks, and to allocate those functions among the automated and human components of the flight deck. The first design path embodies pilot roles and responsibilities that represent the "pilot as pilot". The second design path embodies roles and responsibilities that represent the "pilot as manager". The two design concepts were used to generate high level human factors flight deck design guidelines that apply to both design paths in a future "NextGen" air traffic environment, focusing on those that are new or different from traditional human factors flight deck design guidelines.

Keywords: NextGen, Flight Deck, Function Allocation

1   INTRODUCTION

Emerging Next Generation Air Transportation System (NextGen) operational concepts represent a radically different approach to air traffic management (ATM) and, as a result, a dramatic shift in the tasks, roles, and responsibilities for the flight crew to ensure a safe, sustainable air transportation system. Future operations envisioned in the NextGen involve more complex and precise operations, more available information for the flight crew and air traffic control, more highly automated and complex systems, and increased flight crew tasks and responsibilities. A major issue of the NextGen concept of operations, particularly in a high-density terminal area, is the potential brittleness of the system to disruptions and the reduced potential for humans to supply the needed resilience for non-normal and off-nominal operations. Off-nominal types of situations, while not involving a system failure or major operational incident, can potentially lead to higher pilot workload, higher stress, and require extremely smooth coordination of flight deck automation and flight crew to equal the resilience of today's systems.

To mitigate these potential NextGen issues, it is necessary to identify the functions of a critical element of future aviation systems—aircraft flight decks, and to allocate those functions among the automated and human components of the flight deck. The study involved analyzing operational and functional requirements of the NextGen environment, developing two alternative flight deck design concepts and resulting high-level human factors flight deck design guidelines. It is expected that new flight deck designs supporting new pilot roles and responsibilities will be required to meet NextGen safety and efficiency goals.

2   METHOD

2.1   Operational Requirements

The goal of the operational requirements task was to define the NextGen flight deck operational requirements that have a major impact on what the future flight deck must be able to do. Due to practical limitations, the scope of the flight deck operational requirements considered in this study was restricted to only those requirements that were judged to have a major identifiable impact on future flight decks in the NextGen environment.

A library of NextGen documents from various sources, including the JPDO NextGen Concept of Operations 3.0 (JPDO, 2009), was reviewed to identify NextGen operations relevant to the flight deck. The implications of current flight deck issues on the NextGen environment were derived from current flight safety data. This analysis identified the operational implications and possible human factors issues that need to be addressed in the design of future flight decks. Finally, the most current database of Honeywell's continuously collected Voice of Customer (VOC) data was used to obtain the perspective of airline operators and Original Equipment Manufacturers (OEMs) on future operations and flight deck issues. The requirements were organized in line with the NextGen operational concepts.

2.2 Functional Requirements

The functional requirements identify what the flight deck must do to accomplish the operational objectives. Two important differences between this flight deck functional analysis and prior ones are (a) expansion of the meaning of systems management and (b) inclusion of task management as a high-level flight deck function (Abbott & Rogers, 1993; Funk, 1991). Both systems management and task management functionality relate to the attention and effort that pilots will expend in mission management, information management, collaborative decision making, task management, and other "non-flying" functions resulting from the increased automation complexity and NextGen flight deck operational demands.

The outcome of the functional requirements gathering exercise was not intended to be an exhaustive list of flight deck functional requirements. Instead, it was aimed at a subset of functional requirements that will most significantly differentiate the design paths developed here. For each operational requirement, a functional requirement was derived describing the function(s) that must be performed to fulfill the flight deck operational requirements.

2.3 Scenario Planning

An analysis exercise, referred to here as the "scenario planning" exercise, was conducted for future flight deck designs. The scenario planning exercise identified key uncertainties influencing the "flight deck of the-future" and outlined the implications for a technical approach to future flight deck designs. The scenario planning exercise included the following steps:
- Identification of the focus question: "What functions will the automation and flight crew of the future flight deck need to perform?"
- Identification of the factors and forces that could influence the functions of the future flight deck.
- Categorization, prioritization, and selection of factors and forces that created the most uncertainty and could exert the most influence on design decisions.
- Translation of those factors and forces into key design-influencing dimensions.
- Development and description of scenarios created by the end points of the dimensions.

The forces and factors identified in the scenario planning exercise were combined into several dimensions. Two dimensions emerged as key: (1) the degree to which required functionality was on the aircraft versus on the ground and (2) the degree to which pilots were active aircraft controllers and decision makers versus managers, monitors, and backups to automated systems. The key relevant finding of the scenario planning exercise was the identification of crew roles and responsibilities as a major dimension of uncertainty that will significantly influence future flight deck design. The end points of this dimension provided the basis for the two design paths explored in this study.

2.4   Function Allocation Framework

A significant amount of literature exists on principled approaches to allocating functions between humans and automation systems (e.g., Palmer, Rogers, Press, Latorella & Abbot, 1995 and Billings, 1991). To categorize the two design paths in terms of function allocation, the framework created by Rogers (1996) was used. This framework focuses on four main allocation decisions (i) who performs the task, (ii) what allocation options are available, (iii) what the authority/permission protocol is, and (iv) who has final authority and override capability. It also considers six stages of information processing: sensing, processing data, assessing situation, setting goals, planning actions and responding.

3   NEXTGEN FLIGHT DECK DESIGN PATHS

3.1   Design 1: Pilot as Pilot

Assumptions

The key assumption for this design path is that pilots will retain final authority and responsibility for mission success factors: safety, efficiency, and passenger comfort. The use of hard and soft envelope protections will likely continue to increase, and automation will likely be assigned to perform more functions and tasks. However, the ability of pilots to intervene at any time and take over control of the aircraft must be retained.

Function Allocation

This design path is intended to support the flight crew in their conventional role as pilots—actively engaged in flight control with final authority and responsibility. Generally, pilots will authorize and delegate task performance to the automation. However, they can intervene and take back control and authority at any time. Pilots may not be actively engaged in all functions all the time. They have the ability and authority to intervene at any time in the performance of any function and direct the automation to perform in different modes or at different levels. For this design path, pilots are physically engaged as much as possible. For flight control, the level of active engagement could even increase over that on today's flight decks. This is dependent on an envisioned level of simplicity of the flight control laws, control and guidance modes, and control devices which can be achieved to minimize workload. Pilots will be fully engaged in the tactical execution of flight deck functions as well as in the planning and goal setting related to those functions. The roles of automation, the overall cockpit layout, as well as display functions and formats, are all designed in ways that support the pilot. Automation will need to assist with information and decision functions more than it does today. It should also assist with achieving the needed simplicity in the pilot interfaces for all functions. This will allow pilots to stay fully involved, take on more tasks and responsibilities, without creating workload that is too high.

Pilot as Pilot Themes

Pilot as Pilot (PAP) is the more conventional design path where design supports pilot involvement in all functions, including the ability to control the aircraft with minimal use of automation. Pilots have final authority; they authorize and delegate tasks to the automation. Both control automation and information automation are designed in the context of systems to aid pilots in performance of flight deck functions. The themes that embody PAP design path are listed in Table 1.

Table 1 Pilot as Pilot Design Themes

Design Theme	Example
Flight deck design simplicity	• Intuitive flight controls with simple relationships between inputs (e.g., throttles, stick) and aircraft parameters (e.g., speed, pitch). • Energy management that is achieved automatically through goal-oriented flight control.
Pilot physical engagement	• Manual control that can effectively simplify flying complex operational procedures and in non-normal airplane states (e.g., engine out). • Flight controls that provide physical feedback.
Graphical information display	• Integrated graphical displays for situation awareness. • Graphical presentation of intent/prediction and energy state.
High-bandwidth, low-workload interaction	• Natural human-machine interfaces (speech, touch) to increase pilot input bandwidth, lower pilot workload, and improve pilot input accuracy. • Input checkers to minimize pilot input errors.
Automation as monitor, pilot as final authority	• Automation monitoring of pilot, including monitoring of pilot inputs, pilot state (workload, fatigue), pilot communication, and inference of pilot intent; monitoring of performance requirements. • Pilot has final authority in flight deck decisions and tasks.
Strategic and tactical task support	• Automation aids to support more strategic and tactical tasks performed by pilot; more strategic task aids and mission management aids; more alerts for tactical issues to help pilot transition from strategic to tactical. • Pilot engaged at tactical level, naturally prepared for response to non-normal and off-nominal situations.

Design Theme	Example
Flight deck interaction style	• Automation needs to be more explanatory than today, including communication of automation intent and status of accomplishing pilot intent. • Consistent alerting and notification aimed at bringing situations, events, and states to the pilot's attention.

Implications and Potential Issues

The key implications of the PAP design path are related to the consistency of this design and function allocation approach with traditional human-centered design principles. Namely, humans perform better in non-normal or off-nominal events if they have been actively involved in performing relevant tasks during normal conditions. Humans can also develop and maintain better situation awareness if they are intimately involved in relevant task performance. The PAP design inherently keeps pilots more involved and actively engaged in performance of various tasks and functions than the second design path. As a result, the benefits of that more intimate involvement should be greater for this design path.

Further, pilots traditionally have final authority not only for legal reasons, but also because humans perform better in dynamic, unanticipated situations where knowledge and skill may need to be applied in unforeseen ways beyond the capability of automation. Therefore, this design path would be expected to be more robust and resilient in terms of handling unforeseen situations that are not only unanticipated by the flight crew, but are also unanticipated by automation.

Potential issues likely to manifest themselves with this design path relate to workload and pilot performance. Pilots will be required to perform in an increasingly complex environment. There will be greater demands for precision and efficiency with inherently smaller margins of safety. The types of pilot errors observed today and the human performance constructs underlying those errors could be exacerbated in the NextGen environment. This is because there will be greater demands on pilot resources. Pilots will still need to be engaged in all functions and retain authority and responsibility for all functions.

3.2   Design 2: Pilot as Manager

Assumptions

The key assumptions for this design path are that pilots will share authority and responsibility with automated systems for mission success factors such as safety, efficiency, and passenger comfort. Most functions will be executed by automation and managed by the pilots. This, in turn, assumes that extremely reliable and highly capable automation exists to perform most NextGen flight deck functions. Automation is responsible for the following information processing tasks: sensing, data processing, planning, and response execution. Situation assessment and goal setting are shared responsibilities between the pilot and automation.

Function Allocation

Pilot as Manager (PAM) is the alternate design path where forces and functions push the pilot toward the role of manager. This design path is intended to support the flight crew in the role of NextGen flight deck information managers and collaborative decision makers with automation. Automation is responsible for the majority of aircraft control and navigation tasks. It is also responsible for information processing tasks for which the pilot is responsible in the Pilot as Pilot design path. A key to this design path is how to support the pilots as managers and monitors in an active, engaging way. This might be characterized as intentionally designing the flight deck in a way that encourages the pilot to be a micro-manager (without the negative connotations). The allocation decisions between pilot and automation for task performance, authority, and override capability are all designed to support this design path. Much more analysis will be required to determine exactly how pilots and automation can share authority in a dynamic way and who has final authority in different situations.

Pilot as Manager Themes

Pilot as Manager is the more novel design path in which design supports the pilot in managing, monitoring, and collaborating with automation to effectively perform flight deck tasks. The pilot will share authority and responsibility with the automation for mission success factors such as safety, efficiency, and passenger comfort. The themes that embody the PAM design path are listed in Table 2.

Table 2 Pilot as Manager Design Themes

Design Theme	Example
Shared authority between pilot and aircraft automation	• Automation and pilots can set goals and override each other. • Clearly defined commanded authority transitions from pilot to pilot, pilot to automation, and automation to pilot.
Function allocation between pilot and aircraft automation	• Automate low-level tasks as much as possible since the list of functions that "machines are better at" continues to grow. • Tactical tasks and most aviate/navigate tasks executed by automation and mission management, setting goals executed by pilot.
Design to maintain pilot engagement	• Pilot involved in all functions; Automation provides goal and execution options from which pilot selects. • Status displays to explain automation behavior.
Tactical task support	• Manual intervention mode for emergencies and unexpected tactical events needs to be simple, quick, and error-tolerant.

Design Theme	Example
	• Awareness and decision aids for identifying and responding to failure, anomalous, and off-nominal conditions across the flight deck functions.
Flight deck design simplicity	• Automation needs to integrate information and make diverse and complex operations more understandable to pilots. • Design needs to be easy to train through knowledge-based and rule-based techniques.
Flight deck interaction style	• More active automation, e.g., prompting, challenging, explaining—generally more communicative about what it's doing and why. • Flexibility to change interaction style based on situation.
Strategic, management, and cognitive task support	• Information content needs to support more strategic tasks and goal setting with longer time horizons; more predictive, trend, and "what-if" information. • More strategic alerts and reminders and mission management aids.

Implications and Potential Issues

The main implication of this design path is that the flight crew will have greater bandwidth to manage all aspects of the flight. They will also be able to stay aware of all important events and situations because they are not spending attention and effort performing lower level manual tasks. With the pace, complexity, and diversity of NextGen operations, this ability to maintain a higher level perspective and perform at a strategic level may be essential. However, workload could still be an issue because of the need for the pilot to monitor more automation and manage more flight deck information.

A potential issue with the PAM design is how to keep the flight crew engaged. Actively performing tasks are known to be more engaging than monitoring or managing tasks. The key is how to make pilots behave as active rather than passive managers. They should be fully involved, skeptical, and challenging, even when the automation is highly reliable and non-normal and off-nominal events are rare. The extensive use of reliable automation could mean infrequent and unexpected automation failures. Such failures could be difficult to detect when they occur. Also, complacency could lead to less monitoring of the flight deck system by the Pilot as Manager, resulting in loss of situation awareness. As a result, even when a failure is detected, the pilot could be less likely to deal appropriately with a situation.

Finally, there are many more unknown implications of the PAM design than the PAP design. For example, due to the differences in the roles of the PAM, special training may be required for pilots to effectively perform their new role. New standard operating procedures (SOPs) may also be required for pilot interaction with automation systems, especially in authority sharing and in emergencies.

Additionally, procedures addressing how pilots allocate tasks between them would be very different since both pilots would normally be in "Pilot Monitoring" roles.

Perhaps the largest potential issue with PAM is how to safely share authority in a dynamic way. It is not certain whether a simple, workable protocol can be established that provides an unambiguous means for pilots to have final authority in some situations and automation to have it in others. The legal issues for shared authority also need to be addressed.

3.3 Design Guidelines

The development of the guidelines consisted of a review of current human-factors-focused flight deck design guidelines, particularly those that seemed important for the challenging characteristics of the NextGen operational environment. This review provided context and ideas not only for guideline content, but also for the granularity and wording of the guidelines. The main elements of the two design paths and the associated pilot tasks were also reviewed.

4 CONCLUSION

Two distinct design paths were developed as a result of operational and functional analyses and function allocation. The design paths for the flight deck differ mainly in the allocation of functions between human operator and automation—Pilot as Pilot and Pilot as Manager. For the PAP design path, the pilot performs roles similar to a pilot on a flight deck of today. The main concern of this design path is pilot workload due to increased information processing requirements and task responsibilities. The pilot is primarily responsible for conventional aviate, navigate and communicate tasks. In the PAM design path, the pilot is responsible for management of automated systems and flight desk tasks, with the majority of flight deck functions performed by automation. The major concern for this design path is keeping the pilot engaged and maintaining his or her situation awareness.

The themes that emerged for the two design paths were used to develop conceptual flight deck design descriptions. Function allocation decisions for information processing tasks assigned task performance, flight deck authority, and override capability for the two designs to the pilot and the automation. The descriptions of the design paths were presented in terms of assumptions about (i) nature of information input and output on the flight deck, (ii) cognitive and information demands of pilot tasks; and (iii) the general role of automated systems and the pilot–automation interactions required. Characteristics of each design path were captured in a set of design themes. The two design paths reflect end points of a continuum of function allocation possibilities. They are intended to describe plausible future flight deck designs that account for NextGen flight deck operational requirements, available technologies, and pilot capabilities.

Based on the high-level design concepts developed for the PAP and PAM paths, a key conclusion is that no matter which direction future flight deck designs evolve in, very significant flight deck design changes will be required to meet the safety and efficiency needs of envisioned NextGen operations. The two designs and their accompanying guidelines identify important flight deck design concepts and issues that will need to be addressed to maintain or improve NextGen safety while increasing air traffic capacity and efficiency. The designs represent end points of a continuum of possible human–automation function allocation strategies. The PAM design in particular may not be practical for a number of reasons, including liability uncertainties, unachievable automation reliability levels, and the difficulty in preserving the ability of the flight crew to supply needed resilience for unexpected situations. However, the PAP and PAM design themes illustrate potentially useful design concepts and highlight potential issues that will need to be solved regardless of the direction of future flight deck changes.

It is expected that many of the concepts and guidelines may be difficult to realize, and with further analysis, it may become clear that some of them are not feasible. However, these design concepts should be developed in more detail so that modeling analyses and empirical studies can be conducted to evaluate performance tradeoffs, workload issues, risks and benefits of new technologies, and pilot engagement issues. Ultimately, flight deck designs will likely take a path somewhere between the end points described here. Many of the issues identified and the design concepts described here are expected to still be applicable.

ACKNOWLEDGMENTS

This work was performed under NASA contract number NNA10DE13C. The authors would like to acknowledge Sandy Lozito of NASA, Ames for her engagement and guidance in helping us perform the work

REFERENCES

Abbott, T. S., and Rogers, W. H. (1993). Functional Categories for Human-Centered Flight Deck Design, Proceedings of the 12th Digital Avionics Systems Conference, New York: AMA/IEEE.

Funk, K. (1991). Cockpit task management: Preliminary definitions, normative theory, error taxonomy, and design recommendations, The International Journal of Aviation Psychology, 1(4), 271-285.

JPDO. (2009). Concept of Operations for the Next Generation Air Transportation System, Version 3.0, Washington, DC, 153 pp.

Palmer, M.T., Rogers, W.H., Press, H.N., Latorella, K.A., and Abbot, T.S. (1995). A Crew-Centered Flight Deck Design Philosophy for High-Speed Civil Transport (HSCT) Aircraft, NASA Technical Memorandum 109171.

Rogers, W.H. (1996). A Principled Approach to Allocation of High Speed Civil Transport (HCST) Flight Deck Functions, Draft NASA Contractor Report, 1996.

CHAPTER 30

Effectiveness of Training on Near-Term NextGen Air Traffic Management Performance

Ariana Kiken, Thomas Z. Strybel, Kim-Phuong L. Vu, & Vernol Battiste

California State University, Long Beach
Long Beach, CA
aegkiken@gmail.com

ABSTRACT

NextGen, a planned introduction of automated tools into the National Airspace System, is expected to change the roles and responsibilities of air traffic controllers. Consequently, new methods for training need to be developed to train ATCs in the use of both current-day and NextGen procedures. The present study compared Part-Whole and Whole-Task training methods for student ATCs in a mixed-equipage, near-term, NextGen environment. Participants were 15 students enrolled in a 14-week radar simulation internship. The Part-Whole Lab learned only manual conflict detection and resolution skills during the first 7 weeks of the internship followed by manual skills and NextGen automated tools training for next 7 weeks. The Whole-Task Lab learned manual and NextGen tools together for the entirety of the 14 weeks. Although training did improve performance, situation awareness and workload, the most important factor was individual differences within each training group.

Keywords: NextGen, part-task training, whole-task training, air traffic control

1   NEXTGEN AND TRAINING

An expected increase in air transportation demand over the next two decades has led to the proposal of an overhaul of the air traffic management (ATM) system

(JPDO, 2010). The goal of the Next Generation Air Transportation System, NextGen, is to increase safety, efficiency, and dependability within the National Airspace System (NAS) as traffic density continues to increase. NextGen is expected to incorporate automated tools for air traffic controllers (ATCs) and pilots. These tools are intended to decrease the workload of ATCs so that they may safely manage more aircraft. NextGen may also change the role of the ATC. Current-day ATM procedures require ATCs to interact to some extent with each aircraft that travels through his or her sector. Many of the proposed NextGen concepts at least partially shift responsibility from ATCs to pilots or automated agents, thus making the role of the ATC more concerned with passively monitoring than actively controlling air traffic (Metzger & Parasuraman, 2001). A shift in ATC responsibility, coupled with the increase in air traffic necessitates an evaluation of training procedures that may be used to train both existing and new ATCs in the use of new, automated tools. Of particular concern is the training required to thoroughly prepare ATCs for an airspace that is composed of both NextGen equipped and unequipped aircraft. A mixed-equipage airspace is expected in the near-term as NextGen improvements are gradually introduced (JPDO, 2010).

The present study examines two methods of training ATM with both NextGen and current day procedures for novice controllers. The goal of our research is to determine if training method has an impact on the amount of improvement over the course of training and if either type of training enhances particular aspects of ATM.

ATM is a highly complex task involving many interrelated elements. It can be challenging to train complex tasks without overloading cognitive resources because the learner must understand the task simultaneously on the global and component levels. Often times, this imposes a load on working memory, making it difficult to process all of the task elements (Pollock, Chandler, & Sweller, 2002). It is therefore critical that training help to gradually establish a mental model or schema for the learner that can be stored in long-term memory and serve as a structure on which details can be placed (Pollock et al., 2002).

Training literature is quite dense and spans decades, yet there is ongoing debate over the optimal methods. Two methods have emerged as being well-suited for training highly complex tasks with many interrelated components. These training methods are the part-task method and the whole-task method (Salden, Paas, Broers, & van Merriënboer, 2006). In the part-task training method, tasks are decomposed into constituent subtasks, and each subtask is trained individually. Whole-task training methods train all of the subtasks together as the whole task.

Variations of both methods exist. One variation of part-task training is called forward-chaining. In forward-chaining, training begins with one subtask, and additional subtasks are subsequently added until all of the subtasks are trained together as the whole task. While the concept of decomposing a task into subtasks is quite straightforward, in practice it is not always clear how to best decompose the task in order to train individual subtasks (van Merriënboer et al., 2003). Difficulty parsing subtasks is one of the primary disadvantages of part-task training methods. Moreover, part-task training methods fail to provide practice in coordination of subtasks (van Merriënboer et al., 2003). For these two reasons whole-task training has emerged as a more effective method of complex-task training.

Few studies have assessed part-task training in the ATM domain with NextGen tools. Billinghurst et al. (2011) assessed the order of training of current-day, manual conflict detection and resolution techniques, and NextGen tools. Students studying to become ATCs served as participants during the course of their 16-week training course. One group learned strictly manual conflict detection and resolution skills then NextGen tools, while the other group learned these components in the reverse order. Order of training was not found to have affected ATM performance; however, performance was better in the testing session that immediately followed manual conflict detection and resolution training. Kiken et al. (2011) evaluated the performance of retired ATCs during approximately 18 hours of simulation training. Training progressed according to a forward-chaining training procedure. ATCs were first trained on center sector procedures followed by Next-Gen tools and finally an online probe query procedure. ATC performance initially declined with the introduction of NextGen tools but eventually reached the level of performance seen with manual techniques.

Whole-task training has been proposed as a method to overcome "fragmentation" of subtasks during training (van Merriënboer et al., 2003). Because subtasks are trained together, the primary advantage of whole-task training is its ecological validity in coordination of subtasks (Paas & van Gog, 2009). The ecological validity of whole-task training comes at the cost of an increased cognitive load (Paas & van Gog, 2009). A high cognitive load may lessen the effectiveness of the training, lengthening the amount of time that is necessary for successful learning (Pollock et al., 2002).

The present study investigated the effectiveness of part-task versus whole-task training on ATM in mixed-equipage airspaces. We determined the effectiveness of each training method on performance, situation awareness and workload.

2   METHOD

2.1   Participants

Fifteen students, enrolled in an FAA Collegiate Training Initiative (CTI) program served as participants. The students were enrolled in a 14-week radar-simulation internship at the Center for Human Factors in Advanced Aeronautics Technologies (CHAAT). Eleven males and 4 females with a mean age of 23.4 participated. Time spent studying in the CTI program ranged from less than 1 year to 2 years, with a mode of 1 year. Participants were compensated $10.00 per hour for their participation in the experimental simulation.

2.2   Apparatus

Multi Aircraft Control System (MACS) software, simulating Indianapolis Center (ZID-91), was used for both the training and experimental simulations.

Indianapolis Center includes arrival and departure streams to and from Louisville International Airport as well as overflights.

The NextGen tools included in the present study were: conflict probing, conflict alerting, a trial planner, and Data Comm. Conflict probing is an algorithm that detects conflicts within a 6-minute window. Potential conflicts between 2 equipped aircraft were visually alerted on the radar scope. A trial planner tool was available for equipped aircraft, allowing ATCs to dynamically visualize changes to aircraft routes while simultaneously probing for potential conflicts. Integrated Data Comm allowed commands to be sent directly to the aircraft CDU where they could be reviewed and executed by the pseudopilot.

2.3   Procedure

One time per week, participants received 3 hours of radar simulation training and 2 hours of classroom lecture. The instructor for both training groups was a retired, radar-certified ATC. Students chose to participate in one of two lab groups, each assigned a different method of training by the researchers. During the first 7 weeks of the internship, training method differed between lab groups. One lab was taught only manual conflict detection and resolution skills and verbal communication protocols. This group is henceforth referred to as the Part-Whole lab. The other lab, referred to as the Whole-Task lab learned manual skills and verbal communication procedures in combination with NextGen tools. The Whole-Task lab trained on scenarios populated with 75% NextGen equipped aircraft for the entire 14 weeks of training. During weeks 1 through 7, the Part-Task lab trained on identical scenarios with no NextGen equipped aircraft. NextGen tools were introduced to the Part-Whole lab during week 8 of training, at which point the Part-Whole lab began training on 75% equipped scenarios. The actual traffic density was the same for both groups.

A midterm test was administered to both labs between weeks 8 and 9, and a final test was administered after week 14. Prior to the midterm test, participants in both training groups learned basic ATM movement techniques: altitude, speed, heading, and structure. Participants managed traffic using only one of these methods per scenario until passing the method. After passing all 4 methods, participants were deemed lab "Journeymen" by the instructor. At the time of the midterm, approximately half, 8 participants, were lab Journeymen and 7 were not Journeymen. All participants reached Journeyman status by the time of the final testing session.

Three, 50-minute, experimental scenarios were presented at each testing session. The 3 scenarios differed in percentage of NextGen equipped aircraft, traffic patterns, and conflict locations. Measures were taken during scenario development to equate the scenarios as much as possible on difficulty of conflicts, number of aircraft being managed at any given time, and total number of aircraft in each scenario. The 3 NextGen equipages were 0%, 50% and 100% equipped. Based on the finding by Hah, Willems, and Schulz (2010) that controller workload is substantially reduced when at least half of aircraft are equipped, scenarios in the present experiment were designed to illicit strong differences in workload. The differences between equipages in the present study were designed to be large enough that they would likely be easily detected by the participants and result in different strategies being employed to manage traffic. Each scenario had 32 aircraft that entered the sector over the course of 50 minutes and 8 pre-planned conflicts that were spaced approximately 5-7 minutes apart. Conflict alerting varied based on scenario equipage. The 0% scenario had no conflict alerting, the 50% had half of the conflicts alerted, and the 100% scenario had all 8 conflicts alerted.

In addition to ATM, participants were asked to answer probe questions regarding their SA and workload. Questions were presented on a touch screen adjacent to the participants' simulated radar scope. The Situation Present Awareness Method (SPAM), an online probe query technique was used (Durso & Dattell, 2004). Participants received an auditory signal when a question was ready to be answered. If participant workload allowed, the participant indicated that he or she was ready for a question to be presented by pressing the ready prompt. A question was then presented with multiple choice answer options provided. The ready prompt and the question both timed out after 60s if the participant was unable to respond. Ready latency was used to measure workload; probe question latency and accuracy were used to measure situation awareness.

2.4 Design

The present study was a 2(Training Method) x 2(Testing Session) x 3(Percent of Equipped Aircraft) mixed factorial design. Testing session and Percent of Equipped Aircraft were both within subjects factors. A two-level between subjects blocking variable, "Journeyman" status at midterm was included in the analyses. An additional two-level variable, Question Type, was included in analyses of probe performance. Levels of Question Type were conflict questions and status questions. Conflict question regarded aircraft in conflict in the past, present, or future, and status questions regarded the status of aircraft in the sector.

3 RESULTS

Performance was measured by the number of LOSs in each scenario. Probe ready latency, question latency, accuracy, and responses to workload probes were also measured. Probe ready latency gives an indication of workload. Longer latencies indicate higher workload. Probe question latency and accuracy are measures of situation awareness (SA): longer latency and lower accuracy are associated with lower SA (e.g., Bacon et al., 2011). Workload probes asked participants to rate their workload at four different times per scenario. Participants rated their workload on a Likert-type scale of 1 ("very low") to 7 ("very high"). Additional variables were measured; however, they are outside of the scope of this paper and are not reported. Mixed-factorial ANOVAs were performed. Hyunh-Feldt corrections are reported where appropriate.

3.1   LOS

Scenario Equipage was found to have a significant effect on average number of LOS, $F(2, 22) = 7.17$, $p = .004$. More LOSs occurred in the 0% scenario compared to the 50% and 100% scenarios. Journeyman status was also found to have a significant effect on LOS, $F(1, 11) = 4.88$, $p = .049$. Non-Journeymen had more LOSs than Journeymen. A significant interaction between Scenario Equipage and Journeyman was also found, $F(2, 22) = 6.87$, $p < .004$, as shown in Figure 1. Non-Journeymen had more LOSs than Journeymen in the 0% scenario, $p = .023$. There were no significant differences between Journeymen and Non-Journeymen on the 50% or 100% equipped scenarios, $p > .05$.

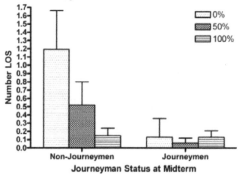

Figure 1 Average LOS by Scenario Equipage and Journeyman status.

3.2   Probe Question Accuracy

A 2 (Training Method) x 2 (Testing Session) x 3(Scenario Equipage) x 2(Question Type) mixed factorial ANOVA with the Journeyman blocking variable was performed. Scenario Equipage was significant, $F(1.86, 20.40) = 14.50$, $p < .001$ with participants answering more questions correctly in the 0% scenario ($M = .85$, $SEM = .02$) than either the 50% scenario ($M = .74$, $SEM = .03$) or the 100% scenario ($M = .74$, $SEM = .02$).

A significant interaction was found between Training and Journeyman status, $F(1, 11) = 4.84$, $p = .05$. For the Whole-Task group, Journeymen answered more probe questions correctly than Non-Journeymen. There was no difference for the Part-Task lab, $F(1, 11) = .02$, $p = .883$.

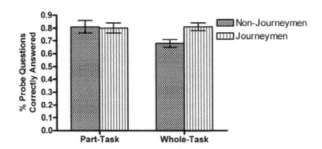

Figure 2 Percent probe questions correctly answered by Training method and Journeyman status.

3.3   Probe Question Latency

A main effect of the probe question latency was found for Scenario Equipage, $F(2, 22) = 9.71$, $p = .001$. It took participants longer to answer questions during the 100% scenario than it did during the 0% and 50% scenarios. A main effect of Testing Session was also found, $F(1, 11) = 6.75$, $p = .025$. probe question latencies were longer at the midterm than at the final.

The main effects were qualified by an interaction between Scenario Equipage and Testing Session, $F(1.83, 20.14) = 4.86$, $p = .021$. As seen in Figure 3, for the 100% scenario, question latency was significantly longer at the midterm than at the final, $p = .013$. Differences at the 0% and 50% levels were not significant, $p > .05$.

An interaction between Scenario Equipage, Testing Session, and Journeyman status was also found when the Question Type variable was included in the mixed-factorial ANOVA, $F(2, 22) = 3.98$, $p = .033$. An interaction between Testing Session and Scenario Equipage was significant for Non-Journeymen, $F(2, 10) = 6.09$, $p = .019$. It was not significant for Journeymen, $F(2, 12) = 1.59$, $p = .245$. As seen in Figure 4, results indicated that at the Midterm, question latencies were significantly longer for the 100% scenario than for the 0% scenario, $p = .005$ and the 50% scenario, $p = .026$. No differences were found for the final, $p > .05$.

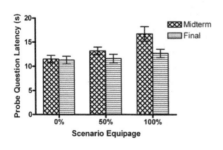

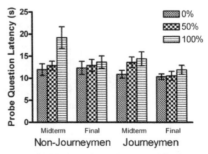

Figure 3 Probe Question latency by Scenario Equipage and Testing Session.

Figure 4 Probe Question latency by Scenario Equipage, Testing Session, & Journeyman.

3.4   Ready Latency

Main effects were found for Scenario Equipage, $F(2, 22) = 4.05$, $p = .032$, Testing Session, $F(1, 11) = 14.33$, $p = .003$, and Journeyman status $F(1, 11) = 5.41$, $p = .04$. Ready prompt latencies were shorter for the 0% scenario ($M = 4.5$s, $SEM = .88$) than the 100% scenario ($M = 6.7$s, $SEM = .88$), $p = .041$. Differences on the 50% scenario ($M = 5.9$s, $SEM = .65$) were not significant, $p > .05$. Ready prompt latency decreased from the midterm testing session ($M = 7.0$s, $SEM = .97$) to the final testing session ($M = 4.3$s, $SEM = .43$). Journeymen ($M = 4.1$ s, $SEM = .90$) were faster at responding to the ready prompt than Non-Journeymen ($M = 7.2$ s, $SEM = 4.13$).

3.5   Probe Workload Ratings

Reported workload was relatively low overall when collapsed across testing sessions. A main effect of Scenario Equipage was found, $F(2, 22) = 5.05$, $p = .016$. Workload was lower on the 100% scenario ($M = 2.38$, $SEM = .15$) than on the 50% scenario ($M = 2.77$, $SEM = .15$), $p = .011$. Workload also tended to be lower on the 100% scenario than the 0% scenario ($M = 2.71$, $SEM = .15$), although the difference was marginal, $p = .056$. A main effect of Testing Session on workload ratings was found, $F(1, 11) = 5.82$, $p = .034$. Workload ratings were lower at the final testing session than at the midterm testing session.

An interaction between Scenario Equipage, Training, and Journeyman status was found, $F(2, 22) = 6.54$, $p = .006$. As seen in Figure 5, there was a significant interaction between Scenario Equipage and Journeymen status for the Whole-Task lab, $F(2, 12) = 5.56$, $p = .02$. The interaction was not significant for the Part-Whole lab, $F(2, 10) = 1.58$, $p = .25$. For Whole-Task lab Journeymen, workload ratings were significantly lower on the 100% scenario than on the 0% scenario, $p = .050$ and the 50% scenario, $p = .027$. Differences were not significant for the Whole-Task lab Non-Journeymen, $p > .05$.

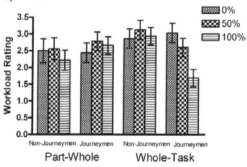

Figure 5 Probe workload ratings by Scenario equipage, Training method, and Journeyman status.

4   DISCUSSION

This preliminary experiment examined Part-Whole and Whole-Task methods of training student ATCs on manual and NextGen ATM procedures. The Part-Whole group received manual procedures training alone followed by training on both manual and NextGen procedures for a mixed-equipage airspace. The Whole-Task group trained on manual and NextGen procedures together for the entire course. Students in the Part-Whole training group had more practice with manual ATM compared to students in the Whole-Task training group. Manual skills training was previously shown to affect student performance more than NextGen training (Billinghurst et al., 2010). Students in the Whole-Task training group had more practice coordinating manual and NextGen procedures. Nevertheless, the most significant differences obtained were not related to training method but to the overall student ability, based on the speed with which participants attained Journeyman status.

Approximately half of the students in each training group attained Journeyman status by the midterm as assessed by the instructor. Overall, Journeymen had fewer LOS, lower workload, and higher SA, based on probe latency. With respect to probe accuracy, Non-Journeymen in the Whole-Task group were less accurate compared with Whole-Task Journeymen, Part-Task Non-Journeymen and Part-Task Journeymen (see Figure 2). Apparently, the lack of manual skills training in the Whole-Task group affected the less-skilled, Non-Journeymen, while Part-Whole training may have increased SA for the less-skilled students.

Training did produce improvements in performance, workload and SA between midterm and final, and in some cases, Non-Journeymen showed greater improvement. The workload results with respect to equipage were contradictory: subjective ratings of workload indicated that workload was lowest in the 100% scenario, yet objective ready latencies indicated that workload was lowest in the 0% scenario. The disagreement between these two workload measures may reflect the students' inexperience in evaluating his/her workload. In summary, performance, workload, and SA metrics indicated that while training method did lead to differences between groups, individual differences may be a more important variable to consider when developing a training program for ATCs to learn manual and NextGen skills. In future experiments on training we intend to examine the cognitive and personality factors that may be responsible.

ACKNOWLEDGEMENTS

This research was supported by the NASA cooperative agreement NNX09AU66A, *Group 5 University Research Center: Center for Human Factors in Advanced Aeronautics Technologies* (Brenda Collins, Technical Monitor).

REFERENCES

Bacon, L.P., T. Z. Strybel, K.-P. L. Vu, J. M. Kraut, J. H. Nguyen, V. Battiste, and W. Johnson. 2011. Situation awareness, workload, and performance in midterm NextGen: Effect of variations in aircraft equipage levels between scenarios. *Proceedings of the 16th International Symposium on Aviation Psychology*: 32–37. Dayton, OH.

Billinghurst, S. S., C. Morgan, R. C. Rorie, A. Kiken, L. P. Bacon, K. P.-L. Vu, T. Z. Strybel, and V. Battiste. 2011. Does the order of introducing NextGen tools impact student learning of air traffic management skills? *Human Factors and Ergonomics Society Annual Meeting Proceedings* 55: 128–132. Las Vegas, NV.

Durso, F. T. and A. Dattel. 2004. SPAM: The real-time assessment of SA. In *A Cognitive Approach to Situation Awareness: Theory, Measures and Application*, eds. S. Banbury & S. Trembley 137–154. New York.

Hah, S., B. Willems and K. Schulz. 2010. The evaluation of data communication for the future air traffic control system (NextGen). *Proceedings of the 54th Annual meeting of the Human Factors and Ergonomics Society* 99–103. Santa Monica, CA.

Joint Planning and Development Office (JPDO) 2010. *Concept of operations for the next generation air transportation system version 3.2.* Accessed September 3, 2011, http://jpe.jpdo.gov/ee/docs/conops/NextGen_ConOps_v3_2.pdf

Kiken, A., R. C. Rorie, L. P. Bacon, S. Billinghurst, J. M. Kraut, T. Z. Strybel, K.-P. L. Vu, , and V. Battiste. 2011. Effect of ATC training with NextGen tools and online situation awareness and workload probes on operator performance. In. *Lecture Notes in Computer Science*, eds. G. Salvendy, & M. J. Smith. Human Interface and the Management of Information, Interacting with Information 6772: 483–492. Berlin, Germany.

Metzger, U., and R. Parasuraman. 2001. The role of the air traffic controller in future air traffic management: An empirical study of active control versus passive monitoring. *Human Factors* 43(4), 519–528.

Paas, F., and T. van Gog. 2009. Principles for designing effective and efficient training of complex cognitive skills. In *Review of human factors and ergonomic,* ed. P. R. DeLucia. 7: 166–194. Santa Monica, CA.

Pollock, E. P. Chandler, J. Sweller. 2002. Assimilating complex information. *Learning and Instruction* 12: 61–86.

Salden, R. J. C. M., F. Paas and J. J. G. van Merriënboer. 2006. A comparison of approaches to learning task selection in the training of complex cognitive skills. *Computers in Human Behavior* 22, 321–333.

van Merriënboer, J. J. G., P. A. Kirschner and L. Kester. 2003. Taking the load off a learner's mind: Instructional design for complex learning. *Educational Psychologist* 38(1): 5–13.

CHAPTER 31

Student Impressions of Different Air Traffic Control Training Methods to Introduce NextGen Tools

Meghann Herron, Sabrina Billinghurst, Tiana Higham, Vyha Ho, Kevin Monk, Ryan O'Connor, Thomas Z. Strybel, Kim-Phuong L. Vu

California State University, Long Beach
Long Beach, CA
meghann.herron@gmail.com

ABSTRACT

As a preliminary step toward identifying the best training techniques for introducing NextGen tools and concepts to air traffic control, the present paper reports data from a debriefing session with 15 students who were trained to manage a sector with NextGen equipped and unequipped aircraft after 14 weeks of practice with a particular training schedule. Participants in the Whole-Task Training group learned manual air traffic control skills in conjunction with potential NextGen tools. Participants in the Part-Whole training group were trained with only manual skills for 7 weeks, and then were subsequently trained with manual skills and NextGen tools during the remaining 7 weeks. In general, students' performance benefited from both training procedures. The students indicated that it was easy to learn NextGen tools and that the use of the tools allowed them the capability to manage more traffic. However, if all the aircraft in their sector were NextGen equipped, the participants indicated that the air traffic management task became boring, and that it was hard for them to pay attention to the task. Moreover, after being exposed to NextGen tools, participants in the Whole-Task Training group felt it would be more difficult to revert back to only the manual mode of air traffic management than participants in the Part-Whole group. Implications of these findings for training are discussed.

Keywords: NextGen, training, air traffic control

1   INTRODUCTION

In the United States, a program called the Next Generation Airspace Transportation System (NextGen) is being implemented to allow the National Airspace System (NAS) to meet the needs of increasing demands in air travel while maintaining safy and efficient operations. NextGen developments are overseen by the Joint Planning and Development Office (JPDO), which includes many key organizations including, among others, the Federal Aviation Administration (FAA), Department of Transportation, and the National Aeronautics and Space Administration (NASA). The FAA (2009) indicates that NextGen will allow more passengers, more cargo, and more types of aircraft to access the NAS. However, to meet the goals of NextGen, new concepts of operation and technologies have to be conceptualized, researched, and tested for implementation. These new technologies and concepts of operation will likely alter the roles of the human operators by providing them with more tools. For example, air traffic controllers (ATCos) may be given tools that detect aircraft on conflicting trajectories and other tools that offer resolutions to these conflicts. These technologies will reduce the ATCos workload, but may also have the unintended consequence of changing the ATCos's task of active control to passive monitoring (see e.g., Prevot et al., 2012; Vu et al., 2012). Being passive monitors can decrease the ATCos situation awareness and performance (see e.g., Pop et al., 2012). Because many NextGen tools and concepts require that human operators interact with the technology, human factors analyses must be conducted.

According to the JPDO (2007), there are three phases of NextGen. Phase 1 (2007-2011) includes the development of avionics technologies to support potential NextGen concepts. Phase 2 (2012-2018), will include the initial implementation of the NextGen concepts that will be followed by continued and more widespread implementation of NextGen technologies, concepts, and procedures in Phase 3 (2019-2025). Much research has been conducted on actual and potential NextGen technologies and concepts of operations (see, e.g., the Special Issue of the International Journal of Human-Computer Interaction on Human Factors and Human-Computer Interaction Considerations for NextGen: http://www.tandfonline.com/toc/hihc20/28/2). The goal of this paper focuses instead on determining the best methods for training new operators on the use of NextGen tools.

As an initial step toward identifying the best training techniques for introducing NextGen tools and concepts, the present paper reports data from a debriefing session with 15 students who were either trained to perform manual air traffic control skills (representative of current day air traffic management techniques) in conjunction with potential NextGen tools (i.e., the Whole-Task training group) or trained with manual skills prior to the introduction of potential NextGen tools (i.e., the Part-Whole training group). The goal of the paper was to identify training themes and students' impressions of the two types of training techniques employed in the larger study. Performance data from the simulation study is reported by Kiken et al. (in press) and will only be referred to in the present paper to provide context for the effectiveness of the different training methods, as indicated by the participants.

The Whole-Task and Part-Whole training paradigms were used in the present study because the task of air traffic management (ATM) is a complex task involving the integration of many individual techniques for managing traffic. Whole-task refers to use of manual techniques in conjunction with NextGen techniques. We consider this the whole task because NextGen equipped aircraft are likely to be introduced into the NAS gradually (i.e., during Phases 2 and 3 of NextGen) and will result in a near-term environment where NextGen equipped aircraft must be managed alongside of NextGen non-equipped (or current-day equipped) aircraft. For the Part-Whole group, we trained the students in manual skills before introducing NextGen ATM techniques. This order of training was selected based on a recent simulation conducted by Billinghurst et al. (2010), where performance was better in the testing session that immediately followed manual ATM training than NextGen training.

Within each group, the specific skills trained to the student participants in this study included identifying conflicting aircraft and estimating time to loss of separation (LOS). A loss of separation is defined as aircraft being within 1,000ft vertical and 5 nautical miles lateral of each other. Students were trained on four types of methods for resolving conflicts using a pure-part, segmented training procedure (Briggs & Naylor, 1963), in which each skill is mastered, one at a time, before all skills are combined. This procedure was employed because it would be challenging to train all the intricacies of ATM to students over the 14-week period, and doing so would likely overwhelm the students. Pollock, Chandler, and Sweller (2002) argue that simplifying the task during training will allow the student to develop appropriate mental models for the task which will facilitate overall learning.

2   METHOD

2.1   Participants

Fifteen students from Mount San Antonio College (Mt. SAC) an FAA CTI institution, were recruited to participate in this simulation. The participants reported having an average of 1.2 years of coursework completed at Mt. SAC prior to internship. Fifteen volunteered to participate in the study. Eleven participants were male and 4 female (M_{age} = 23.53 years, SD = 2.80 years). Participants were compensated $10 an hour for their participation in the experimental simulation.

These students were enrolled in a 14-week radar simulation internship offered by California State University Long Beach's Center for Human Factors in Advanced Aeronautics Technologies (CHAAT), a Group 5 NASA University Research Center. The internship provided radar training with a simulated DSR station in which the students were trained to manage en route air traffic using a variety of procedures, described below. The participants were also taught phraseology for communicating with the aircraft inside their sector (piloted by other students in the class).

2.2 Simulation Apparatus and Environment

The simulation was run using the Multi Aircraft Control System (MACS) software developed by the Airspace Operations Laboratory at NASA Ames Research Center (Prevot, 2002). The simulation environment is considered to be of medium fidelity. The simulated airspace was Indianapolis Center, ZID, Sector 91. The simulated traffic in the airspace included arrivals and departures to and from Louisville Standiford Airport (SDF), as well as overflights.

For the NextGen unequipped aircraft, the students were trained to mentally "sweep" the radar scope to identify aircraft that were on conflicting trajectories. When the students identified conflicting aircraft, they were to resolve the conflict. All communications with unequipped aircraft occurred verbally through a voice communication system using push-to-talk headsets. For the NextGen equipped aircraft, the students were told that these aircraft had conflict alerting, where conflicts between 2 equipped aircraft were visually alerted (flashing red) on the radar scope, with a time estimate to LOS indicated (within a 6-minute window). For NextGen equipped aircraft, the students could use a trial planner tool to alter the altitude or route of the aircraft. The trial planner was coupled with a conflict probe that allowed the students to visualize changes in aircraft route modifications and see whether these route modifications resulted in secondary conflicts. Finally, the equipped aircraft had integrated DataComm, where flight plan changes could be sent to the aircraft and uploaded by the flight deck. The participants interacted with experimental confederates who piloted the simulated aircraft. The pilots were referenced to as "pseudopilots". The pseudopilot "flew" all of the aircraft in the sector, and acted as a "ghost" controller, taking and making handoffs for aircraft entering/exiting the active sector.

In addition to the participants' radar screen, another touch screen computer was located adjacent to the radar monitor. This computer was used to present the participants with workload and situation awareness probes. Although the situation awareness and workload measures obtained through the probes are not the main variables of interest in the present paper, we mention them here because we wanted to know if students felt that the use of probes to capture their workload and situation awareness interfered with their learning and performance.

2.3 Procedure

For the radar internship, participants met once a week. The internship consisted of a 3-hour morning lab, 2-hr afternoon classroom lecture, and 3-hour afternoon lab. Students in the morning lab were trained for the first 7 weeks using only manual ATM skills, where they had to detect conflicts between aircraft and generate their own resolutions. Once these resolutions were generated, the students had to follow verbal communication protocols to communicate the clearances with the pseudopilot that was controlling the aircraft. On the 8th week of the class, this lab was introduced to NextGen tools. The students then managed their sector with a 75% NextGen equipped and 25% NextGen unequipped mixture. More equipped aircraft were used in the sector because the benefits of NextGen tools are evident when the number of equipped aircraft exceeds 50% (Hah et al., 2010). At this point, a mid-term test was administered. For the remainder of the internship, the students managed traffic using both NextGen tools for equipped aircraft and manual techniques for unequipped aircraft. A final test was administered at the end of the internship.

Students in the afternoon lab received Whole-Task Training, where they were exposed to both NextGen equipped and unequipped aircraft from the beginning of the internship, with a 75% equipped-25% unequipped mixture. Although the training method did not change, the students also engaged in the midterm and final test at the same time as the first lab and experienced the same scenarios for testing. The classroom component, involved instruction on the characteristics of the simulated airspace, exercises in identifying conflicting trajectories, and estimating time to LOS, and phraseology and air traffic manage techniques. The instructor for both training groups and the lecture component was a retired, radar-certified ATCo, who has been conducting this internship course for the past three years.

As indicated in the Introduction, the students were trained with specific conflict resolution and air traffic management techniques using a pure-part segmented procedure. Specifically, participants learned to manage traffic solely using one of four separation techniques; vectoring aircraft (i.e., issuing heading clearances), altitude separation (i.e., issuing climb or descend clearances), speed separation (slowing or increasing aircraft speed) and structured traffic flows (i.e., creating "blocks" or "corridors" for specific air traffic patterns that will ensure no LOS if all conventions are followed). Participants managed traffic using only one of these methods and the same training scenario until they were able to complete an entire scenario without any LOS. After passing all 4 methods of air traffic management, the student was considered certified by the instructor as "Lab Journeyman". Lab Journeymen were given harder scenarios and could manage the air traffic using any technique or combinations of techniques. At the time of the midterm, 8 participants were considered Lab Journeymen, four in each lab period; all participants reached the Lab Journeymen status by the end of the internship.

For both the mid-term and final tests, participants performed three 50-minute, experimental scenarios that differed 3 NextGen equipages, where 0%, 50% or 100% of the aircraft in the sector were NextGen equipped. Each scenario had 32 aircraft that entered the sector over the course the scenario, with 8 predetermined conflicts, spaced between 5 and 7 minutes apart. Because conflict alerts only occurred for equipped-equipped aircraft, none of the conflicts were alerted for the 0% scenario, half of the conflicts were alerted in the 50% scenario, and all conflicts were alerted in the 100% scenario. As noted earlier, participants in the mid-term and final test were asked to answer probe questions regarding their situation awareness and workload using the Situation Present Awareness Method (SPAM); Durso & Dattell, 2004).

At the end of the final test, which was also the end of the internship, students were given a questionnaire that asked for their impressions about the training method used, the use of probe questions to assess their situation awareness and workload during the test sessions, and their thoughts and comments about NextGen tools implented in the simulation. Participants were asked to comment on these questions in a group debriefing as well.

3 RESULTS

A content analysis was performed on the students' answers to the debriefing questionnaire to reveal training issues as well as the participants' impression of NextGen tools used in the simulation. Students also provided recommendations for training and indicated that there were a couple of desireable features of the interface display that NextGen researchers may want to examine and incorporate into future research.

3.1 Effects of Training Type

For the morning lab, which received the Part-Whole training, 4 of the 7 participants achieved the lab Journeymen status at the mid-term. For the afternoon lab, which received the Whole-Task Training, 4 of the 8 participants achieved the lab Journeymen status at the mid-term. Thus, in our internship, the training method did not appear to influence the number of participants who achieved the lab Journeymen status.

Participants from both training groups indicated that they were well-trained for the testing scenarios, with only one student from the Part-Whole group indicating that s/he could have benefited from more training with DataComm commands at the midterm. In general, though, the students thought that their performance was better at the final testing than at the midterm, regardless of training group. All students stated that it was easy to learn NextGen tools, and although there were some drawbacks with DataComm (i.e., pilot response was not as efficient), several students in the Whole task group indicated that the conflict alerting with equipped aircraft helped them with their ability to scan the scope and the trial planner made it easier for them to separate aircraft.

When asked how many class periods did the students think they needed to feel comfortable managing mixed equipage traffic in their sector, the students in the Part-Whole group indicated 1-3 sessions (with three students indicating 1 session, three indicating 2 sessions, and one indicating 2-3 sessions), whereas in the Whole Group training condition, five students indicated 3 sessions were needed and the remaining three students indicating 2 sessions needed. Thus, it appears that once the students mastered manual techniques, the addition of NextGen tools did not add much difficulty to their task.

After being exposed to the NextGen tools, five of the seven participants in the Whole-Task training group indicated that it would be more difficult to go back to only manual air traffic management techniques, whereas only two of the eight participants in the Part-Whole training group indicated that it would be difficult to manage traffic with only manual techniques. Thus, the group that started with learning to manage traffic with only manual techniques indicated little difficulty with reverting back to that condition, but those trained with NextGen tools in conjunction with manual techniques thought it would be more difficult to manage traffic without the NextGen tools.

When asked whether each participant thought that s/he would be able to manage more traffic in the simulation tests with the NextGen tools, a majority of the participants said, "yes", especially those in the Whole-Task training group. Most participants also indicated that it was harder to manage the sector when all aircraft were unequipped because of the need to constantly scan the scope. However, participants also indicated that the 100% equipped condition was too boring and it was hard for them to pay attention to the traffic because it only involved monitoring the screen for the conflicts to be alerted.

Finally, when asked whether they would have preferred training with manual skills first or with both manual and NextGen skills simultaneously, all students from the Part-Whole training group said manual skills should be trained first. For the Whole-Task group training condition, five participants also agreed that manual skills should be trained first while the remaining three indicated that manual and NextGen skills should be trained together.

3.2   Influence of Situation Awareness and Workload Probes

Probe questions, administered using the SPAM (Durso & Dattell, 1994) procedure, were used to assess the students' situation awareness and workload during the midterm and final test sessions. Because participants were only given one training session with the probe procedure while managing traffic during the test sessions, we asked participants whether they felt that they had sufficient practice answering the probe questions before the experimental trials began. All but one participant (in the Whole-Task training group) said that one session was sufficient.

Because the probe questions could ask about potential conflicts between aircraft in the sector, traffic flows, or the flight status of particular aircraft, participants were asked whether the probe questions changed their awareness of the traffic. In general, participants from both training groups indicated that the probe questions were more intrusive during the midterm than final testing, and that during the midterm, some questions did cue the participants about potential conflicts. However, during the final testing, the participants indicated that they were able to identify many of the conflicts before the probe questions about the conflicting aircraft were presented, and thus, at the final testing, the probe questions did not change their awareness of the air traffic in their sector.

3.3   Recommendations for Future Interface Design

At the end of the debriefing session, participants were asked to give recommendations on features of the interface that could have made their air traffic management task easier. It should be noted that the recommendations are based on their experience with the interface used in the simulation we performed. The participants came up with two recommendations that were not based on limitations of the simulation environment we used in the study:

1) Implement macros so that common commands can be transmitted with little typing for the NextGen equipped aircraft.
2) Use color coding to distinguish between different types of aircraft (e.g., equipped versus unequipped) or indicate the status of an aircraft.

4   SUMMARY AND DISCUSSION

In general, students benefited from both the Part-Whole and Whole-Task training procedure, as their performance was better at the final testing than at the midterm, regardless of training group (see Kiken et al., in press). All students stated that it was easy to learn NextGen tools and that the use of the tools allowed them the capability to manage more traffic. The students also indicated that learning to use the NextGen tools was not difficult, and that they were able to master the skills in 2-3 class sessions. Training group does seem to influence the perceived difficulty of reverting back to only manual air traffic management techniques once the NextGen tools are removed. The participants in the Whole-Task training group indicated that it would be more difficult to go back to only manual air traffic management techniques compared to participants in the Part-Whole training group. This finding has implications for situations in which automation fails. In the case of automation failure, controllers would have to revert back to the manual mode of air traffic management. Our findings indicate that use of Whole-Task training may decrease the ability of controllers to manage traffic under conditions in which the tool or automation fails.

Although the students appreciated the benefits of NextGen tools, when all aircraft in their sector were NextGen equipped, participants indicated that it was difficult for them to pay attention to the task because the task became too boring. This finding is consistent with the idea that making controllers passive monitors of automation may have detrimental effects (Metzger & Parasuraman, 2001). In fact, most students indicated that they preferred to be trained with manual skills first so that they can master the air traffic management skills before using NextGen tools to help them perform their tasks. Again, this finding suggests that manual skills should be trained first. Once students master the air traffic control task, our findings suggest that it would be easy for them to learn to use NextGen tools to increase their performance, and that having the manual skills will allow students to continue to manage air traffic more effectively in the case of automation failure.

We end by noting limitations of the present paper. This study was a preliminary investigation examining various training methods in which NextGen tools were introduced to students learning air traffic management skills. Given that we only evaluated three specific NextGen tools under a limited training environment, these findings may not be generalizable to other NextGen environments and scenarios.

Also, the data presented in this study reflects the impressions of a small sample of students and may not be representative of the future air traffic control students as a whole.

ACKNOWLEDGEMENTS

This research was supported by the NASA cooperative agreement NNX09AU66A, *Group 5 University Research Center: Center for Human Factors in Advanced Aeronautics Technologies* (Brenda Collins, Technical Monitor).

REFERENCES

Kiken, A., Strybel, T.Z., Vu, K.-P. & Battiste, V. (in press). Effectiveness of training on near-term nextgen air Traffic Management Performance. *Applied Human Factors and Ergonomics.*

Billinghurst, S. S., Morgan, C., Rorie, R. C., Kiken, A., Bacon, L. P., Vu, K.-P. L., Strybel, T. Z., & Battiste, V. (2011). Does the order of introducing NextGen tools impact student learning of air traffic management skills? *Human Factors and Ergonomics Society Annual Meeting Proceedings* 55: 128–132. Las Vegas, NV.

Briggs, G.E. & Naylor, J. C. (1962). The relative efficiency of several training methods as a function of transfer task complexity. *Journal of Experimental Psychology.* 64, 5, 505-512.

Durso, F. T. & A. Dattel. (1994). SPAM: The real-time assessment of SA. In *A Cognitive Approach to Situation Awareness: Theory, Measures and Application,* eds. S. Banbury & S. Trembley 137–154. New York.

Federal Aviation Administration (2009). FAA's NextGen implementation plan 2009.

Retrieved December, 23, 2010, from http://www.faa.gov/nextgen/media/NGIP.pdf.

Hah, S., Willems, B., & Schulz, K. (2010). The evaluation of data communication for the future air traffic control system (NextGen). *Proceedings of the 54th Annual meeting of the Human Factors and Ergonomics Society* 99–103. Santa Monica, CA.

Joint Planning and Development Office (2007). *Concept of operations for the next generation air transportation system version 1.2.* Accessed September 1, 2011, http://www.jpdo.gov/library/nextgenconopsv12.pdf

Joint Planning and Development Office (JPDO) 2010. *Concept of operations for the next generation air transportation system version 3.2.* Accessed September 3, 2011, http://jpe.jpdo.gov/ee/docs/conops/NextGen_ConOps_v3_2.pdf

Metzger, U., & Parasuraman, R. (2001). The role of the air traffic controller in future air traffic management: An empirical study of active control versus passive monitoring. *Human Factors* 43(4), 519–528.

Pollock, E. Chandler, P., & Sweller, J. (2002). Assimilating complex information. *Learning and Instruction* 12: 61–86.

Pop, V.L., Stearman, E.J., Kazi, S. & Durso, F.T. (2012). Using engagement to negate vigilance decrements in the nextgen environment. International Journal of Human-Computer Interaction. (Special Issue). 28(2), 99-106.

Prevot, T. (2002). Exploring the many perspectives of distributed air traffic management: The Multi Aircraft Control System: MACS. *International Conference on Human–Computer Interaction in Aeronautics, HCI–Aero 2002,* 23–25.

Prevot, T., Homola, J.R., Martin, L.H., Mercer, J.S., & Cabrall, C.D. (2012). Tow: automated air traffic control - investigating a fundamental paradigm shift human/systems interaction. : *International Journal of Human-Computer Interacti* 28(2), 77-98.

Vu, K. P. L., Strybel, T. Z., Battiste, V., Lachter, J., Dao, A. Q. V., Brandt, S., Ligda, S., Johnson, W. (2012). Pilot performance in trajectory-based operations under concepts operation that vary separation responsibility across pilots, air traffic controllers, a automation [Special Edition]. *International Journal of Human-Computer Interacti* *28(2)*, 107-118.

CHAPTER 32

StormGen: A Proposed Solution to Weather Simulation in NextGen Research

Joshua M. Kraut*[1], Thomas Quinonez[2], Shu-Chieh Wu[1], Robert Koteskey[1], Walter Johnson[3], and Vernol Battiste[1]

[1]San Jose State University
San Jose, CA, USA
[2]Dell Services Federal Government
Moffett Field, CA, USA
[3]NASA Ames Research Center
Moffett Field, CA, USA
*KrautJosh@gmail.com

ABSTRACT

Next Generation Air Transportation System (NextGen) concepts demand that both predicted and observed weather data are assimilated into air traffic management decision making. Consequently, research that evaluates concepts concerning weather decision making in NextGen requires the use of these weather data in the simulation environments. Current sources of real-world 3D convective weather data are often sparse, leave large coverage gaps, and are not constructed to meet specific research and concept evaluation requirements. As a result, there is a strong need for a simple and versatile tool that can be used for generating tailored, yet realistic weather for simulation-based research. StormGen, the software tool showcased in this paper, has been designed to produce convective weather systems for use in NextGen airspace simulations. StormGen provides a graphical user interface for the construction and placement of storm cells anywhere in a simulated contiguous United States airspace. StormGen functions support morphing of storm cells between different sizes, shapes, altitudes, positions, and intensities over time. The produced weather objects can be exported in multiple formats for use by other simulation components, such as the Multi Aircraft Control System (MACS), the Cockpit Situation Display (CSD), and out-the-window flight simulator views. MACS is an emulation and simulation program which provides a small to large scale airspace environment, and air traffic controller (ATC) display, for current and future air traffic operations in the National Airspace System (NAS). CSD is a 2D and 3D volumetric multifunction interface designed to provide flight decks with a 4D depiction of the interrelationships between surrounding traffic, weather, and terrain within the proximate airspace. Both MACS and CSD are able to utilize the dynamically updated weather delivered by StormGen to display 2D weather information, while CSD is also able to display the 3D weather objects created by StormGen, either in 3D perspective views, or by simulating the 2D scans returned by a simulated airborne radar application. The resolution at which the dynamic weather is updated can be determined by the StormGen software, or the simulation environments displaying the weather information. Thus, it can support six-minute display updates similar to Nexrad, or the virtually continuous display updates found with airborne radars. Finally, depending on the scope and purpose of the simulation environment, the exported weather objects can be used to integrate and simulate the display of predicted (forecast) weather information. This capability is important for the development and evaluation of technologies proposed to utilize such predicted information. There are multiple proposed and planned improvements to StormGen, which would improve the realism of generated weather objects. For example, StormGen presently has a simplistic wind capability that is reflected in the temporal movement of storm cells. Ideally, the editor would support the creation and exporting of wind fields. It would also be advantageous to use publicly available images from ground-based weather radar for the creation of storm systems in StormGen. We envision StormGen to be a continually evolving tool for generating convective weather systems for simulation research in NextGen environments.

Keywords: NextGen, aviation, weather, simulation, storm model

1   BASIS FOR DEVELOPING A WEATHER CREATION APPLICATION FOR USE IN NEXTGEN RESEARCH

The Next Generation Air Transportation System (NextGen) requires the use of new technologies and processes to meet the requirements for increased capacity and efficiency, while maintaining safety and decreasing the impact on the environment. Far from developing isolated enhancements in the National Airspace System (NAS), the goal is to seamlessly combine concepts, procedures, and technologies to create a large cohesive system to meet the goals and demands of future air travel. It has been reported that weather impacts are responsible for 75% of all NAS delays (FAA, 2011). The prospect of increasing efficiency and throughput, while maintaining safety, is the main reason that the assimilation of weather into decision making was identified as one of eight concepts necessary to achieve NextGen objectives.

The weather integration concept identified as necessary for the success of NextGen demands that both predictive and observed weather data be assimilated into air traffic management decision making (JPDO, 2007). Single displays integrating weather and traffic information will be optimized with decision-oriented automation and human decision-making procedures, and all stakeholders will share access to a single authoritative weather source. Improved sharing and integration of weather information into automation and human decision-making capabilities will potentially result in less airspace being constrained by weather, leading to fewer delays and higher traffic throughput, and will allow weather-impacted operations that are at least as safe as those currently available.

Before NextGen visions can be realized, research will be needed to evaluate concepts and tools concerning decision making with relation to weather in simulated environments.

Weather data for use in a simulation-based experiment has the unique requirements that it must be realistic in the eyes of pilot and air traffic controller (ATC) participants, while being deterministic, repeatable, and tailorable to the experimental design. Previous attempts at creating such weather involve searching National Oceanic and Atmospheric Administration (NOAA) archives for weather data that are within range of meeting experimental requirements. The weather data would then be customized by applying translations, rotations, filters, etc., until the weather was close enough to the study's minimum requirements. Not only is this approach labor-intensive, but also relies on the ability to find appropriate weather data in the archives. When the experimental design requires dynamic weather that develops over time in a prescribed fashion, it may not be possible to find real-world weather that progresses as required.

Another disadvantage in using historical weather information is that this "real-world" weather data is not actually weather: it is data generated from weather models or sensors. For example, most convective weather data is weather radar scan data, subject to beam physics, attenuation, blind spots in the scan pattern, and other hardware artifacts. A volume scan from a weather radar can be composited down to a 2D map that is usable in a 2D display. However, for advanced 3D displays, it is impossible to recreate the 3D weather that gave rise to the initial radar data. Once the weather data is created, there is also an issue of how this data will be consumed by simulation components.

A typical airspace simulation facility will comprise multiple components, such as ATC consoles, cockpit simulators, traffic generators, etc., invariably from multiple vendors or organizations. Each component may have its own mechanism for ingesting weather data, with an associated native data format. For a valid simulation, all components must be using the same weather, despite differences in their respective capabilities in terms of dimensionality (2D versus 3D), fidelity, and update rate. Therefore, all weather data, no matter the source, must be modified according to the specifications required by each system, and in a file format required for consumption by each system trusting that the same resultant weather is achieved for all simulation components.

Given that current sources of real-world 3D convective weather data often fail to

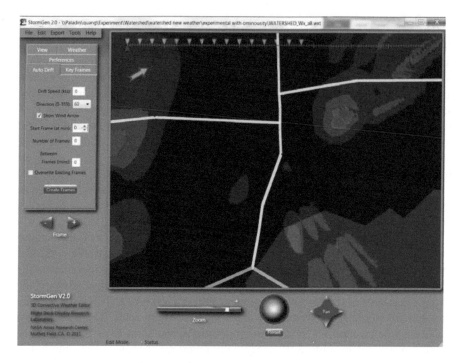

Figure 1 StormGen interface, with sectors.

meet specific research and concept exploration requirements, a strong need was identified for a simple and versatile tool that can be used for generating realistic weather for simulation research, one that can export that data in multiple formats for multiple consumers. StormGen, the software tool showcased in this paper, has been developed to fulfill the need in producing convective weather systems for use in NextGen airspace simulations (see Figure 1).

2   STORMGEN'S CAPABILITIES AND FEATURES

StormGen provides a graphical user interface for the construction and placement of storm cells anywhere in a simulated contiguous United States airspace. Individual cells can be manipulated to tailor both the initial weather state, and the development of weather over time, to exact user requirements. Some of the adjustment options include: rotation, resizing, shifting, shell intensity, shell altitude, cell contour smoothness/roughness, cell propagation, cell replacement, automatic cell dissipation, automatic cell growth, automatic cell drifting, and the ability to adjust the individual polygons making up a cell. Built-in StormGen functions also support the automatic morphing of storm cells between different sizes, shapes, altitudes, positions, and intensities over time. Beyond individual storm cell manipulation, features allow the interaction of multiple storm cells in the simulation environment over a period of time. For example, storm cells can be made to converge, creating a larger cell, or divide, creating multiple smaller storm cells, each with their own attributes and characteristics.

Figure 2 The key frame time-line allows for arbitrary placement of frames for weather manipulation over time.

In addition to placing storm cell objects at different points in space, StormGen is able to place the cell objects at different points in time. Objects located at designated time locations occupy a "key frame," and StormGen can use a morphing function to interpolate frames between the existing frames to change the cell's placement, size, intensity, shape, etc. (see Figure 2). In other words, a user can design the weather in a start frame, and design the weather in a frame 15 minutes later. Then, instruct StormGen to automatically generate intervening frames, in which the weather morphs between frames (see Figure 3).

The time-line has a notional resolution of 1-minute; when a weather frame is created, a marker, the inverted triangle, is inserted on the time-line to represent the new frame's temporal location within the overall weather scenario. Any frame can be manipulated temporally by dragging its marker along the time-line. Individual frames can be exported as files to be consumed by simulation components. This control over frame exporting allows the same weather scenario to be exported as time-synchronized weather to consumers with different update rate requirements: for example ground displays with a 15-minute update rate and cockpit radar displays that can take advantage of the 1-minute resolution data. The temporal resolution at which the dynamic weather is updated can be determined by the simulation environments displaying the weather information. For example, the CSD is able to interpolate between frames to create weather snapshots that are 1-second apart.

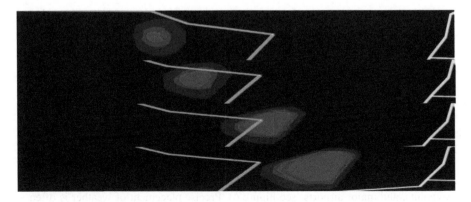

Figure 3 The above time interval series shows the same weather cell as its size, placement, shape, and intensity are automatically morphed by StormGen.

Depending on the design, scope, and purpose of the simulation environment, the created weather objects can be used to integrate and simulate the display of predicted forecast weather information. This capability is especially important for the development and evaluation of technologies proposed for use in a NextGen environment, which will be required to use such predicted information.

In StormGen, each storm cell object is a simple 3D model based on an idealized conceptualization of a typical supercell thunderstorm. A simple modeled cell conceptually comprises three "shells" of intensity which model nominal threshold levels of storm moisture content, which determine levels of radar reflectivity (high, medium, and low). Each of the intensity shells is a set of two-dimensional polygons, each possessing an assigned altitude, intensity value, and object ID. A single polygon represents a 2D slice, at a particular altitude, through a 3D region where the precipitation is of a particular intensity (see Figure 4).

Development of a weather scenario begins with the choice of one of the set of predefined cell objects from the provided template. Additional cell objects can also be added to the existing pallet of choices. Once inserted, cell objects can be merged to create weather patterns of arbitrary complexity.

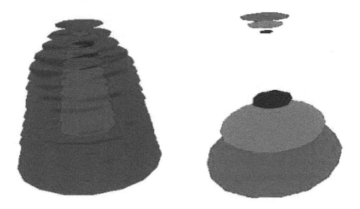

Figure 4 Two distinct views of a single storm cell. Each of the three shells comprises a set of polygons at multiple altitudes. The cell displayed on the left is a 3D view with partially transparent shells. The cell displayed on the right shows only the top and bottom polygons of each shell in the cell.

StormGen includes interface controls, which give the ability to pan, tilt, and zoom within a 3D virtual airspace. To assist in the placement of weather objects, the ability is provided to selectively display key state outlines and airspace sector boundaries.

Additional features to aid in storm cell placement and movement include the display of navigation aids and other background markers, range rings, custom waypoints, and major airports (see Figure 6). Precise placement of weather is often essential to experiment design. For example, cases, such as the one illustrated in Figure 7, in which a storm cell is positioned to block arrivals, can be used to evaluate myriad concepts and tools designed to mitigate the effects of adverse weather.

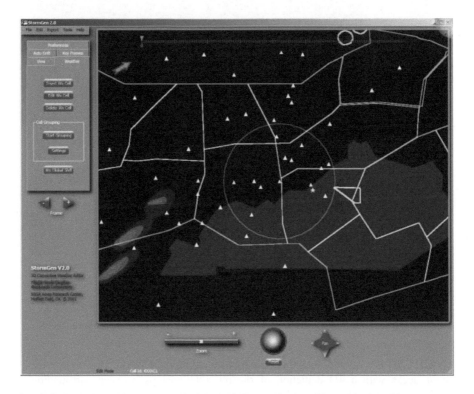

Figure 6 StormGen with sectors, waypoints, and a range ring around a navigation aid.

3 UTILIZING WEATHER CREATED BY STORMGEN IN NEXTGEN RESEARCH APPLICATIONS

The weather objects produced by StormGen can be exported in multiple formats for use by other simulation environments, such as the Multi Aircraft Control System (MACS), the Cockpit Situation Display (CSD), and the out-the-window view, powered by Microsoft® Flight Simulator X, of a medium fidelity, fixed-base 777 flight simulator.

MACS is an emulation and simulation program which provides a small to large scale airspace environment, and ATC display, for current and future air traffic operations in the NAS. Convective weather is displayed on the MACS ATC screen as an emulation of ground-based weather radar, and is a 2D bit-map pattern that is read in from a run-length encoded XML file.

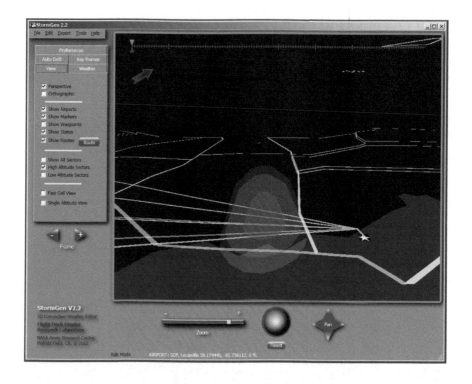

Figure 7 A storm cell positioned to block planned approaches into Louisville International airport. The user has tilted the visual in 3D to verify that the top of the storm is above the air routes

The CSD is a 2D and 3D volumetric multifunction interface designed to provide flight decks with a 4D depiction of the interrelationships between surrounding traffic, weather, and terrain within the proximate airspace. The CSD can display convective weather either in 3D perspective views, or by simulating the 2D scans returned by an airborne radar application. Note that 3D data is essential for a realistic radar emulation, because it allows modeling beam height and tilt. The CSD reads weather data from a binary file containing 2D polygonal regions at multiple altitudes, which define the boundaries of a 3D volume.

StormGen provides an export function that outputs a 3D weather dataset in a format targeted for a specific system. For the CSD, it outputs the required polygonal regions, for MACS, it composites the 3D data to 2D, then encodes the data into XML. It can also export the data as composited 2D polygonal regions. StormGen provides a single place to host all the conversions necessary for heterogeneous simulation components to inter-operate using the same weather. It provides a single tool that a researcher can use to design and evaluate a weather scenario, which can then be used throughout the simulation environment.

StormGen has been used to produce weather for a number of experiments that used multiple simulation components. Often, the individual components have mechanisms to create weather for internal use, however guaranteeing that the same weather is used and displayed on heterogeneous components has heretofore been an intractable problem- one that StormGen has been designed to solve.

4   PLANS FOR FURTHER DEVELOPMENT OF STORMGEN

As with many software programs, the vision, scope, and use of StormGen continues to evolve. As such, the proposed, planned, and undertaken improvements to StormGen have similarly evolved. Some of these advances would improve the realism of generated weather objects. For example, StormGen presently has a simplistic wind capability that is reflected in the temporal movement of storm cells. Ideally, the editor would support the creation and exporting of wind fields, in addition to the weather objects. Another feature that would be advantageous would be the ability to use publicly available images of ground-based weather radar for the creation of storm systems.

Additionally, some weather systems, such as squall lines, have a typical structure and are common in certain geographic areas, during certain seasons. The StormGen user must currently create a larger weather system by placing each cell individually. A possible future enhancement would be to provide the ability to select and insert whole weather systems from a geographically and/or seasonally preferred set of systems.

5   CONCLUSION

Any large-scale advancement to systems and concepts of operation will require a simulation environment to develop and test proposed system components. In the case of integrating weather with decision making processes by users of the NAS, the StormGen program allows the creation and manipulation of convective storm cells over time. Not only does StormGen have the capability to manipulate single cells in regards to shape, intensity, position, size, etc., but it also has the ability to morph and model the interaction of multiple storm cells.

The storm systems and cells that are created by StormGen must be usable by simulation components in order to properly test the hypotheses of a study. To this end, StormGen can export its weather files in multiple formats for use by other software and simulation components. As NextGen concepts continue to evolve to meet the needs of updated requirements, StormGen must also progress to meet the needs of the studies being designed to evaluate NextGen technologies and concepts. As a result, StormGen has evolved to its current state because of improvements that were made over time, and will continue to evolve to meet the needs of integrating weather NAS simulations.

ACKNOWLEDGMENTS

This program development was supported by the NASA Concepts and Technology Development Project Airspace Program.

REFERENCES

FAA, 2011. Retrieved from http://www.faa.gov/nextgen/media/NNEW.pdf March 1, 2012.
Joint Planning and Development Office, 2007. Concept of Operations for the Next Generation Air Transportation System v3.2. Washington, DC.

CHAPTER 33

Human Factors Validation of NextGen Technologies and Procedures

Esa M. Rantanen

Rochester Institute of Technology
Rochester, USA
esa.rantanen@rit.edu

ABSTRACT

Validation has received remarkably little attention in the systems development literature. In particular, there has almost been no progress in the area of human factors validation of complex system in the 20 years since the publication of Wise, Hopkin, and Stager's (1993) seminal volume on the topic. Yet, large-scale modernization efforts of air transportation infrastructure in the U.S. (NextGen) and elsewhere should by all reason have proved to be a boon to research on validation and verification methods. Alas, this appears not to be the case. This session in the 4th International Conference on Applied Human Factors and Ergonomics and the 1st International Conference on Human Factors in Transportation was put together to refocus the attention on verification and validation of NextGen technologies and procedures. The five papers in the session address diverse questions about validation of different components of NextGen. Mosier, Fischer, and Orasanu review the particular challenges of human interaction with highly automated systems. Gore, Hooey, Mahlstedt, and Foyle describe a computational model to investigate pilots' performance in self-separation and closely spaced parallel operations using associated NextGen technologies. Hall, Underhill, Rodriquez, and DeLaura describe validation of a departure route planning tool. Histon, Rantanen, and Alm review past research on aviation communications and offer guidelines for the use of communications analysis in validation of many aspects—also beyond datalink communications—of NextGen. Finally, Archer, Liu, and Wise review the issue of de-conflict algorithms and their role in NextGen. This paper will summarize the contributions of these authors and offer an integrated review of the state of human factors validation and verification of complex systems in the context of NextGen.

Keywords: Human factors, complex systems, validation and verification, NextGen.

1   NEXTGEN VALIDATION CHALLENGES

In the past thirty years, since Wiener and Curry's (1980) seminal paper on the "promises and problems" of cockpit automation, automation research has yielded a tremendous crop of valuable knowledge about human-automation interaction as well as many methodological innovations to study various aspects of automation in diverse domains. For example, the golden anniversary special issue of the journal *Human Factors* contained several articles on automation (e.g., Lee, 2008; Sarter, 2008; Parasuraman & Wickens, 2008), and Sheridan's (2002) book "Humans and Automation" continues to be a best seller. Dedicated chapters on automation are common in human factors textbooks (e.g., Wickens, Lee, Liu, & Gordon-Becker, 2004; Wickens & Holands, 2000) and handbooks (e.g., Salvendy, 1997). A Google Scholar search with keywords "automation" and "human" returns nearly two million hits. Yet, NextGen represents an unprecedented influx of new technologies and procedures to an immensely complex system, the National Airspace System (NAS) and despite all the aforementioned knowledge about automation the final impact of modernization of the air traffic control (ATC) and –management (ATM) remains unknown.

1.1   Measurement Issues

As all scientific endeavors, validation and verification of complex systems rest on validity of the methodologies used for the purpose and in particular the validity of measures. Rantanen (2004; Rantanen & Nunes, 2003) took an emphatically systematic and comprehensive approach to the measurement problem in ATC research. On one hand, it is important to perform periodic and thorough reviews of past and current research efforts and to organize the findings in a manner that facilitates the use of existing knowledge for a basis of future evolvement of ATC measurement, or, to avoid "reinventing the wheel". On the other hand it is prudent to proceed cautiously on an issue as complex as ATC measurement and consider carefully all the constraints, assumptions, and threats to validity that may emerge. A result of this work was a taxonomy and a database for ATC measures.

In all measurement issues, including the ATC measures taxonomy, it is critical to distinguish between direct and indirect measures. Direct measures are those that can be explicitly measured. Examples of such measures include a direct observation of a controller's action, measurement of a response latency, or count of aircraft in a sector at a given time. Indirect measures are those that cannot be measured directly but must be inferred from directly measurable variables. For example, certain actions of a controller may be indicative of his or her performance, response latency can be used to make inferences on some covert cognitive processes, and a number of aircraft in a sector can be used to signify sector complexity.

Altogether 37 different indirect measure classes were identified in the literature (Rantanen, 2004). These, however, rested on very narrow theoretical foundation for making the associations between direct and indirect measures. This is not to say that the measures reported in the literature are not valid; rather, insufficient information was provided to fully assess their validity. Only a minority of all articles reviewed by Rantanen (2004) explicitly justified making inferences about indirect variables based on direct measures by citing past research where such associations have been established. Those articles that did typically cited the same sources, making the body of supporting research literature for ATC measurement remarkably small. Few articles took advantage of the opportunity to add to the foundational body of research in addition to reporting results on their particular topics and thus help validating measures used towards their ends (cf., Vicente & Torenvliet, 2000).

Yet another aspect of classification of measures that warrants discussion is the differentiation between what may be termed primary and secondary measures. Primary measures are those that are measured directly, for example, count of aircraft in a sector, or number of heading changes per aircraft. Secondary measures are those derived from primary measures, for example an average number of traffic in a sector, its variance, or range. In the case of the average, the criteria are implicit (the time duration or interval during which the aircraft were counted and the number of samples) but nevertheless have an impact on the eventual measure. It is clear that descriptive statistics are crucial for reducing and making sense of the data; yet, it is very important to acknowledge how such techniques might obscure some aspects of the data and skew others. These concerns are seldom effectively dealt with in the literature (Vicente & Torenvliet, 2000). This particular facet of measurement will be of especially great consequence when multiple measures representing various aspects of an indirect variable of interest are combined in some sort of an index, such as workload (Hart & Staveland, 1988) or dynamic density (Laudeman, Shelden, Branstrom, & Brasil, 1998).

1.2   Specific Challenges

The entire NextGen system will rely on automation to an unprecedented degree, and therefore most of the validation challenges also pertain to automation and human–automation interaction. Mosier, Fischer, and Orasanu (this volume) bring up many of the unique challenges NextGen present to human operators in the NAS. These include increased demands for deep systems knowledge and coherence skills to stay in control of the myriad automated systems and corresponding requirements for training and maintenance of such knowledge and skills. Complicating the training challenges is the operators' familiarity and ingrained habits with the existing system and procedures and the potentially very subtle and covert changes brought about NextGen.

With NextGen "collaboration and coordination" and the concept of "team" will have a new meaning. As Mosier, Fischer, and Orasanu (this volume) point out, automated systems must be coordinated with and monitored much in the same way operators presently monitor and collaborate with other human operators. No doubt this will also precipitate new behaviors and patterns, and possibly unforeseen problems. Automation support tools may offer an attractive means for investigation of human-automation interaction as NextGen is implemented and as means of validation of the various NextGen components.

Finally, Mosier, Fischer, and Orasanu (this volume) discuss the changing communication patters within teams that include automation. As NextGen tools may facilitate also nonverbal communication between human operators, they may offer novel sources for data on communication patterns and methods of validation of NextGen system components (see also Histon, Rantanen, & Alm, this volume).

2 VALIDATION APPROACHES IN NEXTGEN

This section summarizes four specific approaches to validation of NexGen components detailed in this volume, and discusses the broader challenges to validation and verification as well as some proposed solutions. Specific problems with data analysis and some novel solutions to them will be discussed in the third main section of this paper.

2.1 Validation by Computational Models

As Gore et al. (this volume) discuss in their paper, validation of NextGen technologies prior to their implementation is difficult because as the systems are being built there are no physical prototypes to be used in human-in-the-loop (HITL) simulations and because there are no procedures for their use or definitions for operator roles and responsibilities. Additional constraints for validation of system components in experimental settings are that it is difficult and often impossible to mimic operational task environments in simulations and systematic experimentation requires large experimental designs, which in turn are expensive and time-consuming to run. Computational human performance models (HPM) circumvent many of these constraints and allow for rapid simulation of different characteristics of the model (representing the physical system to be tested) and different inputs representing different operational settings. Because the model only exists in software, it can also be quickly and inexpensively changes to reflect possible changes in the design of the actual system.

As Gore et al. (this volume) also point out, validity of HPM depends on two distinct aspects: validity of the inputs to the model, which must accurately represent the operational scenarios and task environments where the system being modeled will be used, and validity of the model itself and outputs from it. To be valid, inputs to computational models must be representative of operational settings. This can be accomplished in two primary ways.

First, subject matter experts (SMEs) may define the input variables according to their experience. There are two problems with SMEs, however, as with all expert opinions. It is often impossible to separate expert opinions from subjective biases. Different experts may also offer wildly different estimates of the same variables in the same scenarios. Furthermore, SMEs are seldom able to represent the true range of performance and behavior variations encountered in the operational world. The second way is to use data gathered directly from the operational world as inputs to computational models.

There are many feasible sources for operational data of pilot behavior. For example, flight data recorder (FDR) data has proven invaluable in accident and incident analysis to determine the aircraft states and crew actions in abnormal or emergency situations. With advanced technology the quality and quantity of such data have substantially improved, and these data have also become more accessible than before. Consequently, a number of participating airlines have begun to routinely analyze the flight data from regular flights within Flight Operational Quality Assurance (FOQA) programs to identify patterns of irregularities and potentially hazardous situations (Brandt, 1999). Although the current FDRs provide a wealth of data that can be efficiently handled and stored, information extraction from these data is hampered by the lack of proven performance measures. For example, only when predetermined parameter values are exceeded are they automatically flagged in the FOQA data.

Rantanen et al. (Johnson, Rantanen, & Talleur, 2004; Rantanen & Talleur, 2001) developed several novel time series- based performance metrics derivable from automatically collected data from various flight training platforms. These metrics differentiated between pilots of different skill and performance. These results suggest ways in which development of robust and valid FDR-based measures may not only find important applications in quality assurance programs such as FOQA but also as sources of inputs to computational models. Derivation of measures other than simple exceedances would significantly enhance the FOQA program by reducing the labor- intensive manual analyses and potentially allow for detection of trends and patterns that may not manifest themselves as exceedances, but that may nevertheless compromise safety.

2.2   Validation of NextGen tools

Hall, Underhill, Rodriquez, and DeLaura (this volume) describe validation of the Integrated Departure Route Planning (IDRP) tool. This complex technology provides air traffic managers critical departure information reducing their workload in identification, planning, coordination, and implementation of re-routes around convective weather. The validation efforts detailed in their paper are exemplary in their comprehensiveness and careful definition of relevant measures.

In particular, Hall et al. (this volume) bring up the importance of derivation of variables in tool evaluation that have critical human factors implications. System stability, which was expressed using accuracy (correctness) and precision (consistency, reliability) measures of the IDRP forecasts, is such a variable. Forecast volatility was derived from the spread of wheels-off forecasts. These variables are directly experienced by operators using the tool and will impact their use, disuse, misuse, or abuse (cf. Parasuraman & Riley, 1997) of the system.

2.3   Validation Opportunities Afforded by NextGen Tools

The description of evaluation of the IDRP tool by Hall et al. (this volume) suggests also novel opportunities and data sources for validation of NextGen technologies. The IDRP produces departure demand forecast issuances and updates at a rate of once every minute. The IDRP data can be combined with data from ASDE-X systems that provide information about aircraft location on the airport surface and in the immediate airspace. From these data secondary variables may be derived, for example predicted wheels-off error, predicted wheels-off spread, and predicted fix demand spread (Hall et al., this volume). This is great example of the opportunities various NextGen components afford for validation and verification efforts.

2.4 Validation of NextGen by Communications Analyses

Communications between different human operators within the NAS as well as communications between humans and automated systems are fundamental to NextGen. Therefore, communications analysis should be one of the topmost methods used for validation of NextGen components as they are implemented and operationalized. There are many true and tried methods of communications analysis that could be readily employed in NextGen validation efforts, but also promising novel techniques that have yet to be applied in this domain, such as computational linguistics (see Histon, Rantanen, & Alm, this volume).

Communication analysis under NextGen falls into two broad categories. First, NextGen will introduce several novel communications between human operators (e.g., Controller-Pilot Data Link Communications, CPDLC), between human operators and automated agents through various interfaces, and between automated agents. All these communications provide potentially very rich sources of data to be used in validation and verification of NextGen components and the system as a whole. Many existing communications analysis tools may also readily be used to examine these novel communications. The second category consists of the traditional voice communications by radio (between controllers and pilots) and telephone (between controllers) that will persist in parallel of the other means of communications germane to NextGen. These communications will in all likelihood also provide very rich and interesting data for validation and verification purposes in themselves. Perhaps the most important application of communications analyses is investigation of comparable variables and metrics from pre-NextGen operations and operations under NextGen according to the classic pre-treament, post-treatment paradigm.

2.5 Validation of De-Conflicting Algorithms

One of the bottlenecks in the present-day ATC is individual controllers' capacity to monitor and control multiple aircraft to ensure that adequate separation between aircraft is maintained at all times. This task is greatly dependent on the structure of the airspace, perhaps to even greater extent than on the number of aircraft under the controller's responsibility. For this reason ATC sectors are designed around recurring traffic flows to minimize the number of specific points of conflict. With NextGen, not only will controllers need to monitor larger number of aircraft that presently, but direct navigation will make traffic flows much more varied and less predictable than in the current NAS. To allow controllers nevertheless do their job under NexGen, they need automated de-conflicting systems to help ensure conflict-free traffic flows in largely unstructured airspaces.

Validation of the de-conflicting algorithms will be one of the most difficult tasks in NextGen implementation. As Archer, Liu, and Wise (this volume) discuss, the algorithms are mathematically extremely complex and their logic opaque to the operators. In addition to the issues of false alerts (indicating a conflict when none exist) and misses (not detecting a conflict) in the performance of the algorithm, examination of human interaction with de-conflicting tools and issues of trust and reliance are of critical importance.

The performance of de-conflicting algorithms in high-fidelity simulations or operational settings could be validated by analysis of radar data. Fortunately the new technologies associated with NextGen also allow for routine gathering of data that could be stored for a myriad of analyses. Radar data (the lat. and long. coordinates plus altitudes of aircraft) can be recorded at very high frequencies (e.g., every 10 seconds), providing a 4-dimensional picture of all air traffic for accurate reconstruction and analysis using software tools such as SATORI (Rodgers & Duke, 1993) and the Performance and Objective Workload Evaluation Research (POWER; Manning, Mills, Fox, & Pfleiderer, 2000). Today, radar data are increasingly used to evaluate NAS operations through joint Federal Aviation Administration (FAA) and National Aeronautics and Space Administration (NASA) program, the Performance Data Analysis and Reporting System (PDARS; Den Braden & Schade, 2003). As laudable and crucial to the NextGen overhaul of the U.S. air transportation infrastructure as these efforts are, however, the current measures predominantly focus on the system performance with little regard on the human component within it.

The POWER program was never used beyond a couple validation studies (Manning, Mills, Fox, Pfleiderer, & Mogilka, 2002). In a study by Rantanen, Naseri and Neogi (2007) several additional measures of controller performance were derived from operational data obtained from Indianapolis (ZID) ARTCC. For this research, the POWER program was augmented by the Medium Term Conflict Detection (MTCD) algorithm, which allowed for derivation of several critical metrics, including counts of aircraft pairs at the same altitude and time to loss of separation and duration of conflict. Another algorithm (Laudeman, Shelden, Branstrom, & Brasil, 1998; Sridhar, Sheth, & Grabbe, 1998) implemented in the POWER program calculated a dynamic density value for every minute. Given the sophisticated data collection and analysis capabilities or PDARS, this system, too system could plausibly be augmented by a number of additional metrics that focus on human variables. Rantanen (2009) analyzed data collected from the FAA's evaluation simulations of Future En route WorkStation (FEWS) under different air traffic load configurations. The data were first processed into event and trajectory files. The former contained a time line of all controller and pilot actions and events during the run to millisecond accuracy. A separate data processing program was developed to derive temporal task performance variables, the opening and closing of a window of opportunity to perform a given task, and when the task was initiated and completed. From these variables it was possible to determine air traffic controller's task prioritization schemes (first come, first served) as well as the effect traffic density had on their performance.

These studies also bring up an interesting juxtaposition of de-conflicting algorithms to be validated and other conflict detection algorithms used to validate them. Such cross-validation studies might provide valuable insights into the performance and behavior of different algorithmic applications.

3 STATISTICAL CHALLENGES IN NEXTGEN VALIDATION

Contexts such as the NAS provide a very potent field of study for the application of data mining and machine learning techniques, primarily because of the potentially very high number of variables. A justification for the use of these techniques comes from the fact that it is not common to end up with situations where the complexity of the underlying structure of the data renders standard off-the-shelf statistical techniques obsolete and inadequate. Due to the potentially complex nature of the patterns underlying the observed data and considering the fact that in some cases the sample size is not large enough to adequately cover the input space of interest, nonstandard statistical analyses are most likely to be needed to better handle this kind of research. Given the nontrivial, or at least nonstandard, nature of the structure potentially under consideration, it is very likely that expert's knowledge would be of great importance in zeroing in on the patterns underlying the data. Bayesian statistics provide tools and methods for incorporating such knowledge into modeling in order to achieve a better extraction of the true structure.

It would be very surprising if the patterns of workload management and situation awareness under NextGen, for example, were the same (homogeneous) across the whole population under consideration. It is reasonable to hypothesize that there might be clusters of patterns rather than a single pattern that captures the structure of the whole population. This is likely to impact even a regression analysis performed on the data. Modern data mining and machine learning that perform nonlinear regression and pattern recognition along with feature selection and variable selection are likely to be candidates for this kind analysis. Generalized Linear Models, Generalized Linear Mixed Models and modern techniques such Support Vector Machines should be attempted in this kind of studies as these techniques handle complex structures better than standard techniques.

4 SUMMARY AND CONCLUSIONS

Validation and verification of NextGen is a deep and multilayered problem. Separate validation of human performance measures presents further challenges. Undoubtedly many metrics that may me derived from various NextGen outputs will measure approximately the same thing, or closely related things. Therefore, such metrics should be closely correlated when derived from the same data set. Metrics that do not agree with others supposedly measuring the same thing are suspect, warranting closer inspection of their validity. Similarly, comparison of the same metrics from data sets with known differences will allow for assessment of the sensitivity of the metrics (sensitive metrics are obviously preferred). Examination of corroborating evidence will thus be the primary means to validate the measures.

The primary challenge in deriving meaningful, reliable, and valid human measures from operational data lies in the matching of psychological research done in laboratories (controlled experiments) with the operational task environments where air traffic controllers work. This challenge is not insurmountable, however, as has been demonstrated (Rantanen, 2009; Rantanen, Naseri, & Neogi, 2007). The richness of data available from operational ATC through various NextGen system components and programs like PDARS plausibly allows for identification of common variables between the two domains, that is, independent and dependent variables in psychological research literature that are comparable to the task demands and performance indices identifiable in the operational world.

A prime example of such common variable is time. Time is also common to the human, the task, and the environment and thus offers a common unit of measurement of human performance in the context of the task. Time is central to several critical aspects of human performance. Time pressure is a key element of task load and one of the primary drivers of subsequent mental workload (Hendy, 1995; Hendy, Liao, & Milgram, 1997; Hancock & Chignell, 1988; Laudeman & Palmer, 1995; Loft, Sanderson, Neal, & Mooij, 2007). Time is also central to what is arguably the most critical construct of human performance in ATC, situation awareness (cf., Rantanen, 2009). Thus, accurate time stamps associated with myriad of variables collected from operational NextGen provide a promising point of contact with vast amounts of existing psychological research.

REFERENCES

Brandt, M. H. (1999). The next generation FOQA programs. *Presented in the Int'l Symposium on Transportation Recorders.* May 3-5, 1999, Arlington, VA.

Den Braden, W., & Schade, J. (2003). *Concept and operation of the performance data analysis and reporting system (PDARS)* (SAE Technical Paper No. 2003-01-2976). SAE.

Hancock, P. A., & Chignell, M. H. (1988). Mental workload dynamics in adaptive interface design. *IEEE Transactions on Systems, Man, and Cybernetics, 18*(4), 647-658.

Hart, S. G., & Staveland, L. E. (1988). Development of NASA-TLX (Task Load Index): Results of empirical and theoretical research. In P. A. Hancock & N. Meshkati (Eds.), *Human mental workload* (pp. 139-183). Amsterdam: North-Holland.

Hendy, K. C. (1995). Situation awareness and workload: Birds of a feather? *AGARD AMP Symposium on Situation Awareness: Limitations and Enhancements in the Aviation Environment* (21-1-21-7). Brussels, April 24-28, 1995.

Hendy, K. C., Liao, J., & Milgram, P. (1997). Combining time and intensity effects in assessing operator information processing load. *Human Factors, 39*(1), 30-47.

Johnson, N. R., Rantanen, E. M., & Talleur, D. A. (2004). Time series based objective pilot performance measures. *Int'l Journal of Applied Aviation Studies (IJAAS), 4*(1), 13-29.

Laudeman, I. V., & Palmer, E. A. (1995). Quantitative measurement of observed workload in the analysis of aircrew performance. *Int'l Journal of Aviation Psychology, 5*(2), 187-197.

Laudeman, I. V., Shelden, S. G., Branstrom, R., & Brasil, C. L. (1998). *Dynamic density: An air traffic management metric* (NASA-TM-1998-112226). Moffett Field, CA: NASA Ames Research Center.

Lee, J. D. (2008). Review of a pivotal Human Factors article: "Humans and automation: Use, misuse, disuse, abuse". *Human Factors, 50*(3): 404-410.

Loft, S., Sanderson, P., Neal, A., & Mooij, M. (2007). A review of mental workload in en route air traffic control. *Human Factors. 49*(3), 376-399

Manning, C. A., Mills, S. H., Fox, C. M., & Pfleiderer, E. (2000). Investigating the validity of performance and objective workload evaluation research (POWER). *Presented in the 3rd USA/Europe Air Traffic Management R & D Seminar,* Neaples, Italy, Jun 13-16, 2000.

Manning, C. A., Mills, S. H., Fox, C. M., Pfleiderer, E. M., & Mogilka, H. J. (2002). *Using air traffic control taskload measures and communication events to predict subjective workload* (DOT/FAA/AM- 02/4). Washington, DC: FAA OAM.

Parasuraman, R., & Wickens, C. (2008). Humans: Still vital after all these years of automation. *Human Factors, 50*(3), 511-520.

Parasuraman, R., Riley, V. (1997). Humans and automation: Use, misuse, disuse, abuse. *Human Factors, 39*(2), 230-253.

Rantanen, E. M. (2004). *Development and validation of objective performance and workload measures in air traffic control* (AHFD- 04- 19/FAA- 04- 7). Savoy, IL: Aviation Human Factors Division.

Rantanen, E. M. (2009). Measures of temporal awareness in air traffic control. *Proc. 53rd Annual Meeting of the Human Factors and Ergonomics Society* (pp. 6–10). Santa Monica, CA: HFES.

Rantanen, E. M., Naseri, A., & Neogi, N. (2007). Evaluation of airspace complexity and dynamic density metrics derived from operational data. *Air Traffic Control Quarterly, 15*(1), 65- 88.

Rantanen, E. M., & Nunes, A. (2003). Taxonomies of measures in air traffic control research. *Proceedings of the 12th International Symposium on Aviation Psychology* (pp. 977- 980). April 14- 17, 2003, Dayton, OH.

Rantanen, E. M., & Talleur, D. A. (2001). Use of flight recorders for pilot performance measurement and training. *Proceedings of the 2001 International Aviation Training Symposium.* Oklahoma City, OK: FAA Academy.

Rodgers, M. D., & Duke, D. A. (1993). SATORI: Situation assessment through recreation of incidents. *The Journal of Air traffic Control, 35*(4), 10- 14.

Salvendy, G. (ed.) (1997). *Handbook of human factors and ergonomics.* John Wiley & Sons.

Sarter, N. (2008). Investigating mode errors on automated flight decks: Illustrating the problem-driven, cumulative, and interdisciplinary nature of human factors research. *Human Factors, 50*(3): 506-510.

Sheridan, T. B. (2002). *Humans and automation: System design and research issues.* Wiley-Intersicence.

Sridhar, B., Sheth, K. S., & Grabbe, S. (1998). Airspace complexity and its application in air traffic management. *Proceedings of the 2nd USA/Europe Air Traffic Management R&D Seminar.*

Vicente, K. J., & Torenvliet, G. L. (2000). The Earth is spherical (p<0.05): Alternative methods of statistical inference. *Theoretical Issues in Ergonomics Science, 1*(3), 248- 271.

Wickens, C. D., & Hollands, J. G. (2000). *Engineering psychology and human performance.* Pearson.

Wickens, C. D., Lee, J. D., Liy, Y., & Gordon-Becker, S. (2004). *Introduction to human factors Engineering.* Pearson.

Wiener, E. L., & Curry, R. E. (1980). Flight deck automation: promises and problems. *Ergonomics, 23*, 995–1011.

Wise, J., Hopkin, V. D., & Stager, P. (Eds.) (1993). *Verification and validation of complex systems: Human factors issues.* Berlin: Springer Verlag.

CHAPTER 34

Working Towards NextGen in Traffic Flow Management

Tanya Yuditsky-Mittendorf,[1] Bart Brickman[2]

[1]Federal Aviation Administration
[2]TASC, Inc.
Atlantic City International Airport, NJ, USA
Tanya.Yuditsky@faa.gov

ABSTRACT

The fundamental goal of Traffic Flow Management (TFM) is to strategically manage the flow of air traffic to minimize delays and congestion due to system stressors such as weather or equipment outages in the National Airspace System (NAS). Effective traffic management is crucial to realizing the anticipated benefits associated with greater efficiency and fewer delays envisioned for the Next Generation Air Transportation System (NextGen). However, due to the shortfalls inherent in today's system, Traffic Managers currently lack the tools necessary to accomplish the core functions of TFM: analyzing capacity, assessing demand, and managing constraints. As a consequence, the systems and capabilities that are used for TFM today cannot provide an adequate foundation for future system enhancements and NextGen capabilities. In this paper we describe the human performance aspects of the shortfalls identified and tie them to operational impacts through real-world examples of resulting inefficiencies.

Keywords: Traffic Flow Management, NextGen

1 INTRODUCTION

The Next Generation Air Transportation System (NextGen) will modernize domestic air traffic control, and is expected to yield tangible benefits in terms of greater efficiency, fewer delays, and ultimately lower costs associated with flight operations. In anticipation of NextGen, we conducted the current study to examine the current state of Traffic Flow Management (TFM). The purpose of this work was to evaluate the readiness of the current system (i.e., processes and tools) and identify whether any improvements were needed to lay the ground work for NextGen.

1.1   Traffic Flow Management

During peak periods, there are typically 6000 – 7000 aircraft operating simultaneously in domestic airspace, yielding approximately 55,000 flight operations system-wide per day (Somersall, 2011). The fundamental goal of TFM is to strategically manage the flow of this traffic to minimize delays and congestion due to constraints caused by system stressors such as severe weather, excessive volume, or equipment outages. Traffic Managers address these constraints through the execution of the TFM mission to, "balance air traffic demand with system capacity to ensure the maximum efficient utilization of the National Airspace System (NAS)" (FAA, 2011).

Traffic Flow Management is applied throughout the NAS at national, regional, and local levels through the collaboration of Traffic Managers at the Air Traffic Control System Command Center (ATCSCC), each of the 21 Air Route Traffic Control Centers (ARTCCs), and approximately 50 Terminal Radar Approach Control (TRACON) and Air Traffic Control Tower (ATCT) facilities. Ongoing coordination is also required with other stakeholders, including flight operators and the military.

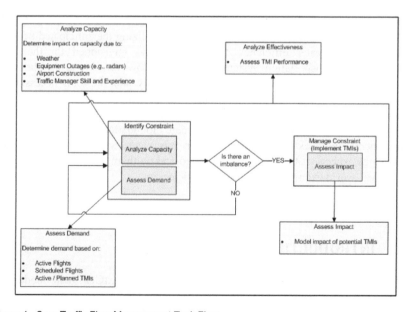

Figure 1. Core Traffic Flow Management Task Flow

In execution of the TFM mission, Traffic Managers conduct three core TFM tasks, as shown in Figure 1. They assess the available capacity of NAS Resources such as airports, routes, airways, and fixes; determine aircraft demand on those resources; and manage constraints that occur when system stressors disturb the balance between capacity and demand.

In order to perform effectively, Traffic Managers must develop and maintain a comprehensive understanding of the current and predicted capacity of NAS resources. That is, they must develop and maintain an awareness of how many flights may occupy or transit a given airspace within a given unit of time. This number is not static and is greatly influenced by system stressors such as severe weather events, equipment outages, construction, and the relative skill of the Traffic Managers. For example, the number of flights that can land at an airport in severe weather will be reduced as compared to the number of flights that can land during clear weather operations. The Traffic Managers' role is to determine how much of an impact these stressors will have on the airspace, and translate that understanding into a capacity value for the airport.

Traffic Managers must also develop and maintain awareness of current and predicted aircraft demand on NAS resources. They need a comprehensive and accurate count of how many planes are expected to occupy or transit a given airspace within a given unit of time. Aircraft demand is based on counts of both airborne and scheduled flights. These counts are subject to change due to variations in airline schedules and can be affected by any traffic management actions impacting the system along the route of flight.

Constraints (i.e., when expected demand exceeds available capacity), are managed through the judicious implementation of Traffic Management Initiatives (TMIs). Examples of common TMIs include Mile-In-Trail restrictions (e.g., aircraft follow one another on a given route by a specified distance), Time Based Flow Metering (e.g., aircraft are given specific times to arrive at a given location), and Ground Delay Programs (e.g., aircraft headed to a common destination are held on the ground to manage arrivals at that destination). Each of these TMIs imposes restrictions on traffic flows to bring the demand back into balance with the available capacity. One of the core principles of TFM is to only impose the minimum amount of control necessary to manage a given constraint. Too much control over-restricts the system and causes flight delays and inefficiency; too little control can put the system into gridlock.

These three components: understanding capacity, maintaining awareness of demand, and managing constraints represent core components of TFM. Success in each of these areas is important in order to realize the TFM mission of utilizing the NAS in a maximally efficient manner. Effective TFM is also crucial to realizing the anticipated benefits envisioned for NextGen. In this study, we evaluated the current state of TFM to prepare for the enhancements coming in NextGen.

1.2   NextGen

According to its Concept of Operations, NextGen will provide an operational environment in which:

"... [Collaborative Air Traffic Management] participants share a common awareness of overall constraints and the impacts of individual and system-wide decisions. Automation tools and system-wide information exchange capabilities improve decision making, enabling participants to understand the prevailing constraints, short- and long-term effect of decisions, and interdependence among national, regional, and local operations. To manage information across all phases of flight, advanced automation is utilized to make the system more agile in responding to changes in environment or demand." (JPDO, 2010)

While the fundamental mission of TFM is not expected to change, the current TFM system must be transformed and enhanced with new tools and capabilities in order to meet the NextGen vision. The NextGen Concept of Operations describes the transformation of TFM as necessary due to the "inherent limitations of today's system, including limits driven by human cognitive processes and verbal communications" (JPDO, 2010). These limitations not only preclude Traffic Managers from utilizing the NAS in a maximally efficient manner today, they are also a source of risk to the success of NextGen. NextGen enhancements will be deployed via incremental improvements to the core functions of TFM today. But weaknesses of the current system would yield an unsuitable foundation to build upon for NextGen. In order to evaluate the delta between where we are today and where we need to be for NextGen, we used a multidisciplinary team to identify, validate, and prioritize TFM Shortfalls in today's system from an operational perspective.

2   METHOD

The Federal Aviation Administration (FAA) established a multidisciplinary team to evaluate the manner in which the 3 core TFM tasks are accomplished today in an effort to determine if there were shortfalls inherent in current operations that needed to be addressed in anticipation of NextGen. The team was comprised of TFM Subject Matter Experts (SMEs) with extensive experience in Command Center, ARTCC, and TRACON domains; four concept engineering teams from academic, industry, and government organizations; and Human Factors Specialists. The SMEs ensured that the team addressed existing problems as expressed by operational users, provided input on the impact of those issues on operations, and provided real world examples. The concept engineering organizations provided input on existing shortfalls derived from their research and industrial experience. The human factors specialists contributed guidance on the impact of the emerging shortfalls in terms of human performance, situation awareness, and workload.

The team conducted two sessions of shortfall analysis. The first session began with a review of shortfalls gathered from previous work including an initial NextGen TFM Concept of Operations (Brennan, et.al, 2011), NextGen Mid-Term TFM Roadmap (Bernard, et.al, 2010), and Gap Analysis (Geffard, Sgorcea, &

Stalnaker, 2010). This review was followed by individual presentations in which the concept engineering organizations identified TFM shortfalls based on their research and unique industrial experience. The SMEs provided input on which problems had the biggest impact from an operational perspective. At the second session, the shortfalls were further refined and expanded to include specific examples of how the shortfalls impact operations. After each of the sessions, all of the participants received a meeting summary for their review and comment. The maturing set of shortfalls was further validated through a series of field visits to FAA facilities during which Human Factors Specialists observed TFM operations and interviewed Traffic Managers.

3   RESULTS

As a result of these data gathering and analytical activities, the team found that Traffic Managers were faced with several operational shortfalls in the system today. They lack basic tools, capabilities, and views of information necessary to effectively execute the core TFM functions of understanding capacity, assessing demand, and managing constraints. In this section, we describe the shortfalls identified in each of these areas in terms of their impacts on human performance and on air traffic operations.

3.1   Capacity

Weather remains the largest driver of capacity constraints in the system today, but interpretation of weather forecasts is not an easy task. Weather is multidimensional and its impact depends on many factors such as the exact location of the impact in relation to routes, the presence of convective activity, and winds. Ideally, all Traffic Managers should be able to view the same forecast and obtain an automated assessment of how the weather will impact their airspace (i.e., the capacity).

Today, capacity is set by manually gathering information on weather, equipment outages, staffing, and any other factors that may impact the flow of air traffic. The impact of weather, for example, is assessed by gathering forecasts from multiple sources. Since there is no authoritative, automated assessment of the impact that weather will have on capacity, the estimates depend greatly on the skills and experience of the individuals involved and, therefore, vary between stakeholders (e.g., flight operators, Traffic Managers at the TRACON, and Traffic Managers at the Command Center). Because the capacity will determine how many flights will be allowed to land at a particular airport or transition a piece of airspace, the stakeholders will often have to negotiate what the capacity will be set to. Overestimating capacity can result in gridlock while underestimating it unnecessarily restricts the flow of air traffic and results in delays.

3.2   Demand

One would think that demand prediction would be a fairly straightforward calculation. If you know when a flight is scheduled to depart, and you know its route and speed, a simple calculation will allow you to predict where the flight is going to be at any point in time. However, there are many factors that affect a flight's departure time, trajectory, and speed. The calculation becomes much more complicated when taking these into account. Ideally, automation should consider all of these factors and calculate a fairly accurate trajectory over time for every flight in the system. The calculations should be continually updated as things change. For example, if a weather system blocks an arrival route at the destination airport, the flight will be delayed on the ground and demand predictions should reflect this information in subsequent updates.

The determination of current and predicted demand is only partially automated today and the algorithms used by the automation are suboptimal. Traffic Managers simply do not have an accurate comprehensive view of the demand on NAS resources. For example, although hundreds of Mile-In-Trail restrictions are implemented daily, current automation does not ingest and reflect these restrictions in the demand counts displayed to Traffic Managers. These TMIs intentionally reduce demand by increasing the separation between the aircraft using a constrained route. So while necessary to reduce demand, the speed, altitude, or directional changes required to achieve the separation imposed by Mile-In-Trail restrictions can substantially impact a flight's trajectory and arrival time. But because these restrictions are not considered in calculations of when an aircraft will arrive at any given point along the flight path, these perturbations from the filed flight plan will throw off the demand counts displayed to the Traffic Managers.

In order to effectively assess true demand, Traffic Managers have to gather information from several sources and mentally adjust the data in order to estimate the combined impact of any TMIs that apply to their facility as well as those that affect flows outside of the facility. Returning to our example above, Traffic Managers have to manually adjust the predicted demand values they obtain from their automated systems to account for the effects of Mile-In-Trail restrictions. The ability to do this well varies with experience and across individuals. Again, poor estimates of these effects can lead to over-restricting the airspace and lost capacity, while underestimating demand for a real constraint can lead to gridlock.

3.3   Managing Constraints

The ultimate purpose of estimating available capacity and assessing demand is to identify and manage potential and existing constraints in the NAS. Ideally, it would be useful to have an alerting system to notify Traffic Managers to conditions in which demand is, or likely will, exceed capacity. However, the alerting system in use today is inadequate at best, and can negatively impact the Traffic Managers' ability to manage constraints. It is based on the same flawed demand calculations described above and thus suffers from inaccuracy. This shortfall also undermines the Traffic Managers' ability to recognize when conditions have improved enough for restrictions to be eased or cancelled. As described earlier, weather data should ideally be ingested by automation and translated into the likely impact on capacity. If, for example, a TMI was implemented to reduce demand due to a constraint, it would be useful if automation could alert the Traffic Manager when the constraint has passed and the TMI may be reduced or altogether removed.

Today's systems do not provide such alerts and these opportunities sometimes go unnoticed resulting in unnecessary restriction on the system. The SMEs reported an example of the operational impact of this shortfall in which an ARTCC in the Northeast continued to reroute flights for two additional hours after routes that were closed due to severe weather were reopened. In this case, the Traffic Manager simply missed a transient verbal notification, so the lack of an adequate alerting system and real-time status information resulted in the continuation of a TMI that was no longer necessary. This can be very costly in terms of additional workload for Traffic Managers (who continued to manually reroute flights), Air Traffic Controllers (who dealt with increased air traffic complexity), and inefficiency due to increased fuel burn caused by longer routes that ultimately resulted in flight delays.

In addition to being aware of the opportunities to take action, in order to effectively manage constraints, it is equally important for Traffic Managers to understand the likely impact that any potential action will have on the system before it is implemented. Ideally, Traffic Managers would be able to model the impact of any potential TMI, or combination of TMIs, against a comprehensive, accurate view of the current state of the system. This would allow Traffic Managers to project their understanding of the system into the operationally relevant near future. This aspect of situation awareness is important because their decisions can have far reaching effects, impacting the entire NAS.

Today however, Traffic Manager's have very few tools for modeling TMIs. They lack modeling capabilities for some of the most commonly used TMIs, and tools for modeling the effects of multiple TMIs in combination are extremely limited. The lack of situation awareness induces overly conservative behavior when managing constraints. Without a clear understanding of the likely impact, the risk associated with trying innovative solutions becomes prohibitive and Traffic Managers fall back on conservative strategies that have been effective in the past, even if they are not the most efficient. It also inhibits the Traffic Managers' ability to fine tune TMIs to be minimally restrictive yet still adequately address the constraint.

Returning to the previous example, although hundreds of Mile-In-Trail restrictions are issued each day, there is no automated support for deciding where to place the restriction or how many miles of separation are needed to have the desired effect. These decisions can have a large impact on operations, but are left to the Traffic Manager's experience and best guess based on how the initiative has performed in the past. Not only are these processes cognitively demanding and workload-intensive, they also make the effectiveness of the outcome highly dependent on the skill of the individual Traffic Manager.

3.4   System View

Most of the deficiencies described above that prevent utilization of the NAS in a maximally efficient manner today can be categorized as a general lack of system view. Traffic Managers do not have an integrated, accurate source of data for performing their critical tasks. This shortfall directly impacts Traffic Managers' ability to assess capacity, determine demand, and manage constraints.

Traffic Managers use many different systems to perform operational tasks, but they function mostly as stand-alone applications. As systems and capabilities in TFM evolved, there was little attention paid to their integration. Figure 2 depicts a typical position in an ARTCC Traffic Management Unit (TMU). The TMUs are crowded with workstations that provide piecemeal operational information and tools, but do not come together to formulate a dynamic, complete view of the operation or to provide optimal support to operational decision-making.

As described in the Capacity, Demand, and Managing Constraints sections above, these core TFM tasks require essential elements of information that are either altogether missing or are not available in an integrated, operationally usable format and ultimately require judgments based on individual skill and experience.

Figure 2 Typical position within an ARTCC Traffic Management Unit

For example, the impact of weather is not integrated with current automation or translated into operationally meaningful information (e.g., impact on capacity), so capacity values are set through an imprecise process of negotiation between stakeholders. In addition, Traffic Managers are required to monitor inaccurate system-generated demand counts, manually collect information on relevant TMIs, mentally integrate the information, and adjust their mental models to ultimately arrive at a more realistic prediction of actual demand. And finally, although hundreds of TMIs are issued every day, the lack of an integrated modeling capability precludes Traffic Managers from knowing what the actual effects of the TMIs will be prior to actually impacting the operational system. Their decisions on which TMI to implement are largely based on experience and manual assessment of impact.

These processes are manifestations of the problems inherent in the current system and are all symptomatic of the lack of an accurate, comprehensive, and integrated system view. They require too many cognitively demanding manual steps, increase the potential for human error, and significantly increase the Traffic Managers' workload. This is especially problematic during severe weather when the situation can change rapidly and the Traffic Managers' workload is very high.

In summary, as one of the SME's described the problem, Traffic Managers do not have an integrated view of all of the decisions that have already been made. They also lack the ability to readily understand the impact of decisions they have yet to make. A more accurate, comprehensive view of the system that affords better responses to existing problems and more proactive responses to potential constraints is needed in order to meet the goals envisioned for NextGen.

4   CONCLUSION

The shortfalls described above have broad implications on human performance of core TFM functions associated with determining current status and predicted demand, assessing the capacity of NAS resources, and managing constraints. These deficiencies lead to real world operational problems such as unnecessary restriction of the NAS, flight delays, and greater costs. Yet this compromised system is expected to provide the foundation for future enhancements and operational improvements as currently defined in the NextGen Concept of Operations (JPDO, 2010). In an effort to move toward NextGen, we used a multidisciplinary team to identify and understand shortfalls associated with the current system from an operational and human performance perspective.

NextGen will modernize air transportation in the United States, with expected benefits to include greater efficiency, fewer delays, and ultimately lower costs. The realization of these benefits, however, is at risk due to the operational shortfalls associated with the current system. The problems related to accomplishing core TFM tasks must be addressed in order for NextGen to be effective. Automation should assist Traffic Managers in the determination of capacity and the assessment of predicted demand. The Traffic Managers should be able to examine and evaluate the potential impacts of their decisions before implementing them. Finally, all of these must be incorporated within the framework of an improved, comprehensive, and more accurate integrated view of the system. These enhancements will afford substantial improvement in Traffic Managers' ability to perform the core TFM tasks. In this manner, TFM can move forward and achieve the benefits envisioned for NextGen.

REFERENCES

Bernard, J., Geffard, M.V., Guensch, C. A., Kamine, S., Katkin, R.D., Keppel, J.G., Levin, K.M., Nguyen, K.L., Nussman, P., & Topiwala, T.N. (2010). *Traffic Flow Management (TFM) Roadmap*. MITRE CAASD report MTR100257R1.

Brennan, M., Ermatinger, C., Hoffman, R., Huegel, C., & Huston, W. (2011). *NextGen Mid-Term TFM Concept of Operations*. Dulles, VA: Metron aviation.

FAA (2011). *FAA Order JO 7210.3W, Facility Operation and Administration, Part 5: Traffic Management System, Chapter 17: Traffic Management National, Center, and Terminal*. Washington, DC: Federal Aviation Administration.

Geffard, M. V., Sgorcea, R., & Stalnaker, S. (2010). *TFM in the Mid-Term: Identification of Business Rules and Gap Assessment Level 2*. MITRE CAASD, MP100312.

JPDO (2010). *Concept of Operations for the Next Generation Air Transportation System, Version 3.2*. Washington, D.C.: Joint Planning and Development Office

Somersall, P. (2011). *TFM Tools: Overview*. Briefing presented to AWS on February 4, 2011.

CHAPTER 35

Expertise, Cognition, and CRM Issues in NextGen Automation Use

Kathleen L. Mosier
San Francisco State University
kmosier@sfsu.edu

Ute Fischer
Georgia Institute of Technology

Judith Orasanu
NASA Ames Research Center

*The HART Group**

ABSTRACT

The aviation NextGen environment will be substantially different from the current National Airspace (NAS) operations. More and more sophisticated automation will enable innovations such as trajectory-based flight plans with precision paths from take-off to landing, aircraft self-separation and traffic avoidance, electronic flight bags, datalink communications, all-weather operations on very closely-spaced parallel runways (VCSPR). Flight crews will be required to interact proficiently with new automated systems and engage them for the flying tasks. Increasingly complex automated systems, however, will pose formidable challenges to crew decision making and resource management. System design and crew training will need to be mindful to the capabilities and limitations of human cognition and decision making to ensure safe operations in the NextGen airspace. In this chapter we describe what experienced crews and automation bring to the decision process in NextGen; define challenges and impediments to cognition,

---

* Frank Durso, Karen Feigh, Ute Fischer, Vlad Popp, & Katlyn Sullivan at Georgia Tech; Dan Morrow at the University of Illinois at Urbana-Champaign; Kathleen Mosier at San Francisco State University.

decision making and human automation interaction; and present guidelines for system design and crew training.

Keywords: expertise, human-automation interaction, decision making

1 THE ROLE OF EXPERTISE IN AVIATION DECISION MAKING

Experts' superior performance in domains such as aviation is directly dependent on their domain-specific knowledge, and that this knowledge enables experts to anticipate future actions and prepare for them more efficiently (Charness & Tuffiash, 2008). Klein's model of expert Recognition-Primed Decision Making (1993), for example, describes expertise as the ability to identify critical cues in the environment, and to understand their structural relationship. According to this model, expert pilots look for familiar patterns of relevant cues, signaling situations that they have dealt with in the past, and base their responses on what they know "works." Importantly, in naturalistic environments experts are able to quickly identify the most ecologically valid cues, that is, the subset of information most critical to situation assessment and to match these cues against patterns in their experience base. Experienced pilots, for example, match the shape, color and size of clouds with patterns in memory to determine whether they should fly through them or around them. The ability to use sensory-driven and intuitive processes in naturalistic environments enables fast diagnosis and decision making as experts are able to quickly size up a problem and know what actions should be taken (Orasanu & Connolly, 1993). This is a critical factor in time-constrained situations.

Experts are also sensitive to dynamically changing conditions and adapt their mental models accordingly. Mosier and Chidester (1991), for example, found that successful flight crews demonstrated an iterative process in solving an oil pressure problem. They made preliminary, modifiable decisions and continued to gather information and monitor results of their actions to refine their diagnoses.

Experts also employ strategies that enable them to cope with ambiguity and time pressure. They are proactive, anticipate potential failures, risks or conflicts, and prepare for them. They know how to 'buy time.' For instance, pilots may request a holding pattern to gather additional decision-relevant information (Orasanu, 1990). To manage time wisely, experts anticipate developments, make contingency plans, prioritize tasks, and use low-workload periods to prepare for upcoming events (e.g., Fischer, Orasanu, & Montalvo, 1993).

2 CHALLENGES TO PILOT-AUTOMATION INTERACTION IN NEXTGEN

The benefits of advanced technology in terms of increased efficiency, data storage and manipulation are self-evident—automated systems can assimilate more information and process it faster than humans. Highly sophisticated decision support tools (DSTs) and information sources will be available to crews. Some shortcomings, however, have been documented in past research and must also be considered. The availability of information, for example, may depend on the crew's ability to locate it through multi-layered systems. Many automated systems are opaque in their operations, and do not enable operators to track their functioning. They cannot take into account context-specific information, and thus may generate faulty recommendations (Gawande & Bates, 2000; Mosier, 2002).

Several researchers have documented problems in the use of advanced automated systems, including mode misunderstandings and mode errors, failures to understand automation behavior, confusion or lack of awareness concerning what automated systems are doing and why, and difficulty tracing the functioning or reasoning processes of automated agents (e.g., Sarter & Woods, 1994a,b). *Automation surprises*, or situations in which operators are surprised by control actions taken by automated systems (Sarter, Woods, & Billings, 1997), may be the result of non-coherent situation assessment processes—that is, when operators have incomplete or inconsistent information about the situation, or misinterpret or mis-assess data on system states and functioning (Woods & Sarter, 2000). Mode errors, or confusion about the active system mode, have resulted in several aviation incidents and accidents (e.g., Sarter & Woods, 1994a,b; Woods & Sarter, 2000). The tendency toward automation bias, or errors resulting from the use of automation as a heuristic replacement for vigilant information search (Mosier & Skitka, 1996), is exacerbated as the logic of systems becomes more difficult to trace.

Additionally, information from automated systems is absorbed primarily via analytic or coherence processes: the *parameters* of the system, the *status* of the flight in terms of flight path and airspeed, the *steps to take* to correct anomalies, the *precise position* of the aircraft. Data and information are specific and highly reliable, but consistent and comprehensive integration of all relevant information for situation assessment, is central as this enables accurate diagnoses and decisions.

A critical component of expertise and pilot-automation interaction in the NextGen cockpit will be knowing when a situation requires analysis as opposed to pattern recognition. Expert pattern matching does not work when data are in digital format, are hidden below surface displays, or mean different things in different modes (Mosier, 2010). Some current displays, however, may in fact be leading operators astray by fostering the assumption that data can be managed in a non-analytical fashion. This is an especially potent trap for highly experienced decision makers, who may be particularly vulnerable to errors if they 'see' patterns before they have sought out and incorporated all relevant information into their assessment.

3 CHALLENGES AND IMPEDIMENTS TO COGNITION AND DECISION MAKING IN NEXTGEN

New tasks, new tools, and new systems. In the early stages of NextGen, flight crews will be unfamiliar with new systems and procedures, and will need to develop expertise in the changed operational domain. The knowledge store that formed the basis for informed heuristics will need to be updated, and the use of shortcuts that work in current operations may lead to errors or mis-assessments in NextGen operations. For example, crews may use old models rather than new automation capabilities when judging the feasibility of a low-visibility VCSPR approach, or may break out of a paired approach based on old standards of safe separation.

Even the most seasoned professional pilots will experience challenges to their expertise as new automated systems and procedures are introduced in NextGen operations. The training issue for NextGen is less a matter of how to teach but rather *what* to teach to enable the development of new expertise. That is, we need to define how expert performance in NextGen will differ from expert performance in current operations. What generalizes? What mental models are likely to lead pilots astray because they do not apply? Where will pilots have skill or knowledge gaps?

Results of a study examining performance in paired simultaneous VCSPR approaches illustrate the trade-offs involved in mixing old and new procedures and automation (Verma et al., 2011). Pilots on a paired approach to SFO were responsible for maintaining a 15sec temporal separation from the lead aircraft using either a position vs. prediction display - and current automation vs. future auto speed control automation. Perhaps the most interesting finding was the interaction between display and automation with respect to workload. When pilots were working with current automation, the position display was rated slightly higher on workload than the prediction display. In the auto speed control automation condition, however, the prediction display produced higher workload than the position display. The researchers posited that because the prediction display resembled the 'green arc' used for altitude prediction, pilots were able to use it relatively easily when combined with familiar automation. Its advantage disappeared when the display was combined with new automation.

These results suggest that pilots will need to be very careful when applying old mental models or 'how to' knowledge to new displays and automation. Experience can be counterproductive if accompanied by the tendency to rely on electronic data and systems in an intuitive manner. Experience can also induce a false sense of coherence, or the tendency to see what one expects to see rather than what is there, as illustrated by the "phantom memory" phenomenon found in part-task simulation data (Mosier et al., 1998; Mosier, Skitka, Dunbar, & McDonnell, 2001). A potential hazard is that pilots may think they 'recognize' and can use a display in familiar ways because it is similar to a current one (e.g., the prediction display is similar to the green arc), and may not be cognizant of differences in interpretation or function.

New automation requires coherent processing - data displays must be examined thoroughly to ensure accurate comprehension, particularly when displays are coupled with new systems and procedures. This represents a shift to knowledge-based mode of problem solving.

Challenges to situation awareness (SA). SA in the NextGen cockpit will require pilots to monitor instruments and to ensure that various indicators and parameters are consistent with each other and the current flying situation. However, the presence of automated systems may lead to biased rather than enhanced information processing and may result in a lack of situation awareness. In a pair of experiments, Mosier and colleagues (Mosier et al., 1998; 2001) identified automation bias as a threat to SA. Pilots overlooked system irregularities that were not detected or indicated by the automation, or alternatively relied exclusively on a single, salient cue automated cue without cross-checking its validity.

SA can also be compromised as a consequence of keeping the user too far removed from the operations of the automated system. As users become more distant from the raw data, and instead rely on system assessments of that information, they may be taken "out of the loop," and more dependent on the automation, which can undermine their SA and action selection (Chidester, 1999). Similarly, a lack of salient feedback signaling a change in the state or mode of the system could lead to misunderstandings and error (Sheridan & Parasuraman, 2006).

Mode awareness problems occur when pilots are unaware of current or future states (or behavior) of automated flightdecks, so that they perform actions consistent with the presumed but not the actual state of the system (Sarter & Woods, 1995). Mode errors reflect attentional processes that interact with inadequate knowledge. Specifically, data-driven factors (salience, access effort coupled with high visual demands from concurrent pilot tasks) and knowledge-driven factors (inaccurate system knowledge) may undermine scanning strategies that support mode awareness (and possibly other dimensions of SA) in automated cockpits (see Bjorklund, Alfredson, & Dekker, 2006). An eye-tracking study of crews interacting with an FMS in a B747 simulator found that pilots failed to fixate on the FMA at critical moments, especially for modes that were initiated by the FMS, suggesting that these mode errors reflected inadequate scanning strategies (Sarter, Mumaw, & Wickens, 2007).

The problem may be exacerbated in NextGen if pilots shift scanning strategies to accommodate new displays at the expense of scanning primary flight data. In this case, they may allocate less attention to the detection of unexpected events than they would otherwise. This scanning shift may reflect data-driven factors such as the salience of new displays (e.g., 3D synthetic displays) or top-down factors such as the (incorrect) expectation that all situational information is represented on these displays, causing them to miss infrequent and unexpected events that threaten safety (e.g., undisplayed traffic; Wickens & Alexander, 2009).

4 CHALLENGES TO CREW COORDINATION AND COLLABORATION IN NEXTGEN OPERATIONS

The safe operation of an aircraft requires crewmembers to collaborate and to coordinate their efforts. Communication is essential as crewmembers need to ensure that they have a common understanding of their situation and action plans. Research consistently showed that well-performing crews discussed in greater detail the problem they faced and their response to it than poorly performing crews (e.g., Orasanu & Fischer, 1992; Sexton & Helmreich, 2000), and were more explicit about their reasoning and intentions (Fischer & Orasanu, 2000). High-performing crews also tended to adhere to normative communication patterns—e.g., questions were answered, and observations acknowledged or elaborated upon (Kanki, Lozito & Foushee, 1989). In so doing they facilitated mutual understanding because conversations had a tight structure and thematically related contributions followed one another in a coherent fashion.

The introduction of automation into the cockpit may change the nature of pilots' communications in one of two important ways. Some (Bowers, Oser, Salas, Cannon-Bowers, 1996; Costley, Johnson, & Lawson, 1989) observed that automation reduced crew communication apparently because pilots could retrieve critical flight information directly from automated systems without input from the other crewmembers. In a related finding Parke et al. (2001) noted that datalink clearances were not consistently shared among crewmembers. Issues of information retrieval and sharing may become even more pronounced in NextGen as pilots interact with more and increasingly complex automated systems.

On the other hand, automation may lead to an increase in pilots' verbal communication. This hypothesis reflects concerns expressed by Wiener (1989) and others (e.g., Segal & Jobe, 1995) who pointed out that automation makes it more difficult for pilots to see and monitor each other's actions, and thus limits pilots' opportunities for non-verbal communication. Pilots may thus compensate for the loss of non-verbal information by talking more. Consistent with this hypothesis several studies showed an increase in crew communication in automated compared to traditional aircraft (Veinott & Irwin, 1993), specifically an increase in pilots' task-related communication (Damos, John, & Lyall, 2005). Similarly, Parke, Kanki, McCann and Hooey (1999) report that pilots adapted their communication behavior to automation–specific affordances. The availability of Electronic Moving Maps during taxi, for example, reduced the number of simple information requests but increased the number of traffic-related observations in pilots' discourse. Technology that provides pilots with shared visual information may facilitate mutual understanding insofar as this information can be presumed as common ground and need not be explicitly addressed in their communications, and in so doing it frees pilots to focus discussions on issues that are not yet part of their common ground.

5   CAVEATS AND CONCLUSIONS

Experts are considered well calibrated if they are aware of the areas and circumstances for which they have correct knowledge and those in which their knowledge is limited or incomplete (Sarter et al., 1997, p. 1929)

Throughout this chapter, we have emphasized the point that the key driver in NextGen decision making will be automation. Decision support systems will provide data about aircraft state and help pilots synthesize it, and may offer recommendations for action. Automated sources such as SWIM (System Wide Information Management) will make up-to-the minute weather information easily accessible. Automated broadcasting systems (e.g., ADS-B, Automatic Dependent Surveillance - Broadcast) will facilitate accurate tracking of traffic. Automated navigational systems will enable flight paths that are precise with respect to both position and time. Sophisticated datalink systems will allow crews to input new clearances or flight changes with the push of a button. Automation will enable all-weather operations that have not been possible in the past, such as VCSPR approaches and landings. In the NextGen environment, electronic data and information will supplant most out-the-window cues, and pilot attention will be much more inside the cockpit than ever before. Although the benefits of NextGen automation are undeniable, some caveats must be acknowledged and addressed in crew training as well as system design.

Pilots need accurate mental models of automated systems. Current pilot automation training focuses on procedural knowledge rather than functional relationships. The 'cookbook,' step-by-step approach to managing automated systems results in large gaps in pilots' understanding of automated systems and limits their ability to cope with any situations outside of normal, standard operations More than a decade ago, Sarter and Woods (1994a,b) called for training on how systems work as well as how to work the systems, and the potential costs of ignoring their call in NextGen operations are enormous. Pilots will always form the last line of defense in anomalies or emergencies, and they will need the resilience provided by a deep level of system knowledge during non-normal operations. This is especially important as new responsibilities are shifted from ATC to flight crews.

Electronic systems require analysis. Electronic data are not amenable to intuitive processes. Even when data are presented in graphic format or seem to present patterns for recognition, their meaning must be interpreted in light of automation mode, flight phase, and system status. Expert pilots may not be able to rely on short-cuts or heuristics, such as pattern-matching, because the architecture of the knowledge space has changed. Expertise in the NextGen cockpit entails coherence skills – seeking out relevant data and information and ensuring that they form a consistent 'story' of the situation. Training programs should include a focus on coherence skills, and system design must facilitate the analytical processing required when dealing with electronic systems.

System awareness is key to situation awareness. SA in the NextGen cockpit cannot be achieved without systems awareness. Because most of the information for SA will reside within electronic systems, an awareness of what systems are doing and why, what mode they are in, and what they will do next is essential to accurate assessment and awareness. Training and design need to facilitate systems awareness as well as the integration of data inside the cockpit with external elements.

Several approaches have been used to support SA in automated cockpits. Peripheral visual and haptic cues for instance may provide effective bottom-up guidance to critical information in visually saturated automated cockpits (Nicolic & Sarter, 2001). Other approaches focus on creating more accurate pilot mental models of how automated systems function, which may support more effective knowledge-based attention strategies (see Sarter, 2008 for review).

What's new and different? Changes must be highlighted. One potentially hazardous aspect of NextGen is that it is an updated paradigm, mixing current and new automation and displays. Changes to systems and procedures are being introduced incrementally, and much of the 'new' automation actually comprises enhancements to current automation. The FMS, for example, will be upgraded to a 4D management system, and will enable autoload of clearances. NextGen systems and displays will be somewhat different from current systems and displays, but not completely so - luring pilots into old patterns of thinking and interacting when these behaviors are no longer appropriate. Pilots with the most experience with old systems may be most susceptible to this trap, as their patterns are more strongly entrenched and they are susceptible to 'strong-but-wrong' erroneous behavior that is more consistent with past practice than current requirements (Reason, 1990). Training and design must include a focus on what is different in systems and displays. Pilots must be aware of nuances of new automation, differences in its functioning, and situations in which it is likely to mislead them.

'Monitor and challenge' also applies to automation. A primary responsibility of pilots in NextGen will be to use new tools wisely in their decision making processes. Crew resource management programs train pilots to monitor each other, and to challenge each other when anything seems awry. A similar approach to automation will increase the probability that the promise of increased safety and efficiency in NextGen operations will be realized.

Design needs to support common ground between humans and automation. Common ground is a central component of human communication. Conversational partners need to have a common knowledge base which, in turn, influences their production and comprehension of messages. Partners' common ground, moreover, is not a static set of knowledge; rather, partners update what is mutually known moment by moment as they converse. These characteristics of communication have at least four implications for the design of cockpit automation, in particular when human-machine interactions are conceptualized as team communication. (1) System design needs to be compatible with pilots' task model and control strategies to ensure that both parties (pilots and automation) come to a shared situation understanding (see also Kaber, Riley, Tan & Endsley, 2001). (2) Automation needs to be transparent to facilitate human-machine communication. This requirement is particularly relevant as automated systems become more complex and autonomous, with higher levels of authority (Sarter & Woods, 2000). One solution is offered by Woods, Roth, and Bennett (1990) who propose that systems clearly indicate their current state, goals, knowledge, hypotheses and intentions. This suggestion is consistent with findings suggesting that superior task performance is related to team members' problem-related contributions, presumably because they facilitate the development of shared situation models (e.g., Orasanu & Fischer, 1992; Sexton & Helmreich, 2000). (3) System design should emphasize the collaborative nature of communication. This suggestion entails a change in pilot-machine interactions - pilots would be required to explicitly accept automated information or directives as understood (see also Miller, 2004). Conversely, pilots' input to automated systems would have to be explicitly acknowledged by the system, for instance by requesting confirmation of the input, such as "Are you sure you want to change the flight level to X feet?" Requiring pilots and systems to provide evidence of their understanding may be especially important when automation presents critical information or pilots propose significant changes to the system status. (4) Related to this, systems should be able to flag pilot input that is inconsistent with or deviates significantly from previous inputs or system states, indicating problems of understanding. Such a feature may prevent tragedies such as the Cali accident (Aeronautica Civil of the Republic of Columbia, 1996). If the FMC had indicated that Romeo was off the programmed flight path and questioned the captain about whether he really wanted to turn toward it, he may have realized that something was amiss.

Design needs to support shared situation models among crewmembers. As mentioned earlier, automation curtails the availability of non-verbal information as system inputs by one crewmember cannot be observed by the other (Sarter, 1995). As a consequence pilots may be required to talk more to compensate for the loss of visual information, or risk jeopardizing mutual understanding. Increased verbal communication may distract pilots from other pertinent tasks and thus may not be a viable solution. Instead, system design should facilitate non-verbal communication between crewmembers. This could be achieved by ensuring that system interfaces are visible to both pilots or by clearly displaying input changes by one crewmember on the interface of the other.

Metacognition will be crucial. The development of expertise for NextGen, as with any new task or system, will require a qualitative shift in processing and organization of knowledge, with an emphasis on developing coherence skills. Metacognition—the ability of experts to monitor their own problem-solving efforts (Druckman & Bjork, 1991)—will thus be an important aspect of pilot expertise in NextGen. Training could target expanding these metacognitive skills for the new context, for instance by providing practice in strategy selection in response to situation events (Cannon-Bowers & Bell, 1997), and by articulating procedures to help pilots monitor their cognitive processes.

ACKNOWLEDGMENTS

This work was supported by funding from the FAA (cooperative agreement DTFAWA-10-C-00084 with Georgia Tech) and NASA (cooperative agreement NNX08AW90A with Georgia Tech and NNX10AK52A with San Francisco State University).

REFERENCES

Aeronautica Civil of the Republic of Columbia (1996). Aircraft accident report: Controlled flight into terrain, American Airlines Flight 965, near Cali, Columbia, December 20, 1995. Santafe de Bogota, D. C., Columbia: Author.

Bjorklund, C. M., Alfredson, J., Dekker, S. (2006). Mode monitoring and call-outs: An eye-tracking study of two-crew automated flight deck operations. *International Journal of Aviation Psychology*, 16, 263-275.

Bowers, C. A., Oser, R. L., Salas, E., & Cannon-Bowers, J. A. (1996). Team performance in automated systems. In R. Parasuraman & M. Mouloua (Eds.), *Automation and human performance: Theory and applications* (pp. 243–263). Mahwah, NJ: Erlbaum.

Cannon-Bowers, J. Bell, H. H. (1997). Training decision makers for complex environments: Implications of the Naturalistic Decision Making perspective. In C. E. Zsambok & G. Klein (Eds.), *Naturalistic decision making* (pp. 99-110). Mahwah, NJ: Erlbaum.

Charness, N., & Tuffiash, M. (2008). The role of expertise research and human factors in capturing, explaining, and producing superior performance. *Human Factors, 50,* 427-432.

Chidester, T. (1999). Introducing FMS aircraft into airline operations. In S. Dekker & E. Hollnagel (Eds.). *Coping with computers in the cockpit* (pp. 153-194). Farnham, Surrey: Ashgate.

Connolly, T. (1988). Hedge-clipping, tree-felling, and the management of ambiguity. In M. B. McCaskey, L. R. Pondy, & H. Thomas (Eds.), *Managing the challenge of ambiguity and change.* NY: Wiley.

Costley, J., Johnson, D., and Lawson, D. (1989). A comparison of cockpit communication B737-B757. *Proceedings of 5th International Symposium on Aviation Psychology* (pp. 423-418). Columbus, OH. The Ohio State University.

Damos, D., John, R. S., and Lyall, E. A. (2005). Pilot activities and level of cockpit automation. *The International Journal of Aviation Psychology, 15,* 251-268.

Druckman, D. & Bjork, R. E. (Eds.), In the mind's eye: Enhancing human performance. Washington, DC: National Academy Press.

Fischer, U., McDonnell, L., & Orasanu, J. (2007). Linguistic correlates of team performance: Toward a tool for monitoring team functioning during space missions. *Aviation, Space, and Environmental Medicine, 78(5),* II, B86-95.

Fischer U., & Orasanu, J. (2000). Error-challenging strategies: Their role in preventing and correcting errors. *Proceedings of the Human Factors and Ergonomics Society 44th Annual Meeting*, San Diego, CA, Vol. 1, 30-33.

Fischer, U., Orasanu, J., & Montalvo, M. (1993). Efficient decision strategies on the flight deck. *Proceedings of the 7th International Symposium on Aviation Psychology*, 238-243.

Gawande, A., & Bates, D. (2000, February). The use of information technology in improving medical performance: Part I. Information systems for medical transactions. *Medscape General Medicine, 2,* 1-6.

Kaber, D. B., Riley, J. M., Tan, K.-W., & Endsley, M. (2001). On the design of adaptive automation for complex systems. *International Journal of Cognitive Ergonomics, 5*(1), 37-57.

Kanki, B. G., Lozito, S. C., & Foushee, H. C. (1989). Communication indices of crew coordination. *Aviation, Space, and Environmental Medicine, 60,* 56-60.

Klein, G. (1993). Recognition-primed decision (RPD) model of rapid decision making. In G. A. Klein, J. Orasanu, R. Calderwoodk & C. Zsambok (Eds.), *Decision making in action: Models and methods* (pp. 103-147). Norwood, NJ: Ablex.

Miller, C. A. (2004). Human-Computer etiquette: Managing expectations with intentional agents. *Communications of ACM, 47*(4), 31-34.

Mosier, K. L. (2002). Automation and cognition: Maintaining coherence in the electronic cockpit. In E. Salas (Ed.), *Advances in Human Performance and Cognitive Engineering Research, Volume 2* (pp. 93-121). Elsevier Science Ltd.

Mosier, K. L., (2010). The human in flight: From kinesthetic sense to cognitive sensibility. In E. Salas & D. Maurino (Eds.), *Human factors in aviation* (pp. 147-173). NY: Elsevier.

Mosier, K. L., & Chidester, T. R. (1991). Situation assessment and situation awareness in a team setting. *Designing for Everyone: Proceedings of the 11th Congress of the International Ergonomics Association*, Paris, 15-20 July, pp. 798-800.

Mosier, K., & Skitka, L. (1996). Human decision makers and automated decision aids: Made for each other? In R. Parasuraman & M. Mouloua (Eds.), *Automation and Human Performance: Theory and Applications.* NJ: Lawrence Erlbaum Associates (pp. 201-220).

Mosier, K., Skitka, L. J, Heers, S., & Burdick, M. D. (1998). Automation bias: decision making and performance in high-tech cockpits. International Journal of Aviation Psychology, 8(1), 47-63.

Mosier, K. L., Skitka, L. J., Dunbar, M., & McDonnell, L. (2001). Air crews and automation bias: The advantages of teamwork? *International Journal of Aviation Psychology, 11*, 1-14.

Nikolic, M. I., & Sarter, N. B. (2001). Peripheral visual feedback: A powerful means of supporting attention allocation and human automation coordination in highly dynamic data-rich environments. *Human Factors, 43*, 30–38.

Orasanu, J. (1990). *Shared mental models and crew decision making.* Technical Report No. 46. Princeton, NJ: Princeton University, Cognitive Science Laboratory.

Orasanu, J., & Connolly, T. (1993). The reinvention of decision making. In G. A. Klein, J. Orasanu, R. Calderwood, & C. E. Zsambok (Eds.). *Decision making in action: Models and methods* (pp. 3–20). Norwood, NJ: Ablex.

Orasanu, J., & Fischer, U. (1992). Distributed cognition in the cockpit: Linguistic control of shared problem solving. In *Proceedings of the Fourteenth Annual Conference of the Cognitive Science Society* (pp.189-194). Hillsdale, NJ: Erlbaum.

Parke, B., Kanki, B. G., McCann, R. S., & Hooey, B. L. (1999). The effects of advanced navigation aids on crew roles and communication in ground taxi. *Proceedings of the 10th h International Symposium on Aviation Psychology* (pp. 804-809). Columbus, OH: The Ohio State University.

Parke, B., Kanki, B. G., Munro, P. A., Patankar, K., Renfroe, D. F., Hooey, B. L., & Foyle, D. C. (2001). *The effects of advanced navigation aids and different ATC environments on task-management and communication in low visibility landing and taxi.* Presentation at the 11th International Symposium on Aviation Psychology. Columbus, OH.

Reason, J. T. (1990). *Human Error.* Cambridge, UK: Cambridge University Press.

Sarter, N. B. (1995). "Knowing when to look where": Attentional location on advanced automated flight decks. *Proceedings of the 8th International Symposium on Aviation Psychology* (Vol.1, pp. 239–242). Columbus: OH: The Ohio State University.

Sarter, N. (2008). Investigating mode errors on automated flight decks: Illustrating the problem-driven, cumulative, and interdisciplinary nature of human factors research. *Human Factors, 50*, 506-510.

Sarter, N. B., Mumaw, R. J., Wickens, C. D. (2007). Pilots' monitoring strategies and performance on automated flight decks: An empirical study combining behavioral and eye-tracking data. *Human Factors, 49*, 347-357.

Sarter, N. R., & Woods, D. D. (1994a). Decomposing automation: Autonomy, authority, observability and perceived animacy. In M. Mouloua & R. Parasuraman (Eds.), *Human performance in automated systems: Current research and trends* (pp. 22-27). NJ: Lawrence Erlbaum Associates.

Sarter, N., & Woods, D. D. (1994b). Pilot interaction with cockpit automation II: An experimental study of pilots' model and awareness of the flight management system. *International Journal of Aviation Psychology, 4*, 1-28.

Sarter, N., & Woods, D. D. (1995). How in the world did we ever get into that mode? Mode error and awareness in supervisory control. *Human Factors, 37*, 5-19.

Sarter, N. B., & Woods, D. D. (2000). Team play with a powerful and independent agent: A full-mission simulation study. *Human Factors, 42*, 390-402.

Sarter, N B., Woods, D. D., & Billings, C. (1997). Automation surprises. In G. Savendy (Ed.), *Handbook of human factors/ergonomics* (2nd ed., pp. 1926-1943). NY: Wiley.

Segal, L. & Jobe, K. (1995). On the use of visual activity communication in the cockpit. *Proceedings of the 8th International Symposium on Aviation Psychology* (Vol.1, pp. 712-717). Columbus, OH: The Ohio State University.

Sexton, J. B., & Helmreich, R. L. (2000). Analyzing cockpit communication. The links between language, performance, error and workload. *Human Performance in Extreme Environments, 5*, 63-68.

Sheridan, T. B., & Parasuraman, R. (2006). Human-automation interaction. In R. S. Nickerson (Ed.), *Reviews of Human Factors and Ergonomics, Volume 1* (pp. 89-129). Santa Monica, CA: Human Factors and Ergonomics Society

Veinott, E. S., & Irwin, C. M. (1993). Analysis of communication in the standard versus automated aircraft. *Proceedings of the 7th International Symposium on Aviation Psychology* (pp. 584-588). Columbus, OH: The Ohio State University.

Verma, S., Lozito, S., et al. (2011). *Integrated Pilot and Controller Procedures: Aircraft Pairing for Simultaneous Approaches to Closely Spaced Parallel Runways.* Presentation at the 9th USA/Europe Air Traffic Management Research & Development Seminar (ATM2011). Berlin, Germany. June 13-16.

Wickens, C.D. & Alexander, A, L. (2009). Attentional tunneling and task management in synthetic vision displays. *International Journal of Aviation Psychology*, 19(2), 1-17

Wiener, E. (1989). *Human factors of advanced technology ("glass cockpit") transport aircraft. Technical Report 117528.* Moffett Field, CA: NASA Ames Research Center.

Woods, D. D., Roth, E. M., & Bennett, K. B.(1990). Explorations in joint human-machine cognitive systems. In W. W. Zachary, S. P. Robertson, and J. B. Black (Eds.) *Cognition, computing, and cooperation* (pp. 123-158). Norwood, NJ: Ablex.

Woods, D. D., & Sarter, N. B. (2000). Learning from automation surprises and "going sour" accidents. In N. B. Sarter & R. Amalberti (Eds.), *Cognitive engineering in the aviation domain.* (pp. 327-353) Mahwah, NJ: Erlbaum.

Section VI

Modeling

Enhanced GOMS Language (E-GOMSL) Tool Development for Pilot Behavior Prediction and Simulation

Guk-Ho Gil[1], David Kaber[1], Sang-Hwan Kim[2]

[1]Department of Industrial and Systems Engineering,
North Carolina Sate University
Raleigh, NC, USA
ghgil@ncus.edu, dbkaber@ncsu.edu
[2]Department of Industrial and Manufacturing Systems Engineering,
University of Michigan-Dearborn
Dearborn, MI, USA
dysart@umd.umich.edu

ABSTRACT

This research developed a new (E)nhanced GOMS (Goals, Operators, Methods, and Selection Rules) Language (E-GOMSL) cognitive modeling tool. EGOMSL integrates capabilities of existing tools and includes new functions to promote usability and learning by aviation system designers involved in the conceptual phase of cockpit automation design. The tool was evaluated by a group of human-computer interaction experts and actual aircraft systems designers using two different types of usability tests, including a cognitive walkthrough (CWs) and system usability survey. The CW revealed specific usability issues, including: (1) a lack of visibility of some interface controls for accessing tool functions; (2) a lack of visibility of a dialog and tool bar for interface prototyping; (3) a lack of user understandability of a flow diagramming capability; and (4) complexity of the method for modifying diagrams. The average system usability score (SUS; 53.13 out of 100 with a criterion of 65) also indicated some usability problems might exist for designers. Recommendations were made to address the usability issues revealed by the CW and to improve the overall SUS from the designers perspective.

Keywords: cognitive modeling, pilot behavior, GOMS, cockpit design

1   INTRODUCTION

One of the main limitations of existing approaches to complex human-machine system design is the requirement for empirical data as a basis for alternative design selection. In addition, design decisions are often based on collections of design guidelines with limited theoretical explanations for why they may be effective from a human performance perspective. This lack of information limits understanding of when and how guidelines can be applied by designers. In order to better support conceptual design of interactive systems, various human performance modeling (HPM) techniques and tools have been developed based on architectures of human information processing (HIP). However, these techniques and tools also have several limitations from a design perspective. Existing tools are not easy to use and designers or developers may need extensive training and practice in use. Furthermore, there is currently no fundamental set of tool capabilities, such as supporting task workload analysis, identifying patterns of HIP (e.g., memory use), simulating visual interface object use (e.g., eye movements), providing interface design support, etc.

Early forms of HPM techniques include the GOMS (Goals, Operators, Methods, and Selection Rules) cognitive modeling technique developed by Card et al. (1993). GOMS models describe user behavior with interactive systems and can be used to evaluate system design from usability and performance perspectives. However, the GOMS Language was limited to representing expert behavior in tasks that could be decomposed to sets of procedures. GOMS models also did not support representation of basic forms of vision (e.g., foveal, peripheral) and motor behaviors as well as parallel processing of these operations. Another major limitation was that time estimates for operations coded in models for predicting task completion times were deterministic. Therefore, model output could not accurately represent individual differences in performance or the stochastic nature of human behavior. Card et al. (1983) previously suggested that variances in operation times could be determined by observation on target tasks and applied in model coding.

Other contemporary modeling tools include the Queuing Network-Model Human Processor (QN-MHP; Liu et al., 2006) and CogTool (John et al., 2004; Teo & John, 2008). Their tools were reviewed in terms of steps to use, methods for user interface modeling, workload analysis and user performance simulation. With respect to modeling steps, the CogTool includes a system interface prototyping capability that allows a designer to mock-up an interface for analysis of potential user behaviors. This is an important capability because the CogTool can automatically generate cognitive models for demonstrations of prototype use. An important feature of the QN-MHP is that it supports user workload analysis in complex systems control by using a graphing technique. With respect to user behavior simulation, the QN-MHP uses computer graphics to represent human visual attention to interface features and the flow of information processing in task processing. CogTool also has the capability to represent patterns of HIP and it does so by using a Gantt-chart to activation times for various visual, cognitive and motor processors.

These capabilities of existing cognitive modeling tools were applied in the development of a new (E)nhanced GOMS Language (E-GOMSL) modeling tool. In addition, the tool includes new functions to promote usability and learning by designers involved in the conceptual phase of the systems design process. Unlike the original GOMS Language, the E-GOMSL tool supports detailed description of low-level visual and motor operations in cognitive models, coding of parallel execution of such operations, the use of stochastic variables for representing operation times, and flexibility in coding different kinds of operations that occur in different work domains, such as aircraft piloting.

The new tool was evaluated through usability tests conducted with a group of six human-computer interaction (HCI) experts and actual aircraft systems designers. Three cognitive walkthroughs (CWs) were performed on various E-GOMSL functions by following a use case scenario. The participants also completed the System Usability Scale (SUS) developed by Brooke (1996). The evaluation approach and results are detailed below.

2   METHOD

2.1   Definition of E-GOMS Language

Kieras (2006) previously developed a computational form of the GOMS Language (i.e., GOMSL). We used GOMSL as a basis for the new E-GOMS Language. We also adopted the concept of flexibility in task method description and operator representation, as used in Natural GOMS Language also developed by Kieras (1997). Each operator in E-GOMSL is defined in terms of four different properties, with reference to a model of HIP (Wickens & Hollands, 1999). These properties include channels, control objects, operator syntax and operator times. There are four processing channels available for modeling behaviors in the E-GOMS Language, including auditory, visual, cognitive and motor (hand, foot and vocal). Each channel has multiple control objects. For example, the cognitive channel includes an information flow control, parallel processing capability, a working memory (WM) store, a mechanism for transactions between WM and long-term memory (LTM), and a decision making control. E-GOMSL operator syntaxes have similar patterns to those used in GOSML and NGOMSL (e.g., Look_for_object_whose <property> is <value>, and_store_under <tag>). With respect to operator times for describing pilot use of cockpit automation interfaces, stochastic variables were determined based on video recordings of actual line pilot behavior in a simulator experiment. A stopwatch was used to record times for specific operations. Due to limitations of this method, times for only five operators were collected, including "Confirm" a visual object, "Look at" a visual object, "Think of" of a cognitive process, "Store" information, and "Recall" information. We used the lognormal distribution as a basis for the stochastic variables since performance times for simple behaviors, like reading dials and entering data by keystroke, are generally skewed (Card et al., 1983, P. 85) and non-normal. The lognormal distribution represents skewed distributions that are common when measurements have low mean and large variance values with all positive data (Limpert et al., 2001). The lognormal distribution has (1) a threshold parameter (i.e., a lower limit) (γ), (2) a scale parameter (θ), and (3) a shape parameter (σ) (Cohen, 1988).

2.2   E-GOMSL Tool Development

The overall concept for the E-GOMSL tool development was to provide an easy to learn and use method and procedure for creating cognitive models of human behavior. To satisfy these needs, six software modules were created including a prototyping module, task flowchart builder, a flowchart to E-GOMSL model translator, E-GOMSL model editor, parser and compiler, and a simulator and report generator. Microsoft Excel Visual Basic for Applications (VBA) macros were used to develop the modeling functions.

Prototyping Module

The prototyping module of the E-GOMSL tool is used to load or draw visual objects as part of a target system (e.g., interface images) and to identify and describe non-visual task objects (e.g., auditory communications from air traffic control (ATC) in the context of piloting). There are three sub-modules including: an XML (extended marked-up language) file loader; (2) a tool for reviewing and modifying visual objects; and (3) a tool for reviewing non-visual objects (task items). In order to define visual objects and execute a simulation, a prototype image must be loaded at the beginning of the E-GOMSL tool use. With the image, a user can identify and modify visual objects in an interface, which are intended for use by pilots in task performance. Figure 1 shows the flow of prototyping modules.

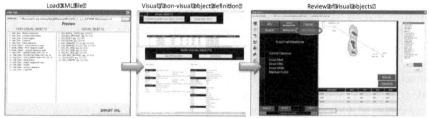

Figure 1. Flow of prototyping modules.

Flowcharting

Figure 2 presents the flowchart module interface. There are two general areas of the interface used for drawing (Figure 2, inset (a)) and coding of an E-GOMSL model (Figure 2, inset (b)). In the drawing area, designers can create a task flowchart (i.e., a pilot activity flow diagram (AFD)) and the corresponding E-GOMSL code is automatically generated in the coding area. In order to create an AFD, designers must use the flowcharting tool bar (Figure 2, inset (c)). The toolbar includes seven buttons. First, a circle icon is used to represent the beginning and end of methods in an E-GOMSL model. Second, a rectangle icon is used to represent steps in a task flow with multiple operators. Third, a diamond icon is used to represent selection rules, as in the GOMS Language. Fourth, a "DEL" button is used to delete a shape or object in the drawing area, and results in removal of a related line of code in the spreadsheet form of the E-GOMSL model. Fifth, a "RELOAD" button is used to refresh information on a visual/non-visual object from system memory. Sixth, an "INFO" button is used to reveal the E-GOMSL code associated with a selected object in the flowchart. Last, a "SIMULATION" button pops-up when the model simulation preparation interface is accessed.

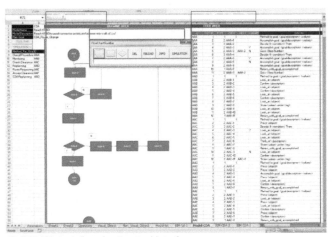

Figure 2. The E-GOMSL task flowcharting tool interface

Simulation

The simulation tool as part of the E-GOMSL software suite includes five display features: a prototype viewer, a window on the pattern of pilot information processing (PIP), a WM graph, a simulation status window, and a navigation control panel. The prototype viewer (Figure 3, inset (a)) presents images of cockpit interfaces when a simulation is running. The PIP pattern window (Figure 3, inset (b)) presents the processors/channels activated by certain operators coded in an E-GOMSL model when the simulation is running. The WM graph (Figure 3, inset (c)) presents the number of chunks of information in WM that are counted by the simulation function at any given time during model execution. The simulation status window (Figure 3, inset (d)) shows the elapsed time, current aircraft location (DME) and text-based real-time simulation information. During model simulation, a designer can look at the simulation status window to observe firing of operators in the cognitive model and to verify the pilot task sequence. The navigation control panel (Figure 3, inset (e)) is used to control the phases of the simulation. A "PLAY" button is used to start the simulation. The simulation "update rate" can be set using a spin button control. When the designer checks the "STEP by STEP" box in the navigation control panel, the PLAY button is automatically disabled and the "STEPS" button is enabled. The STEPS button can be used to increase the simulation time step according to the update rate.

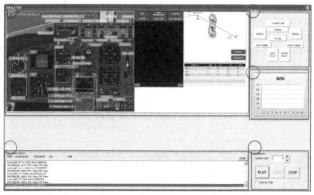

Figure 3. Simulation dialog for visualization of cognitive model execution.

Outputs

Upon completion of a model simulation, the E-GOMSL tool generates three different types of output, including: (1) time estimates, (2) complexity indices, and (3) a Gantt chart representing the extent of activation of various information processing channels. The time estimates (Figure 4, inset (a)) include the simulation time, number of replications, start time, end time, and the percent of total time accounted for by each method in the model during execution. The complexity indices table (Figure 4, inset (b)) include the number of methods in a model, the number of steps in a model, the total number of chunks in WM, the highest number of chunks in WM at any given time, and the number of LTM transactions occurring based on the E-GOMSL model compilation. The Gantt chart (Figure 4, inset (c)) is created based on the task time estimates and complexity indices. It reveals the elapsed time for each method in a model and whether pilot visual, cognitive or motor processing occurred during that time. It also visualizes the passage of task time by using cells in an Excel spreadsheet.

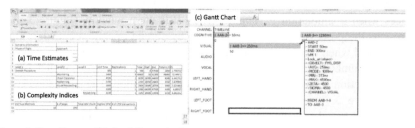

Figure 4. Cognitive model simulation outputs

2.3 Usability test of E-GOMSL tool

Two different methods were used to evaluate the usability and applicability of the new E-GOMSL tool for design purposes. These methods included the CW and a post-trial survey on the usability and utility of the new tool. The CW was applied to three different steps in the use of the E-GOMSL tool. The first step involved user correction of a visual object definition. The function of an ATC clearance rejection button ("REJECT_BTN") in a cockpit interface was incorrectly defined. In the second step, due to the incorrect definition of the button object, an associated pilot behavior flowchart contained an error in a "Replanning" operation as part of a more complicated flight scenario. This user was to correct the definition of the operator in the flowchart. The third step involved user walkthrough of the E-GOMSL simulation and visualization modules. After each step, four usability questions were posed to an evaluator, including: "Is the effect of the action the same as the user's goal at this point?"; "Will the user see that the action is available?"; "Once the user has found the correct action, will they know it is the one they need?"; and "After the action is taken, will the user understand the feedback they get?".

On the usability questionnaire, participants completed the SUS (Brooke, 1996). The survey includes ten statements with Likert scales for users to rate the degree of agreement or disagreement. Some example statements include: "I think that I would like to use this system frequently."; "I found the system unnecessarily complex."; "I thought the system was easy to use." etc. The survey is intended to provide a global assessment of perceived usability (Brooke, 1996).

Following the actual data collection period, another questionnaire was presented to only the avionics designer to assess the perceived utility of the new E-GOMSL tool for design process support. This same designer also participated in the above usability evaluation with the HCI experts and other system designers. The first CW step took 10 min to complete. The second and third CW steps each took 14 min to complete. The post-trial questionnaire took 5 min. The total test time was 42 min.

3   TEST RESULTS AND DISCUSSION

3.1 Cognitive Walkthrough Step 1 (CW1)

For the CW applied to the first step of tool use, all participants were satisfied with the E-GOMSL tool in accomplishing their goal. There was a potential usability problem relating to a "Review Visual Objects" button. For some users, the button was not clearly associated with the goal of correcting the visual object definition. Since designers may need to use the "Visual Object" component frequently, the CW analysis resulted in a recommendation to include a more detailed description on use of the "Review Visual Objects" button in the E-GOMSL tool.

During the first step of the CW, the majority of participants were also able to find correct actions; however, one participant indicated that the visual object list was not visible. The participant expressed some confusion in identifying required actions. In order to increase control visibility, it was recommended that the object list, and "EDIT" and "DELETE" buttons be highlighted. This change in the tool design was expected to allow designers to find appropriate information more easily.

One participant was not able to understand the feedback provided by the E-GOMSL tool because he could not identify visual objects for use before they were highlighted by the application. The reason was the participant did not select a target visual object in a list of objects for modification. The E-GOMSL tool only highlights visual objects for users when a selection is made in the visual objects list. Thus, designers may require more explanation on each component in the prototyping module.

3.2 Cognitive Walkthrough Step 2 (CW2)

During the second step of the CW testing, in general, all participants were satisfied with the design of the flowcharting module for accomplishing goals; however, some evaluator comments suggested potential usability problems. One participant noted that modifying a flowchart required 16 steps at the application interface in order to delete and then add new operators. It was recommended that the method be changed to reduce the number of user actions in representing pilot behaviors to four. Table 1 summarizes an approach to reducing the steps in the flowchart modification method.

Table 2 Comparison of steps for modifying the flowchart.

Previous Steps		New Steps
• Select operator		• Select operator
• Click delete button		• Click modify button
• Click rectangle button		• Modify operator dialog (requires **1** step)
• Click CTRL + Mouse right button	→	• Click OK button
• Setup operator dialog (requires **11** steps)		
• Click OK button		
Total steps: **16** steps		Total steps: **4** steps

Regarding use of the flowchart toolbar, during certain steps of the test, the toolbar was not visible due to user scrolling. When a designer scrolls down an Excel spreadsheet containing an AFD, he or she cannot find the "TOOL BAR". Therefore, it was recommended that the bar be moved to a ribbon at the top of the E-GOMSL interface. Furthermore, based on a videotape analysis of user behavior, one participant indicated the flowchart element labeling method (e.g., AAA-1) was not clear. It was recommended that flowchart element labels provide more information (e.g., AAA-1: Accomplish_Goal) so that designers can better understand the content.

3.3 Cognitive Walkthrough Step 3 (CW3)

In general, all participants were satisfied with their use of the simulation and visualization modules of the E-GOMSL tool during the third step of the CW. One participant commented that the WM graph needed labels for the X- and Y-axes.

3.4 Post-trial Questionnaire

Evaluator responses to the SUS questions were compiled according to the methodology described by Brooke (1996). The average survey score across evaluators was 53.13 (on a 100 point scale). Brooke (1996) said that systems with scores less than a criterion of 65 are likely to have outstanding usability issues. Although this was the case for the E-GOMSL tool, there was substantial variability in the responses among participants. Two of the evaluators (human-computer interaction (HCI) experts) indicated that the overall design was acceptable. Another HCI expert and an avionics designer indicated that there were usability limitations of the interface. It is likely that this response variability was due to different viewpoints on use of the new tool. Some participants focused on how to setup the system (i.e., defining visual objects and non-visual objects) while others focused on using the simulation and analysis tool.

Based on the results of the CW, the usability issues were limited to: (1) a lack of visibility of the "Review Visual Objects" button; (2) a lack of visibility of the "Visual Object" dialog and the "TOOL BAR"; (3) a lack of user understandability of the AFD; (4) a lack of predictability of the key sequence for diagramming; and (5) complexity of the method for modifying AFDs. It is expected that addressing the design recommendations from the CW methodology in a revision of the E-GOMSL tool would produce an overall SUS score (across users) surpassing the problem criteria (65 points) defined by Brooke (1996).

4   CONCLUSION AND FUTURE RESEARCH

In general, the results of the usability and applicability testing support the use of the new E-GOMSL tool for representing pilot behavior with prototypes of cockpit automation interfaces. The tool may serve as an effective basis for making decisions in the conceptual phase of avionics design processes.

Currently, the E-GOMSL tool is not capable of considering certain system (aircraft) and environment factors in modeling pilot behavior. This reduces the accuracy of model-based task completion time estimates and decreases the generalizability of model results to actual flight situations. For example, during a predefined flight scenario, when an aircraft changes ALT (altitude) or SPD (speed), the E-GOMSL tool cannot apply aircraft dynamics to constrain pilot motor behavior. Furthermore, if there are severe weather conditions (e.g., rain, wind, clouds, etc.) in a flight scenario, the E-GOMSL tool cannot apply such conditions to change the state of the simulated airplane and, consequently, pilot behavior. This limitation could be overcome by integrating the tool with a flight simulation application, such as the X-Plane Simulator, which is capable of providing realistic aircraft flight behavior representation (Laminar, 2012). Since the X-Plane simulator provides API functions for developers, the E-GOMSL tool could make use of the capability of X-Plane for modeling airplane dynamics.

ACKNOWLEDGMENTS

This research was supported by NASA Ames Research Center under Grant No. NNH06ZNH001. Mike Feary was the technical monitor. A team from APTIMA Corporation, led by Paul Picciano, executed the usability test that was designed by the North Carolina State University researchers. The opinions and conclusions expressed in this paper are those of the authors and do not necessarily reflect the views of NASA.

REFERENCES

Brooke, J., 1996. SUS: a "quick and dirty" usability scale. In: P. Jordan, B. Thomas, B. Weerdmester and A. McClelland, eds. 1996. *Usability Evaluation in Industry*. London: Taylor and Francis.

Card, S., Moran, T., and Newell, A., 1983. *The psychology of human-computer interaction*. Erlbaum.

Cohen, A. C., 1988. *Three-parameter estimation*, chapter 4, (pp. 113-138). New York and Basel: Marcel Dekker, Inc.

John, B.E., K. Prevas, D.D. Salvucci, and K. Koedinger. 2004. Predictive human performance modeling made easy. *In Proceedings of the SIGCHI conference on Human factors in computing systems* (pp. 455-462), New York, NY.

Kieras, D. 1997. A guide to GOMS model usability evaluation using NGOMSL. In: M. Helander, T. K. Landauer, and P. Prabhu Eds. 1997. *Handbook of Human-Computer Interaction*. Amsterdam: Elsevier Science, pp.733-766.

Kieras, D. 2006. *A guide to GOMS model usability evaluation using GOMSL and GLEAN4* (Tech. Rep.). University of Michigan, Ann Arbor, MI.

Laminar Research, 2012. X-Plane Flight Simulator. [online] Available at: <http://www.x-plane.com> [Accessed 17 February 2012].

Limpert, E., W.A. Stahel, and M. Abbt. 2001. Log-normal distributions across the sciences: Keys and clues. *DioScience*, 51 (5), pp.341-452.

Liu, Y., R. Feyen, and O. Tsimhoni. 2006. Queuing network-model human processor (QN-MHP): A computational architecture for multitask performance in human-machine systems. *ACM Transactions on Computer-Human Interaction*, 13 (1), pp.37-70.

Teo, L. and B.E. John 2008. CogTool-explorer: towards a tool for predicting user interaction. *In Proceedings of 2008 CHI extended abstracts on Human factors in computing systems*, (pp. 2793-2798), New York, NY.

Wickens, C.D. and J.G. Hollands. 1999. *Engineering Psychology and Human Performance*. Prentice Hall.

CHAPTER 37

CRM Model to Estimate Probability of Potential Collision for Aircraft Encounters in High Density Scenarios Using Stored Data Tracks

R. Arnaldo, F. J. Sáez, E. Garcia, Y. Portillo

Universidad Politecnica de Madrid
Madrid, Spain

ABSTRACT

Collision risk estimation in airspace has been studied for many years using different approaches based on different assumptions. The results of this paper supports the fact that the risk is not the same for all the proximate events on which aircraft pass closer than the prescribed horizontal and vertical separation minima. The proposed approach aims at providing an individual probability of potential collision (severity) of each individual encounter, based on the kinematics of the encounter and the minimum lateral and vertical separation at the predicted CPA. The formulation presented allows not only the estimation of the severity for each individual potential aircraft encounter but also the expected probability of collision for all aircraft pairs potentially violating the separation standards, in a given en route airspace scenario and for an aircraft population. In order to evaluate the human factors implications of the research findings, results are applied to stored aircraft tracks that have flown within MUAC Airspace.

Keywords: Risk, CPA, Probability of collision

1. INTRODUCTION

Risk has been defined (Krystul, 2006) as the probability of a particular adverse event occurring during a stated period of time. Applying this definition to an ATC scenario, it is accepted that risk is closely related to situations in which two aircraft are on conflict course and would not only pass closer than the prescribed horizontal and vertical separation minima but, in fact, collide. Collision risk estimation in airspace and mathematical modelling of mid-air collisions has been carried out for over more than 40 years. Initial models (Machol, 1995; Reich, 1964) used information relating to the probabilistic distributions of aircraft's lateral and vertical position, traffic flows on the routes, aircraft's relative velocities and aircraft dimensions to generate estimates of collision risk. Since then, new models have been developed and have been continually refined and improved. (Brooker, P, 2006; Sherali et al, 1998; Burt, 2003; Carpenter and Kuchar, 1997; Kuchar and Winder 2001; Powell and Houck, 2001; Houck and Powell, 2003; Shepherd and Cassell, 1997; Blom et al, 2001; Paielli and Erzberger, 1999).

Following the research line initiated on (Campos, 2001; Garcia, Saez and Izquierdo, 2007; Saez ct At, 2010), the authors are developing a more detailed mathematical model for both components of probability of collision in a radar ATC environment, where recorded tracks can be obtained for all aircraft flying in it from Radar Data Processing systems (RDP) providing indirect information which is closely related to the "human factor response".

2. FUNDAMENTALS BEHIND PROBABILITY OF COLLISION ESTIMATION

The work presented here was originally inspired by the principle (Marks, 1963) that "… the task of relating collision risk to a traffic configuration can be taken in two parts:

- Determining the frequency with which aircraft are exposed to risk by passing close together; and
- Determining what chance of collision is inherent in the passing".

According to this idea, the probability of aircraft collision can be expressed as:

$$P(collision) = FeR * P(pot.coll/pot.conf) * P(coll/pot.coll)$$

where:

- **FeR**, Frequency of exposition to Risk, here is considered as the relative frequency that an aircraft would potentially violate the separation standards defined for the particular situation, here referred to as potential conflict. It is easily seen that this value increases with the traffic density.
- **P(pot.coll/pot.conf)** is the conditional probability of a potential collision (pot.coll) between two aircraft that have previously violated the separation standards (pot. conf). Its value depends on the encounter kinematics and uncertainties associated to predicted positions. It represents the intrinsic severity of the encounter and it is independent of the traffic density.

- **P(coll/pot.coll)** is the conditional probability of collision among potential collisions having failed all the safety barriers (ATC, TCAS) which are in place to mitigate the risk.

A time horizon is established within which all aircraft positions are projected to explore existence of "potential conflicts". In the following discussion 10 minutes look ahead time has been considered. According to that, the relative frequency of potential collisions among potential conflicts F(pot.coll/pot.conf) could be expressed as:

$$F(pot.coll \, / \, pot.conf) = \frac{Num. \, of \, pot. \, collisions}{Num. \, of \, pot. \, conflicts} \approx E[P_a]$$

where Num. of pot.collisions is the number of aircraft that are about to collide (and will do if all safety barriers fail).

An initial expectation for probability of potential collision among potential conflicts, $E(P_a)$, could be obtained as the relative frequency that two aircraft, on a conflict course, would not only pass closer than the prescribed horizontal and vertical separation minima, but would in fact collide. This expression provides an expected, or global, value and does not assess the severity of each individual potential encounter itself. This paper proposes an approach to estimate the severity of the encounter using the conditional probability of a potential collision P_a for each particular aircraft encounter. This proposed approach aims at improving the previous works by:

- Providing an individual probability of collision of each individual encounter based on the: (1) geometry and kinematics of the encounter, (2) the minimum predicted lateral separation at the CPA, and (3) the minimum predicted vertical separation at the CPA.
- Taking into consideration the radar data errors and the segmentation errors.
- Allowing the severity of each individual potential encounter or any subsets of traffic samples and/or airspace parts.

3. DERIVATION OF A GENERAL EXPRESSION FOR PROBABILITY OF COLLISION (PA).

In order to obtain a general expression of Pa an impact plane is defined as a generic projection plane containing the centre of reference aircraft ACi (assumed as static) and perpendicular to $\overrightarrow{v_{ji}}$ (relative velocity vector between the two aircraft i and j involved in the proximity event). Additionally, the collision area is defined as the projection of the collision cylinder ($2\lambda_{xy}$, $2\lambda_z$). If the conflict cylinder would be settled in ACi, being its centroid the one of the cylinder as well, it could be also defined the conflict area as the projection of the conflict cylinder (2R, 2H). The CPAP (Closest Point Of Approach Projection) is a point with coordinates y1p and z1p obtained by projecting intruder aircraft. Figure 1 shows that a conflict will occur if ACj encounters the stationary conflict area, that is, if the CPAp coordinates (y1p, z1p) are inside the conflict area. In the same way, a collision will occur if ACj encounters the stationary collision area, that is, if the CPAp coordinates are inside the collision area.

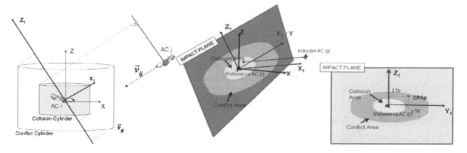

Figure 1. Impact Plane, Collision Area, Conflict Area and Projected CPA. definition

Considering the changes in the CPA coordinates due radar and radar data segmentation errors, the probability of potential collision for an intruder aircraft that has violated the separation standards and whose projection consequently hits within the conflict area can be calculated as:

$$P_a\left(y_{1p}, z_{1p}\right) = \int_{S_{PCF}} dP_1 P_2 =$$

$$\approx S_{PCOL} \cdot \int_{S_{PCF}} f_1\left(y'_{1p} - y_{1p}, z'_{1p} - z_{1p}\right) \cdot f_2\left(-y'_{1p}, -z'_{1p}\right) dy'_1 dz'_1$$

This equation provides and individual probability of collision based on: (1) geometry and kinematics of the encounter, (2) the minimum predicted lateral separation at the CPA, and (3) the minimum predicted vertical separation at the CPA. It takes into consideration the two probability density functions stating segmentation lateral and vertical errors and the projection lateral and vertical errors characterization. As a result the bi-dimensional probability density function of the CPAPs can be derived from previous equation as:

$$f_a\left(y_{1p}, z_{1p}\right) = \int_{S_{PCF}} f_1\left(y'_{1p} - y_{1p}, z'_{1p} - z_{1p}\right) \cdot f_2\left(-y'_{1p}, -z'_{1p}\right) dy'_1 dz'_1 \quad (10)$$

Both expressions estimate the probability of potential collision, having a potential separation violation (potential conflict), for each aircraft encounter, provided that uncertainties in the projection of segmented trajectories and in the segmentation process have been characterised by associated pdfs, f1 and f2, respectively.

4.   RESULTS AND DISCUSSION

The previous mathematical formulation is supported by previously mentioned ad-hoc software developed by the authors for Eurocontrol in the framework of the

3D-CRM programme. This software is intended to measure the collision risk in high density ATC en route airspace, based on an analysis of the stored aircraft tracks that have flown in it within a given time frame.

For the purpose of evaluating the mathematical expressions to estimate the probability of collisions, the mentioned software tool has been applied to a radar data sample from the Maastricht Upper Area Control Centre (MUAC). EUROCONTROL's Maastricht Upper Area Control Centre (MUAC) is a regional air traffic control centre providing seamless air navigation services in the upper airspace (above 24,500ft) over a large (approximately 700,000 square kilometres) multinational airspace in Europe. An advanced and complex ATC automated system named MADAP (Maastricht Automated Data Processing and Display System) is the technical enabler responsible for managing, processing and presenting in real time information relating to the air traffic flows in the whole area. MADAP performs centralized multi-radar tracking using the information provided by a large number of radars and computes a high quality air traffic situation. In MUAC, a unique horizontal separation standard of 5 NM is used throughout the total area of responsibility. The vertical separation minimum of 1000 ft. is used.

4.1   Empirical Estimation for Pa

The general expression of expected Pa is calculated numerically from the relative frequency of potential collisions among all potential conflicts using the following equation:

$$E[\text{P}(\text{pot.coll}/\,pot.conf)]= E[P_a] \approx \frac{Num.\,of\ pot.\,collisions}{Num.\,of\ pot.\,conflicts} = \frac{19}{35166} = 5.4*10^{-4}$$

Figure 2 illustrates the obtained bi-dimensional histogram of the projected horizontal and vertical separations at the CPA for the whole data period analysed. As it is shown, the number potential conflicts are higher when encounters are between aircraft established at the same flight level (0ft vertical separation) and, as well, between aircraft having 2.5 and 5NM of lateral separation. It is also noticed that number of encounters having 1000ft separation is higher than for any other vertical separation except the 0ft. This is easily understood when taking into account that within the en-route airspace most of the time aircraft are in level flight (nominally always 1000ft apart between contiguous flight levels). If safety barriers have not been applied the number of collisions to happen would have been 19. The area used to compute the number of potential collisions is shown as remarked by a red dotted circle.

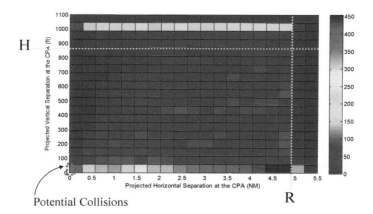

Figure 2. 2D histogram of projected horizontal and vertical separations at the CPA (31 days of radar data)

4.2   Pa estimation for each aircraft encounter

Once the empirical general or expected value for Pa has been obtained, Pa was estimated for each particular encounter by the next expression.

$$P_a(y_p, z_p) = 4\lambda_{xy}\lambda_z \cdot \frac{v_x}{\sqrt{v_{x+}^2 v_z^2}}\left[1 + \frac{\pi}{4} \cdot \frac{\lambda_{xy}}{\lambda_z} \cdot \frac{v_z}{v_x}\right] \cdot f_2^{\cdot}(-y_{p'}, -z_p) =$$

$$= 2\lambda_{xy} \cdot f_{2y}(-y_p) \cdot 2\lambda_z \cdot f_{2z}(-z_p) \cdot \frac{v_x}{\sqrt{v_{x+}^2 v_z^2}}\left[1 + \frac{\pi}{4} \cdot \frac{\lambda_{xy}}{\lambda_z} \cdot \frac{v_z}{v_x}\right]$$

This equation provides and individual probability of collision based on: (1) kinematics of the encounter (ratio vz,to,vx), (2) the minimum predicted lateral separation at the CPA (yp), and (3) the minimum predicted vertical separation at the CPA (zp). It also takes into consideration the segmentation lateral and vertical errors (f_{2y} and f_{2z}).

A result for Pa estimation for level flight encounter is shown in the upper part of figure 3. For this case when CPAp coordinates (yp,zp) are very close to the reference aircraft (ACi), Pa estimated value reaches 3*10-2. This value has a magnitude of two orders higher than the empirical expected result ($5.4*10^{-4}$), but strongly decreases when predicted CPAp lays apart from ACi, resulting values much lower than the empirical one. In the lower part of this figure, the graphs shown when one or both aircraft are climbing/descending but having vz/vr ratio close to zero, it could be seen that in spite of the decrease of the maximum value for Pa (7*10-3) it is still greater than the empirical expected result for Pa. Furthermore, the probability of collision for CPAp for which yp coordinates close to zero but zp coordinates separated from the ACi remains significant. Pa estimation for encounters having two different aircraft climbing/descending (vz / vx) ratios is shown in figure 4.

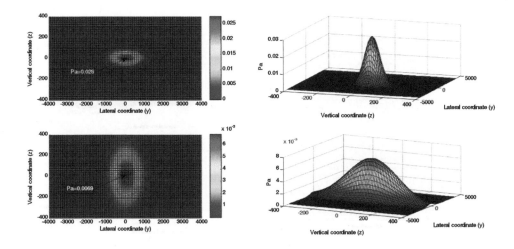

Figure 3. Pa estimation for different CPAp. Aircraft established at a defined flight level or vz equals to zero (upper) and aircraft with vz close to zero (lower).

Despite the fact that shape of both functions for Pa are similar to the one obtained in the lower part of figure 13 (aircraft climbing/descending and vz/vx close to zero), the maximum values for Pa are different in both cases (9*10-3 for vz/vx=0.1, and 2*10-2 for vz/vx=20), showing that Pa maximum values for CPAp close to reference aircraft (ACi) has a decreasing trend when vz/vr ratio increases. The following table summarises the results obtained from empirical and estimated Pa for the worst case, that is to say Pa for predicted CPAp=(0,0).

The results clearly shows that it is unrealistic to assign the same probability for potential collisions to all potential conflict, independently of the predicted coordinates for CPA, no matter how these coordinates have been derived.

Table 1 Worst case Pa estimation

Empirical result for expected P_a, $E[P_a]$	$5.4 * 10^{-4}$
Estimated P_a for CPA_p=(0,0) and level flight	$3*10^{-2}$
Estimated P_a for CPA_p=(0,0) and $v_z/v_x \approx 0$	$7*10^{-3}$
Estimated P_a for CPA_p=(0,0) and v_z/v_x =0.1	$9*10^{-3}$
Estimated P_a for CPA_p=(0,0) and v_z/v_x =20	$2*10^{-2}$

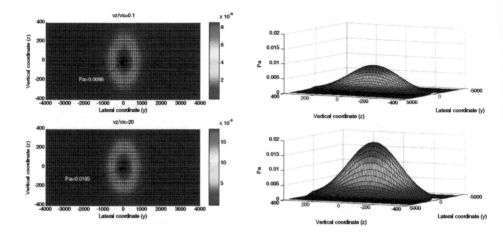

Figure 4. Pa estimation for different CPAp . Aircraft climbing/descending and different vz/vx ratios.
vz / vx=0.1(upper), vz / vx=20 (lower)

4.3 Expected Pa estimation for a given scenario and traffic sample

When a collision risk analysis is applied to a representative aircraft population, using segmentation of their stored radar tracks, a 2D histogram of projected horizontal and vertical separations at the CPA can be obtained, as it is shown in figure 3. This histogram provides a first approach for expected Pa using equation (4), which is the way used to obtain E[Pa]= 5.4*10-4, this value can taken as reference value for Pa. If the histogram exhibits a close to uniform distribution, it can be understood that any "generic" potential conflict would became a potential collision with the same probability. It is also possible to propose a different approach to establish the expected value for Pa in a given scenario and for a given aircraft population, discussed below.

$$E[P_a] = \frac{1}{N}\sum_{ji} P_a\left(y_{ji}, z_{ji}, r_{ji}\right) = \frac{\lambda_{xy}\lambda_z}{N}\sum_{ji}\left[1 + \frac{\pi}{4}\cdot\frac{\lambda_{xy}}{\lambda_z}\cdot r_{ji}\right]f_{2y}\left(y_{ji}\right)\cdot f_{2z}\llbracket(z\rrbracket_{ji})$$

$$= \frac{\lambda_{xy}\lambda_z}{N}\sum_{ji}\left[1 + \frac{\pi}{4}\cdot\frac{\lambda_{xy}}{\lambda_z}\cdot r_{ji}\right]f_{2y}\left(y_{ji}\right)\cdot f_{2zji}\llbracket(z\rrbracket_{ji})$$

Where Pa (y_{ji}, z_{ji}, r_{ji}) is the individual probability of each potential collision being rji=vz/vx the between vertical and horizontal relative speeds, and f2zji the probability density function applied to each aircraft encounter (between each pair of aircraft, i and j). When this equation is applied to previous MUAC data sample, expected value for Pa results 8.2*10-4, slightly higher than empirical.

5.   CONCLUSIONS.

This paper analyse in detail the inherent collision risk involved on each aircraft proximity event by assessing the conditional probability Pa of a potential collision between aircraft that are exposed to risk, that is to say, they are potentially going to violate the separation standards defined for a specific airspace if it is no corrective action is taken. The proposed approach allows the determination of the severity of each aircraft encounter as the probability of potential collision of each individual aircraft encounter in high density ATC en route airspace, based on an analysis of the stored aircraft tracks that have flown in it within a given time frame. The authors propose the mathematical formulation to characterise the severity of each aircraft proximity event by the convolution of the bi-dimensional probability density function of the predicted Closest Point of Approach between the involved aircraft and the distribution of aircraft lateral and vertical error in the projected position The presented work aims to provide an individual probability of collision based on the geometry and kinematics of the encounter and the minimum lateral separation and the minimum vertical separation at the predicted Closest Point of Approach or CPA. The formula takes into consideration uncertainties introduced by the radar data error and the segmentation error. The results of this paper shows that there is no the same severity for all the proximity events on which aircraft pass closer than the prescribed horizontal and vertical separation minima, but also that the expected severity for given scenario and traffic sample can also vary depending upon the kinematics characteristics of involved aircraft within this scenario. It is also considered that collision risk for high air traffic density can be analysed from the estimation of three different factors:

- Relative frequency of exposition to risk (FeR). The value of this factor can be easily obtained from any radar data sample and strongly depends on the minimum applied horizontal and vertical separations standard and increases with air traffic density,
- Expected severity E(Pa). This value can be directly derived from individual probabilities of potential collision (Pa). Furthermore, having individual severities, it also permits additional assessment on safety (hot spots identification, etc.).
- Expected probability of failure of safety barriers (ATC, TCAS, etc.)

As the two first factors can be derived from the stored tracks of the traffic sample, using the software tool developed by the authors [38], further work is now devoted to develop of the probability of failure of the ATM safety barriers. Once the probability of failure were stated and validated, it will be possible to estimate the collision risk for individual encounters, scenarios and air traffic samples. Results obtained for MUAC, with data sample used in previous discussion, exhibits a rounded value for frequency of exposition to risk of FeR=0.3. Probability of potential collision among encounters exposed to risk, Pa or its expected value E(Pa) for the same sample, oscillates between $8.2*10-4$(expected) and $2*10-2$(worst case). Previous results demand a probability of "safety barrier failure" lower than between $0.4*10-5$ and $1.7*10-7$ respectively, to reach an ATM en route target level of safety of TLS=10-9. This last value is normally the one used as TLS. For instance, in reference (Eurocontrol, 2006) mid-air collision given as accident frequency (per flight) is 5.4*10-09, specifying that, among them, the frequency of fatal accident, directly caused by ATC (per flight), is 3.5*10-09.

REFERENCES

Blom, H.A.P., et al, 2001, 'Accident Risk Assessment for Advanced ATM,' *Air Transportation Systems Engineering*, 463-480.

Brooker, P, 2006, Longitudinal Collision Risk for ATC Track Systems: A Hazardous Event Model. Journal of Navigation, Vol. 59-1: 55-70.

Burt L , 2000, 3-D Mathematical Model for ECAC Upper Airspace, Final Report

Campos, L. M. B. C., 2001, Probability of collision of aircraft with dissimilar position errors , Journal of aircraft, vol.38-4: 593-599

Carpenter, B., Kuchar J., 1997, Probability-based collision alerting logic for closely-spaced parallel approach. AIAA-1997-222 Aerospace Sciences Meeting and Exhibit, 35th, .

Garcia, E., Saez, F.J., Izquierdo, M.I. 2007, Identification and analysis of proximate events in high density en route airspaces ,7th USA – EUROPE ATM R&D Seminar.

Eurocontrol, 2006 , Main Report for the: 2005/2012 integrated risk picture for Air Traffic Management in Europe EEC Note No. 05/06, Project C1.076/Eec/Nb/05.

Houck, S. W., Powell, J D., 2003 Probability of Midair Collision During Ultra Closely Spaced Parallel Approaches. *Journal of Guidance, Control, and Dynamics*, Vol.26-5: 702-710

ICAO, 1988, Review of the General Concept of Separation Panel, 6th meeting, Doc 9536.

Kuchar J., Winder, L. F., 2001, Generalized philosophy of alerting with applications to parallel approach collision prevention. *AIAA Guidance, Navigation, and Control Conference and Exhibit.*

Krystul, J. 2006, Modelling of stochastic hybrid systems with applications to accident risk assessment.

Machol, R. E. , 1995, Thirty Years of Modelling Midair Collisions, Interfaces 25: 151-172.

Marks B. L, 1963, Air traffic control separation standards and collision risk. Royal. Aircraft Establishment Technical Note No. 91.

Paielli, R.A. and H. Erzberger, 1999, Conflict Probability Estimation Generalised to Non-Level Flight', Air Traffic Control Quarterly, Vol. 7: 195-222,.

Powell, J. D. Houck, S. W. , 2001, Assessment of the possibility of a midair collision during an ultra closely spaced parallel approach. *AIAA Guidance, Navigation, and Control Conference and Exhibit.*

Reich, P.G., 1964, A theory of safe separation standards for Air Traffic Control, Technical Report 64041, Royal Aircraft Establishment, UK

Saez F, Arnaldo R, Garcia E, McAuley G, Izquierdo M., 2010, Development of a three dimensional collision risk model tool to asses safety in high density en-route airspaces. Proceedings of the Institution of Mechanical Engineers, Part G: Journal of Aerospace Engineering, vol. 224 – 10: 1119-1129.

Shepherd, R. Cassell, R. 1997, A reduced aircraft separation risk assessment model, AIAA Guidance, Navigation, and Control Conference.

Sherali H. D, Smith C., Trani A.A., Sale. S.and Chuanwen Q., 1998, Analysis of Aircraft Separations and Collision Risk Modeling. NEXTOR - National Center of Excellence for Aviation Operations Research.

CHAPTER 38

Evaluation of the Big-airspace Operational Concept through a Fast-time Human Performance Computation Model

Wenbin Wei, Amit Jadhav, Damir Ceric, and Kevin Corker

Human Automation Integration Lab (HAIL)
College of Engineering
San Jose State University
San Jose, CA 95192

ABSTRACT

To reduce congestion and delay for the air traffic around the metropolitan areas, FAA has developed a concept of operations for an integrated Arrival/Departure Control Service, call the "Big Airspace" (BA) concept. The objective of this study was to apply the fast-time human performance model, Air MIDAS, to evaluate the feasibility and effectiveness of the BA concept on human operators, who are responsible for executing and managing airspace operations.

The human performance model operates in an integrated mode with the fast-time airspace simulation system representing the external world, and the Air MIDAS system representing the human operators in that world. The two simulation models (airspace and human) synchronize their operations through their respective internal clocks.

In order to judge the impact of "Big Airspace" (BA) procedures and other related conditions such as traffic level, weather and communication processes, we conducted an analysis of variance (ANOVA) to see if there was an effect of the experimental conditions on the controllers' performance. Our analysis indicated that "Big Airspace" procedures were operationally feasible. BA operations produced reduced workload overall when compared with present operations under similar experimental conditions. Our model also indicated that data link communications decreased operator workload especially under the BA procedural paradigm.

1. INTRODUCTION

There have been great needs in evaluating human performances for the new technologies or procedures that are proposed to be applied in air traffic control or management. The fast-time human performance model, Air Man-Machine Integrated Design and Analysis System (Air MIDAS), filled a gap in the standard methodologies of both the fast-time logistical studies and the human-in-the-loop studies. Standard approaches use an incremental increase in testing complexity, fidelity and cost from empirical results, to prototype, to full mission simulation and then field testing. Such a process is, of course, a valid paradigm. However, the rate of development of ATM systems, the tremendous economic pressure to implement and reap immediate benefits from technologies, and the significant complexity and cost of large scale distributed air-ground tests suggest the development of other, more cost effective methods of human factors research. Prominent among these is computational human performance modeling (Laughery, Archer and Corker, 2001). In this paradigm the human and the system elements of interest are represented as computational entities (or agents). These agents interact as the system elements would in actual field operations, and behaviors can be observed. The benefit accrued is that system and human characteristics can be quickly varied (e.g. based on an assumed technology change or procedural changes) and the impact of those changes can be identified in the full-system context. Such models help focus the expensive and complex simulation and field tests. In addition, performance at the edge of system safety can be explored in the computational human performance modeling paradigm.

To reduce congestion and delay for the air traffic around the metropolitan areas, FAA has developed a concept of operations for an integrated Arrival/Departure Control Service, call the "Big Airspace" (BA) concept. The graphical descriptions of "Big Airspace" (BA) with respect to the baseline space are provided in detailed context in Section 3 below. The objective of this study was to apply the fast-time human performance model, Air MIDAS, to evaluate the feasibility and effectiveness of the BA concept on human operators, who are responsible for executing and managing airspace operations.

2 METHOD

The fast time human performance model, Air MIDAS, was linked to a fast time airspace simulation to represent the airspace, air traffic and human controller performance in BA operations. This simulation replicated, in part, the scenarios undertaken in fast time simulation and in the human-in-the-loop simulations reported herein. Figure 1 illustrates the Air MIDAS model's organization and the flow of information among the model's components and its linkage to the Reconfigurable Flight Simulator for airspace representation (Shah et al., 2005). The model is based on an agent architecture that is illustrated in the figure, with the boundary of each of the hierarchically structured agents represented by a boxed enclosure. Data are transformed in each agent and passed to other agents in a cyclic process. The "cost" of the transformation in terms of operator resources required and time are calculated and archived for "workload" analysis.

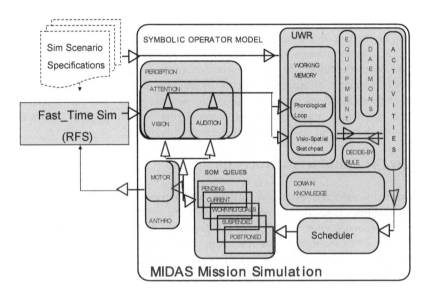

Figure 1. Air MIDAS component organization and information flow.

The human performance model operates in an integrated mode with the fast-time simulation system representing the external world in which the operator models are required to interact, and the Air MIDAS system representing the human operators in that world. Communication among the modules coordinates their performance. The models (airspace and human) synchronize their operations through their respective internal clocks.

2.1 Functions Represented in the Model

Specific elements of the model are as follows: 1) Perceptual process models are provided for visual and audition actions of the air navigation service providers (ANSPs). 2) An Updateable World Representation (UWR) represents the operator's understanding of the world (i.e., information about environment, equipment, physical constraints and procedures). The declarative information about the world is represented in a semantic net. The procedural information is held as decomposition, interruption and completion procedures for goals, subgoals and activities. Specific values in the world information serve to trigger activities in the simulated operator. 3) Active decision-making processes are represented as rules in propositional structure, as heuristics, or as software triggers, daemons, which serve to trigger action in response to specific values in the environment. 4) A scheduling mechanism imposes order on activities to be performed depending on their priority and the available resources to perform those activities. The scheduling mechanism also incorporates a switching mechanism that selects among control "modes" in which an operator can perform. These modes are a computational implementation of the COCOM model (Hollnagel, 1993; Verma & Corker, 2001) in which the model uses qualitatively different action sequences to perform required activity depending on the resource availability, time and competence level of the operator. 5) A set of queues manage current activities, activities that are interrupted or waiting to be rescheduled. 6) A representation of motor activity represents the process of performance of the selected action in the simulation.

Each of these functions is represented in independent agents that communicate with each other in a message-passing protocol and that keep track of those interactions as a source of data about the functions of the model.

2.2 How the Simulation Proceeds

The model is a discrete-event simulation in which events are defined as temporal increments (called "ticks") of a clock that sends a message to all agents to proceed with their functions at each event. The events/time base is a variable that can be set by the analyst using the model. The resolution of this event-base was varied in the Big Airspace simulation so that the model could concentrate its data collection in epochs of high intensity operation, and reduce its collection in times of reduced activity. Its lower bound is a 100 msec step-size.

An activity is triggered in the model in two ways. Either by decomposition of goals to be performed, or by occurrence of a specific value in the environment for which there is a daemon to respond and identify that value as significant and requiring action. When an activity is triggered in service of the mission goal or in response to environmental stimuli, the activity, prior to its having an effect in the simulation, is managed by the scheduler.

Activities are characterized by several defining parameters that include the conditions, under which the activity can be performed, its relative priority with respect to other activities, an estimate of its duration for scheduling, its interruption specifications, and the resource required to perform that activity. The scheduler is a blackboard process that evaluates these parameters and develops a time for the activities performance in the ongoing simulation. Following a multiple resource assumption, activity load is defined for Visual, Auditory, Cognitive, and Psychomotor dimensions (McCracken & Aldrich, 1984). Activities also require information for their performance. The information requirement is identified as knowledge either held in the operator's memory or available from the environment. If the information necessary for activity performance is available, and its priority is sufficient to warrant performance, then the scheduler operates according to heuristics that can be selected by the analyst. In most cases, the heuristic is to perform activities concurrently when that is possible, based on knowledge and resource constraints.

Behavior in our model is generated by the simulated operator's knowledge of his/her primary goals (e.g., traffic flow control), and by values of attributes in the updateable world representation. These goals are used to select action to be performed from a library of available actions. Action is performed either through a decomposition of the mission goals to tasks (the planned mode of action) or action is initiated by rules that match the incoming perceptual information to required action when some parameter values in the world are in that state. The action required is then "scheduled" to represent both human strategic behavior management and the limitations of capacity that human operators bring to the performance of tasks. The model also provides for activities to be interrupted and resumed (according to specification in the activity's definition). Activities that are interrupted (or aborted or forgotten) are placed in queues to either support their initiation or resumption or to document as data their abortion and/or forgotten state. Once activities are initiated in the simulation, the transformation and information exchange that are defined in those activities takes place in the simulated world. The simulation world is then changed and the cycle of performance continues.

3 SIMULATION ARCHITECTURE

The objective of the simulation study is to study procedure and workload issues of the controller by evaluating various traffic flows and manipulating controller tasks, communication processes and procedures associated with Big Airspace.

All systems that interact with the physical world are usually hybrid systems containing both discrete and continuous time behavior. To model hybrid systems it is necessary to have a distributed simulation and modeling environments. The High Level Architecture discipline is used to stimulate the distributed, interoperable models. Figure 2 illustrates various components associated in integration.

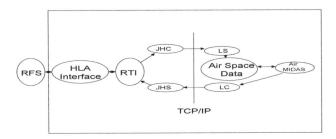

Figure 2. Software architecture.

The TCP/IP link to the external Reconfigurable Flight Simulator (RFS) moves through High Level Architecture (HLA) links to a LISP socket configuration to link with air MIDAS.

The Reconfigurable Flight Simulation System (RFS) that represents the operational airspace and was developed at Georgia Tech University (Shah et al. 2005).

The airspace used in this simulation is provided in figures 3 and 4.

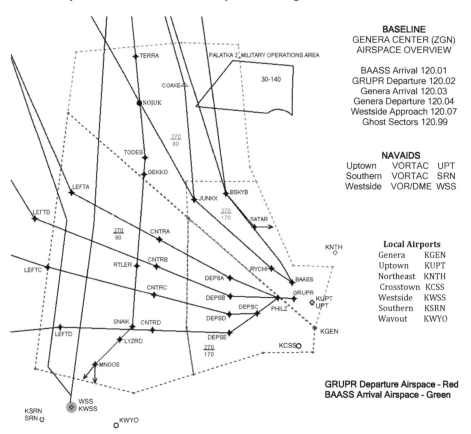

Figure 3. Simulation airspace: Baseline airspace

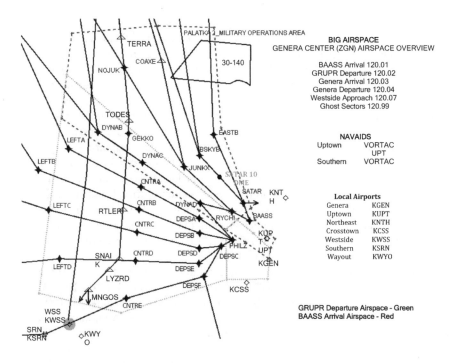

Figure 4. Simulation airspace: Big airspace

4 EXPERIMENT DESIGN

The Human Performance Model (HPM) was configured with controller roles and responsibilities for a central Florida ARTCC and TRACON set of sectors. Each sector has two controller representations: one for R-Side and the other for D-Side. The controllers had automation tools, URET and TMA as well as DATALINK, in one set of conditions.

In applying the Air MIDAS model, we performed the following steps: 1) Specified and encoded airspace to apply to the human performance model. In assessing human performance, researchers concentrated on the following airspace elements: the GRUPR1 SID, the BASS1 STAR, the transition airspace sectors, and traffic associated with west-north operations (i.e., NSDLR1 SID; and SNAPR1, LZARD1, DARBW1, CALMP1, and DARNE1 STARS). 2) Filtered traffic files to find aircraft associated with the airspace and routes selected in the time period specified. 3) Identified activities for each human operator to perform in the scenarios to be simulated. A definition of procedures for the airspace operations was decomposed into activities for the models. An activity is triggered in the model in two ways: either through decomposition of goals to be performed or through occurrence of a specific value in the environment for which there was a daemon to respond to and identify that value as significant and requiring action. 4) An activity was characterized by several defining parameters, including conditions under which the activity can be performed, its relative priority regarding other activities, an estimate of its duration for scheduling, its interruption specifications, and the resources required to perform that activity.   Assuming multiple resources, researchers defined activity load for Visual, Auditory, Cognitive, and Psychomotor dimensions (McCracken & Aldrich, 1984). 5) Simulated human operators by including a radar controller and data controller in each sector. This was needed to identify workload impact, and specifically the impact of changes in communication processes (voice vs. data link). The western region of the airspace was modeled with two TMU supervisors to decide and communicate weather response strategies and dynamic airspace shifts. 6) Developed two terminal airspaces for the fast-time simulations: the BL condition and BA condition. 7) Included two major airports and three satellite airports in the simulation to show multiple arrival and departure routes for the terminal area.   The major airports included Orlando International Airport (MCO) and Tampa International Airport (TPA).   The satellite airports included Orlando Sanford International Airport (SFB), Orlando Executive Airport (ORL), and Melbourne International Airport (MLB).

The model was tested under the following conditions: 1) Three levels of traffic: Traffic was provided at three levels; simulated air traffic loads at BL (2012 BL), BL plus 50 percent and BL doubled.   Both BL operational procedures and BA procedures were run at each traffic level with and without weather. Additionally, the BA operations were run in a voice communication and data link communication mode. 2) Procedures and rules: With and without Big Air Procedures in place. This includes the expanded ARTCC airspace in Big Air operations, three-mile separation throughout the airspace and dynamic response in sector structure as a function of weather in the airspace. 3) Weather event: The simulation was run with and without disruptive weather events in the scenario. 4)   Communications media: The simulation was run with and without the use of data link communications between the controller and flight crew. All conditions were tested across all runs in a fully crossed design.   That is to say that the simulation was run in all conditions and levels of those conditions crossed with all others.

The major derived results from the simulation output included: 1) Workload measures from the human performance model.   The human performance model outputs workload in four-dimensions or resource types. These are visual, auditory, cognitive and motor loads. 2) Number and type of clearances delivered in the scenario. 3) Number of tasks undertaken and successfully completed (or aborted/deferred).

In this paper, we focus on the studies of workload analysis.

5 RESULTS

In order to judge the impact of Big Air, traffic level, weather and communication processes several analyses were undertaken on the data generated by the models of human controllers in arrival, departure and surrounding airspace. We conducted an analysis of variance (ANOVA) to see if there was an effect of the experimental conditions on the controllers' performance. In this analysis the dependent variable "cognitive workload" was used, and the data are from the Radar controller model.

5.1 2x3x2 ANOVA Repeated Measures

Cognitive workload is a measure that the model generates to estimate the loading for task performance in the cognitive dimension for each activity that the model performs. As indicated above, the cognitive load for a given activity is an estimate that is based on subject matter expert opinion (McKracken & Aldrich, 1984). Table 1 provides the means and standard deviations of that workload for all the tasks performed by the radar controller across the simulation scenarios in the arrival airspace. The arrival airspace had the highest workload for the controllers.

Table 1. Cognitive workload means and standard deviations for experimental conditions.

	Mean	Std. Deviation	N
Big100Wx	2.4595	1.40439	20
Big100NoWx	2.5250	1.55109	20
Big150Wx	2.4430	.45966	20
Big150NoWx	1.8745	.65528	20
Big200Wx	3.8850	.25082	20
Big200NoWx	2.1620	1.15123	20
BL100Wx	1.3385	.11160	20
BL100NoWx	1.3025	.22985	20
BL150Wx	1.3435	.09354	20
BL150NoWx	1.3120	.20086	20
BL200Wx	1.4675	.16964	20
BL200NoWx	1.4835	.12926	20

In order to determine whether the conditions that we manipulated (control mode, weather, traffic density), an analysis of variance was performed. Table 2 provides the "F" statistics for the cognitive workload analysis of variance across all conditions and the interaction of those conditions. All main effects and interactions are significantly different from at an alpha level of 0.05. This indicates that the cognitive workload is statistically different under the varied conditions of the simulation.

Table 2. Tests of Within-Subjects Effects: Number of Clearances Issued

Source	Type III Sum of Squares	Df	Mean Square	F	Sig.	Partial Eta Squared
BigAir_BL	84.052	1	84.052	128.597	.000	.871
Level	10.684	2	5.342	6.753	.003	.262

Weather	8.645	1	8.645	26.439	.000	.582
BigAir_BL * Level	5.145	2	2.573	3.542	.039	.157
BigAir_BL * Weather	7.881	1	7.881	30.143	.000	.613
Level * Weather	7.729	2	3.864	19.462	.000	.506
BigAir_BL * Level * Weather	8.733	2	4.367	24.023	.000	.558

Figure 5 illustrates the cognitive workload for the radar controllers in the simulation. As can be seen in the figure the cognitive workload is higher overall (that is across all sectors in the simulation) in the BL operating conditions of the experiment as compared with the Big Air Operating conditions.

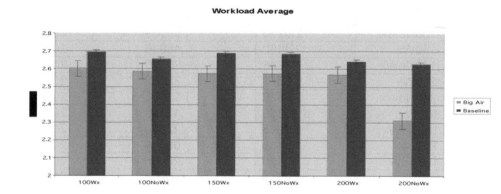

Figure 5. Cognitive workload average across conditions of traffic density, weather and standard BL verses BA operating conditions. (Bars represent one standard error.)

The conclusions drawn from this analysis is that on average the baseline operations imposed a higher cognitive load on the controller model than did the operations using the Big Air procedures, and the increased workload level was consistent across conditions (weather and traffic load).

5.2 Data Link versus Voice/Radio Communications

In addition to the workload under the conditions of traffic increase, weather events and Big Air procedures, a separate analysis was performed taking into account the use of data link (controller to pilot data link communications). In this set of conditions, the activities needed to communicate a clearance to the flight deck were simulated to represent the activities needed to compose and send a message to the flight deck that included a heading, speed, vector or route change. The data link procedures represented transfer of control and communication at the sector boundary and the delivery of clearance information. In the alternative condition to this, the controller used voice/radio communication procedures to provide the flight crew with clearances. A "t" test for differences was performed and the data link versus voice clearance process were found significantly different from each other t= 2.99 (dof, 14) at $\alpha=0.05$. This difference is seen as a reduction in cognitive workload during Big Air procedures as opposed to BL procedures. Figure 6 illustrates these differences.

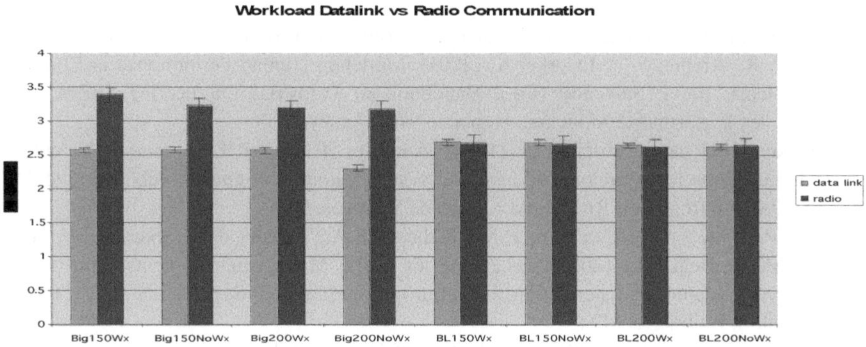

Figure 6. Workload: data link versus radio communication.

The conclusion drawn form this analysis is that in Big Airspace operations data link communication consistently resulted in reduced workload on average when compared with standard operations. The effect of data link communications in baseline conditions did not have an average net reduction. This result (a significant interaction) means that the use of data link communications is differentially more effective in the Big Air procedure process, making Big Air operations even more efficient. Further analyses would need to be performed in order to be sure why this differential effect is seen, but a hypothesis that seems likely is that Big Air procedures allowed route changes and approach RNP/RNAV selection which were efficiently communicated (single action selection of the route from a controllers' preferred route list and single action delivery of this clearance to the aircraft).

6 SUMMARY

A fast time human performance model was developed and run under conditions that simulated Big Airspace operational concept procedures. The model indicated that Big Airspace procedures were operationally feasible. Big Airspace Operations produced reduced workload overall when compared with present operations under similar experimental conditions. The model also indicated that data link communications decreased operator workload especially under the Big Air procedural paradigm.

ACKNOWLEDGMENTS

The authors would like to acknowledge the financial support from FAA to complete this study.

REFERENCES

Hollnagel, E. (1993). Human reliability analysis: Context and control. Academic Press

Laughery, R., Archer, S. and Corker K. (2001). Modeling Human Performance in Complex Systems. In G. Salvendy (Ed.), Handbook of Industrial Engineering. Interscience Human Performance Modeling, Hoboken, New Jersey: Wiley.

McCracken, J. H. and Aldrich, T.B. (1984). Analysis of selected LHX mission functions: Implications for operator workload and system automation goals (Technical note ASI 479-024-84(b)). Fort Rucker, AL: Anacapa Sciences, Inc.

Shah, A.P., Pritchett, A.R., Feigh, K.M., Kalaver, S.A., Jadhav, A., Corker, K.M., Holl, D.M., & Bea, R.C. (2005). Analyzing air traffic Management systems using agent-based modeling and simulation. USA/Europe International Symposium in Air Traffic Control, Baltimore, MA.

Verma, S. and Corker, K. (2001). Introduction of Context in human performance model as applied to dynamic resectorization. In Proceedings of the 11th International Symposium on Aviation Psychology, Columbus, Ohio: Ohio State University.

CHAPTER 39

Extending Validated Human Performance Models to Evaluate NextGen Concepts

Brian F. Gore[1], Becky L. Hooey[1], Eric A. Mahlstedt[1], David C. Foyle[2]
[1]San Jose State University at NASA Ames Research Center

[2]Human-Systems Integration Division / MS 262-4
Moffett Field, CA 94035-0001 USA
E-mail: Brian.F.Gore@nasa.gov; Becky.L.Hooey@nasa.gov;
Eric.Mahlstedt@nasa.gov; David.C.Foyle@nasa.gov

ABSTRACT

To meet the expected increases in air traffic demands, the National Aeronautics and Space Administration (NASA) and the Federal Aviation Administration (FAA) are researching and developing Next Generation Air Transportation System (NextGen) concepts. NextGen will require substantial increases to the data available to pilots on the flight deck (e.g., wake projections) to support more precise and closely coordinated operations (e.g., closely spaced parallel operations, CSPOs). These NextGen operations, along with the pilots' roles and responsibilities, must be designed with consideration of the pilots' capabilities. Failure to do so will leave the pilots, and thus the entire aviation system, vulnerable to error. A validated Man-machine Integration Design and Analysis System (MIDAS) v5 model was extended to evaluate changes to flight deck and controller roles and responsibilities in NextGen approach and land operations. Compared to conditions when the controllers are responsible for separation on descent to land phase of flight, the output from these model predictions suggest that the flight deck response time to detect the lead aircraft blunder will decrease, pilot scans to the navigation display will increase, and workload will increase.

Keywords: NextGen CSPO, MIDAS v5, Human Performance Model

1   INTRODUCTION

The National Airspace System (NAS) in the United States is currently being redesigned because it is anticipated that the current air traffic control (ATC) system will not be able to manage the predicted two to three times growth in air traffic in the NAS (JPDO, 2011). To meet the expected increases in air traffic demands, the National Aeronautics and Space Administration (NASA) and the Federal Aviation Administration (FAA) are researching and developing Next Generation Air Transportation System (NextGen) concepts to alleviate bottlenecks caused by the anticipated growth.

One such bottleneck is anticipated to be in the decent, approach, and landing phases of flight. Closely Spaced Parallel Operations (CSPO) are expected to enable paired approaches to minimum runway spacing in instrument meteorological conditions (IMC) while maintaining an acceptable level of risk (Cox, 2010). The current requirement for landings in IMC is at least 4300 ft of lateral runway spacing (as close as 3000 ft for runways with a Precision Runway Monitor) whereas operations in visual meteorological conditions (VMC) require lateral runway spacing to be equal to or greater than 750 ft. It is feasible for aircraft to perform both arrival and departure operations in IMC using VMC parallel separation standards as advanced navigation technology, sophisticated wake avoidance algorithms, 4-D flight management systems, and advanced flight deck displays become more widely available (Rutishauser et al., 2003).

In the highly automated CSPO environment envisioned by NextGen, a paradigm shift might be required that would transfer the responsibility for separation from ATC, as is currently the case, to the flight deck. As flight decks are modified to accommodate the new suite of automation tools and displays required to support this, research must be conducted to ensure that they are designed and implemented in a safe manner without leaving pilots vulnerable to errors or excess workload. These NextGen procedures and operations, along with the pilots' roles and responsibilities must be designed with consideration of the pilots' capabilities. Failure to do so will leave the pilots, and thus the entire aviation system vulnerable to performance inefficiencies caused by error. This is a particular concern in the CSPO environment where wake threats become an important issue for trailing aircraft on approach and landing operations.

1.1   Using HPMs to Evaluate NextGen Concepts

There are large challenges associated with evaluating novel NextGen concepts such as CSPO and changes to pilot / ATC roles and responsibilities. Because NextGen concepts are still in the early stages of the design lifecycle, operator roles and tasks are often not well defined, and NextGen technologies have not necessarily reached a level of sufficient maturity to allow for physical prototypes. These factors limit the feasibility of full-mission human-in-the-loop (HITL) simulations. However, human performance models (HPMs) can be used to make meaningful contributions early in the design lifecycle, particularly for concepts that have high consequences associated with their failure. Models can be advantageous because they are cost effective, and eliminate concerns often associated with HITL testing of new concepts such as novelty and training effects. Furthermore, models are advantageous as compared to HITL simulations where you have to build physical prototypes because HPMs represent the location and nature of information symbolically (i.e. one can classify flight deck information as text or symbol, without defining the exact text phraseology or designing the symbol) and therefore allow rapid prototypes of concepts to be generated and tested early in the design phase. One such HPM tool, the Man-machine Integration Design and Analysis System (MIDAS), is discussed next.

1.2   The Man-Machine Integration Design and Analysis System (MIDAS)

NASA's MIDAS is a dynamic, integrated HPM that facilitates the design, visualization, and computational evaluation of complex man–machine system concepts in simulated operational environments (Gore, 2008). MIDAS symbolically represents many mechanisms that underlie and cause human behavior including the manner that the operator receives/detects information from an environment, comprehends and registers this information in a memory store, decides on a response, and responds to the information within the context of operational rules and human performance capacities. MIDAS combines these symbolic representations of cognition with graphical equipment prototyping, dynamic simulation, and procedures/tasks to support quantitative predictions of human-system effectiveness, and improve the design of crew stations and their associated operating procedures. MIDAS provides an easy to use and cost-effective means to conduct experiments that explore "what-if" questions about domains of interest.

2   A PROCESS FOR EVALUATING NEXTGEN CONCEPTS USING HPMS

One challenge associated with developing valid models of NextGen concepts is the lack of HITL data with which to validate the models. In this paper we propose a candidate process for developing and validating HPMs for the evaluation of NextGen concepts. The process includes four steps: 1) Develop a baseline (current-day) model; 2) Validate the baseline (current-day) model; 3) Extend the baseline scenario to NextGen; and 4) Conduct iterative what-if scenarios to explore early design concepts. This process addresses the validation challenge by first developing models of known, current-day, operations, which are well-defined and proceduralized, and for which HITL data exist to enable validation.   Then, the validated model platform is modified by integrating assumptions about likely NextGen changes that will be made to the flight deck equipage and pilot tasks. Confidence is attained that the validity of the model is preserved through the documentation of assumptions and through the small iterative model changes to the validated model. This process, as applied to the CSPO concept, is discussed next.

2.1   Develop the Baseline (Current-Day) Model

A MIDAS v5 high-fidelity model of a two-pilot commercial crew flying current-day area navigation (RNAV) approach and landing operations was developed using a methodical, multi-dimensional approach (Gore et al., 2011). The model represented a Boeing 777 flying from 10,000 ft to touchdown at Dallas Fort-Worth (DFW) airport. The modeled scenario began with the aircraft at an altitude of 10,000 ft and 30nm from the runway threshold (see Figure 1). The cloud ceiling was 800 ft, with a decision height (DH) of 650 ft at which point the modeled pilots disconnected the autopilot and manually flew the aircraft to touchdown. The model assumed that the "pilot flying" (PF) was in the cockpit's left seat and the "pilot monitoring" (PM) was in the right seat. The model scenario included communications with DFW Regional Approach Control, tower, and ground control, as well as intra-cockpit communications.   In total, over 970 pilot tasks were included in the model.

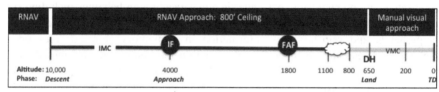

Figure 1. Baseline Current Day RNAV Model of Approach and Landing. Notes: DH = Decision Height; FAF = Final Approach Fix; IF = Initial Fix; IMC Instrument Meteorological Conditions; RNAV = Area Navigation; TD = touchdown; VMC = Visual Meteorological Conditions.

A task network model architecture such as the one embedded in MIDAS contains top level tasks (e.g., land) that are subsequently decomposed into finer-grained tasks, generally to the button-press level for physical control input tasks, scan fixation points for the visual system, and verbal strings for the vocal communication output. The tasks in the network are then tied to a set of behavioral primitives. These tasks then wait for release conditions to be satisfied by the environment, the operators, the controls, or the displays. The behavioral tasks in MIDAS are termed the operator primitives. The task network model illustrated in Figure 2 illustrates a subset of the entire task network model, a snapshot of the flight deck's flap-setting procedure required when landing the aircraft in the simulation.

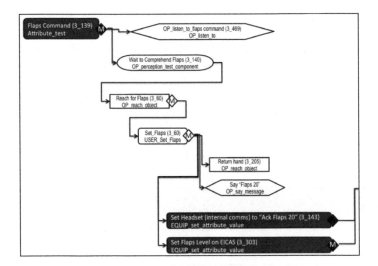

Figure 2. Task Network Model Implementation of a Set Flaps Sequence.

2.2   Validate the Baseline (Current-Day) Model

The model inputs, including the task trace and input parameters, were validated using focus group sessions comprised of a total of eight commercial pilots with glass-cockpit aircraft and RNAV flying experience. The pilot-centric scenario-based cognitive walkthrough approach captured the context of operations from 10,000 ft to touchdown and enabled pilots to assess the modeled tasks and identify tasks that were missing, or in the wrong sequence. Out of 74 tasks in the MIDAS RNAV application presented to the focus group pilots, 12 tasks were identified that should be removed, reordered, or added. The pilots also completed quantitative rating scales, which were used to validate the model input parameters for workload and visual attention. Thirty-nine tasks were rated on the visual, auditory, cognitive, and motor workload dimensions (as relevant for the task).   The workload of four of the pilot tasks was modified and three new primitives were created based on the pilots' ratings. The model was refined based on the results of this input validation process. Next the model outputs, workload and visual attention, of the refined model were statistically compared to existing HITL data (Hüttig, Anders, and Tautz 1999 Mumaw, Sarter and Wickens, 2001; Anders, 2001; Hooey and Foyle, 2008). The workload model output correlated (r^2=.54) with a comparable HITL study with for overall workload. The individual workload dimensions also correlated positively with the HITL study (r^2=.55 to .94). Visual scan time correlated with three HITL studies (r^2 = .99). These results provide confidence that the model validly represents pilot performance.

2.3.   Extend Validated Model to CSPO Scenarios

The RNAV scenario was modified to reflect the CSPO concept based on assumptions about changes to: 1) flight deck equipage (e.g., the addition of data communications, augmented wake and traffic information on the Primary Flight Display (PFD) and Navigation Display (Nav), and visual and auditory wake threat alerts; and, 2) flight crew tasks (e.g., identifying and tracking paired traffic, receiving and accepting datalink, monitoring wake displays). This model assumed an operational environment consistent with NextGen goals of reduced landing minima, specifically, a cloud ceiling of 200 ft and a DH of 100 ft (see Figure 3). The assumptions were made based on interviews with NextGen concept developers and scenario-based focus groups with pilots experienced with current-day Simultaneous Offset Instrument Approaches (SOIAs), which are similar to CSPO but conducted in VFR conditions and with larger runway separations.

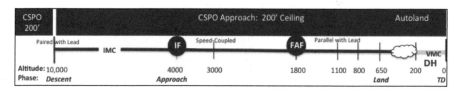

Figure 3. CSPO 200 ft Ceiling Scenario Timeline. Notes: IF = Initial Fix, FAF = Final Approach Fix, DH = Decision Height, IMC = Instrument Meteorological Conditions, VMC = Visual Meteorological Conditions, TD = Touchdown

In order to ensure the verifiability and validity of the CSPO model, the specific CSPO task changes and input parameters were validated using the same pilot focus group sessions described previously. In the focus group sessions, after the pilots completed the task trace and input parameter rating scales for the RNAV model, the CSPO concept was introduced. The pilots were briefed on the goals of NextGen, expected changes to flight deck equipage, and pilot procedures. Examples of the wake displays on both the PFD and Nav and the visual and auditory wake warnings and alerts were presented. A video of two pilots completing CSPO procedures from a HITL simulation (Verma et al., 2008) was also presented. Pilots completed the task trace and input parameter rating scales for the new tasks of the CSPO model.

2.4   Conduct "What-if" Scenarios

Using the validated CSPO model, the model was then exercised to explore a number of CSPO design concepts including varying the flight crew task allocation, pilot-ATC roles and responsibilities, and the format and location of wake and spacing information. In total, 26 model-based scenario manipulations were completed. Analyses of pilot performance measures, including time required to complete tasks, pilot workload, pilot scan patterns and response times to off-nominal events were used to draw conclusions regarding the information requirements necessary to support NextGen CSPO concepts (see Hooey, Gore, Mahlstedt, and Foyle, 2012).

3   CSPO CONCEPT EVALUATION RESULTS

In this section, results of one of the CSPO 'what-if' evaluations are presented. Specifically, the model was modified to evaluate the effect of delegating separation to the flight deck for CSPO operations. Two scenarios were modeled: 1) **Pilot-responsible separation**, a scenario to represent NextGen CSPO conditions in which the pilots (PF and PM) are responsible for separation delegation and for detecting and initiating emergency escape maneuvers; and, 2) **ATC-responsible separation**, a scenario to represent current-day conditions in which the ATCs are responsible for separation and for detecting and initiating emergency escape maneuvers. As the aircraft was on final approach (1800 ft), a wake threat occurred in which the wake of the lead aircraft extended into the ownships' trajectory requiring a missed approach by the following aircraft. In the pilot-responsible for separation scenario, pilots were provided with a dynamic wake display and a two-stage alert system that first issued a visual and auditory warning as the wake threat developed and a final alert commanding an immediate take-off / go-around (TOGA) procedure. The dynamic breakout maneuver was shown on the Nav. In the ATC-responsible separation scenario, ATC received the two-stage wake warning and alert, then issued a verbal 'Go-Around' command to the aircraft, with a missed approach path that accounted for metroplex traffic, terrain, and wind conditions.

Ten Monte Carlo simulation runs were generated to evaluate the pilots' time to initiate the emergency escape maneuver in response to the wake threat alert, the pilots' scan (percent dwell time, PDT) performance, and the pilots' workload.

3.1   Response Time to Wake Threat Alert

The response time to the wake threat alert as a function of current-day ATC-responsible or NextGen Pilot-responsible for separation can be found in Figure 4. This figure illustrates that pilots were only negligibly faster to initiate the emergency escape maneuver in the pilot-responsible condition (2.9s) than in the ATC-responsible condition (3.2s); $(t(9)=.07, p>.05)$.

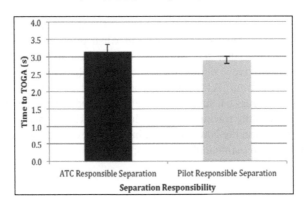

Figure 4. Time to TOGA as a Function of Crew Responsibility for Separation.

3.2   Flight Crew Scan Performance

The output of the flight crew scan performance during the alert phase was analyzed to determine the impact that the change from ATC-responsible to Pilot-responsible had on the pilots' scanning during the wake threat event. Scan performance was measured by the percent dwell time (PDT) that the pilot spent on three main areas of interest (AOIs): Primary Flight Display (PFD), Navigation Display (Nav), and Out the Window (OTW). As shown in Figure 5, when pilots were responsible for separation they spent more time monitoring the Nav containing wake information than when ATC was responsible for separation ($t(19)=27.59$, $p<.01$). As a result, *both* pilots spent less time monitoring OTW and critical flight performance data on the PFD in the pilot-responsible paradigm.

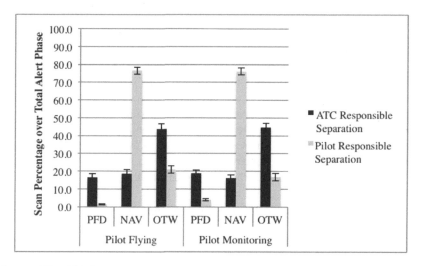

Figure 5. PF and PM Scan over the Total Alert Phase

3.3   Mean Workload

Figure 6 presents mean workload predicted by MIDAS over the total alert phase when ATC was responsible for separation and when pilots were responsible for separation. It is evident that the workload is predicted to be higher for both members of the flight crew during the alert phase when the responsibility for separation is shifted from ATC to the cockpit ($F(1,18)=31.48$, $p<.01$). This increased workload occurs because the PF and PM spend more time on workload-inducing tasks like visually tracking the lead aircraft, making mental comparisons to displayed information (on the Nav), determining the missed approach response, and making control inputs to maintain the correct spacing.

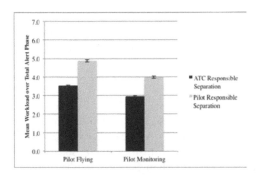

Figure 6. PF and PM Mean Workload Ratings Across the Total Alert Phase.

4 DISCUSSION AND IMPLICATIONS

A model of commercial airline pilots conducting approach-and-landing procedures was created using the MIDAS software following a methodical development and validation approach. The premise that guided the current work was that model validity is a process, not solely a single value at the conclusion of a model development effort. Valid inputs lead to valid outputs. It is therefore necessary to follow an iterative input validation process as well as an iterative output validation process. Conducting only one of these validation processes may lead to invalid models. This is especially true as the complexity of the operational environment and tasks increase.

The current research has evaluated proposed changes to flight deck technologies, pilot procedures, operations, and roles and responsibilities that are likely with NextGen CSPOs through the use HPM outputs of time required to complete tasks, pilot scan performance, and pilot workload. The three measures of pilot performance (blunder response time, scan time, and workload) provide insights into potential costs of transitioning to a CSPO paradigm in which pilots are responsible for separation. CSPOs are predicted to influence the flight deck by directing a greater scan percentage to the Nav. Blunders in CSPOs are predicted to be detected faster (negligibly) when the flight deck is given responsibility for separation but this comes at a cost of increased workload for the PF and the PM and a reduced spread of visual attention across cockpit and out-the window. The flight-crew scan performance output reveals that in CSPOs, both pilots attend to the Nav during the alert phase the majority of their time. This is important because neither of the pilots is looking at the PFD, the primary display used for flight. This has important implications for NextGen CSPOs because it may be better for one of the pilots to focus on the Nav display, while the other pilot focuses his/her attention on the PFD. The appropriate allocation of attention is something that needs to be defined.

Future research, using HITL methods with both pilots and ATC, is required to better quantify the actual response-time benefit offered by the pilot-separation concept. In addition, system studies are required to determine if the response-time

416

benefit provides a practical advantage in operational environments. Furthermore, HITL research is required to fully assess the workload and visual attention decrements associated with the pilot-separation scenario.

ACKNOWLEDGEMENTS

The composition of this work was supported by the Federal Aviation Administration (FAA)/NASA Inter Agency Agreement DTFAWA-10-X-80005 Annex 5. The authors would like to thank Connie Socash, Christopher Wickens, Marc Gacy, and Mala Gosakan (Alion Science and Technology) for their model development support and Nancy Haan (Dell Perot) for the supporting the validation.

REFERENCES

Anders, G. 2001. Pilot's Attention Allocation during Approach and Landing: Eye- and Head-Tracking Research in an A330 Full Flight Simulator. Paper presented at the 11th International Symposium on Aviation Psychology, Columbus, OH.

Cox, G. 2010. Closely Spaced Parallel Runway Operations (CSPO). Presented at the CSPO Working Group Meeting, June 3, 2010. Available at: http://www.faa.gov/about/office_org/headquarters_offices/avs/offices/afs/afs400/afs440/cspo/meetings/media/cspo_working_group.pdf

Gore, B.F. 2008. Chapter 32: Human performance: Evaluating the cognitive aspects. In. *Handbook of digital human modeling*. ed. V. Duffy. Boca Raton: CRC Press Inc. pp. 32:31-32:18.

Gore, B.F., B.L. Hooey, and C. Socash, et al. 2011. Evaluating NextGen closely spaced parallel operations concepts with human performance models. Human Centered Systems Laboratory (HCSL) Technical Report (HCSL-11-01). Moffett Field, CA: NASA Ames Research Center.

Hooey, B.L. & D.C. Foyle. 2008. Aviation safety studies: Taxi navigation error and synthetic vision system operations. In. Human performance modeling in aviation, eds. B.L. Hooey and D.C. Foyle. Boca Raton: CRC Press/Taylor & Francis. pp. 29-59.

Hooey, B.L., B.F. Gore, E. Mahlstedt, and D.C. Foyle. 2012. Evaluating NextGen Closely Space Parallel Operations Concepts with Validated Human Performance Models: Flight Deck Guidelines (Part 1 of 2). Human Centered Systems Laboratory (HCSL) Technical Report (HCSL-12-01). Moffett Field, CA: NASA Ames Research Center.

Hüttig, G., G. Anders, and A. Tautz. 1999. Mode Awareness in a modern Glass Cockpit – Attention Allocation to Mode Information. Paper presented at the 10th International Symposium on Aviation Psychology, Columbus, OH.

Mumaw, R.J, N. Sarter and C.D. Wickens. 2001. Analysis of pilots' monitoring and performance on an automated flight deck. Paper presented at the 11th International Symposium on Aviation Psychology, Columbus, OH.

Joint Planning Development Office (JPDO), 2011. Targeted NextGen Capabilities for 2025. JPDO: December.

Rutishauser, D., G. Lohr, D. Hamilton, and R. Powers, et al. 2003. Wake Vortex Advisory System (WakeVAS) Concept of Operations NASA/TM-2003-212176, April.

Verma, S., S. Lozito, and T. Kozon, D. et al. 2008. Procedures for off-nominal cases: Very closely spaced parallel runways approaches, Digital Avionics System Conference, St. Paul, MN.

CHAPTER 40

Model Individualization for Real-Time Operator Functional State Assessment

Guangfan Zhang, Roger Xu, Wei Wang, and Aaron A. Pepe

Intelligent Automation, Inc.
{gzhang, hgxu, wwang, apepe}@i-a-i.com

Feng Li, Jiang Li, and Frederick McKenzie

Old Dominion University
feng.odu@gmail.com; jli@odu.edu; rdmckenz@odu.edu;

Tom Schnell, Nick Anderson, Dean Heitkamp

University of Iowa
thomas-schnell@uiowa.edu;nicholas-anderson@uiowa.edu; dean-heitkamp@uiowa.edu

ABSTRACT

Proper assessment of Operator Functional State (OFS) and appropriate workload modulation offer the potential to improve mission effectiveness and aviation safety in both overload and under-load conditions. Although a wide range of research has been devoted to building OFS assessment models, most of the models are based on group statistics and little or no research has been directed towards model individualization, i.e., tuning the group statistics based model for individual pilots. Moreover, little emphasis has been placed on monitoring whether the pilot is disengaged during low workload conditions. The primary focus of this research is to provide a real-time engagement assessment technique considering individual variations in an aviation environment. This technique is based on an advanced

418

machine learning technique, called enhanced committee machine. We have investigated two different model individualization approaches: similarity-based and dynamic ensemble selection-based. The basic idea of the similarity-based technique is to find similar subjects from the training data pool and use their data together with the limited training data from the test subject to build an individualized OFS assessment model. The dynamic ensemble selection dynamically select data points in a validation dataset (with labels) that are adjacent to each test sample, and evaluate all the trained models using the identified data points. The best performing models will be selected and maximum voting can be applied to perform individualized assessment for the test sample. To evaluate the developed approaches, we have collected data from a high fidelity Boeing 737 simulator. The results show that the performance of the dynamic ensemble selection approach is comparable to that achieved from an individual model (assuming sufficient data is available from each individual).

Keywords: Operator Functional State, Engagement, EEG, Machine Learning, model individualization

1   INTRODUCTION

In the past decade, a wide range of research has been undertaken regarding Operator Functional State (OFS) assessment. Many existing studies utilized psychophysiological measurements to index the level of cognitive demand associated with a task (Boucsein and Backs, 2000), fatigue (Trejo, Kochavi and Kubitz, 2005), engagement (Stevens, Galloway and Berka, 2007) and other functional states (Hockey, 2003).

Individual differences, commonly referred to idiosyncratic regularities of the physiological reaction or Individual Response Specificity (IRS) (Marwitz & Stemmler, 1998), need to be considered in the OFS assessment model. It is often difficult, if not impossible; to train an OFS assessment model for each individual pilot due to lack of sufficient training data from each individual. On the other hand, a generalized OFS assessment model built on the data from a large amount of training subjects often yields poor performance when being applied to a new subject. A possible solution is to perform model individualization that adapt the trained generalized model to an individual. Olofsen et al. (2010) categorized two model individualization techniques. The first involves basic research, which aims to discover the individual differences of physiological responses reflecting the OFS. Based on the knowledge and understanding of the physiological response, model parameters can be adjusted accordingly to address individual variations. The second approach doesn't rely on the understanding of the physiological differences and only involves statistical analysis/modeling approaches. Currently, the understanding of the nature of OFS is limited and many fundamental issues, including individual variations in physiological response, are not well understood. Thus, most of the current research focuses on the statistical approaches. Rajaraman et al. (2009) has developed a method for developing individualized biomathematical models for predicting cognitive performance impairment of individuals subjected to total sleep loss. Their method systematically customizes the parameters in the two-process model of sleep regulation for an individual by optimally combining the performance information with a priori performance information using a Bayesian framework. Zhang et al. (2009) presented a similarity-based approach for model individualization. This approach finds similar subjects from the training data pool and uses their data together with the limited data from the test subject to build an individualized OFS assessment model. The idea is based on the assumption that if the test subject's data is similar to the data of a training subject in one or more specific conditions, he or she will have similar behaviors to the training subject in other conditions. Thus all the training data from this training subject can be used to build a model for the test subject.

This paper presents a framework for real-time individualized OFS (engagement, more specifically) monitoring in an aviation environment. Two major components are included. The first component is a base OFS assessment component, which is based on an advanced machine learning technique, enhanced committee machine. A committee machine is a strategy to improve classification/regression performance by combining responses from multiple committee members. The enhanced committee machine integrates a bootstrapping technique, an advanced feature selection method and a Neural Networks-based classification method to build base classifiers/committee members (Zhang et al., 2009). By aggregating outputs from the committee members, the final engagement decision can be more robust and accurate. The second component is the model individualization component, which performs individualized engagement assessment. We further explore the similarity-based individualization technique by comparing different similarity measures and investigate a novel dynamic ensemble selection-based approach for model individualization.

The dynamic ensemble selection framework dynamically selects a subset of committee members/classifiers from tens/hundreds (or more) of committee members. The selection is based on the performance of the committee members on the limited validation data (with labels). The rational of this design is based on the assumption that if the test sample is similar to its local validation samples in the feature space and we select a subset of trained committee/members classifiers who perform well on the local validation samples, we may achieve a good engagement assessment results for the test data point. If there is limited data from the test subject with labels (known engagement states), it can also be included in the validation dataset to evaluate the performance of each classifier. The first few top classifiers are then selected to classify the test sample by majority voting (Giacinto, 2001).

To evaluate the developed approach, we have collected data from a high fidelity Boeing 737 simulator, including flight technical data, psycho/physiological signals and behavioral measures. An extensive feature study has been performed to extract promising and robust features. The results show accurate engagement/disengagement detections, making it suitable for real-time assessment. Both of the model individualization techniques have been evaluated and the results show that the performance of the dynamic ensemble selection approach is comparable to that achieved using an individual model assuming sufficient data is available from each individual for training. We have also shown that the similarity-based technique has the potential to individualize the OFS assessment model. However, the performance achieved based on the limited subjects doesn't show an improvement comparing to the results achieved from a generalized model. It may be due to the limited number of subjects available for performance evaluation. We will further investigate this method in the future.

The remainder of the paper is organized as follows. Section 2 describes model individualization based on enhanced committee machine for real-time engagement assessment. Section 3 describes the flight simulation configuration and experiment design. Section 4 shows performance evaluation results. Section 5 concludes the paper.

2   MODEL INDIVIDUALIZATION BASED ON ENHANCED COMMMITTEE MACHINE

In this research, based on the enhanced committee machine framework (Zhang et al. 2009), we have developed two model individualization techniques: similarity-based and dynamic ensemble selection-based.

2.1   Enhanced Committee Machine

A committee machine is a strategy to improve classification/regression performance by combining responses from multiple committee members. Different algorithms can function as committee members, for example, Neural Networks (NNs), Gaussian Mixture Models (GMMs), and Support Vector Machines (SVMs). If committee members have the diversity property, i.e. they are unlikely to make errors in the same feature space, the errors from individual committee members will be canceled by each other to some extent. Furthermore, since the committee machine "averages" its individual member's estimation, the variance of the committee machine can be significantly reduced. As a consequent, the performance of the combination of the estimation from each committee member is often superior to that of its committee members (Zhang et al., 2009). In this paper, we implemented the committee machine using a Multi-Layer Perceptron (MLP) neural network trained by the standard Back Propagation (BP) algorithm as the base classification model.

2.2   Model Invidualization

It is known that physiological signals from different individuals usually have different characteristics. Therefore, a generalized model built on data from other individuals may not perform well. When an individual's data is not sufficient to build an individual's model, an alternative option is to individualize the generalized model for each individual. We have investigated two approaches for model individualization in this research: similarity-based and dynamic ensemble selection.

The architecture of the two model individualization techniques based on enhanced committee machine for engagement assessment is shown in Figure 1. It is common that both methods perform model individualization by selecting a subset of committee members, from similar subjects (similarity based) or best performing committee members based on performance evaluation results using a validation dataset (dynamic ensemble selection). We will introduce both methods in this section.

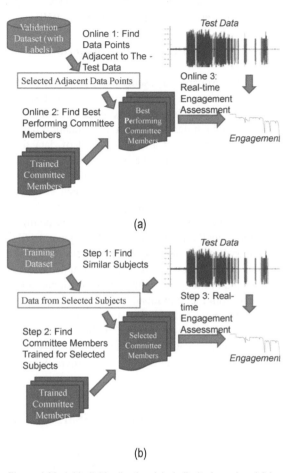

(a)

(b)

Figure 1 Model individualization:(a) similarity-based and (b) dynamic ensemble selection

2.2.1 Similarity-based Model Individualization

In many cases, training data for an individual is limited and is expensive to collect. It may be infeasible to train an individualized OFS monitoring model using only the individual training data. Therefore, we developed an approach to build an individualized OFS monitoring model by identifying training subjects that have similar physiological responses and extracting their data for model individualization.

Ideally, for each individual being tested, we assume that we can find one or more "similar" subjects among all the subjects with training data. In each functional state that the individual has experienced, there may exist one or more subjects that have similar training data in the same functional state. The purpose of computing subject distance is to find such subject(s) in a specified functional state. After scanning all the functional states of the individual, we can select a subset of subjects that are similar to the individual in some functional states based on the similarity metrics. All the trained committee members from these subjects can be extracted to form an individualized committee machine.

The similarity can be measured by different metrics. In this paper, we investigated five different similarity measures implemented in Matlab: t-test, entropy, Bhattacharyya distance, ROC and Wilcoxon test (MATLAB 7.11, The MathWorks Inc., Natick, MA, 2010).

2.2.2 Dynamic Ensemble Selection

It is reasonable to assume that an individual's signals may carry similar patterns to other persons in a specific context when they are performing a similar task. For a particular person for whom we want to build an engagement assessment model, if we can find similar patterns in the validation data (with labels; such as limited data from the test subject and/or a subset of data from other subjects) and select the classifiers that perform the best on the validation data, we may obtain good results by using these selected classifiers. This technique is different from the similarity-based individualization technique, which is based on a static selection procedure. Comparing to the similarity-based approach, we utilized a dynamic ensemble selection framework to dynamically individualize a generalized model to adapt to an individual's characteristics. The generalized model is built using available training datasets, which consists of a set of neural network classifiers as committee machine members. In dynamic ensemble selection, each classifier's accuracy is estimated in the local feature space surrounding the data similar to an unknown test sample from the individual. The first few top classifiers are then selected to classify the test sample by majority voting (Giacinto, 2001). The rational of this design is based on the assumption that if the test sample is similar to its local validation samples in the feature space, we may achieve a good assessment result for the test data point by utilizing those classifiers which perform well on the adjacent validation data points.

3   ENGAGEMENT ASSESSMENT EXPERIMENT DESIGN

In order to study engagement, we conducted experiments in a fully equipped Boeing 737 simulator involving commercial pilots (Ellis, 2009). The functionality of the simulator can be described as a fully functional flight deck with full glass cockpit displays, five outside visual projectors, functioning mode control panel with autopilot and autothrottle, and standard Boeing 737 controls. Several subjects participated in the pilot engagement study. Pilots had varying levels of experience with different types of aircraft. The experiment involved a flight from Seattle Tacoma International Airport to Chicago O'Hare International Airport. The details of the flight have been extracted from an actual American Airlines flight which took place on May 10th, 2010 (Zhang, et al., 2011). Details can be retrieved from on flightaware.com.

The experiment procedure has been previously described in our paper (Zhang et al., 2011). In order to study engagement, all pilots were scheduled to arrive at 5:30pm and were asked to avoid drinking caffeinated beverages such as coffee on the day of the experiment. An orientation video was shown to the subjects before the simulated flight. The video contained a description of the experiment as well as a Control & Display Unit (CDU) programming training section. The video included a description of the sensors and video recording devices used, as well as the responsibilities that the pilots would have during the experiment. The details shared with the subjects did not include information on the probes used to measure engagement levels so that pilots would not anticipate these probes throughout the experiment. During the flight simulation, one of the staff controlled the simulation computer to play pre-recorded audio files mimicking ATC transmissions. An experimenter was in charge of tagging the data to make sure that proper labels were added to the data sheets to identify the phases of the experiment as well as the times when the pilot responded to ATC. At the end of the experiment, the subjects filled out a subjective survey (such as NASA TLX) to assess their workload, fatigue and situational awareness during different phases of flight (Zhang, 2011).

In addition to the subjective surveys, objective data was collected from flight technical data (altitude, speed, *etc.*); three types of psycho-physiological sensors, including eye tracking cameras, an electroencephalogram (EEG) net, and electrocardiogram (ECG); and performance data such as response time to ATC calls or pump failure.

4   EXPERIMENTAL RESULTS

The developed techniques were evaluated with the experimental data collected through a Boeing 737 flight simulator. In this study, two engagement states and their time durations were first identified by evaluating videos of subjects. A pilot's state during takeoff or while handling a pump failure was considered as 'engaged' The pilot's state during level flight without any manipulation or if napping was recognized as 'disengaged'. Calculated features can then be labeled with these states by aligning with the identified time information.

We focused on the EEG sensor signals in this paper. The EEG data was first preprocessed, including removal of environmental and DC artifacts, removal of EEG datasets with unreasonable measurements based on standard deviation (such as 0 indicating no signal collected), selection of EEG channels of interest, identification of spikes/excursions/amplifier saturation, removal of eye blink/body movement induced artifacts, and calculation of Power Spectrum Density (PSD) (Li et al, 2011). Based on existing research studies on EEG (Berka et al. 2007) we selected a subset of EEG signals, including Fz-Oz, Cz-Oz, C3-C4, F3-Cz, F3-C4, Fz-C3, P7-Oz and P8-Oz. In this study, EEG absolute PSD variables for each 1-s epoch were computed and for each bipolar pair, the power spectrum within each band was summed up as a feature. All the features were analyzed and selected based on an advanced feature selection algorithm (Li et al., 2006) before being fed into the committee machine.

4.1   Similarity-based Model Individualization Evaluation

To study the effectiveness of the similarity-based model individualization technique, we first investigated the effectiveness of each similarity measure: t-test, entropy, Bhattacharyya distance, ROC and Wilcoxon test using the function provided in MATLAB 7.11 (the MathWorks Inc., Natick, MA, 2010). The hypothesis is that, if we can identify a similar subject in the training data pool, we can use his/her data to build a model and achieve a good performance when applying to the test subject.

We have evaluated the similarity–based individualization technique on 8 subjects. Among the different similarity measures, the Wilcoxon test can better represent the subject similarity based on the cross-subject evaluation performance, and was chosen as the similarity measure for similarity-based model individualization. The similarity-based engagement assessment performance results are shown below.

Table 1 Performance: generalized vs. similarity-based individualization

Subject	1	2	3	4	5	6	7	8	Avg.
Generalized	93.98	99.28	74.17	83.8	98.24	63.2	43.75	80.98	79.67
Similarity-based	90.65	97.93	88.76	82.48	97.88	61.7	49.31	71.77	80.05

The performance by the model individualization does not show an improvement on the performance comparing to the performance achieved using a generalized model (trained using data from other subjects). Further investigation is needed with a large data corpus.

4.2   Dynamic Ensemble Selection for Model Individualization

We used eight subjects' data collected from the Boeing 737 flight simulator and evaluated the dynamic ensemble selection technique for model individualization. There are seven scenarios, as described below, to test the engagement assessment performances.

- Scenario 1. We utilized the first 20% of data from one subject for training (26 committee members in total) and the remaining 80% data for testing the same subject, thus obtaining the individual model performance.
- Scenario 2. For each subject, we trained 5 committee members/models using data from each of other subjects. Since there are 8 subjects in total, 5*7=35 committee members were trained for the testing subject. This will give us the baseline generalized model performance.
- Scenario 3. The same as Scenario 2 except that the dynamic ensemble selection technique was applied using data from other subjects for validation.
- Scenario 4. The same as Scenario 3 except that the validation dataset for dynamic ensemble selection was from the testing subject (first 20%).

- Scenario 5. For each testing subject, there were committee members trained using the testing subject's first 20% of the data and another 5*7 models trained using data from other subjects.
- Scenario 6. The same as Scenario 5 except that we used the dynamic ensemble selection technique and combined the data from other subjects and the first 20% of data from the testing subject as the validation dataset.
- Scenario 7. The same as Scenario 6 except that the validation data only contained the first 20% of dataset from the testing subject.

The performance for all the subjects is shown in Table 2.

Table 2 Experiment results

Scenario / Subject	1 (Individual Model)	2 (Generalized Model)	3 (Dynamic Ensemble Selection)	4 (Dynamic Ensemble Selection)	5 (Individual Model)	6 (Dynamic Ensemble Selection)	7 (Dynamic Ensemble Selection)
2	88.74	93.98	89.51	96.31	94.56	91.26	92.82
4	96.12	99.28	96.55	97.99	97.55	96.26	96.98
5	86.35	74.17	79.7	85.24	85.98	81.55	90.04
6	90.51	83.8	79.27	86.24	93.21	85.37	91.29
18	92.84	98.24	94.82	97.36	97.47	91.85	96.04
20	99.53	63.2	75.82	82.48	94.86	81.54	99.65
21	94.41	43.75	61.04	84.44	87.9	74.87	94.41
22	88.09	80.98	79.48	80.66	88.33	84.32	87.38
Average	92.07	79.67	82.02	88.84	92.48	85.87	93.57

We found that 20% of the data from each subject is sufficient to train a reasonably good model for engagement assessment (Scenarios 1 & 5), which performs better than the generalized model (Scenario 2). The dynamic ensemble selection strategy boosts the performance of the model even without data from the individual (Scenario 3). If we use data from a test subject to dynamically select committee members, the performance can be improved (Scenarios 4, 6 and 7).

5   DISCUSSIONS

In this research, we have explored two different model individualization techniques for engagement assessment in aviation environments. Future tasks include enhancing the model with additional sensory information (flight technical data and ECG, for example), further improving the model individualization technique and continuing to further verify and validate the real-time assessment technique with additional participants' data.

ACKNOWLEDGMENTS

This project was funded by the NASA (Contract No: NNX10CB27C). We thank Dr. Alan T. Pope, Mr. Chad L. Stephens, and Dr. Kara Latorella for their comments and suggestions as we performed this research.

REFERENCES

Berka C., D. J. Levendowski and M. N. Lumicao et al. 2007. "EEG Correlates of Task Engagement and Mental Workload in Vigilance, Learning, and Memory Tasks," Aviation, Space and Environmental Medicine, Vol. 78, No. 5, Section II, May 2007

Boucsein W. and R. W. Backs, 2000. "Engineering psychophysiology as a discipline: historical and theoretical aspects," in Boucsein, W., Backs, R.W. (Eds.), Engineering Psychophysiology: Issues and Applications. Erlbaum, London.

Ellis K., 2009. "Eye Tracking Metrics for Workload Estimation in Flight Deck Operations, Master Thesis, University of Iowa, 2009.

Giacinto G. and F. Roli 2001. Dynamic Classifier Selection Based on Multiple Classifier Behaviour. Pattern Recognition 34, 179-181

Hockey R., 2003. "Operator Functional State: The Assessment and Prediction of Human Performance Degradation in Complex Tasks," NATO ASI SERIES Vol 355

Li F., J. Li and F. McKenzie et al. 2011. "Engagement Assessment Using EEG signals," MODSIM World Conference, 2011, Virginia Beach, VA

Li J., J. Yao and N. Petrick et al. 2006. "Hybrid Committee Classifiers for a Computerized Colonic Polyp Detection System," Medical Imaging 2006: Image Processing, vol. 6144, pp. 1701-1709, 2006.

Marwitz, M. and G. Stemmler, 1998. On the status of individual response specificity. Psychophysiology, 1998. 35(1-15).

Olofsen E., H. Van Dongen and C. G. Mott et al. 2010. "Current Approaches and Challenges to Development of an Individualized Sleep and Performance Prediction Model" The Open Sleep Journal, Volume: 3, Pages: 24-43

Rajaraman, S., A. Gribok and N. J. Wesensten, et al. 2009. An Improved Methodology for Individualized Performance Prediction of Sleep-deprived Individuals with the Two-process Model. Sleep. 2009 October 1; 32(10):1377-1392

Stevens R. H., T. Galloway, and C. Berka 2007. "EEG-Related Changes in Cognitive Workload, Engagement and Distraction as Students Acquire Problem Solving Skills," Proceedings of the 11th international conference on User Modeling.

Trejo, L. J., R. Kochavi and K. Kubitz 2005. "EEG-based Estimation of Cognitive Fatigue," Proceedings of SPIE: Bio-monitoring for physiological and cognitive performance during military operations. v5797. 105-115.

Zhang, G., R. Xu and W. Wang, et al. 2009. "Individualized Cognitive Modeling for Closed-Loop Task Mitigation, ModSim World Conference, Virginia Beach, VA

Zhang, G., W. Wang and A. Pepe, et al. 2011. "A Systematic Approach for Real-Time Operator Functional State Assessment," MODSIM World Conference, Virginia Beach, VA

Section VII

Safety and Ethics

The Applications of a Human Factors-Based Safety Management Model in Aviation Industry

Yu-Lin Hsiao[1]   Colin G. Drury[2]

[1]Chung Yuan Christian University
Chung Li, Taiwan
yhsiao@cycu.edu.tw
[2]State University of New York (SUNY) at Buffalo
Buffalo, USA
drury@buffalo.edu

ABSTRACT

At present the move towards modern Safety Management Systems (SMS) is being advocated in aviation industry. Although SMS emphasizes the integration of human factors and quality management process to contribute to safety satisfaction, the practical techniques to do so still leave as an open question. In this paper, we explore several aspects of how to integrate and apply a human factors-based model, Human Factors Analysis and Classification System – Maintenance Audit (HFACS-MA), within the SMS function to enhance the capability of safety management. The proposal includes aspects such as information fusion of various data sources, sensitivity analysis, and root cause analysis. Issues regarding improvement of the current risk analysis matrix and implementation of Six Sigma are also discussed. All of these practical applications can be integrated to establish a comprehensive safety risk management approach which incorporates human factors concepts in multiple features for SMS. Extensions of the work to domains beyond aviation maintenance, and indeed beyond aviation, are also appropriate as many industries use proactive measures as obligatory tools for their safety management.

Keywords: SMS, HFACS-MA, Sensitivity Analysis, Root Cause Analysis, Risk Matrix, Six Sigma

1   INTRODUCTION

In our lifetime we have seen the change from basing safety on accident investigation to using more proactive safety measures, for example safety audits. These proactive functions are carefully designed to focus attention on what their designers expect to be the detectable precursors of accidents or incidents, such as work environment, manual contents and training performance. It is well accepted by the industry that if we could effectively fix those existing or potential problems found from proactive techniques such as audits, their remediation will contribute to the risk mitigation of erroneous activities. This will eventually benefit future accident or incident prevention.

However a causal relationship between proactive measures as precursors and accident/incident as safety performance must be valid to support the above assumption. In fact, it is crucial to assure that safety audits perform as they are intended, i.e. as a proactive measure of the inherent safety of an objective system. Do the findings of maintenance audits really predict the future safety performance of a Maintenance, Repair and Overhaul (MRO) organization or an airline's maintenance department? Amazingly, this question does not appear to have been answered for aviation maintenance field or indeed any other system where proactive measures are in use.

In our previous study (Hsiao 2011), we tested the above question using existing maintenance audits, and found that audit findings, when suitably analyzed via human factors/ergonomics (HF/E) methods, could indeed predict the following month's safety performance for two different airlines' MRO departments. A human factors-based classification model, Human Factors Analysis and Classification System – Maintenance Audit (HFACS-MA), was developed to analyze safety audit findings in aviation maintenance. Its development was based on the original HFACS model (Shappell and Wiegmann 2001) and Reason model (Reason 1990), modified to fit the aviation maintenance domain and to include specific organizational and supervisory influences. In fact, when HFACS-MA had been developed, it was found that none of the items referred exclusively to aviation or to maintenance, so that future expansion to other domains should be quite straightforward.

In HFACS-MA, errors are classified on a three-level hierarchy, with the top level having four main factors progressing from what Reason (1990) calls the "sharp end" (*Unsafe Acts*) through immediate influences (*Preconditions for Unsafe Acts*) to more distant influences (*Unsafe Supervision* and *Organizational Influences*) as Fig 1. Each lower level gives progressively more detail of the errors. Using Unsafe Acts as the example, the middle level has two factors: unintended actions (*Error*) and intended but wrong actions (*Noncompliance*). It should be noted that in this paper we use "Noncompliance" replacing the wording "Disobedience" in our previous study, to allay misconceptions in the MRO industry. However, its definition remains the same.

Both the HFACS-MA model's measurement reliability and prediction validity have proven statistically significant (Hsiao 2011). In the reliability test, the Kappa values were at least acceptable for levels of analysis and the specific factors. For Unsafe Acts, Preconditions and Unsafe Supervision, Kappa was highest at about 0.8 (on a scale of 0 to 1), while for Organizational Influences it was somewhat lower at about 0.5. We utilized a Neural Network method to establish the prediction model, and the correlations predicting future safety performance from audit findings were 0.76 and 0.59 for two independent airlines, both statistically significant at $p < 0.01$.

Based on our study, the classifications of HFACS-MA could be used as reliable and valid predictors of an MRO's future safety performance, i.e. monthly incident rate. The result not only provided quantitative evidence for the causality between human error and safety performance from the social science perspective, but was also the first validation of an audit analysis model against a widely accepted outcome measure of safety. In addition to the Neural Network analysis, we also tested multiple regression models with the same data, but their predictive power was not as good as the Neural Network models. This was expected when attempting to use non-normally distributed data to predict potentially non-linear effects. Note that prior incident data did not contribute to the quality of prediction: predictions based on audits were far stronger. Note also that use of the HFACS-MA was required to achieve valid prediction: merely counting audit findings without analyzing their human factors implications did not predict future outcomes.

2   APPLICATION AND INTEGRATION WITH SMS

Since we have proven that prediction of future safety performance from audit data is possible in MROs now, the next question is how to incorporate this academic achievement into on-going operations in the aviation industry. At present the move towards modern Safety Management Systems (SMS) is being advocated and indeed mandated by the International Civil Aviation Organization (ICAO). The idea behind SMS is that data from various sources need to be integrated to provide a comprehensive view of the safety of a system, leading to actions based on more than mere response to the latest audit or incident. Although SMS emphasizes the importance of human factors (includes human performance and organizational factors) as the foundation of safety risk management and the necessity to integrate into quality management techniques and processes to contribute to the achievement of safety satisfaction (p. 7-10) (ICAO 2009), it still leaves the human factors concepts in the theoretical phase. A clear guideline or a practical method to apply HF/E in operational aspects is still missing for the industry and waiting for researchers to fulfill. In the following section, we will introduce several proposals for how to integrate our HF-based model within the SMS function and more.

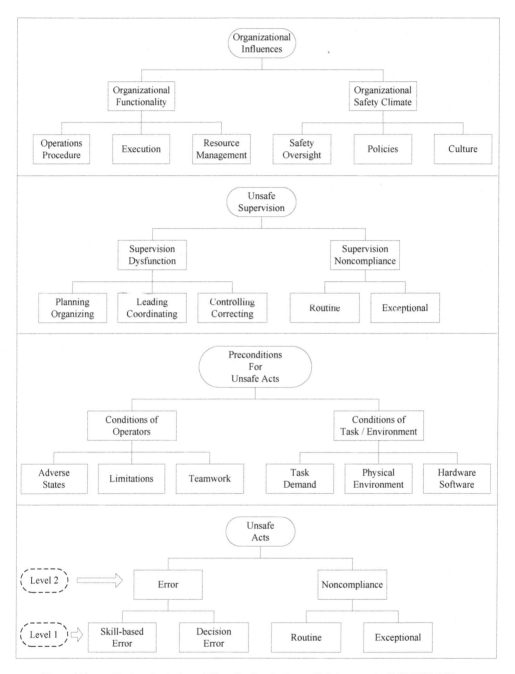

Figure 1 Human Factors Analysis and Classification System – Maintenance Audit (HFACS-MA)

2.1   Information Fusion

First, in addition to the safety audit data used in our study, there are other proactive data sources already being collected in the aviation maintenance field, e.g. the Aviation Safety Action Program (ASAP), the Maintenance Error Decision Aid (MEDA), the Line Operations Safety Assessment-Maintenance (LOSA-Mx), Self Audit or the Continuing Analysis and Surveillance System (CASS). All of these existing safety functions have provided effective improvements to flight accident prevention. And one of the common features of these data is that they are generally, at least in part, related to human factors. Thus if these other sources could be integrated with the audit findings and be analyzed by the HFCAS-MA simultaneously under the same classifying criteria, the result would provide a more comprehensive picture of the human error status in a MRO. Different weights might be applied to different data sources based on the consideration of importance or severity of the data, or upon their predictive validity. Furthermore, the integration of human error related data would also improve the perfection of prediction capability of future safety performance in some ways.

If we consider a MRO or an airline as a medical patient, the output of this kind of information fusion would be similar to an on-condition health report following a comprehensive physical checkup, revealing systematic weaknesses or potential hazards. Modern Safety Management Systems indeed require such an incorporation of data from different sources to detect existing or potential problems. In Fig 2, the frequencies of the error categories provides a continuing check of the annual human error status, and can help by pointing to other data sources that could be usefully integrated into the SMS. For example, human error reports could be incorporated with the internal monitor and correct process of maintenance deficiencies of airlines' CASS to help illuminate that one specific area.

2.2   Sensitivity Analysis

Although a neural network provides a good platform to solve complicated problems, it still leaves open the question of the precise relationships between input and output. Even though successful convergence of estimation results implies the existence of specific input-output relationships; however these remain implicit features of the network, i.e. a black box. To go further implies that we should conduct a sensitivity analysis, which provides an approach in gauging certain aspects of these inherent input-output mapping 'rules' (Azoff 1994). In sensitivity analysis, researchers examine the sensitivity of the key output to all initial input values, to be sure knowing at least the largest of these sensitivity coefficients (Werbos 1994). The measure of effects of the input variations on the output quantities can provide us an indication of the importance of the inputs. The result of sensitivity analysis could be utilized to benefit other subsequent analysis activities (e.g. root cause analysis). For example, airlines should be more attentive to factors with higher sensitivity results.

434

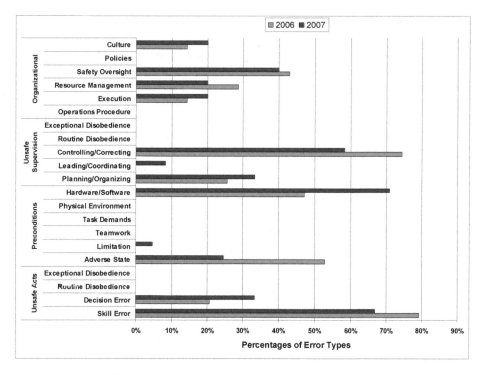

Figure 2 Example of annual checkup report of human error status

2.3   Root Cause Analysis

Even though this valid prediction of future incident rate can be used to point out the risk level in the near future, the analysis still leaves a question to managers: what issue or potential problems they should be aware of? One alternative to answer this question is root cause analysis. For instance, in Fig 3 while the predicted incident rate represents a high predicted incident rate at the seventh month, root cause analysis of the underlying data sources should be conducted to evaluate the specific influences or contributions of each related error input to the output. The result of sensitivity analysis can also be applied to assist the evaluation process to determine the importance of each error category or the priority of intervention and resource allocation. A quantitative warning message and detailed root cause can be sent to all related managers to help maintain situation awareness and to help develop improvement actions to mitigate the potential risk of a high incident rate. Briefly, the prediction of future incident rate should not only be treated as a monthly risk indicator, but also be considered as a decision-making aid for direct human error management.

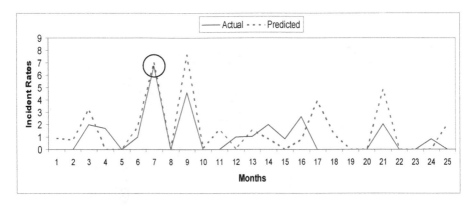

Figure 3 Sample prediction results showing the timing to conduct a root cause analysis

2.4   Risk Analysis

In Risk Analysis, the risk level is usually determined by both the probability and severity magnitude (or called consequence), usually in a multiplicative relationship as shown in Fig 4. However, typically both are subjective interpretations of risk level (Cox 2008). Our study provides a more objective and quantitative method for risk evaluation: overall accuracy of prediction (R^2) represents the likelihood of risk and severity level can be determined by the future incident rate. Alternatively, the measure of the accumulated incident rate using "the accident pyramid" (Heinrich 1950; Bird 1975) or the "worst credible effect" (Pasquini, Pozzi et al. 2011) could help determine the severity outcome. Using the prediction result of two independent airlines as an example, the R^2 of Airline A was 0.58 and 0.35 for Airline B; we could adopt the likelihood as "Probable" for Airline A and "Remote" for Airline B respectively. For the severity, it is practical to set up at least two complementary conditions to decide its rank: the historical average of monthly incident rate and the cumulative annual limit of incidents. For example, the condition for Major level could be set as "exceed the three years average of monthly incident rate".

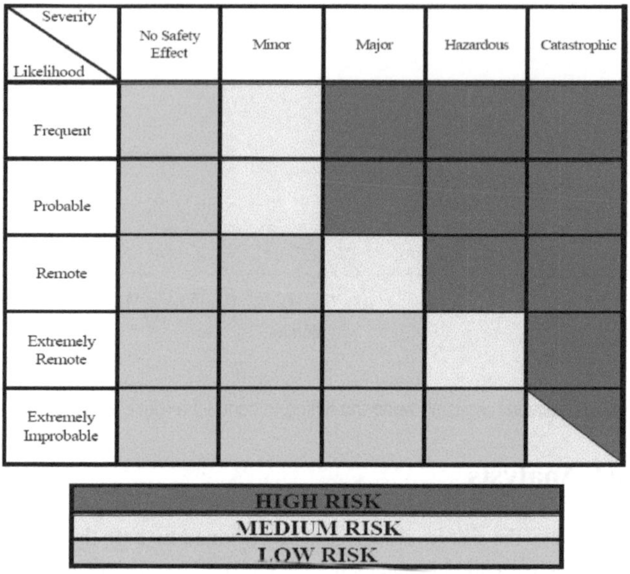

Severity / Likelihood	No Safety Effect	Minor	Major	Hazardous	Catastrophic
Frequent					
Probable					
Remote					
Extremely Remote					
Extremely Improbable					

HIGH RISK
MEDIUM RISK
LOW RISK

Figure 4 Sample of Risk Matrix (FAA 2007)

2.5 Six Sigma

From a quality management perspective, Six Sigma has become a popular alternative for safety management since one of its purposes is to decrease the human error rate (e.g. defects per million opportunities). According to the definition of Six Sigma's DMAIC process (Define, Measure, Analyze, Improve, Control), our HF-based model provides the first three: clear human error definitions, reliable and continuous data collection sources, and a validated analysis method. If we use root cause analysis, the Improve step of DMAIC is also covered. Our inclusion of management and supervision functions in the analysis helps provide logical means to implement the final category: Control. It should be noted that the prediction result could also be considered as a performance measurement.

Human error elimination is an ongoing goal of quality management. But one of the difficulties of implementing Six Sigma in the aviation industry is the low frequency of specific human error behavior followed by considerable severity. This situation increases the complexity of quality control. One advantage of our HF-based model to solve this issue is information fusion which would combine and analyze every singular but similar event or finding into same error category, i.e. noncompliance, to provide a continuous and sufficient database for quality management purposes.

3   DISCUSSION

As has been repeated many times, about 70% of aviation accident causes are related to human factors. Beyond conducting human factors audits or evaluation, what other practical ways can we do to prevent human error from happening? Following the industry trends from passive safety improvement (e.g. accident investigation) to proactive measures (e.g. safety audits), our study provides an advanced alternative to monitor safety performance in a timely manner from a human factors viewpoint.

In fact, most airlines or MROs have met the requirements or philosophy of the SMS in their own ways, and many safety databases have been integrated to provide information to assist managers' decision-making. However, detailed analysis or explanation of the connections between different data sets is still incomplete, especially from HF/E perspectives. HFACS-MA provides a valid and necessary intermediate step between collecting findings and implementing effective countermeasures.

Many different audit systems exist in the aviation industry, e.g. ATOS, but we do not know whether other audit systems would have been just as effective as the one use in our analysis. Also, we have not explored the model and data in enough detail to determine whether any aspects of the audit fail to contribute to predicting the future incident rate, and thus could be omitted. Are standard audits used by regulatory authorities the best ones to use, or should airlines and MROs develop their own audit system to assure safety? Clearly, there are questions that need to be answered subsequent to our positive findings if our techniques are to become useful and standard tools for the aviation industry.

All of these practical applications help establish a comprehensive human factors risk management approach for SMS in aviation maintenance. The methodology exists now, but can be further developed for use by any airline or third-party MRO to check whether *their* proactive measures or audit results predict *their* future safety performance. Note that the data we use is already being collected, so that we can use existing data from different sources in a more integrated manner, as required by SMS. Note also that HFACS-MA proved to be non-maintenance-specific and indeed non-aviation-specific, so can be readily extended to other domains.

4   CONCLUSION

The HFACS-MA classification system worked well and is a valid contribution to human factors in airline safety independent of the prediction model. The prediction worked for two specific airlines using the data already being collected by a regulatory authority (which had requested to remain anonymous). In this paper, we discuss specific ways to integrate HFACS-MA into SMS and to explore the possibility of enhancing modern safety risk management methods. Although the study currently is only in the aviation maintenance field, extensions of the work to domains beyond aviation maintenance, and indeed beyond aviation, are also appropriate. Many industries use proactive measures such as audits as obligatory tools for their safety management.

5   REFERENCES

Azoff, E. M. (1994). Neural network time series forecasting of financial markets, John Wiley & Sons, Ltd.

Bird, F. E. (1975). Control total de pérdidas. New Jersey, Consejo Interamericano de Seguridad.

Cox, A. L. (2008). "What's Wrong with Risk Matrices?" Risk Analysis 28: 497-512.

FAA (2007). AC 150/5200-37, Introduction to Safety Management Systems (SMS) for Airport Operators.

Heinrich, H. W. (1950). Industrial Accident Prevention A Scientific Approach. New York, McGraw-Hill Book Company, Inc.

Hsiao, Y. (2011). An Audit-Based Prediction Model for Aviation Maintenance Safety. Industrial and Systems Engineering, University at Buffalo, The State University of New York. Ph.D.

ICAO (2009). Safety Management Manual (SMM), ICAO.

Pasquini, A., S. Pozzi, et al. (2011). "A critical view of severity classification in risk assessment methods." Reliability Engineering & System Safety 96(1): 53-63.

Reason, J. (1990). Human Error. New York, Cambridge University Press.

Shappell, S. and D. Wiegmann (2001). Applying the human factors analysis and classification system (HFACS) to the analysis of commercial aviation accident data. 11th International Symposium on Aviation Psychology, The Ohio State University, Columbus, OH.

Werbos, P. J. (1994). The roots of backpropagation: from ordered derivatives to neural networks and political forecasting, John Wiley & Sons, Inc.

CHAPTER 42

"I did something against all regulations": Decision making in critical incidents

Katherine L. Plant and Neville A. Stanton

Transportation Research Group
Faculty of Engineering and Environment, University of Southampton, SO17 1BJ
k.plant@soton.ac.uk

ABSTRACT

This paper explores naturalistic decision making when dealing with critical incidents in the cockpit. The aim of the paper was to examine the extent to which the Perceptual Cycle Model is evident in the decision making process. The Critical Decision Method was utilized to provide critical incident data from five helicopter pilots. The data were thematically analyzed according to the elements of the Perceptual Cycle Model and a phase-of-incident coding scheme. Further analysis explored the diagnostic decision types pilots were employing. A variety of different decision types were found and preliminary differences were identified between appropriate and inappropriate decision making. Additional data analysis with a larger sample is intended to verify the claims made here and answer the questions generated.

Keywords: Diagnostic decision making, perceptual cycle model, critical incidents.

1   INTRODUCTION

It is argued that in order to establish a causal explanation of behavior in the cockpit it is necessary to understand why a pilot's actions and assessments made sense to them at the time (Dekker, 2006). Whilst incidents usually arise from a combination of systemic, technical and human factors, what stands out is the human element, usually as there is a desire to ascribe blame (Woods et al., 2010). The emphasis on the human element however, is not without reason. Pilot judgment in decision making during the handling of emergency situations is usually the deciding factor as to whether an incident will become an accident (McFadden and Towell, 1999). Similarly, Stanton et al. (2010) argue that success is ultimately contingent on efficient decision making. The analysis of performance however, has traditionally given limited consideration to the processes leading to it and the context in which it occurred (Maurino, 2000). Dekker (2006) argues that understanding decision making requires an appreciation of the work situation (context) and the acknowledgement that people change this context by the actions they take; this evolving situation can also alter behavior. This is fundamentally the explanation provided in the Perceptual Cycle Model.

The Perceptual Cycle Model (PCM: Neisser, 1976) presents the view that human thought is closely coupled with a person's interaction in the world; both informing each other in a reciprocal, cyclical relationship. World knowledge (schemata) leads to the anticipation of certain types of information; this then directs behavior (action) to seek out certain types of information and provides a way of interpreting that information. The environmental experience (world) results in the modification and updating of cognitive schemata and this in turn influences further interaction with the environment. Schemata are a fundamental part of the model; these are defined as organized mental patterns of thoughts or behaviors that help organize world knowledge (Bartlett, 1932).

The PCM provides a human-in-the-system approach to understanding decision making; individual schemata are considered in the wider systemic context in which they occurred. Smith and Hancock (1995) argue that the usefulness of the PCM explanation lies in the interaction between operator and environment, which is an appropriate approach to take when analyzing socio-technical systems such as those in the aviation domain (Maurino, 2000). Owing to the strengths of the approach, recent research has applied the principles of the PCM to incidents with a high decision making component, including the Ladbroke Grove rail crash (Stanton and Walker, 2011), the Stockwell shooting (Jenkins et al., 2011) and the Kegworth plane crash (Plant and Stanton, 2012). The approach taken here is to analyses decision making during critical incidents within the context of the Perceptual Cycle Model (Neisser, 1976).

2   METHOD

2.1   Design and Participants

This study is exploratory research to determine whether the PCM can be used to understand decision making in the cockpit. The study makes use of five case studies. Case studies relating to critical incidents are a way to increase understanding of cognitive processes through the production of rich, qualitative data (March et al., 2010). A case study approach is appropriate for exploratory research, particularly in the context of the aviation domain, in which access to large samples is difficult.

The participants were male helicopter pilots aged between 41 and 50 years old.

They were voluntarily recruited through an advert for participants placed on the British Helicopter Association website and word-of-mouth. Face-to-face interviews were conducted with each pilot at their places of work. The interviews lasted approximately an hour. All participants were relatively experienced; flying hours ranged between 1150 and 4700 hours (mean = 3270). Ethical permission for the study was granted from the Research Ethics Committee at the University of Southampton.

2.2   Critical Decision Method Procedure

The Critical Decision Method (CDM; Klein et al., 1989) was utilized here in an attempt to elicit schematic processing that underpins the PCM. As argued by Walker et al. (2011), *"gaining insight into mental representations...is experimentally and conceptually challenging"* (p. 879). Schemata are mental constructs and therefore cannot be directly measured but only inferred through the empirical consequences of them. As part of ongoing research, a review of methods for schema elicitation was conducted. It is beyond the scope of this paper to discuss the review in depth; however the CDM was rated highly on the method criteria and a previous case study has tested the proof of concept (Plant and Stanton, 2011).

Knowledge elicitation is achieved with the CDM through the use of cognitive probes to understand decision making during non-routine incidents via a retrospective semi-structured interview. The interview schedule for this study included demographic questions and the probes provided by Klein and Armstrong (2005). The probes cover a range of decision making areas including situation awareness, expectancy and planning.

The nature of the interview was explained to each participant (i.e. they were required to recall an incident and the interview would last approximately an hour depending on length of answers provided). In line with Klein and Armstrong's (2005) recommendations it was emphasized to the participants that answers should state what was actually done, rather than what should have been done. The participants were asked to provide a high level overview of a non-combat incident they had been involved with during which they were the primary decision maker. A critical incident was defined as being 'a non-routine or un-expected event that was highly challenging and involved a high workload'. The incident was broken into distinct phases, physical and mental events that occurred during each phase were recorded and finally the CDM probes were asked.

3   DATA ANALYIS

The method of data analysis employed here was based upon the principles of thematic analysis; the process of classifying text data into meaningful themes. The coding scheme was based on the three categories of the PCM; schema, action and world. Additionally, seven generic incident phases were identified and the data coded according to these: *(1)* pre-incident *(2)* onset of problem *(3)* problem identification *(4)* immediate action *(5)* decision making *(6)* subsequent actions *(7)* incident containment. To assess the reliability of this coding scheme, four colleagues coded the data from one pilot according to the incident phase coding scheme. Agreement between the four coders and the criterion coder averaged ninety-three percent. Literature suggests that an acceptable level of agreement is generally defined as eighty percent or above when reliability is calculated as the number of agreements divided by total number of segments (Marques and McCall, 2005).

Klein (1992) proposed eight diagnostic decision making strategies. Six of these are shown in table 1. Story building is excluded as it did not represent any of the data in the case studies (it is a form of mental stimulation) and Klein's original two decision types of step-by-step belief updating and global belief updating were combined into one type; belief updating. This is based on Klein's (1992) assertion that it is not possible to distinguish between the two belief updating types, particularly in retrospective accounts. This paper seeks to explore the types of diagnostic decision making evident in the CDM data as a way to understand critical incidents in the cockpit. The data from each case study which had previously been coded as relevant to the decision making phase was coded in terms of diagnostic decision making type.

4   RESULTS

Figure 2 represents the average spread of data over the seven incident phases and the average spread of the three PCM elements over the incident phases. For example, the majority of data related to the decision making phase and data coded as schema was generally found in this phase.

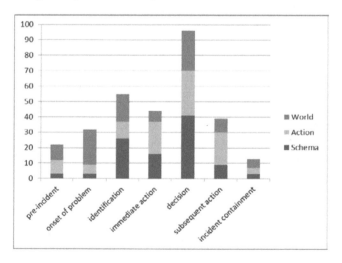

Figure 2. Occurrence of data in each phase of incident according to element of PCM.

Each pilot made one primary decision, however, each employed a number of decision points, using a variety of decision types, to reach this overall decision. Nearly three quarters of the decision types were defined as feature matching (48%) or analogical reasoning (23%). Table 1 describes the decision types, the number of times each type was recorded in the data and an example of each from the pilots' transcripts.

5   DISCUSSION

The majority of the data was coded as 'decision making', which is initially not surprising considering that the data was gathered within the context of a critical decision making interview; the interview schedule was therefore biased to eliciting data relating to decision making. However, standard operating procedures (SOPs) and the rigorous training practices in place for pilots are there to remove this need for extensive decision making by offering a limited set of choices (Simpson, 2001). It would appear however, that when faced with a critical incident, pilots still engage in a substantial decision making process. Therefore the type of decision making that occurs during critical incidents was explored further.

The study sought to establish whether Klein's (1992) diagnostic decision types were an appropriate classification of the decision types found in critical incident data and whether these types shed light on the nature of individual decision making during critical incidents. With the exception of story building, all of the diagnostic decision making types were found in the decision making data and so it would appear that pilots adopt a combination of decision making strategies. Feature matching was most common. This was also found by Klein (1992) in his original study of anti-air warfare decision making in the maritime domain. The second most commonly used strategy in this study was analogical reasoning; this was not reflected in Klein's results. This finding suggests pilots actively employ the strategy of matching to prior cases. This is already acknowledged in the aviation domain as a successful decision making strategy in the form of case-based remembering (O'Hare et al., 2010).

Stanton et al. (2010) also utilized Klein's decision making types in their analysis of distributed decision making in multihelicopter teams. They noticed an interesting dichotomy between formal rules and regulations versus prior experience. The same dichotomy was noticed here in data from Pilot A and D. For example, Pilot A used his experience (of flying in mountainous terrain and knowledge of how crucial the mission was) to make a decision (to continue down through cloud without radar cover) which he stated was *"something that was against all sorts of regulations"*. Therefore in this instance analogical reasoning outweighed feature matching elements of the situation (cloud cover) against known procedures. Interestingly, Pilot A was the only pilot to employ the 'information seeking' strategy. It can be speculated that it was important for him to seek confirmation for the decision he knew was against rules. Pilot D's decision making was influenced by his inexperience, which resulted in him not doing something that he should have done (climb into cloud). Pilot D used feature matching (i.e. bad weather, poor visibility, legal height limits reached) to determine that the the only decision left was to climb into cloud and perform an 'IF abort'.

Table 1. Six diagnostic decision types, occurrence in the data and examples from the case studies (adapted from Klein, 1992).

Decision Type	Description	Number recorded in data (%)	Example
Feature matching	Situation is judged by matching features of the situation against features stored in memory about previous situations or prototypes	15 (48%)	"Amber caution light on, you check the temperature and pressure gauge and they were normal, so I wasn't losing pressure and the temperature wasn't increasing so that puts a bit of safety blanket on and there was no vibration or any other issues with the aircraft" (Pilot C)
Holistic matching	Involves generating inferences about the situation that lead to a meaningful whole in terms of coverage and coherence	4 (13%)	"… it wasn't a case of picking an ideal field. It was the case of picking the best field directly in front of me in a mile range" (Pilot E)
Information seeking	Seeking more information regarding the situation when unable to make a clear diagnosis	1 (3%)	"…no gap down through cloud…talking to the mobile ops. guy on the ground and was told he could see into a certain valley North of where we were, but couldn't see the top of the mountains. Also trying to use the maps to see if the GPS is as accurate as I need it to be" (Pilot A)
Belief updating	Way in which people change their belief judgments as they become aware of new information	2 (6%)	"There is a point when you realize you can't see anything. The weather is worse than forecast, still couldn't see lights" (Pilot D)
Mental simulation	Inferring a causal sequence, imaging how a course of events might have unfolded. Where data is integrated as opposed to just checked off.	2 (6%)	"Neither of us had done it before. Didn't really know whether it was going to work …whether we were going to get shouted at…annoy someone. Cold… might go into icing" (Pilot D)
Analogical reasoning	Retrieval of a match to a prior case that can serve to identify the dynamics of the situation	7 (23%)	"…The severity of the vibration, I had experienced vibration in other aircraft and knew what was normal and what was abnormal. This was extremely abnormal" (Pilot E)

Pilot D stated he had *"trained for and practiced for* [IF abort]*... simulated dozens of IF aborts in training, there is an SOP for it"*. However he had never actually performed one and used mental simulation to infer what the outcomes might be, including *"get shouted out by ATC"*, *"go into icing"* or *"whether it was going to work the way it said it would in the book"*. Even though the pilot had trained for this situation, he ignored the rules and regulations due to a lack of experience with actually performing the manoeuvre.

The examples presented here highlight how both pilots' decision making was influenced by experience at the expense of following rules and regulations (in two very different ways). This dichotomy was originally highlighted by Stanton et al. (2010) and therefore this repeated finding suggests there is conflict in the decision making process between rules and regulations versus experience. Further research is required to explore this further; however rules, regulations and SOPs are intended to limit the decision making choices available (Simpson, 2001). This research has highlighted the fact decision making may not be as clear cut as simply following rules or SOPs, even when an appropriate response is prescribed. Simpson (2001) states that the limited choices afforded by the use of rules and regulations are intended to reduce cognitive expenditure. However, cognitive expenditure is likely to be greater if faced with conflict between those regulations and personal experience, especially in an emergency situation. Training and procedures need to reflect this conflict. Simpson (2001) argues that little has been done to formulate techniques and strategies to teach intuitive decision making. Studies such as this one highlight the types of strategies pilots' employ and areas of conflict, which can be used as the foundation for decision training and aiding techniques, in order to develop specific components and skills for effective decision making.

Inter-rater reliability for the coding of data into decision types was only forty-seven per cent (one colleague coded fifty per cent of the data). This has identified issues within the classification scheme for decision types, in that they are not mutually exclusive. From the description of each type presented in table one; it is clear that feature matching, holistic matching and analogical reasoning all refer to prior or pre-defined cases in some form. For example, the inter-rater coder took analogical reasoning to be any decision based on literal previous experience (i.e. having landed in fields before). The criterion coder however, took analogical reasoning to be either literal past experience or prior experience gained from vicarious experience such as discussing scenarios with colleagues or the experience gained from training. This discrepancy is one example of the ambiguity in the classification scheme. Additional issues include the fact that 'information seeking' appears to be an action based on other decision making types, rather than an actual decision type and mental simulation is hard to identify as it impossible to mentally simulate without relating to some form of, or lack of, experience. Regardless of the issues associated with the coding scheme, the main finding still stands; pilots engage in a variety of decision making types throughout a number of decision points when dealing with critical incidents. As part of on-going research, additional critical decision interviews were undertaken (n=20) and therefore future work intends to analyze the data generated from these in terms of decision making types in light of the issues noticed in this exploratory study. This may lead to the development of a new classification scheme to fully account for decision making during critical incidents. Such a scheme will be strongly process orientated i.e. focusing on the perceptual-action processes involved with decision making. An understanding of the processes involved helps understand why decisions are made which in turn increases understanding when things go wrong and has implications for training and aiding decision making.

This exploratory study has identified numerous paths for further research. For example, one way of treating the data in this study was to see if there were differences in the decision types utilized by pilots who made appropriate versus inappropriate overall decisions (this was defined by the answer given to the question 'were you uncertain about the appropriateness of your decision?' 'no' was classified as an appropriate decision, 'yes' as an inappropriate decision). Two pilots were classed as making appropriate decisions and three made inappropriate ones. The key difference noticed was that the decision types of information seeking, belief updating and mental simulation only occurred in inappropriate decision data. These differences can be explained in terms of the variants of Klein's (1993) Recognition Primed Decision (RPD) model. The RPD approach models decisions for which alternative courses of action are derived from the recognition of critical information and prior knowledge. The variants of the model concern decision complexity. In the simplest case a situation is recognized and the obvious reaction is implemented (e.g. Pilots C and E in their appropriate decision of landing immediately). A more complex model involves the decision maker having to mentally simulate the reaction to evaluate it (e.g. Pilot D and A when deciding whether to fly into cloud and pilot B when deciding whether to go to an airport or ditch on the beach. Pilot B and D stated they would make different decisions if the situation arose again and Pilot A's decision was inappropriate in that it violated regulations). The decision types which were utilized by the pilots who made inappropriate decisions (i.e. mental simulation, belief updating and information seeking) are evident in the more complex RPD model. The RPD approach suggests that effective decision making is most likely to occur when a pilot can match what they are currently experiencing to what they have previously experienced or learnt (Simpson, 2001). This was found to be the case here and whilst only modest assertions can be made, based on the exploratory nature of the study, the finding has implication for the way pilots are trained to ensure that the opportunity for experiential learning (literal or vicarious) is optimized. The differences between appropriate/inappropriate decisions or successful/unsuccessful decisions will be an interesting avenue to explore with a larger sample size.

The thematic analysis using the PCM coding scheme showed that pilots engage in multiple iterations of the PCM when dealing with critical incidents. The PCM reflects the role of schemata, suggesting that pilots are utilizing past experience and integrating them with the environmental information they are presented with in order to produce appropriate actions and decisions. It is widely acknowledged that people use prior encounters to deal with current judgment decisions (Simpson, 2001). However, the PCM reflects the interaction between these prior encounters and current environmental experience and the modifying effect each can have on the other. Therefore, by exploring the use of the PCM in decision making scenarios, a distributed approach to decision making can be conceived in which the decision making neither resides in the individual or the system, but in the interaction of the two. This approach fits with the wider trend of the increased emphasis placed on the collective behaviour of systems as a whole, as opposed to the behaviour of individuals within that system (Stanton et al., 2009). This distributed approach has previously been conceived with regards to situation awareness by Stanton and colleagues (2006; 2009). However, it would be interesting to apply similar principles to general decision making and thus the consequences of decision making such as the study of systems error.

6   CONCLUSION

To conclude, this exploratory study demonstrated that pilots engage in the perceptual-action cycle when dealing with critical incidents in the cockpit. This preliminary investigation paves the way for conceiving a distributed model of decision making in the cockpit. In providing a distributed model of decision making casual explanations of the consequences of the decision making, such as error, will be achieved as an understanding of why a pilot's actions and assessments made sense to them at the time will be gained.

Furthermore, the study looked at diagnostic decision types when dealing with the incidents. It was found that a variety of decision types were employed, preliminary findings suggest that conflict occurs between following regulations versus experience and that inappropriate versus appropriate decision outcomes are associated with utilizing different decision types. The value of research of this nature lies in the potential for improving the instruction of NDM to optimize the occurrence of appropriate decision making. As expected with exploratory studies, more questions were raised than answered. Additional data analysis with a larger sample is intended to verify the claims made here and answer the questions generated.

ACKNOWLEDGMENTS

The authors would like to thank the pilots who took part in the research and provided these interesting case studies and the colleagues who assisted with inter-rater coding assessments.

REFERENCES

Bartlett, F.C. 1932. *Remembering: A Study of Experimental and Social Psychology.*
      Cambridge University Press: Cambridge.
Dekker, S. 2006. *The Field Guide to Understanding Human Error.* Aldershot: Ashgate.

Jenkins, D., Salmon, P. M., Stanton, N. A., Walker, G. H. & Rafferty, L. 2011. What could they have been thinking? How sociotechnical system design influences cognition: a case study of the Stockwell shooting. *Ergonomics,* 54 (2), 103-119.

Klein, G.A. 1992. *Decision making in complex military environments* (Prepared under contract N66001-90-C-6023 for the Naval Command, Control and Ocean Surveillance Centre, San Diego, CA). Fairborn, OH: Klein Associates.

Klein, G.A. 1993. A recognition primed decision model of rapid decision making. In Klein, G.A., Orasanu, J., Calderwood, R. & Zsambok, C.E. (Eds). *Decision Making in Action: Models and Methods.* Norwood, NJ: Ablex Publishing.

Klein, G.A. 2008. Naturalistic Decision Making. *Human Factors,* 5: 456-460.

Klein, G. A. & Armstrong, A. 2005. Critical Decision Method. *In:* Stanton, N. A., Hedge, A., Brookhuis, K., Salas, E. & Hendrick, H., eds. *Handbook of Human Factors and Ergonomics Methods.* Boca Raton, FL: CRC Press, 35.1-35.8

Klein, G.A., Calderwood, R., & Macgregor, D. 1989. Critical Decision Method for Eliciting Knowledge. *IEEE Transactions on Systems, Man and Cybernetics*, 19: 462-472.

March, J. G., Sproull, L. S. & Tamuz, M. 2010. Learning from samples of one or fewer. *Quality and Safety in Health Care,* 12, 465-471.

Marques, J. F. & McCall, C. 2005. The application of inter-rater reliability as a solidification instrument in a phenomenological study. *The Qualitative Repot,,* 10: 439-462.

Maurino, D. E. M. 2000. Human Factors and Aviation safety: what the industry has, what the industry needs. *Ergonomics,* 43: 952-959.

McFadden, K. L. & Towell, E. R. 1999. Aviation human factors: a framework for the new millennium. *Journal of Air Transport Management,* 5: 177-184.

Neisser, U. 1976. *Cognition and Reality.* San Francisco: W.H. Freeman Co.

O'Hare, D., Mullen, N., Arnold, A. 2010. Enhancing aeronautical decision making through case-based reflection. *The International Journal of Aviation Psychology*, 20: 48-58.

Plant, K. L. & Stanton, N.A. 2011. A critical incident in the cockpit: Analysis of a critical incident interview using the Leximancer tool. In *Proceedings of the 3rd CEAS Air&Space Conference,* October 2011, Venice, Italy.

Plant, K.L. & Stanton, N.A. 2012. Why did the pilots shut down the wrong engine? Explaining errors in context using Schema Theory and the Perceptual Cycle Model. *Safety Science,* 50: 300-315.

Simpson, P.A. 2001. *Naturalistic Decision Making in Aviation Environments.* Report No. DSTO-GD-0279. DSTO Aeronautical and Maritime Research Laboratory, Australia.

Smith, K. & Hancock, P. A. 1995. Situation Awareness is adaptive, externally directed consciousness. *Human Factors,* 37: 137-148.

Stanton, N.A., Stewart, R., Harris, D., Houghton, R.J., Baber, C., McMaster, R., Salmon, P., Hoyle, G., Walker, G., Young, M., Linsell, M., Dymott, R. & Green, D. 2006. Distributed situation awareness in dynamic systems: theoretical development and application of an ergonomics methodology. *Ergonomics*, 49(12), 1288-1311.

Stanton, N.A., Rafferty, L.A., Salmon, P.M., Revell, K.M., McMaster, R., Caird-Daley, A., Cooper-Chapman, C., 2010. Distributed decision making in multihelicopter teams: case study of mission planning and execution from a non-combatant evacuation operation training scenario. *Journal of Cognitive Engineering and Decision Making*, 4 (4), 328–353

Stanton, N. A. & Walker, G. H. 2011. Exploring the psychological factors involved in the Ladbroke Grove rail accident. *Accident Analysis and Prevention,* 43, 1117-1127.

Stanton, N.A., Salmon, P.M., Walker, G.H., Jenkins, D., 2009. Genotype and phenotype schemata and their role in distributed situation awareness in collaborative systems. *Theoretical Issues in Ergonomics Science*, 10 (1), 43–68.

Walker, G. H., Stanton, N.A. & Salmon, P. M. 2011. Cognitive compatibility of motorcyclists and car drivers. *Accident Analysis and Prevention*, 43: 878-888.

Woods, D.D., Dekker, S., Cook, R., Johannesen, L., & Sarter N. 2010. *Behind Human Error*. Aldershot: Ashgate.

CHAPTER 43

Evaluating Professional Aircrew Safety Related Behaviour In-flight

Sue Burdekin

University of New South Wales
Australian Defence Force Academy
Canberra. ACT Australia
s.burdekin@adfa.edu.au

ABSTRACT

Complex safety critical organizations, such as nuclear power, oil and gas, health-care, and various modes of transport, place a great deal of reliance on the skills and behaviour of their professional workforce in order to safely conduct everyday high-risk operations. These organisations generally operate at the leading edge of advanced technology, yet incidents and accidents, involving major damage to assets and sometimes loss of life, are still a major issue, particularly when investigations reveal that highly skilled and well trained personnel are involved. Following a series of major accidents in the aviation industry during the 1970s, a behavioural and flightdeck risk management training intervention, now known as Crew Resource Management (CRM), was introduced. CRM has been successfully adapted to address the needs of many high-risk, high-reliability organisations on a global basis, but in order to ensure that the CRM training targets developing issues, and remains relevant, it is essential to evaluate the resulting trainee safety performance in the workplace and feed those results back into the training continuum.

This paper will discuss two experimental research projects that involved the development of an aircrew in-flight evaluation methodology. The evaluation philosophy was based on the multiple assessment approach of Kirkpatrick (1959), which espouses four levels of training program evaluation: reaction, learning, behaviour and results. However, the study focused on gathering self-reported behavioural data across a range of categories of behaviour designed by subject matter experts and tailored to suit the requirements of the operation.

The first study was conducted with Royal Australian Air Force military pilots using a single pilot F/A-18 Hornet flight simulator. Participants were asked to rate their own behaviour after 'flying' a medium or high workload mission, according to the pre-determined behavioural categories, using a Likert scale and open comments. The results across both conditions of workload were highly correlated with the ratings of expert observers.

In the second study, the self-report methodology was further developed to incorporate a crossed design in a multi-crewed civil airline environment in Europe. Sixty flight sectors were observed and the ratings by the observers were found to be significantly correlated with the self-report ratings from the captain and co-pilot. However, further analysis of the data indicated that there was a deeper story to tell. This tale and future plans to validate the results of this behavioural self-report methodology will be discussed.

Keywords: Self-reporting, behavioral categories, non technical skills

1 MISSION OPERATIONS SAFETY AUDITS

Operations in high-risk, high-reliability industries depend upon well trained and skilled personnel performing at peak levels in order to safely and successfully achieve a desired operational output. However, it must be acknowledged that other issues, such as environmental conditions, organisational context, together with national, organisational and professional culture can also influence the performance and productivity of personnel (Flin, O'Connor & Crichton, 2008; Reason, 2008). Systemic safety investigations, operational experience, and research have shown that most incidents and accidents involve some form of human failure that incorporates errors, violations of standard operating procedures, or flawed decisions. In many cases, these human errors and violations occur despite the technical competence of the operators involved (Dismukes, Berman & Loukopoulos, 2007; Dekker, 2002 & 2006; Reason, 1990).

One way to potentially reduce and manage these human factors effectively is to introduce error and violation management processes in the form of individual and team training interventions. A human factors training program that has demonstrated some success in correcting these processes in aviation and other complex safety dependent industries, such as, healthcare and transport modalities including rail, is known as Crew Resource Management (CRM) (Helmreich, 2000). However, in order to determine the content of that training, and to ensure that the training remains relevant and effective, it is necessary to identify what is actually occurring, in terms of human performance, in the workplace.

There are a variety of well established workplace audit and training program evaluation models. However, the use of a structured combination of methodologies, including the assessment of reactions to the intervention, learning, behaviour, and performance results is more likely to provide robust data that can be cross referenced for validity and reliability (Kirkpatrick, 1959 & 1975: Kirkpatrick & Kirkpatrick, 2006).

Whilst it is relatively straightforward to design protocols for the collection of much of this assessment data, it is more challenging to objectively assess human behaviour because of individual differences between people. These differences can be influenced by variations in factors such as, workload, fatigue, as well as physical and psychological stressors (Ackerman & Cianciolo, 2000; Matthews, Davies, Westerman & Stammers, 2000). Traditional approaches to the assessment of human performance include observation, and peer and supervisor review. Although each of these methods has been criticized for perceived bias and inconsistency, a meta-analysis conducted by Harris & Schaubroeck (1988) suggested that self-report accuracy was positively related to the vocational skill of the self-reporter.

A further consideration is that in some professions it is not possible to observe specialists in their working environment - for example, in military aviation single pilot operations. Additionally, the employment of observers/assessors, in a multi-crewed environment can be labour intensive, time consuming, and costly.

It would be extremely useful to safety specialists if it could be demonstrated empirically by scientific experimental means that professional operators were capable of reliable and valid reporting and self assessment of their own behaviour whilst interacting with people and technology in performing their operational tasks in the workplace. As the aviation industry was the first to recognize the importance of human factors (non-technical skills) training to compliment technical skills training, and it was becoming increasingly accepted that these skills needed to be evaluated; aviation was identified as a suitable safety critical industry in which to test a new approach to aircrew performance evaluation.

This paper will discuss two experimental research projects that involved the development of an aircrew in-flight evaluation methodology known as Mission Operations Safety audits (MOSA). The aim was to validate behavioural self-reported information from professional operators working in high-risk, high reliability organizations. Based on a meta-analysis of self-report studies (Harris & Schaubroeck, 1988), it was hypothesized that professional pilots would be able to effectively self-report on their own performance in-flight across a range of pre-determined categories of behaviour.

The first study was conducted in a military F/A-18 Hornet flight simulator where conditions of 'cognitive workload' (Wickens, Lee, Liu & Becker, 2004) could be controlled. The second study was carried out in-flight under normal operating procedures, in collaboration with a European low cost carrier and an aircraft manufacturer. Both studies were tailored to address the operational environment of the individual organizations involved by the design of relevant, structured categories of behaviour (Flin, O'Connor & Crichton, 2008) that were derived from the content covered in contemporary CRM training courses. For example, one category of pilot behaviour was 'automation management', where it could be demonstrated that the interaction between the pilots and the automation was managed to balance situation and workload requirements by effective set-up, the anticipation of recovery from anomaly techniques, and briefing of these plans and actions to other crew members.

Study One – Military Fast Jet Self-Reporting

Experienced pilots from 81 Wing at Royal Australian Air Force (RAAF) Base Williamtown were invited to participate in the study. Thirty-one male F/A-18 rated, active military fighter pilots volunteered for the simulated flights, each of which lasted no longer than one hour. Confidentiality of their data was assured.

Method

The study was designed with three independent variables: rater (observer or pilot), workload (medium or high), and phase of flight (departure, on-task or recovery). Workload was manipulated within subjects so that each pilot flew a mission that included a low, medium and high workload scenario. The difficulty level of the departure phase was low and was invariant across participants. As this study was primarily looking at whether pilots were able to recall and self-report their own behavioural performance maintaining agreement with an independent observer under conditions of high workload, the departure phase of flight in both profiles was not analysed. The remaining flight phases were combined with two levels of workload difficulty – medium and high – to create two flight profiles: 'on-task' phase/high, 'recovery' phase/medium for the first flight profile, and the reverse for the second. Each flight profile was executed by half of the participants and the self-reported ratings were completed immediately after the simulated 'flight' had concluded.

Both flight profiles were designed in consultation with experienced operational personnel. It was important to match the degree of workload difficulty across the phases of flight in order to balance the design so that 'phase of flight' itself could be controlled.

On-task phase - The first profile was designed with the high workload. It required the pilot to make a low level tactical intercept of an unidentified aircraft from a height of 28,000 feet in instrument meteorological conditions (IMC). At the bottom of this high speed, controlled descent the pilot was given a right airframe mounted auxiliary drive (AMAD) failure, which constituted an emergency and required him to break off the intercept and immediately take action to secure the safe return of the aircraft to the airfield. The pilot had to identify the emergency, refer to the emergency manual, decide on a course of action, initiate that action and make an emergency transmission. As a consequence of the emergency, the pilot had to consider dumping fuel in marginal weather so that the aircraft was not landing in a heavy configuration.

The second flight profile was designed with a medium workload in the 'on-task phase' of the flight. It also required pilots to conduct a tactical intercept to identify intruder aircraft but there was no emergency, and it was considered to be a straightforward, rehearsed manoeuvre, which was classified as medium workload.

Recovery phase - In first profile the recovery back to the home airfield was a straightforward, trained-for recovery and was considered to be a routine exercise.

The second profile was designed with a high level of workload for the 'recovery phase' of flight. It required the pilot to recover the aircraft at short notice to another air force base using a difficult and rarely used instrument approach. This approach was considered to provide a high level of workload because it required the pilot to achieve a demanding descent profile into the airport. The pilot would also have to consider the weight of the aircraft, and decide whether to dump fuel in marginal weather conditions.

Rating protocols - The rating protocols were identical for both flight profiles, and specific instructions were given to ensure that pilots and observers knew where one phase of flight ended and another commenced. Provision for a written comment was provided under all phases of flight. The Likert rating scale was one to five, with one representing unsatisfactory and five representing highly effective performance. A comprehensive explanation of each performance rating was included on the front cover of the rating form. The categories of behaviour were designed in consultation with subject matter experts and included: monitor/cross check, workload management, situational awareness, automation management, evaluation of plans, inquiry, assertiveness and emergency/abnormal. The behavioural category definitions were relevant to a military operational environment and were a reflection of the content of the Australian Defence Force (ADF) aircrew CRM training modules.

Results

A *preliminary statistical analysis* was conducted on the ratings of the first 10 participants in order to establish empirically the reliability of the experimenter's (Observer 1) ratings. To measure this reliability, a second observer (Observer 2), who had extensive experience as a flight instructor on the Hornet, also rated the performance of the first 10 participants against the eight behavioural criteria. The preliminary analysis was conducted to determine the ability of Observer 1 to judge accurately each participant's performance against the behavioural categories. As neither observer was subjected to a condition of excessive workload, all phases of flight—including medium and high workload—were combined and the total mean ratings per subject across each behavioural category for observers 1 and 2 were compared statistically.

To determine the degree of agreement between the ratings made by the two observers, the Pearson product moment correlation between the mean ratings for Observer 1 and Observer 2 on each behavioural category was calculated. This process was also used to screen out any criterion upon which there was no statistically significant agreement between the ratings of the two observers. (See Table 1).

The main purpose of this exercise was to ensure that the ratings of pilot performance made by the experimenter, who was a civilian commercial pilot and a trained CRM facilitator (Observer 1), would be reliable and accurate for the remainder of the experimental trials.

Except for 'Inquiry', there was a positive correlation between all of the Observer 1 and Observer 2 total means for each category of behaviour. These correlations were all significant at the $p < 0.05$ level. Therefore, both observers rated seven out of the original eight behavioural markers with acceptable agreement. The 'Inquiry' criterion was eliminated from the main analysis. On the basis of this analysis, Observer 1 only rated the performance of the pilots on the remainder of the simulator trials.

Table 1 Correlation of total means between observers

Criterion	Correlation
Monitor/Cross check	.540 *
Workload Management	.722**
Situational Awareness	.769**
Automation Management	.750**
Evaluation of Plans	.570**
Inquiry	.379
Assertiveness	.510 *
Emergency/Abnormal	.547 *

**	Correlation is significant at the 0.01 level (2-tailed)
*	Correlation is significant at the 0.05 level (2 tailed)

In the Main analysis, following a data cleansing procedure, the observer ratings and each of the remaining pilot participant ratings were initially collapsed into total medium and total high means and standard deviations across each of the behavioural categories (see Table 2) - excluding 'Inquiry'. The means were then subjected to a Pearson product-moment correlation.

Table 2   Means and Standard Deviations for Total Medium and High Workload

Workload	Mean	Std. Deviation	N
Medium-Observer	3.66	.55	14
Medium-Pilot	3.22	.51	14
High-Observer	3.53	.58	14
High-Pilot	3.12	.45	14

To address the first question – can pilots achieve a high level of agreement with independent objective observers by recall and self-report of their own behavioural performance - the mean pilot ratings in the <u>medium</u> workload phase of flight were compared with the mean ratings of the observers (ie. the benchmark), which produced a positive correlation (r = .618; n = 14; $\underline{p}$ < 0.05). These results showed that the pilot's ratings had a significant positive relationship with the ratings of the observer. Therefore, pilots did achieve a high level of agreement with the observer when recalling and self-reporting their own behavioural performance. There was some difference in the magnitude of the mean ratings (t (26) = 2.99; $\underline{p}$ < 0.05) with pilots being more critical of their own performance than the observers (see Table 2).

The second experimental question of interest asked if pilot ratings were affected by workload. The ratings for the pilots and observer under <u>high</u> workload were positively correlated (r = .881; n = 14; $\underline{p}$ < 0.01). The results showed that the pilots' ratings shared a very strong positive relationship with the ratings of the observer. Therefore, under both conditions of workload, there was a significant positive relationship between pilot and observer ratings. Again the magnitude of mean total ratings differed between pilot and observer (t (26) = 2.061, p < 0.05) with pilot ratings slightly lower than the observers.

A point to make in relation to ratings is that pilot and observer scores appeared to be more highly correlated under high workload (r = .881) than under medium workload (r = .618). This observation was confirmed by a test of significance between correlation coefficients (z = 2.46, $\underline{p}$ < 0.01).

Qualitative Data - Comments written by the pilots on the rating protocols were reviewed to check for relevant observations. They mainly referred to explanations and justification for decisions made during the flight, for example, "I was behind the aircraft on recovery - I should have dumped fuel earlier", and "I allowed the altitude to wander occasionally due to high workload". An example of recognised good practice included, "I was happy with my decision to retain fuel due to the uncertainty of the weather."

Discussion

The first MOSA study was designed to test an alternative means of collecting reliable and valid behavioural information from military pilots who fly aircraft that cannot accommodate an observer. The aim was to gain greater visibility of behavioural performance in the field and to develop a CRM evaluation process that would determine if CRM training had transferred from the classroom into the cockpit.

The results of this study demonstrated that F/A-18 pilots are able to recall and self-report their own behavioural performance immediately following an operational flight in the simulator and that their self-reports correlate significantly across the sample with ratings of the same behaviour made by independent observers. The finding supports the Harris and Schaubroeck (1988) meta-analysis of self-reporting, which suggested that the level of vocational skill of the self-reporter is positively related to the accuracy of the report.

RAAF pilots are trained to be critical of their own performance, which may explain their highly discriminating self-reports. Pilots' written comments were also consistent with the realisation that they were well aware of which behaviour was worthy of harsh criticism, and conversely which behaviour was deserving of praise.

To explain why the F/A-18 pilots' self reports were more highly correlated with the observers' ratings in the high workload condition than during the medium workload condition; we need to consider 'normal' RAAF flying. Developing strategies to cope with abnormal and emergency procedures is a major component of RAAF pilot training and daily routine. During the high workload phase of flight, pilots would have been extremely focused on the task at hand, but because of their extensive experience in high workload situations, this focus might not have exhausted their cognitive capacity. The abnormality or emergency might have even enhanced the pilot's ability to remember and make a qualified judgment of his own behavioural performance, because he would have devoted conscious information processing capacity to solving the emergency, compared to what he perceived as, the routine behaviour experienced in the medium workload condition.

Study Two - Civil Airline Pilot Self-Reporting in-flight

On learning of the results from the MOSA study, the Flight Operations Monitoring group from a major aircraft manufacturer was interested to determine whether pilot self-reporting would be a reliable means of gathering operational data from crews in a commercial airline environment. With the assistance of one of their customer airlines, a low cost carrier in Europe, an airline pilot behavioural self-reporting study was conducted. Early in the planning stage, airline management and the research team successfully sought cooperation from the pilot's union and stake holders. All participants were volunteers, and individual pilot confidentiality was guaranteed.

Method

The design of the study was developed in conjunction with airline pilots and Safety Department personnel. Although the protocols were influenced by the MOSA military study, the nature of the information collected was customized to be of use to the airline's Safety Department, with a view to highlighting early detection of safety related issues. The reporting protocol additionally asked pilots to assess their interaction with air traffic control, airports, ground support and passengers. Pilots were asked to rate their own performance across eight categories of behaviour: briefing, contingency management, monitor/cross check, workload management, situational awareness, automation management, communication, and problem solving/decision making. Each category of behaviour was given a comprehensive descriptor by the researcher and subject matter experts , and in order for the airline pilots to make a more informed rating choice, a series of specific 'word pictures' was given to each category of behaviour ranging from a grading of 1 (unsatisfactory) to 5 (exceptional) (refer Table 3).

Table 3   Grading/Word Pictures for the Behavioural Category 'Briefing'

1.	Unsatisfactory briefing standard. Briefing duration and crew interaction minimal. Available company resources not utilized to a satisfactory standard. SOPs not adhered to.
2.	Basic briefing conducted with limited crew interaction. Incomplete use of available resources and workload allocation limited. SOP briefing structure loosely adhered to.
3.	Crew operates in accordance with SOP briefing structure. Interactive briefing conducted in a timely manner, utilizing available resources to an adequate standard.
4.	Effective crew briefing conducted utilizing all company/non-company information. Proficient time and workload management with clear interaction and allocation of duties amongst crew.
5.	Comprehensive and operationally thorough briefing conducted to a high standard. Excellent crew interaction, participation and understanding. All available briefing resources utilized and clear and concise workload allocation amongst crew.

All data collecting missions originated from the airline's home base and were flown in its fleet of A319 aircraft during normal revenue raising flights. The network of destinations was restricted to eight UK and European airports.

In addition to reporting on their own individual performance during the flight, the captain and the first officer were asked to confidentially report on each other's performance using the structured protocols provided. The observers used the same protocols to report on both the captain and first officer. In an effort to be as unobtrusive as possible to regular operations, the pilot volunteers were not given a training program to explain the experimental study but rather a single page of explanatory notes before the flight in addition to a prior letter from management requesting their cooperation.

Results

Sixty flight sectors were observed by trained observers from the jump seat. Following a data cleansing procedure, the ratings of the observer/captain, observer/first officer, and captain/first officer were compared to measure the degree of correlation. The results were found to be statistically significantly correlated, indicating that commercial airline pilots were able to effectively self-report on their own behaviour across a range of structured categories when compared to the reports of an observer and each other (refer Table 4).

Table 4   Ratings across all behavioural markers

Rater	Mean	sd	N	r
OBS CPT	3.99 4.00	.36 .58	432	.344 **
OBS F/O	3.99 3.73	.36 .53	376	.206 **
CPT F/O	4.00 3.73	.58 .53	376	.169

**   Significant .01

The data were then compared by behavioural category in order to determine how individual categories of behaviour contributed to the overall results (refer Table 5).

Table 5 Correlations by behavioural category

Behavioural Marker	OBS/CAPT	OBS/FO	CAPT/FO
Briefing	.630 **	.427 **	.275
Contingency Mgt	.361 **	.056	.000
Monitor-x-check	.278 *	-.098	-.043
Workload Mgt	.364 **	.164	.132
Situational Awareness	.271 *	.368 *	.115
Automation Mgt	.109	.301 *	.195
Communications	.457 **	.441 **	.335 *
Problemsolve/Decision Making	.265	-.232	-.116

*   *Correlation is significant .05*
**   *Correlation is significant .01*

This breakdown revealed that the category of behaviour selection appeared to be understood and rated consistently by the captain and the observer with the exception of 'automation management' and 'problem solve/decision making'.   These categories would need to be reconsidered for future use. It might be that they simply need a better definition. Whilst the captains appeared to be very good at interpreting the categories and rating their own performance, the first officers seemed to have some difficulty. The observers and first officers found agreement on four of the eight behavioural categories, but the captains and first officers only agreed with the ratings for the 'communication' category.

A closer inspection of the data indicated that the ratings made by first officers were more discriminating. They did not seek to present their performance in an artificially inflated score; rather they were more critical of themselves. The most likely explanation for this result was suggested by airline check and training captains who explained that when first officers complete their training they will have been trained to, and assessed at, a very high standard. Because first officers are not generally involved in the formal assessment of other pilots, the higher than average standard expected in a training environment may be the only exposure that they have had in rating operational performance. This may mean that if a pilot self-reporting system were to be introduced comprehensive training in the use of it would need to be an integral part of the package, as would be expected for any new program.

Conclusion

Both the military and the civil airline studies indicate that professional pilots are able to effectively self-report on their own performance across a range of predetermined categories of behaviour. Furthermore, the self-reports from pilots and trained observers were found to be statistically significantly correlated, in both the simulator and under normal 'in-the-field' operating conditions in-flight. These results support the hypothesis, derived from published research, that the level of vocational skill of the self-reporter is positively related to the accuracy of the report.

The aim of this research was to validate behavioural self-reported information from professional operators working at the 'sharp end' of high-risk, high-reliability organisations. In doing this, safety critical issues can be dealt with in a timely manner once management can establish confidence in the ability of these highly trained, highly skilled workers to assess their own performance, and the behaviour of those with whom they interact, within the workplace. Additionally, the information that could be regularly captured in this structured 'category of behaviour' manner can be fed back into the design of human factors safety training courses, such as CRM and other areas of operational training. In this way intervention strategies could be more precisely targeted based on relevant, contemporary issues – in other words providing the basis for a closed loop system.

A further study is planned to test the MOSA methodology in a regional airline operating turbo prop aircraft to islands in the Indian Ocean. This research will determine if the MOSA evaluation approach is acceptable in another national and organizational culture. To validate the data, the research will compare the behavioural self-reports from pilots not only to reports from trained observers but also to safety data that have been collected from other safety management sources within the same organisation, over the same period of time. These data sources will include, for example, flight data analysis from quick access recorders, safety reports, cultural surveys and incident reports.

REFERENCES

Ackerman, P. L., & Cianciolo, A. T.(2000). Cognitive, perceptual speed, and psychomotor determinants of individual differences during skill acquisition. *Journal of Experimental Psychology: Applied, 6* (4), 259-290

Dekker, S. (2002) *The Field Guide to Human Error Investigations.* Ashgate, Aldershot.

Dekker, S. (2006) *The Field Guide to Understanding Human Error.* Ashgate. Aldershot.

Dismukes, K., Berman, B., and Loukopoulos, L. (2007) *The Limits of Expertise: rethinking pilot error and the causes of airline accidents.* Ashgate. Aldershot.

Flin, R., O'Connor, P., & Crichton, M. (2008) *Safety at the Sharp End: a guide to non-technical skills.* Ashgate. Farnham

Harris, M., & Schaubroeck, J. (1988) A meta-analysis of self-supervisor, self-peer and peer-supervisor ratings. *Personnel Psychology*, 38, 43-62.

Helmreich, R. (2000) On Error Management: lessons from aviation. *British Medical Journal*, 320, 781-785.

Kirkpatrick D. L. (1959). 'Techniques for evaluating training programs.' *Journal of American Society of Training Directors*, 13 (3): pp21–26.

Kirkpatrick, D. L. (1975). 'Techniques for Evaluating Training Programs' in *Evaluating training programs* D. L. Kirkpatrick (ed.) Alexandria, VA: ASTD.

Kirkpatrick D.L., & Kirkpatrick J.D., (2006) *Evaluating training programs: the four levels* 3rd Ed. Berrett-Koehler Publishers Inc., San Francisco.

Matthews, G., Davies, D.R., Westerman, S., & Stammers, R. (2000) *Human Performance: Cognition, Stress and Individual Differences.* Taylor & Francis. Philadelphia.

Reason, J. (2008) *The Human Contribution: unsafe acts, accidents, and heroic recoveries.* Ashgate Publishing Limited. Farnham.

Reason, J. (1990) *Human Error.* Cambridge University Press. Cambridge.

Wickens, C., Lee, J., Liu, Y., & Becker, S. (2004) *An Introduction to Human Factors Engineering* 2nd Ed. Pearson Prentice Hall. New Jersey.

CHAPTER 44

How Well Are Human-related Hazards Captured by Multi-agent Dynamic Risk Modelling?

Sybert H. Stroeve[1], Henk A.P. Blom[1,2]

[1]National Aerospace Laboratory NLR, Air Transport Safety Institute
Amsterdam, The Netherlands
[2]Delft University of Technology, Faculty of Aerospace Engineering,
Delft, The Netherlands
stroeve@nlr.nl, blom@nlr.nl

ABSTRACT

Multi-agent dynamic risk modelling is used to assess accident probabilities of air traffic scenarios involving multiple entities, disturbances and their interactions. In this context human performance models should be able to capture a wide variety of hazards. This paper provides an analysis of the coverage of a broad set of hazards that can be attained by model constructs in multi-agent dynamic risk modelling. It shows that multi-agent situation awareness is a prime construct for understanding and analysing hazards in air traffic scenarios.

Keywords: safety, hazards, multi-agent dynamic risk modelling, human models

1   INTRODUCTION

A recent overview of human performance models in aviation (Foyle and Hooey, 2008) concluded that existing human performance models can effectively support system design and evaluation in aviation, especially in combination with human-in-the-loop simulations. This overview shows the capabilities of five human performance modelling frameworks (ACT-R, IMPRINT/ACT-R, Air MIDAS, D-

OMAR and A-SA) with respect to a variety of human performance characteristics, including error prediction, crew interactions, external environment interactions, scheduling and multitasking, memory, visual attention, workload, situation awareness, learning, and resultant and emergent behaviour. Model results for each of these characteristics were derived for two studies on taxi navigation errors and synthetic vision system operations respectively. The results show that each of these five models have their specific strengths in supporting design by testing nominal and off-nominal scenarios, determining the potential for error-precursor conditions and conducting what-if system redesigns efficiently.

Human performance modelling also plays a key role in multi-agent dynamic risk modelling (DRM) of the TOPAZ (Traffic Organization and Perturbation AnalyZer) safety assessment methodology (Blom et al., 2001a, 2006). Multi-agent DRM captures the performance of interacting agents (humans and systems) in safety-relevant scenarios and uses Monte Carlo simulations to assess rare event probabilities of potential safety consequences such as aircraft collisions. In contrast with the five methods mentioned above, the objective of TOPAZ thus is safety risk analysis rather than human performance analysis.

Corker et al. (2005) compared Air-MIDAS and TOPAZ on differences regarding human performance modelling. Both frameworks embrace the generic human information processing model (Wickens and Hollands, 2000) as basis and both incorporate approaches to capture situation awareness (SA) of an agent, human errors and cognitive modes. Both also make use of a systematic specification language and validation means. The specific submodels used by Air-MIDAS and TOPAZ to cover these aspects typically are complementary rather than exclusionary. This was illustrated for a runway incursion case studied by both approaches, where the more detailed Air MIDAS human performance modelling supported tuning of the TOPAZ multi-agent DRM, and the TOPAZ results showed the accident risk implications of these human performance aspects (Corker et al., 2005).

In (Stroeve et al., 2009, 2011c) the TOPAZ multi-agent DRM approach is systematically compared with a classical event sequence-based safety analysis. The findings confirm that the level of safety in sociotechnical organizations depends on the interactions between the organizational entities in their contextual conditions. This coincides with the psychological perspective that the flexibility and system oversight of human operators are often essential to maintain efficient and safe operations in the context of uncertainty and disturbances. Within this context, human performance models should be able to capture a wide variety of hazards, i.e. anything that may influence safety. This safety-driven need has formed an important guidance for the selection and integration of various human-related submodels within TOPAZ multi-agent DRM.

The objective of this paper is to provide an assessment of the types of human-related hazards in air traffic that are represented by submodels currently in use by TOPAZ multi-agent DRM. The hazards are obtained from a large hazard database that has been maintained at NLR for over 15 years. The research questions are: to what extent can hazards be covered by the current submodels and which of these submodels are most powerful in hazard coverage?

The paper is organized as follows. Section 2 presents the hazard database used. Section 3 describes submodels used in TOPAZ multi-agent DRM. Section 4 presents a hazard coverage analysis by these submodels. Section 5 draws conclusions.

2   DATABASE OF HAZARDS IN AIR TRAFFIC

One of the steps in a safety risk assessment cycle is the identification of hazards of the operation under study (AP15, 2007). Over a period of 15 years NLR has collected hazards that were identified in a broad range of air traffic safety assessments. The prime means by which these hazards were gathered is by brainstorm sessions with pilots, controllers and other experts. These hazard brainstorm sessions aim to push the boundary between functionally imaginable and functionally unimaginable hazards (De Jong et al., 2007). Consequently, considerable parts of these hazard brainstorm sessions address human behaviour, conditions and technical systems that influence human behaviour and interactions between humans. Overall, the resulting hazard database includes a broad set of hazards, addressing the performance of interacting humans, technical systems and contextual conditions for a large variety of air traffic operations.

The collection of hazards in the hazard database includes equal or similar hazards and hazards that refer to a study-specific context, e.g. airport layout and route structure. Recently, we analysed the hazards in the hazard database by selecting all unique hazards and formulating them in a generalized way (Stroeve et al., 2011a). These generalized hazards were structured into various hazard clusters, describing classes of technical systems (e.g. aircraft systems, navigation systems, air traffic control systems), human operators (e.g. pilot performance, controller performance) and contextual conditions (e.g. weather, traffic relations, infrastructure and environment). Examples of the hazards are provided as part of the hazard coverage analysis in Section 4.

3   SUBMODELS IN TOPAZ MULTI-AGENT DRM

Several human-related submodels have been developed for TOPAZ multi-agent DRM (Blom et al., 2001b, 2003; Stroeve et al., 2003). At a syntactic level, models in TOPAZ multi-agent DRM are specified by Stochastically and Dynamically Coloured Petri Net, which is a Petri net extension that is able to represent general stochastic hybrid processes in a form that allows stochastic analysis (Everdij and Blom, 2010). At a semantic level, the following submodels are used.

(1) Human information processing (HIP): The human information processing submodel considers processing of information from the environment that leads to actions that may influence the environment during tasks that the human operator uses for the fulfilment of his/her work. The submodel is based on a task analysis, which takes into account the Multiple Resources model (Wickens and Hollands, 2000). In particular, subtasks and resources are aggregated in a fixed pattern, according to time-critical task/resources combinations (Blom et al., 2001b; Blom et al., 2003).

The human information processing submodel includes task identification, task scheduling, task load, decision making, and task execution. *Task identification* considers the ways that the human operator identifies the tasks that need to be performed at a particular time instance, e.g. for a monitoring task it may be based on regular updating at deterministic or stochastic intervals, and for a communication task it may be based on the situation awareness (SA) of the operator. The *task scheduling* process determines which tasks may be performed concurrently, as well as a priority among the tasks that cannot be performed concurrently. The *task load* describes the number of tasks to be performed and/or the resources required by tasks at the level of visual, auditory, cognitive and motor performance. The task load influences the cognitive control mode of the human operator. *Decision making* processes are based on decision rules dedicated to the safety-relevant scenario considered, e.g. describing when aircraft are in a conflict situation, or how to react to a conflict situation. *Task execution* considers the dynamic and stochastic performance characteristics of tasks. The duration of a task may be influenced by the cognitive control mode and the execution of a task may include errors. The task execution may have effect on the SA of the human operator considered as well as on states of other humans or technical systems.

(2) Cognitive control mode (CCM): This submodel considers that humans can function in a number of cognitive control modes, such as Strategic, Tactical, Opportunistic and Scrambled (Hollnagel, 1993). The cognitive control mode may depend on human performance aspects such as the range of tasks to be done and the situation awareness of the human. It influences human performance aspects such as the planning horizon and the accuracy of task performance.

(3) Human error (HE): The human error submodel considers that the execution of a task by a human operator may include large deviations from normal practice and that such deviations may be expressed as 'errors'. The HE submodel includes slips, lapses, mode errors and knowledge and rule-based mistakes that may occur during the human information processing steps (Reason, 1997; Wickens and Hollands, 2000). It does not represent in detail the mechanisms that may have given rise to the error, but it considers the behaviour resulting from these mechanisms at a probabilistic level for a specific task. The error probability is thus task specific and it may be influenced by other submodels, such as the cognitive control mode. For instance, the probability of an error may be higher in the Opportunistic control mode than in the Tactical control mode. Examples of types of errors are the lack of observing an aircraft during a visual monitoring task, or call-sign confusion during a communication task.

(4) Multi-agent situation awareness (MASA): The concept of situation awareness (SA) addresses perception of elements in the environment, their interpretation and the projection of the future status (Endsley, 1995). For a single human, situation awareness components are commonly included in human information processing stages (Wickens and Hollands, 2000). In an air traffic environment with multiple human operators, these aspects and associated errors of SA depend on various human-human and human-machine interactions. The multi-agent SA submodel describes the SA of each agent in the multi-agent system as time-dependent

information of all other agents, where the agents include humans as well as technical systems (Stroeve et al., 2003). The components of the SA of an agent about another agent include the identity (e.g. call-sign), mode variables (e.g. flight phase or alert status), continuous state variables (e.g. aircraft position and speed), and intent variables (e.g. intent of a pilot about the course). Achieving, acquiring and maintaining SA depend on situation assessment processes as observation, communication and reasoning, which are related to human information processing tasks. The timing of the situation awareness updates depend on the initiation and duration of related tasks (e.g. monitoring, communication). The situation assessment processes may be afflicted by stochastic noise or error, e.g. observation noise for visual monitoring of aircraft positions by a controller, or misunderstanding of a clearance by a pilot.

(5) System mode (SM): An abstract submodel that describes the behaviour of a technical system by modes. These modes are discrete states for the functioning the technical systems, such as failure conditions, system settings, etc. These modes have particular durations or modes changes occur instantaneously.

(6) Dynamic variability (DV): An abstract submodel that describes the variability of states of agents due to dynamic processes. For instance, it can describe the movements of an aircraft according to differential equations relating states such as position, velocity, acceleration and thrust.

(7) Stochastic variability (SV): An abstract submodel that describes the stochastic variability in the performance of human operators and technical systems. For a human operator it specifies the variability in task aspects, e.g. duration, start time, accuracy, etc., in a contextual condition, i.e. given the state of other human performance model constructs, such as SA, cognitive control mode and other human modes. The variability is represented by probability density functions with moments that may be functions of the contextual condition. Similarly, the variability of system functioning is described by context-dependent probability density functions.

(8) Contextual condition (CC): A collection of submodels capturing the context of the operation, such as weather, route structure, environmental conditions and airport infrastructure.

4 HAZARD COVERAGE ANALYSIS

On the basis of the hazard database described in Section 2 and the TOPAZ multi-agent DRM submodels described in Section 3, it was determined for each hazard whether it can be covered by the submodels (Stroeve et al., 2011a). This analysis was done qualitatively by performing a mental simulation about the way that the mechanisms of each of the submodels can represent the hazard considered, where a hazard may be covered by none, one or multiple submodels. The result of the analysis for a particular hazard may be that it is

- Well covered, meaning that on the basis of our multi-agent dynamic risk modelling experience we assess that it can be well represented by the current set of submodels;

- Partly covered, meaning that there are aspects of the hazard that are not covered by the current set of submodels; or
- Not covered, meaning that the current set of submodels cannot represent the hazard.

A number of examples of the hazard coverage analysis are shown in Table 1; the results for all hazards are in (Stroeve et al., 2011a). It should be noted that in this table the hazards are mostly described at a local level without considering effects in a multi-agent system. For instance, the hazard "radar not working" might imply that a controller does not have aircraft position data on screen, or the hazard "pilot reports wrong position" might imply that the receiver (e.g. a controller) attains a wrong SA about the position of the aircraft.

Table 1: Examples of hazard coverage analysis.

Hazard	Submodel	Explanation
Well covered		
Radar is not working	SM	Discrete mode *Not Working* with a particular probablity and duration.
VHF R/T frequency is blocked	SM	Discrete mode *Delaying* with a particular probablity and duration, leading to a delay in communication between agents.
Pilot report wrong position	MASA HE	Wrong component about aircraft position in SA by pilot, or pilot makes an error while reporting.
Pilot taxies too fast	HE SV	Pilot agent may make an error in setting the taxi speed. The speed may be chosen from a probality distribution, potentially leading to exceedance of a threshold.
Controller makes a reading error	MASA HE	Controller wrongly updates a SA component by reading.
Controller spends too much time monitoring	HIP	Controller agent frequently schedules monitoring tasks and/or uses long times for monitoring tasks.
Controller is overloaded with coordination messages	HIP CCM	A large number of coordination messages lead to scheduling and execution of tasks to handle them, to a high task load and may lead to a transfer to an *Opportunistic* cognitive control mode.
Resolution of conflict leads to other conflict(s)	HIP MASA DV SV	Conflicts leading to other conflicts depend on the stoachastic dynamics of agents and interactions interactions between agents. Resolution of conflicts may lead to other conflicts depending on the SA and decision making of the involved controllers.
Partly covered		
Alert causes attention tunneling	MASA HIP Others...	An alert influences the SA of an agent and in combination with other SA components it induces the agent to identify particular tasks needed and the scheduling of these tasks. As a result other tasks may be neglected. Other constructs, e.g. for stress-produced tunneling, may be needed.

Change in ATC procedures leads to confusion by pilots	MASA Others...	Specific procedures may be modelled by decision making rules in the SA updating of the pilot agent's SA. However, the confusion due to changes is not covered.
Controller ignores an alert (no evaluation)	MASA Others...	The lack of evaluation has impact on the SA of the controller agent. The background of this behaviour (e.g. lack of trust) is not covered.
Avoiding bad weather leads to higher traffic density	MASA DV CC Others...	Multiple pilot agents being aware of a bad weather situation may lead to correlated aircraft dynamics. Bad weather manifestations are well not covered.
Not covered		
Pilot is fatigued and sleepy	None	Conditions leading to and effects of fatigue and sleepiness are not covered.
Cultural differences between airlines	None	Cultural differences between organizations are not covered.
Complacency of controller	None	Complacency of human operators is not covered.
Controller performance is affected due to alcohol, drugs or medication	None	Effects of alcohol, drugs or medication on human performance are not covered.

Table 2 shows the numbers of hazards for each hazard cluster that are well covered, partly covered or not covered. It follows from these results, that overall 58% of the hazards are well covered, 11% are only partly covered and 30% are not covered. For hazards in the clusters pilot performance and controller performance about 30% to 45% of the hazards are not yet covered. Further details on the types of hazards that are not yet covered are given in (Stroeve et al., 2011a, 2011b). To this respect, it should be noticed that the TOPAZ methodology evaluates the safety effect of non-covered hazards through a bias and uncertainty analysis (Everdij et al., 2006).

Table 3 shows the relative use of the various submodels for modelling hazards in each of the hazard clusters, i.e. the percentage of the use for well covered or partly covered hazards. It follows that the multi-agent SA (MASA) submodel applies most frequently (59%) in the hazard coverage. Especially high percentages are attained for modelling human-related hazards in the clusters pilot performance, controller performance and ATC coordination. In addition, multi-agent SA also covers significant percentages of hazards in the technical systems clusters. The submodel system mode (SM) applies to 29% of the well or partly covered hazards, and mostly focuses on clusters related to technical systems. The submodel human error (HE) is applied in 25% of the cases, which typically concern errors in the tasks of a single human operator. The human information processing (HIP) submodel is applied for 21% of the hazards in general and to larger extents for human-focused hazard clusters. The other model constructs are applied to significantly smaller extents.

Table 2: Results of the hazard coverage analysis per hazard cluster.

	Hazard cluster	Total number of hazards	Hazard coverage					
			Well covered		Partly covered		Not covered	
A	Aircraft systems	14	11	79%	2	14%	1	7%
B	Navigation systems	8	7	88%	0	0%	1	13%
C	Surveillance systems	14	14	100%	0	0%	0	0%
D	Speech-based communication	19	13	68%	2	11%	4	21%
E	Datalink-based communication	10	9	90%	0	0%	1	10%
F	Pilot performance	62	31	50%	13	21%	18	29%
G	Controller performance	55	23	42%	7	13%	25	45%
H	ATC systems	13	7	54%	2	15%	4	31%
I	ATC coordination	12	8	67%	0	0%	4	33%
J	Weather	14	2	14%	4	29%	8	57%
K	Traffic	17	13	76%	0	0%	4	24%
L	Infrastructure & environment	12	11	92%	0	0%	1	8%
M	Other	16	6	38%	0	0%	10	63%
	Total	266	155	58%	30	11%	81	30%

Table 3: Relative use (%) of submodels (columns) for modelling hazards in the hazard clusters (rows).

	Hazard cluster	(1) HIP	(2) CCM	(3) HE	(4) MASA	(5) SM	(6) DV	(7) SV	(8) CC
A	ACS	0%	0%	0%	15%	85%	15%	0%	0%
B	NS	0%	0%	0%	29%	86%	0%	0%	0%
C	SS	7%	0%	7%	43%	71%	14%	0%	0%
D	SBC	7%	7%	60%	47%	40%	0%	0%	0%
E	DBC	11%	0%	33%	56%	67%	22%	0%	0%
F	PP	27%	7%	36%	91%	7%	0%	2%	0%
G	CP	53%	17%	40%	70%	3%	0%	0%	0%
H	ATCS	11%	0%	11%	56%	56%	11%	0%	0%
I	ATCC	38%	25%	25%	75%	0%	13%	0%	0%
J	WT	0%	0%	0%	33%	0%	33%	0%	100%
K	TF	8%	0%	0%	38%	8%	85%	54%	8%
L	IE	0%	0%	9%	45%	27%	9%	0%	73%
M	OT	33%	0%	17%	50%	17%	17%	0%	50%
	Total	21%	6%	25%	59%	29%	12%	4%	10%

5   CONCLUDING REMARKS

TOPAZ multi-agent DRM is used to assess rare event probabilities of safety-relevant scenarios of air traffic sociotechnical systems. Given the importance of human contributions to air traffic safety and the diversity in the ways that human-related aspects may impact safety, there is a need for a sufficiently broad repertoire of human-related submodels in multi-agent DRM.

As a structured approach towards evaluating the richness of the current set of model constructs in TOPAZ multi-agent DRM, we analysed the coverage of a broad set of hazards by submodels. The results show that 58% of the hazards are well covered, 11% is partly covered and 30% is not covered by the current submodels.

The multi-agent SA submodel applies to almost 60% of the covered hazards, especially for the representation of human-related hazards (e.g. in about 90% for the pilot performance cluster). Submodels that apply in 20% to 30% of the covered hazards are system mode, human error and human information processing; other submodels apply at much lower percentages. This shows that multi-agent SA is a key submodel for understanding and analysing hazards in sociotechnical air traffic scenarios.

The above findings form a sound motivation to continue the hazard-guided development and integration within multi-agent DRM of complementary submodels for hazards not yet fully covered. This is already ongoing in follow-up studies of the SESAR WP-E project MAREA.

ACKNOWLEDGMENTS

Part of this research has been done in the SESAR WP-E project MAREA (Mathematical Approach towards Resilience Engineering in ATM). MAREA is co-financed by Eurocontrol on behalf of the SESAR Joint Undertaking (SJU). The views expressed in this research paper are those of the authors. The paper does not purport to represent views or policies of NLR, Eurocontrol or SJU.

REFERENCES

AP15. (2007). *ATM Safety Techniques and Toolbox*: Federal Aviation Administration & Eurocontrol.

Blom, H.A.P., Bakker, G.J., Blanker, P.J.G., Daams, J., Everdij, M.H.C., & Klompstra, M.B. (2001a). Accident risk assessment for advanced air traffic management. In G. L. Donohue & A. G. Zellweger (Eds.), *Air Transport Systems Engineering* (pp. 463-480): AIAA.

Blom, H.A.P., Daams, J., & Nijhuis, H.B. (2001b). Human cognition modelling in ATM safety assessment. In G. L. Donohue & A. G. Zellweger (Eds.), *Air Transport Systems Engineering* (pp. 481-511). Washington D.C., USA: AIAA.

Blom, H.A.P., Stroeve, S.H., Everdij, M.H.C., & Van der Park, M.N.J. (2003). Human cognition performance model to evaluate safe spacing in air traffic. *Human Factors and Aerospace Safety, 3*, 59-82.

Blom, H.A.P., Stroeve, S.H., & De Jong, H.H. (2006). Safety risk assessment by Monte Carlo simulation of complex safety critical operations. In F. Redmill & T. Anderson (Eds.), *Developments in Risk-based Approaches to Safety: Proceedings of the Fourteenth Safety-citical Systems Symposium, Bristol, U.K.*: Springer.

Corker, K.M., Blom, H.A.P., & Stroeve, S.H. (2005). *Study on the integration of human performance and accident risk assessment models: Air-MIDAS and TOPAZ.* Proc. International Symposium Aviation Psychology 2005 (pp. 147-152), Dayton (OH), USA.

De Jong, H.H., Blom, H.A.P., & Stroeve, S.H. (2007). *How to identify unimaginable hazards?* Proc. 25th International System Safety Conference (ISSC2007), Baltimore, USA.

Endsley, M.R. (1995). Toward a Theory of Situation Awareness in Dynamic Systems. *Human Factors: The Journal of the Human Factors and Ergonomics Society, 37*(1), 32-64.

Everdij, M.H.C., Blom, H.A.P., & Stroeve, S.H. (2006). *Structured assessment of bias and uncertainty in Monte Carlo simulated accident risk.* Proc. 8th International Conference Probabilistic Safety Assessment & Management, New Orleans, USA.

Everdij, M.H.C., & Blom, H.A.P. (2010). Hybrid state Petri nets which have the analysis power of stochastic hybrid systems and the formal verification power of automata. In P. Pawlewski (Ed.), *Petri Nets* (pp. 227-252). Vienna, Austria: I-Tech Education and Publishing.

Foyle, D.C., & Hooey, B.L. (Eds.). (2008). *Human performance modeling in aviation.* Boca Raton (FL), USA: CRC Press.

Hollnagel, E. (1993). *Human reliability analysis, context and control.* London, UK: Academic Press.

Reason, J. (1997). *Managing the risk of organizational accidents.* Aldershot, England Ashgate.

Stroeve, S.H., Blom, H.A.P., & Bakker, G.J. (2003). *Multi-agent situation awareness error evolution in accident risk modelling.* Proc.5th USA/Europe Air Traffic Management R&D Seminar, Budapest, Hungary.

Stroeve, S.H., Blom, H.A.P., & Bakker, G.J. (2009). Systemic accident risk assessment in air traffic by Monte Carlo simulation. *Safety Science, 47*(2), 238-249.

Stroeve, S.H., Everdij, M.H.C., & Blom, H.A.P. (2011a). *Hazards in ATM: Model constructs, coverage and human responses.* SESAR Joint Undertaking, report E.02.10-MAREA-D1.2.

Stroeve, S.H., Everdij, M.H.C., & Blom, H.A.P. (2011b). *Studying hazards for resilience modelling in ATM.* Proc. First SESAR Innovation Days, Toulouse, France.

Stroeve, S.H., Blom, H.A.P., & Bakker, G.J. (2011c). *Contrasting safety assessment of a runway incursion scenario by event trees and agent-based dynamic risk modelling.* Proc. 9th USA/Europe Air Traffic Management R&D Seminar, Berlin, Germany.

Wickens, C.D., & Hollands, J.G. (2000). *Engineering psychology and human performance.* Upper Saddle River (NJ), USA: Prentice Hall.

CHAPTER 45

The Safety and Ethics of Crew Meals

Immanuel Barshi, Jolene Feldman

NASA Ames Research Center
Moffett Field, CA
Immanuel.barshi@nasa.gov

San Jose State University
San Jose, CA
Jolene.M.Feldman@nasa.gov

ABSTRACT

Glucose is the fuel of the brain. When blood glucose levels drop, so does cognition. Interestingly, it's the higher cognitive functions, such as decision making and the ability to perceive risk, that go out first. Motor skills go last, and so the pilots' hands may go to the right switches and levers, but his mind will be lagging behind. Data also shows that cognitive performance degrades long before people are aware of any symptoms such as a sense of hunger or having a headache. Thus, low blood glucose levels represent a serious safety risk in complex operations, and especially in aviation where pilots may have to go for many hours with little or no food. Part of the reason for this lack of food is the elimination of crew meals in many domestic operations. We review data showing the effects of low blood glucose on pilot performance, and discuss the ethical and practical issues surrounding the problem.

Keywords: blood glucose, cognition, aviation, pilots, risk, performance, food

1   BACKGROUND

All of us experience the sensation of being hungry from time to time and most of us have experienced some of the acute effects of hunger, such as a sense of lowered energy levels, fatigue, and changes in mood. Moreover, some have experienced the severe effects of hunger, a state that impairs cognitive and decision-making processes. All effects of hunger may be most apparent when job performance is demanding and of a highly sensitive nature. Flight operations often involve busy days, quick turn-arounds, and no food provided onboard the aircraft for the crew. Such operations can create situations in which pilots may not get the food that they need, because they are scheduled to start their duty day before food services at restaurants or airport terminals are open, or end their day after such facilities are closed. In the field of aviation, cognitive difficulties arising from hunger may result in situations that are costly and dangerous.

1.1   Blood Glucose, Cognition and Performance

The brain uses glucose (sugar) as its source of energy. The brain is reliant on a continuous supply of glucose through the blood to function properly, as it cannot store glucose. The primary source of glucose is carbohydrates, and their consumption affects the rise and/or fall of blood glucose levels, based on the quality (glycemic index) and quantity (glycemic load) of the carbohydrate. Other factors also affect blood glucose levels; some can lower levels (e.g., alcohol, exercise), raise levels (e.g., stress, illness), or have variable effects (e.g., medication, vitamins, nicotine). Adequate hydration may also play a role in studies of blood glucose and measures of performance; Liebermann and colleagues (2012) state that single-day food and fluid deprivation studies (which have shown decrements in cognitive function) may induce mild dehydration since a substantial proportion of daily water intake is taken in food (as cited in Sawka, Cheuvront, & Carter, 2005, p. S32).

Blood glucose and performance are investigated in a myriad of ways; individuals in experiments may be diabetic or non-diabetic, induced to hypoglycemic (low blood glucose) states or are fasting (e.g., not eating prior to the experiment), provided with a glucose-drink or not, engaged in everyday tasks and/or administered laboratory measures, and so on. Due to the complexities of studying the effects of blood glucose levels, other factors that affect blood glucose, inherent individual differences in blood glucose regulation, and lack of a clear and standardized methodology within the field, there are often unclear and conflicting research results.

Despite these inconsistencies, research and experience demonstrate that blood glucose levels affect cognitive performance. Furthermore, low blood glucose levels (from fasting or insulin-induced hypoglycemia) impair performance of individuals on tests that assess memory, decision-making, fine motor skill, verbal fluency, reaction time, and auditory/visual processing speed (see review by Feldman & Barshi, 2007).

Importantly, laboratory research on the effects of low blood sugar has demonstrated three critical findings. First, individuals do not need to be hypoglycemic to experience symptoms of low blood sugar (Owens & Benton, 1994; Donohoe & Benton, 1999). Simply skipping one meal can lower blood sugar enough to cause measurable decrements. Second, individuals may not necessarily recognize the symptoms of low blood sugar (Evans, Pernet, Lomas, Jones, & Amiel, 2000). Pilots may be able to maneuver an aircraft and perform all of their physical tasks, but may not realize that their underlying cognitive abilities are impaired. By the time a pilot realizes his or her physiological impairment, their cognitive functions are likely to have already been affected for some time. Third, individuals need time to recover from low blood glucose levels (Holmes, Koepke, & Thompson,1986). It can take up to thirty minutes after eating for an individual to recover from low blood glucose levels. The effects of low blood sugar are not always evident to the individual and can be insidious. However, the potential effects and implications of low blood sugar on pilot performance are clear.

Evans and colleagues (2000) addressed the insidious nature of low blood sugar and its influence on mental capabilities with healthy (non-diabetic) male participants. Blood sugar levels were controlled to induce hypoglycemic states, and participants were assessed through physiological measures, cognitive tests, and a questionnaire. The questionnaire assessed symptoms such as trembling, confusion, headache, and irritability, and also required participants to rank feelings of abnormality from a mild state to a maximal state. The speed of onset of the symptoms of acute hypoglycemia was investigated and researchers concluded that these symptoms may preclude awareness. They found that symptom generation, including the subjective feeling of abnormality, was delayed for up to twenty minutes after the onset of the hypoglycemia nadir (the lowest blood sugar level) and after the onset of detectable cognitive dysfunction.

Further, Holmes and colleagues (1986) found that low blood sugar levels affected higher cognitive functions, like decision-making, or visual and auditory comprehension, earlier than they affected lower functions such as motor skills. Hence, by the time a pilot becomes aware of impairments to motor functions such as a marked slowness in completing a task, or shaking or trembling, higher cognitive functions may have already been negatively affected for some time, without the pilot's awareness.

It is important to note that in experimental studies such as the Evan et al. (2000) study, participants' blood sugar was artificially set at hypoglycemic levels. In routine flight operations, even though pilots are not expected to be hypoglycemic, their blood sugar levels may drop to a level at which cognitive performance is affected. Thus, it could take pilots longer than twenty minutes to recognize such cognitive degradation, if it is recognized at all.

More relevant studies have focused specifically on pilots' flight performance and carbohydrates. A study conducted by researchers from the United States Air Force Academy (Johnson, Carlson, Veverka & Self, 2007) investigated recovery performance of pilots from a simulated unusual attitude after either a high or low carbohydrate drink. Although no significant differences were found, data suggested a trend toward increasing reaction times (initial stick input) and time to straight and level for the lower blood glucose group. Researchers attribute the lack of significant findings to the magnitude of difference in the amount of carbohydrates in the drink conditions (needed to be greater), the differences in level of pilots' simulator experience, and insufficient time to process the higher carbohydrate drink.

Similarly, work to increase aviation safety conducted by Lindseth and colleagues (Lindseth, et al., 2011) investigated the effects of diet on cognition and flight performance of pilots. Researchers found that pilots consuming high-fat and high-carbohydrate diets had significantly better overall flight performance scores (airspeed control, heading control, and altitude control) than pilots consuming high-protein diets. Pilots on the high-fat diet also scored significantly better on a short-term memory test (Sternberg test) than on the high-protein diet. Pilots on the high-protein diet deviated from their assigned altitude significantly more often than did pilots on the other diets.

In yet another study, researchers from the US Army Medical Research (Lieberman et al., 2008) did not find food restriction to negatively affect performance of healthy individuals on various measures. In this highly controlled two-day laboratory experiment, blood glucose levels were monitored using a Continuous Glucose Monitoring System, sleep activity was monitored via wrist-worn monitors, mood was assessed (e.g., Profile of Mood States) and a battery of cognitive tests was administered (e.g., Four-Choice Reaction Time Test, Psychomotor Vigilance Test). Healthy participants consumed three different diets: carbohydrate, carbohydrate-fat, and calorie-deprivation. Individuals' performance was evaluated. While lower glucose concentrations were associated with the calorie-deprivation diet, performance on various cognitive tests, and assessments of sleep and mood did not show adverse effects. Lieberman asserts that this study does not contradict single day studies of total food and fluid restriction that find decrements in cognitive function, as participants in his study were adequately hydrated. This study shows that in the case of young healthy soldiers, the caloric deprivation condition didn't drive blood glucose levels low enough to show performance effects. However, the general airline pilot population may be very different from these young, well conditioned soldiers.

These laboratory studies illustrate the complex effects of food consumption on pilot performance, and the need for further systematic investigation. Regardless, in many of the reports filed with NASA's Aviation Safety Reporting System (ASRS), pilots report "feelings of hunger", and "lack of food", and relate these to impaired flight performance (Bischoff & Barshi, 2003). It is often only in hindsight that the pilots recognize that hunger and low blood sugar levels might have been causal in their incidents; at the time of the incident flight, they might have been aware of feeling hungry, but were unlikely to have been aware of their degraded cognitive performance.

2 PILOTS' REPORTS

To determine the extent of the food availability problem in routine operations and to investigate the reported effects of hunger on pilot performance, we conducted a search of the ASRS database (Bischoff & Barshi, 2003). Several common situations appear in the descriptions of the effects of hunger on pilot performance. Primarily, physiological difficulties, cognitive difficulties, and crew coordination difficulties were described. Pilots reported incidents such as altitude deviations, crossing active runways, misidentifying and landing on the wrong runway, unauthorized landings, or other non-adherence to clearances from air traffic control (ATC).

These ASRS reports are consistent with results from laboratory driving simulator performance studies of individuals (progressively induced to hypoglycemic states), who demonstrated significantly impaired driving – off-road driving (driving across the midline), driving fast, and applying brakes on the open road (Cox, Gonder-Frederick, Kovatchev, Julian, & Clarke, 2000). Furthermore, impaired individuals were reluctant to take corrective action by either pulling off of the road or taking an available glucose drink – 43% severely impaired individuals did not take corrective action. This study is particularly telling because the driving simulator scenario was relatively simple and lasted only five minutes.

The National Transportation Safety Board (NTSB) has also identified and documented lack of food, low blood sugar and/or hypoglycemia as contributing factors (either directly or indirectly) in three commercial airline accident reports (NTSB, 1976; 1979; 1986) and four general aviation aircraft accident reports (NTSB, 1989; 1992; 1997; 1998). It should be noted that an abnormal blood glucose level resulting in hypoglycemia cannot be determined from postmortem specimens using existing procedures (Canfield, Chaturvedi, Boren, Véronneau, & White, 2000). Whether or not a pilot has experienced an abnormal hypoglycemic episode during flight cannot be clearly determined. Also, tests for blood glucose levels, or detailed questions about food intake are not currently a required part of the NTSB accident investigation process, even when the crew survives the accident. Furthermore, adverse effects on pilots' cognitive performance may occur at blood sugar levels that would not be clinically considered as hypoglycemic. Thus, the full extent of the impact of pilots' low blood glucose levels on aviation safety may be greatly underestimated.

3   EFFECTS OF BLOOD GLUCOSE LEVELS ON PILOT PERFORMANCE

A study by Barshi, Mauro, Feldman, Geven, Currin, et al. (2012) is currently investigating the effects of blood glucose level on pilot flight performance, ATC communication, and cognitive performance. Additional mood and physiological assessments are also included. Using scripted scenarios and a flight simulator (piston single/twin aircraft Flight Training Device), pilots fasted overnight and then were provided juice the next morning and prior to beginning their flight legs. Pilots were fitted with a continuous blood glucose monitor (MiniMed CGMS) that averages blood glucose levels every five minutes. This device was worn by the pilots throughout the experiment, and was calibrated through finger sticks (One Touch). In a within-subjects design, pilots prepared their flights, completed checklists and weight and balance sheets, and were administered cognitive tests and mood assessments. They flew four similar flight legs on each of two different days. On both days, after the second flight leg, pilots were either provided or not provided with a snack. Water was available to the pilots throughout the experiment. Participants served as their own controls and the conditions were counterbalanced. Preliminary data analysis of one subject shows that overall blood glucose levels fell from fasting and rose due to a carbohydrate drink/snack, as expected (Figure 1). In the no-snack condition, blood glucose levels fell or were flat, as expected (Figure 2). Note, in both figures, blood glucose level is shown in red; activities, such as juice/snack, flight legs, and breaks are shown in black.

When we evaluated preliminary error data of two subjects, the effect of not eating on performance became clear. Subjects made three times more major errors during the simulated flight mission in the no-snack condition, compared to their performance in the snack condition. The wealth of data from audio and video recordings, and from the simulator, as well as from the various cognitive and mood instruments will provide new insight and a new understanding of the effects of blood glucose levels on pilot performance, and specifically in the areas of flight performance (e.g., altitude deviations), ATC communication (e.g., read back accuracy), and cognitive performance (e.g., decision-making).

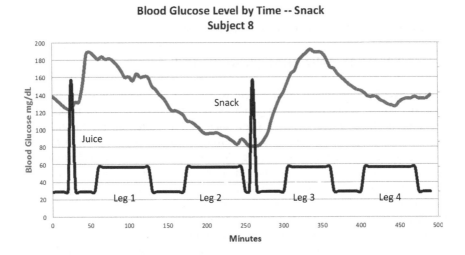

Figure 1. Blood glucose level while flying in the snack condition

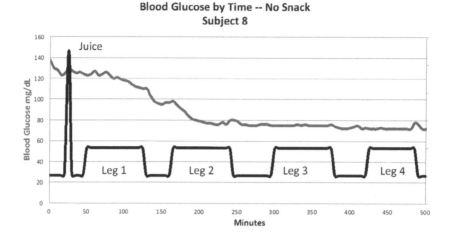

Figure 2. Blood glucose level while flying in the no-snack condition

4   ETHICAL AND PRACTICAL ISSUES

The aviation operational environment is complex and demanding, yet business concerns motivate choices that may not support the basic needs of pilots while operating in these critical settings. Crew members are often unable to obtain the food that they need, to remain well-fed throughout their duty day, or to even be aware of the consequences of not eating. Yet crew meals are gone, and production pressures push the schedules such that there may not be time to buy food between flights. What's more, since the events of September 11th, 2001, the cockpit door is locked, and it's a very complicated process for flight attendants to bring food or beverages into the cockpit. Thus, it should come as no surprise that pilots may not even be aware of their impaired performance resulting from not eating

It is well know that most aircraft incidents and accidents are due to failures in decision-making, in attention, and in memory retrieval. What is not know is how many of these failures are the result of compromised blood glucose levels. Errors such as incorrect taxiing and runway over runs are very similar to those seen in the driving simulator studies – off-route driving, speeding, and inappropriate stopping. These are expensive errors in many more ways than just money. Yet avoiding them is relatively simple.

Blood glucose awareness training is currently administered to diabetic individuals, showing sustained benefits in insulin-treated actively-flying pilots (Grossman, Barenboim, Azaria, Goldstein, & Cohen, 2005; Cox et al., 2001). It is possible to adapt a similar training method and implement this program throughout the aviation industry. Education can increase pilots' self-awareness of food consumption (or lack thereof) and its effect on their flight performance. Through education and training, individuals can learn to become aware of and recognize the early signs of low blood glucose levels, when and what impairments happen specifically to them, and when they might need to take corrective action. Moreover, companies can provide their employees with nutritional snacks, that are easy to carry around over multiple days, at a much lower cost than the cost of incidents and accidents.

5 CONCLUSION

Laboratory studies, pilots' reports, and NTSB investigations show a common thread of serious issues relating lack of food and pilot performance. These issues are relevant for general aviation, commercial, and military pilots – in environments where metabolic demands increase and peak performance is critical. It is likely that the adverse effects of low blood glucose levels on aviation safety is grossly underestimated, particularly so when other employee groups are taken into account such as maintenance workers and ramp crews.

Low blood glucose levels affect performance in all of us – whether we are the pilot or the passenger. Everyday tasks (e.g., remembering to do an errand, performing mental arithmetic, driving or flying safely) are likely impaired from compromised blood glucose levels due to a lack of food – whether we realize it or not. In the case of airline pilots, it is a safety issue, an ethical issue, and a business issue that can and should be resolved.

ACKNOWLEDGMENTS

The authors would like to acknowledge Dr. Robert Mauro of Decision Research, Senior Research Associate Captain Richard Geven of Perrot Systems, and Jeff Currin, Dan Ziller, and Bill Troyer of the University of Oregon at Eugene.

REFERENCES

Barshi, I., R. Mauro, J. Feldman, R. Geven, J. Currin, and D. Ziller (2012). *Effects of blood glucose levels on pilot performance*. Manuscript in preparation.

Bischoff, J. and I. Barshi. 2003. *Flying on empty: ASRS reports on the effects of hunger on pilot performance*. Proceedings of the International Symposium on Aviation Psychology, Dayton, Ohio, 125-129.

Canfield, D.V., A.K. Chaturvedi, H.K. Boren, S.J.H. Véronneau, and V.L. White. 2000. *Abnormal glucose levels found in transportation accidents*. (FAA Publication Report No. DOT/FAA/AM-00/22: 1-11). Washington, D.C.: Office of Aviation Medicine.

Cox, D.J., L. A. Gonder-Frederick, B.P. Kovatchev, D.M. Julian, and W.L. Clarke. 2000. *Progressive hypoglycemia's impact on driving simulation performance*. Diabetes Care, 23(2): 163-170.

Cox, D.J., L.A. Gonder-Frederick, W. Polonsky, D. Schlundt, B. Kovatchev, and W. Clarke. 2001. *Blood glucose awareness training (BGAT-2)*. Diabetes Care, 24(4): 637-642.

480

Donohoe, R. and D. Benton. 1999. *Declining blood glucose levels after a cognitively demanding task predict subsequent memory.* Nutritional Neuroscience, 2: 413-424.

Evans, M.L., A. Pernet, J. Lomas, J. Jones, and S.A. Amiel. 2000. *Delay in onset of awareness of acute hypoglycemia and of restoration of cognitive performance during recovery.* Diabetes Care, 23(7): 893-898.

Feldman, J. and I. Barshi. 2007. *The effects of blood glucose levels on cognitive performance: A review of the literature.* NASA Technical Memorandum 2007-214555. Moffett Field, CA: NASA Ames Research Center.

Grossman, A., E. Barenboim, B. Azaria, L. Goldstein, and O. Cohen. 2005. *Blood glucose awareness training helps return insulin-treated aviators to the cockpit.* Aviation, Space, and Environmental Medicine, 76(6): 586-588.

Holmes, C., K.M. Koepke, and R.G. Thompson. 1986. *Simple versus complex performance impairments at three blood glucose levels.* Psychoneuroendocrinology, 11(3): 353-357.

Johnson, O.C., J.B. Carlson, D.V. Veverka, and B.P. Self. 2007. *Effects of low blood glucose on pilot performance.* BIOS, 78(3): 95-100.

Lieberman, H.R., C.M. Caruso, P.J. Niro, G.E. Adam, M.D. Kellogg, B.C. Nindl, and F.M. Kramer. 2008. *A double-blind, placebo-controlled test of 2 d of calorie deprivation: effects on cognition, activity, sleep, and interstitial glucose concentrations.* American Journal of Clinical Nutrition, 88: 667-76.

Lindseth, G.N., P.D. Lindseth, W.C. Jensen, T.V. Petros, B.D. Hellan, and D.L. Fossum. 2011. *Dietary effects on cognition and pilots' flight performance.* The International Journal of Aviation Psychology, 21(3): 269-282.

National Transportation and Safety Board. 1976. Aircraft Accident Report. *Alaska Airlines, Inc. Boeing 727-81, N124AS, Flight 60, Ketchikan International Airport, Ketchikan, Alaska, April 5, 1976* (Report No. NTSB-AAR-76-24). Washington, D.C.

National Transportation and Safety Board. 1979. Aircraft Accident Report. *Air New England, Inc. deHavilland DHC-6-300, N383EX, Flight 248, Hyannis, Massachusetts, June 17, 1979* (Report No. NTSB-AAR-80-1). Washington, D.C.

National Transportation and Safety Board. 1986. Aircraft Accident Report. *Pan American, Boeing 747-121, N751PA, Flight 362, Miami, Florida, December 21, 1986.* (Report No. NTSB-MIA-87-IA- 054). Washington, D.C.

National Transportation and Safety Board. 1989. Aircraft Accident Report. *Cessna 150, N704SM, Orlando International Airport, Orlando, Florida, December 26, 1989* (Report No. NTSB-MIA-90-L- A041). Washington, D.C.

National Transportation and Safety Board. 1992. Aircraft Accident Report. *Luscombe 8A, N71183, Snoqualmie, Washington, August 24, 1992* (Report No. NTSB-SEA-92-LA-188). Washington, D.C.

National Transportation and Safety Board. 1997. Aircraft Accident Report. *Beech G35, N4552D, Jackson, Wyoming, July, 6, 1997* (Report No. NTSB- SEA-97-LA-159). Washington, D.C.

National Transportation and Safety Board. 1998. Aircraft Accident Report. *Extra Flugzeugbau EA- 300, N301NL, Chenoa, Illinois, April 12, 1998* (Report No. NTSB-CHI-98-LA-121. Washington, D.C.

Owens, D.S. and D. Benton. 1994. *The impact of raising blood glucose on reaction times.* Neuropsychobiology, v30: 106–113.

Sawka, M.N., S.N. Cheuvront, and R. Carter. 2005. *Human water needs.* Nutrition Reviews, 63(6):S30-S39.

CHAPTER 46

Naming of Taxiways on a Large Airport to Improve Safety through Simpler Memorization

Vilmar Mollwitz, Thorsten Mühlhausen, Helmut Többen, Jörn Jakobi,
Michael Kreuz

DLR German Aerospace Center
Institute of Flight Guidance
Braunschweig, Germany
michael.kreuz@dlr.de

ABSTRACT

This study analyzes two different alternatives to designate the taxiways of a newly build airport. These two different layouts are named X and Y (proposed by airport operator and commercial pilots) in the following. The aim of this study is to determine which proposal is more intuitive and therefore easier to remember by pilots. An empirical study was designed and executed with two groups of pilots from different airlines.

All pilots were randomly split into two groups X and Y and got a short training phase for their layouts respectively. The task was to reproduce logical taxiway clearances by drawing the taxi-paths in blank airport charts. In addition they should provide their subjective opinion of this taxiway naming.

The results showed a clear preference for one of the two naming layouts. Pilots also reasoned their preference. The analysis showed that both naming layouts revealed room for improvements.

Therefore, a meeting with the main stakeholder at this airport (ANSP, Airport authority and major airlines) was held to get the best possible layout for all of them. This meeting took place with the collaboration of DLR and a satisfactory solution could have been found for all participants.

Keywords: taxiway, airport, safety, memorization

1 INTRODUCTION AND MOTIVATION

The optimization potentials for the naming of taxiways at a newly build airport are examined. Objective was to get an intuitive solution for pilots and ATC in the frame of ICAO recommendations for taxiway designation. Intuitive in terms of memorability and a related safety gain. Two different variants were favoured in the beginning of this study, one from the airport operator and one from airline pilots. In the following, these two attempts are made anonymous by using the labels X and Y.

2 EXPERIMENTATION

2.1 Background

The naming layout X was designed in accordance to the desires of the local ATC provider, naming layout Y by a local airline supported by DLR.

Thereafter, both layouts were proved for their conformity with ICAO recommendations, Annex 14 "Aerodromes, Volume I – Aerodromes Design and Operations". Conformity was particularly checked with recommendations set in chapter 5.4.3.36 not to mention the letters I, O, X nor the words "inner" and "outer" (ICAO, 2004; Horonjeff et al, 2010): *"5.4.3.36 Recommendation – When designating taxiways, the use of the letters I, O or X and the use of words such as inner and outer should be avoided wherever possible to avoid confusion with the numerals 1, 0 and closed marking".* Both naming layout were updated and brought in line regarding all ICAO recommendations.

2.2 Setup of the experiment

In order to check the memorability of the two naming layouts an experiment was designed by DLR. The experiment itself consists of three phases.

In a first phase one of the two naming layouts is presented to the test subject for a period of five minutes with the instruction to memorize the naming. In order to avoid learning effects or confusions between the two designs, every test subject is presented of either naming layout X or Y, not both. After this "phase of memorizing" a distraction task was given: Each candidate had to play a Sudoku game or to underline a defined letter out of a letter soup for the same period of time (five minutes). Doing so, the naming of taxiways should be erased from the short-term memory.

Third phase was the real experimental task: 12 different taxi instructions were being read to the test subjects. Instructions were read slowly and just one single time. After each instruction the test subject was requested to reproduce the instructions by marking it in an unlabeled layout of the airport. This unlabeled layout is identical to the one provided in the "phase of memorizing" but does not include any naming of taxiways (gates, apron parking positions are not included either). In accordance to real working habits pilots were allowed to take written notes. Working time per taxi instruction was limited to one minute each. Turning back to instruction samples that have already been worked on was not allowed. In both groups taxi instruction and their sequence of presentation were identical spatially but different in naming caused by a different naming layout.

The recruitment as well as the execution of the experiment took place during a professional training. Pilots were unprejudiced and unprepared. Due to lack of time demographically data could not be assessed. Assigning pilots to the layouts Y or X was decided randomly by drawing lots. Finally, 23 subjects worked on layout X and 19 on layout Y. The experiment was conducted in a separate and quiet room, one for each group.

3 RESULTS

In total 42 pilots participated in this study. They were randomly allocated to taxi layout X = 23 test subjects and layout Y = 19 test subjects. Each of them was instructed to listen to 12 different taxi clearances and to reproduce it by a drawn line in a blank airport map. Deviations between the actual taxi clearance and the drawn line were recorded as "mistakes". A complete data set with 42x12 = 504 answers, without missing value, could be assessed.

Every not recognized or wrongly marked taxiway in the layout is evaluated as one mistake. Due to differences in the length of instructions the maximum amount of possible mistakes differs among the instructions. In average 1.96 mistakes per reproduced taxi clearance were observed. With group X every test subject causes 27.04 mistakes in average over all 12 taxi clearances, 2.25 (SE = .021) mistakes per taxi clearance. Instead group Y causes a total of only 19.47 mistake in average, 1.62 (SE = .023) mistakes per taxi clearance. Data were proven for significant differences by a 2x12 mixed-model ANOVA. Mixed-model because one of the factors is repeated-measures and the other is not. The calculated F-value of $F_{(1, 40)} = 1.669$ proved statistically significance (p = .49). Taxi instructions based on layout Y are reproduced more reliable than naming layout X. Figure 1 gives an overview of the average number of mistakes per taxi instructions for both tested layouts.

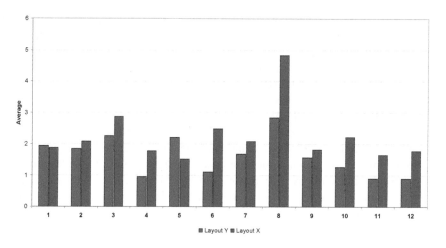

Figure 1 Average mistake per taxi instruction

To be seen in figure 1, except of item 1 and 5, all taxi instructions are performed with fewer mistakes on basis of layout Y than on basis of layout X. The maximum amount of average mistakes and also the greatest difference in average mistakes were observed with taxi instruction number 8 (taxiing from the North to take-off position on runway 07R) representing the greatest length and complexity.

Following Figure 2 gives an overview of the accumulated average mistakes per layout for all taxi instructions, 19.47 for layout Y respectively 27.04 for layout X.

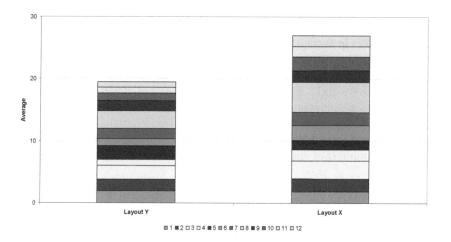

Figure 2 Accumulated average amount of mistakes per test condition

In order to analyze possible results for the deviation in the results shown above, pilots were given the possibility to provide their comments on both layouts, mainly in terms of the felt intuition. Their statements are documented in following chapter 4.

4 COMMENTS BY THE PILOTS

In addition to the execution of the test described in chapter 3 pilots were given the chance to provide comments on the naming layout to find out optimizations criteria with regard to the naming of taxiways. Comments on both layouts are summarized in the following.

4.1 Comments on layout X

The majority of the pilots disapproved layout X (see Appendix B). The reasoning of structure in naming the taxiways is not obvious and in consequence can not easily be memorized by the pilots. It is merely the naming of the four main taxiways (A, B, C, D) being parallel to the runway that is judged as "comprehensive". The naming of the taxiways diverging from the main taxiways (V1, VC, V2, P3) is judged as "not logical" and not comprehensible for most pilots.

Apart from that, the lack of taxiway K2 despite the existence of taxiways K1 and K3 is assessed critically. This "inconsistency" represents a risk in terms of safety especially for landing operations as pilots expect a taxiway named K2 instead of taxiway K3 right after a taxiway K1. Moreover, the naming of taxiways at the take-off position on runway 25L in the south part of the airport with the letter "N" is criticized as this one is associated easily with the direction of North.

Furthermore, the layout of Munich International airport is said to be a positive example for the naming of taxiways at airports. Due to the similar spatial arrangement of the terminal and the two runways, a naming in accordance to the one at Munich airport could lead to a significant improvement in memorability and thus to a gain in terms of safety.

4.2 Comments on layout Y

Concerning the layout Y (see Appendix C) there are heterogeneous opinions. Especially the naming of the taxiways according to the geographical direction (taxiway N in the north and taxiway S in the south of the airport) are judged "very good". According to the pilots asked, another positive characteristic about this layout is that focus is laid on the view of pilots landing/approaching. It is mentioned that the pilot's orientation on ground is supported by the numbering of taxiways (main taxiways labeled with N in the North as well as S for those in the South being parallel to the runway are defined with ascending numbers from runway in the direction of the terminal). The structure of naming the taxiways in the West (marked with W and E for those ones being in the East) is judged by a part of the interviewed pilots as "not obvious" in connection with the parallel taxiways (marked with C) in between. A possible reason for this might be that the given layout represents the first phase of construction and does not take the terminals of the final construction stage into account whereby the targeted system of naming of taxiways could get more obviously.

The naming of taxiways in the north of the airport is judged critically. The use of the letters T and U as well as the naming of taxiways H3 and H4 in the area of the former airport is mentioned as "not obvious" and "confusing". This can be used as an explanation for the higher amount of mistakes concerning taxi instruction number five (taxiing in the northern part of the airport, see Figure 1). Apart from that, using two letters for exit taxiways per runway is considered negatively. Naming these taxiways with only one letter and ascending numbers is mentioned to be a better variant.

In general, the pilots mentioned that the reference point can hardly be recognized. Some of the taxiways are numbered from the West to the East and some from the East to the West. In comparison to this irritating naming of taxiways, the naming of the taxiways at the Airport in Munich is mentioned as a positive example.

4.3 Differences in the naming of taxiways concerning the proposed layout Y and the of taxiway naming at Munich Airport

The geographical similarities between the airport layout of the new airport and Munich Airport (ICAO, 2012) have already been mentioned in chapter 4.1. In the following, the differences concerning the naming of taxiways between the layout offered by party Y and the layout carried out in Munich shall be discussed briefly.

Exit taxiways of the runway in the north at Munich Airport are labeled with the letter A and numbered from the West to the East in ascending order. The same approach has been realized with the exit taxiways of the runway in the South that are labeled with the letter B.

The taxiways being in parallel to the terminals at Munich Airport are named according to the geographic direction with the letters "W" (for the German word "Westen") and "O" (for the German word "Osten" that "East" is English for) and also from West to East in ascending order. At this, it needs to be mentioned that with regard to the recommendations set by ICAO the letter O shall not be used.

4.4. Miscellaneous comments

In general, pilots prefer a standard taxiway that can be used as long as possible without being given any additional instructions. In this regard, Amsterdam airport is mentioned as a good example several times.

Other suggestions such as the naming of taxiways with two letters as being implemented at Vienna airport are discussed controversially by pilots but do not represent any preferences.

5 RECOMMENDATIONS

In accordance to the experiment carried out there is statistically significant higher reproducibility of taxi instructions that are based on layout Y, which is designed according to the geographical direction.

These statistically proofed facts are supported by the comments being summarized in chapter 4. With regard to these comments the layout Y is judged as "easier recognizable" but still with room for improvements. The critical comments of the pilots concerning the layout X coincide with the objective measurable worse result of layout X compared to the one proposed by party Y.

Nevertheless, both variants of naming taxiways do not represent the optimum. Taxiways at Munich airport is indicated as a better layout by the pilots. This might be explained by the higher publicity of Munich airport.

According to this, coordination between the affected airlines, the airport authority as well as the national air navigation service provider were encouraged by the pilots. With the collaboration of DLR a successful conclusion as well as satisfactory solutions for all parties could be achieved in a final summit.

REFERENCES

Horonjeff, R. et al. (2010). *Planning and Design of Airports*, McGraw Hill, Fifth Edition,
ICAO (2012). AIP Germany, Aerodrome Chart AD2, Munich EDDM 2-25
ICAO (2004). *Annex 14 to the Convention on International Civil Aviation, Volume I, Aerodromes Design and Operations*, Fourth Edition, 2004

APPENDIX A: TAXI INSTRUCTIONS

Instructions layout X

1. Landing 25R, taxi to P3 via L1, V2, C
2. Departure P3, taxi to runway 07L via C, Q1, D, L4
3. Landing 25L, taxi to E via M1, R, B, V1
4. Departure V2, taxi to runway 07R via B, Q1, A, M4
5. Landing 07L, taxi to H3 via H1, T
6. Departure R, taxi to runway 25L via B, P3, A, N5
7. Landing 07R, taxi to R via N3, V2, B
8. Departure H3, taxi to runway 07R via T, K3, L4, D, S, B, M5
9. Landing 25R, taxi to H4, via K3, T
10. Departure E, taxi to runway 25R via V1, C, J2, D, J5
11. Departure V1, taxi to runway 25L via E, B, N4
12. Departure F, taxi to runway 07L via V1, C, R, L4

Instructions layout Y

1. Landing 25R, taxi to C2 via L1, E3, N2
2. Departure C2, taxi to runway 07L via N2, C3, N1, L4
3. Landing 25L, taxi to S3 via G1, W1, S2, E1
4. Departure E3, taxi to runway 07R via S2, C3, S1, G4
5. Landing 07L, taxi to H2, via H1, T
6. Departure W1, taxi to runway 25L via S2, C2, S1, F5
7. Landing 07R, taxi to W1 via F3, E3, S2
8. Departure H2, taxi to runway 07R via T, K2, L4, N1, W2, S2, G5
9. Landing 25R, taxi to H4, via K2, T
10. Departure S3, taxi to runway 25R via E1, N2, J2, N1, J5
11. Departure E1, taxi to runway 25L via S3, S2, F4
12. Departure N3, taxi to runway 07L via E1, N2, W1, L4

APPENDIX B: LAYOUT PROPOSED BY PARTY X

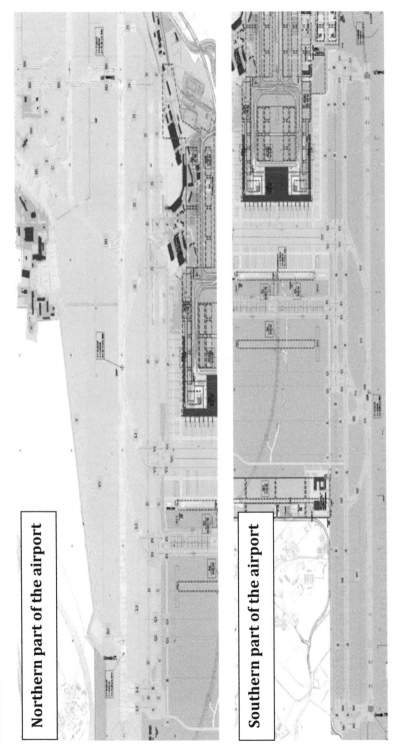

Northern part of the airport

Southern part of the airport

490

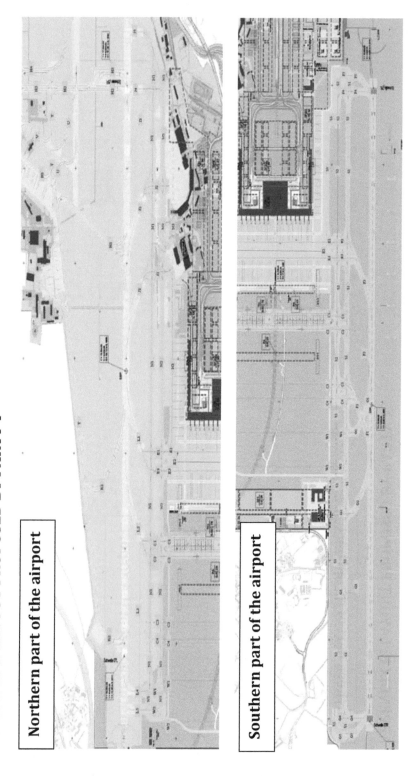

APPENDIX C: LAYOUT PROPOSED BY PARTY Y

Northern part of the airport

Southern part of the airport

The Application of Line Operation Safety Assessment (LOSA) and Flight Data Management Technologies to General Aviation Flight Training

Richard J. Steckel, M.S., Manoj S. Patankar, Ph.D.

Saint Louis University
St. Louis, Missouri, USA
rsteckel@slu.edu

ABSTRACT

This study explored the use of Line Operations Safety Assessment (LOSA) (also known as Line Operations Safety Audit) incorporating the Threat Error Management (TEM) model and flight data management technologies to determine the prevalence and quantity of threats, errors and undesired aircraft states (UAS) that occur during a normal flight in a collegiate aviation flight training program. Digital image, voice and flight data recorders were installed in two PA-28-201 Piper Arrow aircraft used for university FAR Part 141 flight training.

By reviewing the recorded data, the study examined the ability (or inability) of student pilots/instructors to recognize, mitigate and/or "trap" threats, errors and undesired aircraft states that occur. This study also examined the applicability and effectiveness of implementing losa program within a collegiate aviation training program by reviewing the current state of losa in all flight operations, investigating the technologies available to facilitate losa in a general aviation setting, assessing the current state of collegiate aviation training programs and examining possible organizational and safety culture issues associated with implementing a LOSA.

Keywords: LOSA, TEM, flight training, image, voice, data, recorder

1   INTRODUCTION

1.1   Line Operations Safety Audit (LOSA)

LOSA collects data using the Threat Error Management (TEM) model, which focuses on the type and quantity of threats to the flight (e.g., weather, airport design), errors committed by the flight crew (e.g., incorrect radio phraseology, level off at an incorrect altitude, missed checklist item) and undesired aircraft states (e.g., too slow on approach, incorrect aircraft configuration) and how the flight crew react and manage these anomalies that may not be reportable or adversely affect the flight when managed correctly.

The FAA uses a medical metaphor to describe a LOSA. It is similar to an annual checkup at the doctor's office. The purpose of a checkup is to use diagnostic measures to detect health concerns before they manifest themselves in a more serious manner. A LOSA is a diagnostic picture of the health of an air carrier and assessing its strengths and weaknesses in normal operations.

Klinect (2005) defined the ten operating characteristics of a LOSA:

- Jump seat observations of regularly scheduled flights
- Voluntary flight crew participation
- Anonymous, confidential and non-punitive data collection
- Joint management/union sponsorship
- Secure data collection repository
- Trusted and trained observers
- Systematic observation instrument
- Data verification roundtables
- Data derived targets for enhancement
- Feedback of results to line pilots

The Threat and Error Management (TEM) framework focuses simultaneously on the operating environment and the humans working in that environment. Because the TEM taxonomy can also quantify the specifics of the environment and the effectiveness of performance in that environment, the results are also highly diagnostic. The origin of the Threat Error Management model was tied to the development of LOSA (Merritt & Klinect 2006). In a cooperative effort in 1994, Delta Airlines and the University of Texas Human Factors Research Project (UTHFRP) developed Line Operations Safety Audit (LOSA) as a field observation method to assess crew resource management skills (Klinect, 2005). Klinect likens TEM to defensive driving for pilots. TEM does not teach a flight crew how to fly, but how to minimize and successfully manage the risks associated with any flight.

Commercial air carriers found LOSA as an effective tool. For example, after the initial implementation of LOSA in 1996 at Continental Airlines, pilots accepted error management and used it in every day operations. A LOSA conducted four years later in 2000 discovered improvement in crew performance (e.g., checklist usage) and a 70% reduction in non-conforming or unstabilized approaches (Gunther, 2002).

1.2   Transitioning LOSA to general aviation

General aviation accident investigation/prevention still relies primarily on a post-accident/incident investigative process to collect and analyze physical forensic evidence found at the accident site, witnesses to the accident or air traffic control recordings. General aviation continues to have the highest aviation accident rates within U.S. civil aviation: about 6 times higher than small commuter and air taxi operators and over 40 times higher than air carrier operations (NTSB 2011). Since the majority of the 228,683 active general aviation aircraft in 2008 (BTS 2011) were not turbine powered, multi-engine aircraft with more than 10 passenger seats, they were not required by federal regulations to have an operable flight data or cockpit voice recorder (FAA 2011).

This study explored the use of LOSA for improving safety performance and risk management in a general aviation collegiate aviation training environment using current flight digital data, cockpit voice and image recorders in lieu of human observers used in previous LOSAs.

In support of this exploration, this study examined the following research questions:

- What is the current state of technology that will enable LOSA to be accessible, effective and affordable to collegiate GA training programs?
- What are the benefits of and barriers to implementing LOSA in a collegiate general aviation flight training program?
- What types and prevalence of threats, errors and undesired aircraft states occur during a typical collegiate GA training flight (dual or solo) that could affect safety performance and risk management?

1.3   Limitations of human observers

While the LOSA data collected in operations and previous research were ground breaking and very effective, the use of human observers and the analog method of recording data (pencil, clipboard, and appropriate forms) meant that: a) any threats or errors missed by the observer were lost, b) it would be difficult to simultaneously compare data gathered by the observer, FDR or CVR after the flight c) there could have been observation reactivity by the pilots caused by the observer being in the cockpit and d) if there was no provision for a jump seat, an observer could not be used.

Klinect's (2005) research focused on the airline industry where most aircraft come equipped with a jump seat in the cockpit for the observer. A review of the FAA's advisory circular on LOSA (FAA, 2006) and the ICAO LOSA document

(2002) both show their respective processes using a human observer to collect data.

1.4   Use of cockpit image recording

This study used image recorders as a substitute for a human cockpit observer, which is a controversial subject within the aviation community.

The National Transportation Safety Board (NTSB) has implementation of cockpit image recorders on its "10 most wanted list" (NTSB 2011a). While the NTSB has advocated image recorders in the cockpit, airline pilot unions are opposed to this technology. The Airline Pilots Association (ALPA) has said "Current technology already provides investigators with the tools they need to determine the cause of airline accidents....ALPA is opposed to any use of video recording in the cockpit." (ALPA 2011 p. 1)

Miller (2004) states air carrier pilots oppose image recorders in the cockpit today because promises made in the 1960's that cockpit voice recorder tapes would only be used for accident investigations were broken when the cockpit voice recording for a 1989 Delta Airlines accident in Dallas was released to and played by the media.

1.5   Safety Culture Issues

Patankar & Sabin (2010), state there are 4 states of safety culture that can exist in an organization, secretive, blame, reporting and just. LOSA depends on a just culture where individuals are encouraged/rewarded for providing safety information, safety behaviors become more important that outcomes, risk taking behaviors are discouraged regardless of outcome, reckless behavior is penalized regardless of outcome and employees and management are accountable for safety, resulting in a higher employee-management trust.

1.6   Learning Culture Issues

In addition to having a safety culture that will allow a LOSA program to be effective and successful, any organization implementing LOSA should be a transformational learning organization. In a transformational learning organization, undesirable events are used to lead the development of a proactive system to identify hazards, evaluate risks and mitigate potential for a failure, maximizing the knowledge gained from all previous undesirable events.

1.7   A review of current collegiate aviation training program safety initiatives

An informal email survey was conducted among several universities with flight training programs. The university flight departments were asked if they had the following: a safety reporting system, a LOSA program, a FOQA program and/or a safety management system.

All universities had a safety reporting system and all were working towards implementing a safety management system (Table 1). Very few universities had a FOQA program and none had a LOSA program in place.

Table 1 – University safety programs

Institution	LOSA	FOQA	Safety Reporting System	SMS
Saint Louis University	Yes	No	Yes	In work
Embry-Riddle	No	Yes	Yes	In work
University of North Dakota	No	In work	Yes	In work
Western Michigan State University	No	Yes	Yes	In work
Southern Illinois University Carbondale	No	No	Yes	In work

2   METHODOLOGY

To validate the effectiveness of the LOSA in this setting, the study conducted descriptive quantitative research using an observational study as the instrument to gather data on the prevalence and quantity of threats, errors and undesired aircraft states that occur during a normal flight in a collegiate aviation training program. Appareo GAU 2000 digital flight data recorders and ChaseCam digital audio/image recorders were installed in two Saint Louis University Piper Arrow PA-28R-201 aircraft and served as a virtual LOSA observer, recording parameters that might not be captured in FOQA data. By observing crew actions in the cockpit, subtle hand gestures between the flight crew and/or actions taken that might not be mentioned verbally on the checklist and that might cause a threat, error, or undesired aircraft state were collected/recorded. The image recorder had an advantage over a human observer, being able to faithfully replay the flight without the memory or observational lapses that a human observer might have.

Participants were recruited from the instructors and students at Parks College of Engineering Aviation and Technology, Saint Louis University. Participants had the option of not recording a flight if they so chose, and they could leave the study at any time. Forty flights of approximately 2 hours in duration were recorded. Thirty two flights yielded usable data, meeting the study target of recording and analyzing a minimum of thirty flights.

Threats, errors and undesired aircraft states were defined using a modified checklist from Klinect. Observed threats, errors and undesired aircraft states on recorded flights were captured on a data collection sheet.

3   RESULTS

The data were extracted from the data collection sheets and input into IBM SPSS version 19.

The total number of threats, errors and undesired aircraft states were counted. There were a total of 109 threats, 118 errors and 10 undesired aircraft states recorded during the study period.

All of the codes for each event involving a threat, error and undesired aircraft state, the response to the event threat, error or undesired aircraft state (Correct, incorrect or none) and the outcome to the event (corrected, expired, new code) were entered for each flight. Since no incidents or accidents occurred during the study period, that block of data was omitted from the study.

Next, the average number of threats, errors and undesired aircraft states per flight were calculated. There was an average of 3.28 threats, 3.56 errors and 0.34 undesired aircraft states (UAS) per flight, or a total average of 7.18 events per flight. Klinect (2005) in his study only reported average threats per flight, which he found at 4.00 per flight, very close to the results of this study.

A Pareto chart was then created to determine the most prevalent threats, errors and undesired aircraft states that occurred during all of the recorded flights (Figure 1).

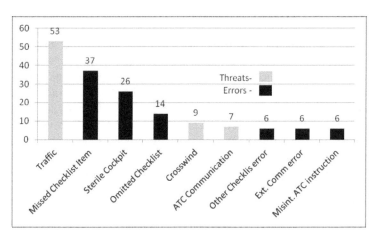

Figure 1 Pareto chart of the most prevalent threats and errors

The most prevalent threat was other aircraft "traffic" and the most prevalent error was missed checklist items.

A cross tabulation was run on SPSS 19 using Type (threat, error, undesired aircraft state), response (correct, incorrect, none) and Outcome (corrected internally, expired, new code, corrected externally) to determine how well students and instructors detected threats, errors and undesired aircraft states, how they responded to them and the outcome to their response or lack of response.

The SPSS cross tabulation format made the data difficult to interpret, so it was transposed onto charts showing the flow of the threats, errors and undesired aircraft states. Each chart (Figures 2-4)) graphically shows how threats, errors and undesired aircraft states were managed.

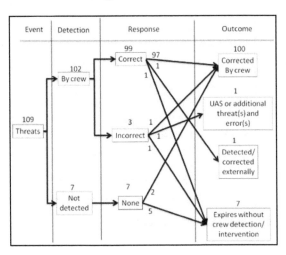

Figure 2 Flow chart of how threats were managed

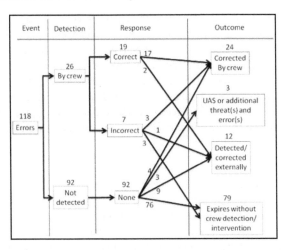

Figure 3 Flow chart of how errors were managed

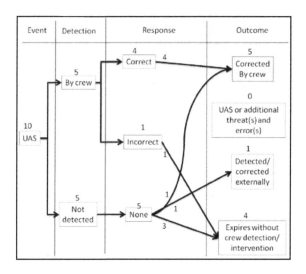

Figure 4 Flow chart on how UAS were managed

The participants successfully detected and responded to threats that occurred. Ninety three percent of threats were detected and 97% of those detected had correct responses.

The participants did not do so well with errors. Seventy seven percent of errors committed by the participants went undetected. Of these, 85.8 percent of these were expired and were never detected. Fourteen percent of undetected errors were corrected externally (by air traffic control), created an undesired aircraft state or were corrected by the participants without knowing an error had been committed.

Of the ten undesired aircraft states, five were detected and four of those had the correct response. The five undetected undesired aircraft states expired, were inadvertently corrected by the participants or were corrected externally by someone outside of the aircraft.

5   DISCUSSION

Enrolling participants in the study was an unexpected challenge. Initially, a personal presentation was made to prospective participants. Flyers were also distributed at the university's airport facility. This resulted in a disappointing two students agreeing to participate.

Interviews conducted by the investigator revealed that the perceived safety culture among students and instructors was one of blame. There were two main concerns; that the flight school management would obtain the data and punish students and/or instructors for any infractions, intentional or unintentional and that the FAA would be able to obtain identifiable data and issue violations to the students and instructors. This was particularly troubling to the students and instructors working towards a career in the airlines, as any FAA issued violation could seriously jeopardize their chances to be employed by an airline. Additional information sessions addressing these concerns of trust resulted in an adequate number of participants to gather the requisite flights.

This study was limited to a structured collegiate flight training program. The majority of pilots either own or rent the aircraft they fly and may not be in a formal safety program. Future research could be conducted to determine how LOSA might be effectively implemented in this pilot demographic. Would rental pilots allow rental agencies to monitor their flights? Would aircraft insurance companies find value in performing LOSAs on their clients? Progressive Insurance (2012) is currently doing something very similar to a LOSA with their automotive clients, allowing them to install "Snapshot" sensors onto their vehicles to monitor their clients' driving habits. Aviation insurance underwriters and their clients may want to investigate this further if it allows the underwriters to more accurately determine client risk.

4   CONCLUSIONS

The study showed that there is technology currently available, affordable and small enough to be equipped in general aviation aircraft to collect usable LOSA data. The results showed that general aviation pilots are exposed to many of the threats, make errors and create undesired aircraft states much like major air carriers.

The number of undetected errors pilots committed, knowing that they were being recorded was surprising. An effective employee error reporting system will not capture errors the employee is not aware of committing.

This study also showed if pilot trust can be gained and the pilots perceive the information will be used to improve safety and not used as a reason for punishment, they will allow voice and image recorders into the cockpit.

REFERENCES

Air Line Pilots Association (ALPA), (2011). *Investigating airline accidents: Cockpit Video is not the answer.* Retrieved August 13, 2011 from http://www.alpa.org/portals/alpa/ pressroom/inthecockpit/CockpitVideo.htm

Bureau of Transportation Statistics (2011). *Table 1-11: Number of U.S. Aircraft, Vehicles, and Other Conveyances.* Retrieved September 10, 2011 from http://www.bts.gov/publications/national_transportation_statistics/html/table_01_11.ht ml

Federal Aviation Administration, (2011). *Federal Aviation Regulation 91.609(e) Flight data recorders and cockpit voice recorders.* Retrieved September 10, 2011 from http://ecfr.gpoaccess.gov/cgi/t/text/text-idx?c=ecfr&sid=ff7df264d40756f97562c877ebce6c97&rgn=div8&view=text&node=1 4:2.0.1.3.10.7.7.5&idno=14

Federal Aviation Administration (2006). *Line Operations Safety Audits. Advisory Circular AC 120-90.* Retrieved July 9, 2011 from http://flightsafety.org/files/AC-20120-9011.pdf

Gunther, D. (2002). Threat and error management training counters complacency, fosters operational excellence. *International Civil Aviation Organization Journal, 57,* No. 4, pp. 12-13.

International Civil Aviation Organization (2002*). Line Operations Safety Audit, Doc 9803 AN/761.* Retrieved July 9, 2011 from http://icao.int/anb/humanfactors/LUX2005/Info-Note-5-Doc9803alltext.en.pdf

Klinect, J.R. (2005). *Line Operations Safety Audit: A cockpit observation methodology for monitoring commercial airline safety performance.* Retrieved July 9, 2011 from http://www.lib.utexas.edu/etd/d/2005/klinectj57340/klinectj57340.pdf

Merritt, A., Klinect, J.R. (2006*). Defensive flying for pilots: An introduction to Threat Error Management.* The LOSA Collaborative. Retrieved July 10, 2011 from http://homepage.psy.utexas.edu/homepage/group/HelmreichLAB/Publications/pubfiles/TEM.Paper.12.6.06.pdf

Miller, L. (2004). *Pilots oppose call for in-cockpit cameras.* The Associated Press, July 27, 2004. Retrieved August 13, 2011 from http://archives.californiaaviation.org/airport/msg31325.html

National Transportation Safety Board (NTSB). (2011*). General aviation safety: What is the issue? Retrieved September 10, 2011 from* http://www.ntsb.gov/safety/mwl-2.html

National Transportation Safety Board (NTSB). (2011a). *NTSB unveils new "most wanted list.* NTSB press release. National Transportation Safety Board, Office of Public Affairs. Retrieved August 13, 2011 from http://www.ntsb.gov/news/2011/110623.html

Patankar, M.S., Sabin, E. (2010). The Safety Culture Perspective. In E. Salas & D Maurino (Eds), *Human Factors in Aviation* (pp. 95-122). Salt Lake City, UT; Academic Press.

Progressive Casualty Insurance Company (2012). *Find out how to save with usage-based insurance.* Retrieved February 25, 2012 from http://www.progressive.com/auto/snapshot-how-it-works.aspx

CHAPTER 48

Action Research & Change Management System in Aviation

M.C. Leva[1], N. McDonald[1], P. Ulfvengren[2], S. Corrigan[1]

[1]Aerospace Psychology Research Group - School of Psychology Trinity College Dublin Ireland
[2]KTH Industriell Teknik och Management | Address SE-100 44 STOCKHOM (Sweden)
levac@tcd.ie

ABSTRACT

The main objective of MASCA is to deliver a structure to manage the acquisition and retention of skills and knowledge for managing change across the air transport system. The project includes different stakeholders in a common operational system (airlines, airports, maintenance companies, etc.) in the common effort to identify critical areas to change the shared operational system to deliver a better service, especially in the area of performance management in safety. To meet the objectives within the Project it was decided to follow an action research framework that concerns the implementation and evaluation of change, as well as the analysis and measurement of operational parameters needed to identify, plan and implement a successful organizational change initiative like the one related to performance and risk. The present paper reports about its initial phases and of the research hypothesis followed in guiding its development.

Keywords: aviation, management of change, action research

1   THE PROBLEM OF MANAGEMENT OF CHANGE AND CULTURE IN AVIATION

Within the aviation industry the need for sustainable change is becoming more and more imperative. Change is in fact something being imposed on the industry from a number of sources (e.g., Regulations, increasing commercial pressure and the introduction of new technologies) and the need for change management skills and capability within organizations is required in order to meet these challenges.

The literature on organizational change demonstrates that, against different criteria and outcomes, only a minority of major change initiatives (typically between 30% and 50%) have a positive outcome (Dent and Powley, 2001, Kotter, 1995). Change is necessary, but it is risky. It is vital therefore that the reasons for organizational change are understood by those responsible for its enactment. Where change involves people in organizational processes it becomes primarily a 'human factors issue' and involves complex, multidimensional solutions. This is made difficult by the fact that change often results in "resistance" by employees and this requires careful management in terms of hearts and minds.

Organizational Culture has a dynamic role in maintaining system stability. Qualitative change in collective understanding may only come after a cumulative aggregation of many minor shifts in the way a social group make sense of their situation; the shift may then be rapid and volatile; this is why the importance of consolidating and embedding change is so often emphasized.

It has also been observed that when management is distant or strategies are not clear employees do not share their competence, their knowledge and keep on working on the basis of what they already know without proposing new ideas, or directions. Instead, when management has a clear strategy and a way to communicate and pursue this strategy with all the employees, involving them towards a common sense of direction the possibilities of establishing an internal process of change for business improvement are greatly increased (Kotter 1996).

As culture cannot be directly 'managed' or controlled, attempts to do so often create an unofficial counter-culture (Kunda 1992). However culture can not be ignored and it is important to be able to focus on those aspects and phenomena that are relevant and upon which company have an influence such as goals, vision and values; the perception of the system and its functions, often called climate; subcultures differentiated by roles and boundaries; the quality of engagement between people and organization, including dimensions like trust or alienation. For a cultural measure to perform this function effectively it has to be grounded in a demonstrated relationship to the functioning of the system it represents. Culture can be seen as the active engagement of a collective in a process of change. The cultural analysis then can become part of the change process. A classic approach to cultural analysis is to seek to establish fundamental meanings and values. Here the aim is to construct a rich in-depth interpretation from a broad range of material. While this may not be particularly useful from a short-term perspective in managing change, it can be extremely valuable in developing a strategic view of some of the challenges that need to be faced in changing an organization in the long run.

1.1   The MASCA Project

The main objective of MASCA is to deliver a structure to manage the acquisition and retention of skills and knowledge for managing change across the air transport system. The project includes different stakeholders in a common operational system (airlines, airports, maintenance companies, etc.) in the common effort to identify critical areas to change the shared operational system to deliver a better service, especially in the area of performance management in safety. The management of safety needs to be integrated into the overall management process if safe out-comes are to be assured. Standard safety performance indicators for instance are necessary, but not sufficient. They need to be part of a greater process able to carry on the indications into possible improvements initiatives; therefore participating in a generic organizational capability for managing or implementing change. Starting from these premises MASCA is concentrating on extending this state of the art by developing a dedicated Change Management System that will facilitate change initiatives through greater transparency, mentoring and information support in a sustainable manner and enhancing the commitment of organizational stakeholders toward its success. This will also involve training and competence development for change management skills and providing stakeholders with a better understanding of the various other aspects of the aviation system that their work impacts on and impacts on them.

Figure 1 illustrates the main elements of the Process of Change proposed in MASCA. It is organized around three dimensions strategy (direction and goals), process (functional system) and competence (making sense of the system). Interpolated in the diagram are some other dimensions that are important in the management of change: the role of internal and external change agents; the extent to which the organization can act in a coherent integrated way; and the use of data to track change and validate strategy. The central premise of the diagram is that change is a process, from diagnosis of needs to evaluation of outcome, which engages these different elements at every step of the way.

2   THE ACTION RESEARCH APPROACH IN MASCA

Action research can be described as a methodical, interactive approach embracing problem identification, action, planning implementation, evaluation and reflection. The insight gained from the cycle feed into further continuous improvement cycles of further collaborative enquiries performed by reflective practitioners (Zuber-Skerritt, 1992). In this sense " the approach followed in the current paper stem from the belief sustained by Torbert (2001) that "Knowledge is always gained through action and for action. From this starting point, to question the validity of social knowledge is to question, not how to develop a reflective science about action, but how to develop genuinely well-informed action — how to conduct an action science". In approaching the problem in MASCA the practitioners involved performed a 'forensic and in-depth' insight into each of the industrial partners involved in the project regarding their current strategic and operational approaches so as to identify, understand and assess their needs in implementing and effectively evaluating change.

The various visits of the members of the research team in the industrial partners involved:

- Structured interviews and observations with key stakeholders
- Documentation Analysis

- Introduction Focus Groups aimed at building an internal MASCA improvement team
- Triggering ideas by providing examples of possible support tools in a MASCA Change Management System

The structured interviews were used to perform a preliminary analysis of the industrial partners around three main areas that a change management system should be able to address:

1. Strategy. An insight the organization's relationship with its environment, the goals it has set of itself, and how those goals are also communicated and used with the member of the organization. In starting a change initiative the first element is the needs for change. This come from the awareness of the key strategic goals each of the companies are pursuing in order to meet the needs that current business and economic challenges are presenting. This is a vital motivational element and needs to be communicated thought-out the company. The management has to commit to a common agenda that deliver benefits also on field or task performance. Further this commitment should also translate in deeper understandings of recommendations for change, ensuring that they are not just 'quick-fixes' and that they are not independent and applicable only to certain 'operating silos'. It is also advisable to try and link up a new project with the current change initiatives that the company may already have in place to facilitate and ensure that overall, as well as local or departmental, strategic objectives are met.

2. Process. What are the minimal conditions for being able to purposefully change a functional system? It is necessary to know how that system works or functions as well as to be able to track what it is doing-understand its processes. The process model derived also needs to be sufficiently well defined and grounded in the operation to be meaningfully linked to real operational data measuring, concurrently, system inputs, activity and outputs.

3. Competence. Where change involves people in organisational processes it becomes primarily a 'human factors issue' and involves complex, multidimensional solutions. This is made difficult by the fact that change often results in "resistance" by employees and this requires careful management in terms of hearts and minds.

It is therefore necessary to address the following three aspects:

- Participation, which consists in including end users in the project from the beginning until the end. It ensures taking into account their opinion and their wishes. Their knowledge of the trade is particularly useful for project managers who will be alerted if the project is going off track.
- Communication will help all employees to understand the aim of the projects and the reasons why changes will happen. Well informed about project development, end users will have a better understanding of its importance for the company and of the impacts it will have on them.
- Learning must ensure that end users have all the keys (theoretical and practical knowledge) to start successfully with a new system or within a new organisation. Facilitating knowledge transformation and transfer

throughout the organisation, organisational learning and organisational memory. Support staff in their efforts to improve performance providing the training, education and mentoring required by change and the delivery of the appropriate training to sustain new operational realities as well as the crucial skills for managing the change.

Figure 1: Main elements of the Change Process in MASCA

In the action research approach adopted within the project the main scope to be fulfilled was to identify an intervention that would actually be of use for the industrial partners. To be able to do so, which means identifying areas in need of improvement the researchers needed to start establishing as much as possible a relationship based on trust, their interaction with the workers in fact will change the way people think about the future and will create expectations. By pointing to certain issues the researcher in an organization will stir things up a little and it is only by trust that an organization and the individual in it may let you access weak spots that are normally left in the dark.

The evidence suggests that the particular combination of internal and external change agents can be critical to the success of change initiatives (Altricher et al 2002). So it makes sense to ask: Where are the influences for change coming? Who are important actual or potential change agents in the organization? What external influences are there and how do they operate? These can include academic and research collaborations, industry networks and associations, industrial innovation districts, communities of practice. In what ways can these influences provide guidance for a change program? European collaborative projects provide the opportunity for a variety of types of collaboration that can stimulate innovation:

- Learning between operational companies
- Innovative collaboration between companies along a business process (e.g. flight operations and maintenance)
- University/ research institute collaboration with industrial partners for RTD development, action research, education and training.

Operational organizations, unlike large design and manufacturing organizations do not have research and development departments – thus collaborative research and development can provide an effective innovation process to compensate for this. However, for this to work, the collaboration has to deliver benefits to meet the short-term operational or organizational goals of the participating organizations, as well as the medium to long-term research goals of a project. This fits well within an action research framework that concerns the implementation and evaluation of change, as well as the analysis and measurement of operational parameters needed to identify, plan and implement a successful organizational change initiative like the one related to performance and risk.

2.1 The Research Assumptions for MASCA CMS.

In defining the Research hypothesis to be tested in MASCA the main issue we wanted to tackle was to provide an evidence base that can justify acting (better than a pragmatic rule of thumb or an armchair academic analysis). Building on the aspects considered in the MASCA CMS some Axis of intervention were identified. In planning an intervention there are two aspects of the system that can be directly influenced – the way the system works, and how people understand and make sense of it (see Figure 2). If both of these aspects are addressed in a congruent manner, then the performance of the new system (as a result of people's collective actions) can be measured; in time, collective experience with the new system will be embedded and consolidated in the collective understanding and values (the culture) of the organization.

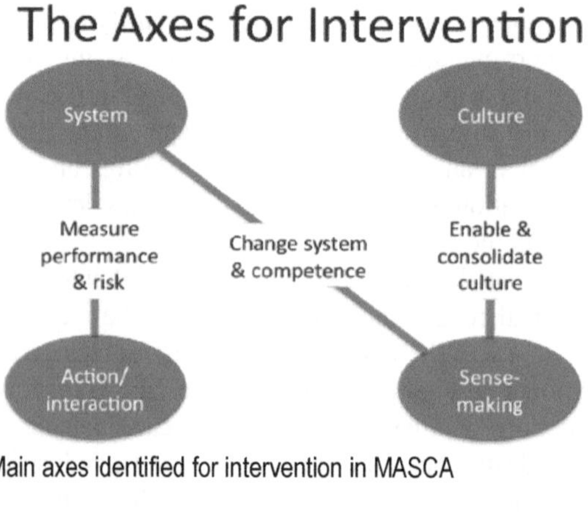

The Axes for Intervention

Figure 2: Main axes identified for intervention in MASCA

The research hypothesis that we are going to test in action therefore can be expressed in terms of the needs that a change initiative needs to address to work out.

- 1: The need for creating a common picture.
  This assumption is based on theories on involvement, participation and motivation as to meet a common problem of resistance to change or how it is done (Maurer 1996). Reasons for the lack of a common picture are different domain expertise and local objectives that in part may be explained by organizations working in "silos" and need to divide operations into result units with different accountabilities as part of a performance management system. An adequate level of understanding of a working operational system requires as a basis a functional knowledge of the relationships between inputs, tasks and activities and outputs in the actual practices.
- 2: The need for measuring performance not only at the sharp end (targets) but also in terms of antecedents.
  Change is often driven by a need to improve performance and the program may rely heavily on performance indicators to express targets and measure progress. Given the enormous power of performance management systems to distort the desired functionality of the system, it is important to understand the ways in which measurable behaviours of people or performance indicators are actually related to the underlying functionality of the system. There are two basic terms in a causal or probabilistic analysis of performance that are necessary to be considered – antecedents and consequences.
- 3: The need for risk assessing an initiative before it starts. Linked to the previous assumption performance in a safety critical domain like aviation has also another element not to be overlooked: risk. Doing a risk assessment as part of a change initiative on the process lay out to achieve the goals will increase the capacity of the project to meet unexpected issues and make the change more effective (as well as being in conformity of the requirements of ICAO SMS and EASA guidelines. Further the capability of managing system risk opens up the possibility of managing the process in a more integrated way – by making explicit the interdependencies between different organizational functions supplying and supporting the operation.
- 4: The need for feedback and reports on progress.
  This is based on theories on motivation and the need for incentives to get the required sense of achievement and ownership to make the change worthwhile.
- 5: The need for developing an agency between research and practice.
  This consideration is based on examples from research in innovation and industrial management. The need for both knowledge from research and practice is needed in successful change. However not enough applied successful research has been made applicable in practice and researchers will never gain enough domain knowledge from the outside. Many industries lack their own research departments.

3 INTERIM RESULTS AND EVALUATION

As already said the main objective in MASCA is to develop and deploy a Change Management System based on assumptions from change management theory and the MASCA model. With respect to evaluation both the system developed and its effect on operational outcome needs to be assessed. The main stakeholders are the industrial partners and there interest in MASCA's overall contribution to outcome of change implementation. The other stakeholders are the researchers interested in validating research questions and evaluating how well the developed CMS also may meet the research objectives. Each partner can provide feedback and evaluate a specific subset of objectives formulated for each case study so that they can be verified and used for evaluation purposes. In the context of the action research approach the researcher are aiming at explicitly verifying the main hypothesis used as a bases of the interventions. Table 1 reports the strategy identified to test all of them. Some of those strategy have already been implemented in the form of a pre-MASCA survey used in the initial site visits.

Table 1: Evaluation strategy identified for each research hypothesis

Research Hypothesis	Verification Method
There is a need for creating a common picture	Masca survey results regarding commitment and understanding o the need for change at the pre-MASCA intervention compared to survey post MASCA intervention regarding the understanding and the sharing of the need for change and perceived usefulness of the initiative.
There is a need for measuring performance not only at the sharp end (targets) but also in terms of antecedents	Comparison of the "As is situation" for follow up initiatives and identification of corrective actions in a KPI driven system with the MASCA proposed intervention regarding performance management (based on antecedents and consequences identification)
There is a need for risk assessing an initiative before it starts	Actual benefits potential and effective delivered to the initiative through the use of a risk assessment at an early stage on the core aspects of an imitative such as effects on resources and allocated time.
There is need for feedback and reports on progress	Survey performed on perceived usefulness of initiative and understanding of data management within the company pre-MASCA and after. Concrete results description.
There is a need for developing an agency between research and practice.	Evaluation of Agency support provided by Industrial partners

4   CONCLUSIONS AND NEXT STEPS

The outcome of the initial phase lead to pave the way to the future direction of the action research being carried out in the project aimed at supporting the following areas:

a)   The role of Competitiveness, regulation and technology induced change in the aviation process: a system of systems perspective for a concrete intervention. The intervention chosen as a test bed is the introduction of Collaborative Decision Making (CDM) in a major European Airport (Swedavia). This entails the establishment of better communication connectivity within each organization (airline EFB and Datalink to ATM), the use of serious games to train people in a new system perspective.

b)   A model for Management of change: preparedness, assessment and training that foresee the use of Process Mapping Serious games and process assessment to prepare and assess possible major changes in a regional airline. This initiative also needs to have a link to the approaches proposed for Performance Management as the outcome of a good model and training for management of change as to be translated in terms of monitoring of performance.

c)   The importance of performance management for continuous improvement driven changes. The interventions to be used as test beds for this topic are going to be: (i) the introduction and use of Safety Performance Indicators in a Safety Management system as a way of identifying and driving improvements in a major airline. (ii) The introduction of a more holistic performance management approach in a small regional airport and its role for continuous and sustainable development. (iii)

d)   A Learning training framework to support the above initiatives
       -    HF training for different aviation stakeholders
       -    Serious games
       -    Mentoring of Masca internal teams
       -    A Master program on managing change in human systems.

Previous experiences shows developing and sharing innovative ideas about people in complex systems, and even exchanging staff between the research group and an industrial company can transfer research-based knowledge into practice and by doing so stimulate process innovation and increase competitiveness (Ward et al 2010) we are hoping to increase the knowledge based previously acquired about management of Change and demonstrate it by finalizing more successful concrete interventions.

ACKNOWLEDGMENTS

The above-mentioned research has received funding from the European Commission's Seventh Framework Programme FP7/2007-2013 under grant agreement 266423 "MASCA".

REFERENCES

Airport Council International (2006) Airport Benchmarking To Maximise Efficiency. Published By ACI World Headquarters Geneva – Switzerland.

Altricher H., Kemmis S., McTaggart R., Zuber-Skerritt O. (2002) The concept of Action Research, The Learning organization Volume 9 Number 3 pp 125-131.

Button, K.J., McDougall, G. (2006) Institutional and structural changes in air navigation service–providing organizations. Journal of Air Transport Management, 12, 236–252.

Clarke, S., 1998. Organizational factors affecting the incident reporting of train drivers. Work & Stress 12, 6–16.

Dent and Powley (2001). Employees Actually Embrace Change: The Chimera of Resistance. Journal of Ap-plied Management and Entrepreneurship, September 26, 2001

Heinrich, H. W. (1980). Industrial Accident Prevention: A Safety Management Approach, (5th ed.). ISBN 0 07 028061 4.

ICAO (2009) Safety Management Manual (SMM) Doc 9859 AN/474.

Kotter, J. P. (1995, Mar-Apr). Leading change: Why transformation efforts fail. Harvard Business Review, Vol. 73, No. 2, March/April, pps. 59-67.

Kotter, J. P. (1996) Leading Change. Harvard Business School Press.

Kunda G. 1992 Engineering Culture. Temple University Press.

Maurer, R. (1996). Beyond the wall of resistance: Unconventional strategies that build support for change. Austin, TX: Bard Books, Inc

McDonald N. 2007 Human Integration in the Lifecycle of Aviation Systems In D. Harris (Ed.): Engin. Psychol. and Cog. Ergonomics, HCII 2007, LNAI 4562, pp. 648–657, Springer-Verlag Berlin Heidelberg

Pransky, G., Snyder, T., Dembe, A., Himmelstein, J., 1999. Under-reporting of work-related disorders in the workplace: a case study and review of the literature. Ergonomics 42, 171–182.

Probst, T.M., Brubaker, T.L., Barsotti, A., 2008. Organizational under-reporting of injury rates: an examination of the moderating effect of organizational safety climate. Journal of Applied Psychology 93 (5), 1147–1154.

Scotti D. (2011) Measuring Airports' Technical Efficiency: Evidence from Italy PhD dissertation University of Bergamo Department of Economics and Technology Management.

Tye, J. (1976). Accident Ratio Study, 1974/75. London: British Safety Council.

Ward M., McDonald, N., Morrison, R., Gaynor, D., & Nugent, T. (2010) 'A Performance Improvement Case-Study in Aircraft Maintenance and its implications for Hazard Identification' in Ergonomics, Special Edition: Human Factors in Aviation. Vol. 53, Issue 2, Pages 247 – 267.

Zohar, D., (2003). The influence of leadership and climate on occupational health and safety. In: Hoffman, D.A., Tetrick, L.E. (Eds.), Health and Safety in Organizations: A Multilevel Perspective. Jossey-Bass, San Francisco, CA.

Zuber-Skerritt O. (1992) Professional Development in Higher Eduation: A theoretical framework for Action Research. Kogan Page, London.

CHAPTER 49

Airline Framework for Predictive Safety Performance Management

P. Ulfvengren[1], M.V. Leva[2], N. McDonald[2], M. Ydalus[3]

[1]KTH Royal Institute of Technology, INDEK, 100 44 Stockholm, Sweden
[2]Aerospace Psychology Research Group - School of Psychology, Trinity College
Dublin Ireland
[3]SAS Scandinavian airlines, STOOB, 195 87, Stockholm, Sweden
pernilla.ulfvengren@indek.kth.se

ABSTRACT

In an ongoing project, MAnaging System Change in Aviation, MASCA (EU-FP7) a model for system change is proposed. In this action research project the objectives are both to develop and to deploy a change management system, CMS, in real change interventions with industrial partners in accordance to this model. This paper reports from one of the planned case studies in an airline in development of an improved PM framework in an advanced SMS. The purpose of this research is to develop, implement and evaluate a performance management framework that will enable the airline to let reliable and relevant indicators drive change to improve the operational process and the airline's safety performance. Initial result from and planned roadmap for future research is presented. A review of earlier research identifies the need and recommendations for new PM based on HILAS and MASCA research. The airline's existing tools and processes are analyzed in relation to the MASCA approach and CMS to identify where contributions are most beneficial and what qualities are there already. Consolidation of existing parts and parts of work in progress results in a model for a proactive PM framework and a brief on planned future research.

1   INTRODUCTION

Can you measure safety? If safety within an organization is more than absence of accidents, at a given time, it is not sufficient to measure safety in numbers of accidents or incidents. In safety critical systems such as the Aviation Transport System, ATS, there is a need to assure safety. There are safety goals set for the future Air Traffic Management, ATM, systems of NextGen and Single European Sky, SES. The European Flight Plan, a European Safety plan 2011-2014 (EASA, 2011) has set a target to increase safety by reducing the accident rate by 80%. To know that the goal has been achieved or that progress is made there is a need to measure safety, both on organizational and system level. Stakeholders' competence and understanding of safety, and what safety model(s) are used will affect which approach is taken and how much effort is made to contribute to that goal. The European commission has produced a list of actions with the aim of improving safety (EC, 2011). The actions are within categories such as systemic issues, operational issues, emerging issues and finally human factors and human performance. It is recognized that the last category will bring improvement across all the other areas.

There is an on-going change in the way regulators and operators manage performance with regard to safety introduced by a new regulation that is performance-based. The International Civil Aviation Organisation, ICAO has issued a Safety Management System, SMS, manual (ICAO, 2008). Safety will be assessed based upon how well an organization manages to change and improve what is considered to facilitate safe operational outcomes, in other words its change capability.

Change capability is one of the most desired qualities in any business. It facilitates innovation and adaption to sudden trends in markets. In the Air Transport System, ATS, two major on-going changes are a new Air Traffic Management system and new regulation. It is evident that a change capability is needed to prepare for these changes, as well as internal improvements for sustainable safety and profitability. This work contributes to the need for an overall change management system. Managing system change in aviation is an extreme challenge considering the safety aspects and the fact that the reported success rate of major change initiatives are low (Dent and Powley, 2001).

There are many challenges with developing a safety performance management framework in aviation. One is to define a safety model that describes what is considered to increase safety in operations. There exist numerous models in the field of safety science and human factors. Integrating safety in other management systems is another challenge, so the framework needs to support integration across departments and, at system level, across organizations.

A new human factors model has been proposed and implemented in steps in two consecutive EU-projects, HILAS and MASCA. This paper reports from on-going MAnaging System Change in Aviation, MASCA (EU-FP7). A model for system change is proposed based on the new human factors model, building on HILAS research and accepted change management theories. In this action research project the objectives are both to develop and to deploy a change management system, CMS in real change interventions with industrial partners in accordance to this model. The scope of this paper is part of one of the planned case studies (Ulfvengren et al., 2012) in an airline in with an on-going development of an improved PM framework in an advanced SMS. The purpose of this research is to develop, implement and evaluate a performance management framework that will enable the airline to let reliable and relevant indicators drive change to improve the operational process and the airline's safety performance. In this paper initial result from this development and planned roadmap for continuation is presented.

2 THEORETICAL FRAMEWORK

Here follows a brief overview from a literature review of earlier research. It presents the scope of theories that have been considered such as general performance management (PM), performance management in relation to safety, safety management systems (SMS), development of indicators and earlier research in HILAS. Ongoing work including the present MASCA model and latest version of the Change Management system is also presented.

2.1 Performance Management and Safety

Today's business environment is constantly changing. Pressure has increased system wide on capacity, safety, quality, security, cost, environment, regulations and competition. This makes the conditions under which processes are operated constantly changing. Management control aims to have intentional impact on operations, executives and personnel towards certain objectives. Support for these tasks is: formal controls such as budgets and performance measures; organisational structure such as allocation of responsibilities and accountabilities and incentives and less formalised control such as organisational culture, learning and authorization (Ax, Johanson and Kulvén, 2009). The overall goal with PM is to deliver improvement of performance, i.e the operational process outcome. Using key performance indicators at various departments allows aggregation and collection of measures to gather an overall summary of results of the business as well as to identify performance of specific areas and need for improvement. An indicator is an observable measure that provides insights into a concept that is difficult to measure directly (Harms-Ringdahl, 2008). Safety is a concept known to be difficult to measure directly. The model(s) of safety in an organisation reflect the understanding of how safety emerges and will also determine what indicators that are identified to be essential for improved safety (Wreathall, 2008). If an indicator is valid or not has more to do with the causal model that is being assumed than with the indicator per se (Hopkins, 2009). SPI's are used primarily to be support of and control in the risk management process, but also support the relation with over-sight authorities. It is argued (Mawdsley, 2010) that an internal goal for safety departments also is to present performance dashboard of key performance metrics to impact decision for safety budgets, resources and training for safety related activities!

An airline is built up by many departments and production areas. Departments are given directions and goals and work out their own strategies to meet them. The core airline has an overall strategy to reach business goals. A strategic decision-making process (Beauchamp, 2008) needs to be in place to ensure this alignment. A lack of strategic decision-making may lead to a failure to balance safety risks with intense production pressures which may result in accidents (Woods, 2003). Many airlines struggle as a business to earn sustainable profit. Management approaches such as Lean (Womack and Jones, 2003) helps a very pressed industry to survive and stay in a cost-cutting mode. In human factors and safety research there are many models explaining how weaknesses in technology, organizations and culture contributes to safety failures leading to incidents and accidents (Norman, 1986; Wickens et al., 2004; Reason, 1997; Hollnagel, 1998, Perrow, 1984). To keep balance between protection and production aviation should be able to manage lean and safe performance in an integrated system. Specifically the new SMS and safety performance regulation is advocating an integrated approach (ICAO, 2008). There are two novel approaches in this initiative. Safety will be integrated in other management systems and safety performance will be assessed, instead of traditional safety outcomes such as frequency of undesired events. Similar to a Quality Management Systems, QMS, safety performance targets are to be set, safety performance indicators will be monitored to follow progress and safety performance requirements will guide mitigation strategies. It is clear that safety will be assessed based upon how well an organization manages to change to facilitate safe operational outcomes, in other words its change capability. In order to implement such management system a new model that has not traditionally been included under the rubric of safety management is needed.

Historically safety has been extremely dependent on quality of technology that constitutes the aircrafts. The need for rigorous control for certification of any detail connected to the aircraft remains today. Later safety outcomes were ascribed to other areas of the system like the "human factor" and human error (Reason, 1990). Operators have been blamed for not using the technology correct and making errors. Still the only measure of this performance was number of incidents and accidents. In human factors research much emphasis has been on humans' interaction with technology, i.e human factors engineering (Wickens et al., 2004). Instead of blaming the operator for using technology wrong usability and cognitive engineering strive for developing technology that when used makes it difficult to do wrong and easy to do right, technology "that make us smart" (Norman, 1993). A combination of various performance improvements in the past has made aviation the safest way to travel. Today, however, aviation is argued to no longer becoming safer (Learmont, 2009). Even if a systemic view has been proclaimed in HF and safety for some time to address this but the ruling model seams still to be reactive attempts to reduce numbers of incidents and undesired events. However a model improving only the traditional measures in operations will not take aviation from the safety plateau (Amalberti, 2001). Eurocontrol (2010) summarizes the scope of influence of Human Performance (HP), leading to human errors, in ATM, i.e. designing the right technology and selecting the right people. In this research approach it is argued that analysing human performance will not necessarily improve the process that produces the outcome. Humans are part of the system but are not producing the operational outcome per se, but surely their performance is influenced by what the system offers them. The process that produces the operational outcome is influenced by more than aircraft performance, flight operations and crew performance. Planning, dispatch and duty officer, maintenance ground as well as airport services and air traffic management and control will build in risk before departure. It was in the field of maintenance that phenomena were found that were not possible to explain by current human factors models which derived in to a new human factors model.

2.2 HILAS

The Aviation Psychology Research Group at Trinity College in Ireland has for many years conducted human factors research in the maintenance domain (McDonald, 1999, 2001 a, b; McDonald et al., 2000). They have identified some main phenomena which are not possible to explain by common theory or HF models: *Double standard* describing the gap between strategic management views of operations and the real "normal" operations as performed by operators, *WIPIDO,* meaning Well Intended People In Dysfunctional Organisations, describes how operators do their best to overcome an actual or perceived lack of support in operations, a strong *professional culture* seams to compensate for weaknesses in the organisation and *cycles of stability* describes that even after failures organizations do not easily learn and change. As an answer to this research a New Human Factors model has been suggested (McDonald, 2009). Four key theoretical themes; System, Action, Sense-making and Culture are identified and combined into this model, believed to have both relevance and leverage in real life applications. "It is proposed that these terms represents or refer to different aspects of the same underlying reality…" (Mc Donald, 2009). In planning an intervention two of these aspects of the system can be directly influenced – the way the *system* works, and the *sense-making,* how people understand it. If both of these aspects are addressed in a congruent manner, then the performance of people's collective *actions* can improve performance, and in time, collective experience with the new system will be embedded and consolidated in the *culture,* the collective understanding, of the organization.

With this background the HILAS project (Human Integration in the Lifecycle of Aviation Systems)(EU-project, 2005-2009) was launched and run and developed a platform for the integration of human requirements across the industry lifecycle (HILAS, 2005). Socio-technical systems theory on the other hand has been strong on general principles for optimizing both technical and social aspects of systems, but weak on specific methodologies for analyzing system functioning. Central to this is a capacity to strategically assess the risks confronting the system and to transform this into a programme of change which can be evaluated in terms of effective risk control. Prerequisites for this are to (McDonald, 2006) model operational risk using quantitative operational data and simulating operational transformation as well as to develop organisational processes for managing multiple goals in an integrated way, processes for managing change, and the role of culture.

An essential tool which supports these prerequisites is the Operational Process Model and Knowledge Space Model (OPM/KSM) (Morrison, 2009). It is clear that the New Human Factors model also has several design implications on performance management directed by a set of policies describing the logic for the new human factors model with regard to performance and safety. Performance management here has two main focuses of balancing control and support in order to both capture relevant data to measure, monitor and evaluate performance from the system as well as enabling and supporting performance. This applies to both individual and system level. At an individual level it is people producing quality and safety but also those building capacity and managing the system. At a system level it is supporting operational processes facilitating and supporting the operational outcome, i.e. innovative reporting tools were developed linked to the Operation Process Models (Leva et al., 2010) which serve the purposes of both individual and system level performance management.

2.2.1 Operational Process Model-Knowledge Space Model

Process Mapping is commonly referred to as the process of collecting information about the activities that an organization needs to perform in order to achieve a certain goal. It is used to identify and define the activities and events that constitute the production process and further to highlight deficiencies and weaknesses in the flow of information and resources in order to rectify these. Within the HILAS project the OPM is used as part of a broader tool for human factors analysis of processes based on what is known as the Knowledge Space Model (KSM). The KSM is a tool which allows an analyst to interrogate operational processes using a serious of questions designed to highlight the most important information, material & human inputs, the relationship between different tasks in the process, the goals, contextual factors and key process enablers.

The KSM framework can be described in several ways: as a model of a system or operation which incorporates rich layers of knowledge about how that system really works, as a HF analysis methodology of a process or task, as a methodology for analysing the current functioning of the system as well as the potential impact by introducing changes as to understand a process, accident, incident, event or risk analysis, as guideline for the knowledge transformation process into a common understanding of the operational implications of potential changes and it represents processes graphically which help understanding the logic of what to measure and why and to identify indicators. For these reasons the OPM/KSM is an ideal tool for monitoring and evaluation.

2.2.2 Prospective risk analysis- PRA

Prospective risk analysis has certain organizational and data requirements as well as required relationship between different actors and tools (Cooke and McDonald, 2009). Currently the standard approach to safety depends almost entirely on identified safety performance indicators which are compiled as a result of experience from past events. However, this is not enough for anticipating the effect of change and to explore risk of operations in advance of the occurrence of incidents so that the reduction in risk may be achieved proactively. To explore risk involves the analysis of data along with the use of the OPM/KSM for anticipating the effects of the operational processes. The need for a specific prospective risk analysis may be triggered by parameter exceedance, i.e., when trends and boundaries being routinely monitored are observed to go beyond acceptable levels. There should be a direct link between the low level measures and the high level KPIs with a dataflow enabling aggregation bottom-up and disaggregation top-down between operational, tactical and strategic management layers. This emphasises the need for the integration of all performance data and the avoidance of data silos. Only by grounding the strategic high level KPIs in relevant operational performance aspects, it is possible to direct the organisation in a meaningful way. Often KPI's are derived top down and thus fail to effectively monitor how the operation delivers its outcomes. As a result interventions are based on misleading assumptions that might increase gap between the actual operation and its strategic oversight. As a complement to this bottom up approach of aggregating indicators, in a similar way the KSM approach facilitates a top down approach. When the strategic overview identifies a requirement for a change of direction, this can be translated and cascaded down from a high level KPI into lower level PIs that are grounded in the actual process. This ensures that any change that is implemented can actually be monitored and evaluated appropriately. This helps to ensure safety as a priority. PRA is dependent on the following conditions being in place:

1. **Safety performance indicators** (SPIs) and low level key performance indicators KPIs and a clear link between low level SPIs and high level SPI/KPIs as identified in the balanced scorecard;

2. Availability of tools **gathering of operational data** i.e the reporting tools and data structure proposed in HILAS and the sourcing of information from other parts of the organisation and the rest of the industry;

3. Availability of tools for **storing, retrieving and analysing** these data with the capability of providing targeted and intelligent reports on performance allowing the identification of action and change requirements;

4. Along with technical support for data management there is a requirement for an **operational process model** (OPM) and employment of the **Knowledge Space Model** (KSM) as a tool for facilitating the populating of the OPM, the qualitative analysis of the processes and the prescriptive identification of needs for change.

5. An **organisational structure and climate** for facilitating staff engagement at all levels of the organisation including human factors training. A "culture of fairness" (Swedish airline concept) is essential for capturing low-level operational data the buy in from managers to work across department on safety issues.

The HILAS model was validated in a real change case. When proven successful the rational was proven also to the industrial partners (Ward et al., 2010). In relation to PM the HILAS approach was proven more effective than a Lean approach which analyses tend to be weak in relation to people and safety functions in complex systems. In relation to understanding processes it was felt that the OPM provided a much richer and safer means of modelling than the Lean 'value stream mapping', which focuses on what adds value to the process and what does not (waste).

2.4   MASCA

The MASCA project (Leva et al, 2011) is further developing the HILAS platform. The MASCA project is organized around two complementary objectives, to both develop and deploy a Change Management System (CMS) that support specific change initiatives. This also involves training and competence development for change management skills, providing stakeholders with a better understanding of the various interdependent aspects of the aviation system that is relevant to them. Building on the new human factors model with System, Action, Sense-making and Culture aspects MASCA will support influences related to improvement in change capability and operational performance. What influences change capability is summarized in the MASCA model and what influences the process and operational performance is captured in the SCOPE model. The MASCA model (figure 1) is organized around core dimensions like: Strategy, Process, and Competence as well as Agency, Integration and Analysis. As central motif is change as a process: Assess, Plan, Implement and Evaluate, engaging all the other parts.

Strategy. Strategy encapsulates the organization's relationship with its environment. Sensing what the competitive challenge requires needs to be matched to internal intelligence on how the organization is performing so that it may be understood how to change the operational system in order to match that challenge.

Process. It is necessary to understand what the minimal conditions are for being able to purposefully change a functional system. A process model needs to be sufficiently well defined and grounded in the operation to be meaningfully linked to real operational data. A process for managing change is also essential and part of the CMS.

Competence. Competence in relation to change may be: the insufficient competence within organizations to manage complex change involving human, social and technical systems; and the lack of a good practical theory to support professional and managerial practice in this area. Change involving people in organisational processes becomes a 'human factors issue' and involves complex, multidimensional solutions and associated training.

Integration towards a joined-up organization or the extent to which the organization can act in a coherent integrated way is facilitated by creating a common picture. Earlier research prior and in HILAS suggests particular 'knowledge transformation' activities, i.e. process modeling in OPM/KSM. These aim to develop common understanding among diverse stakeholders in a change or innovation process. Reasons for the lack of a common picture are different domain expertise and local objectives that in part may be explained by organizations working in "silos" and need to divide operations into result units with different accountabilities as part of the PM.

Analysis represents the need for managing performance and risk and the importance to have capability to give feedback and report on progress in both operations and change processes. New technology for diagnosis has enabled data integration. New theory applied to this field may develop new practice to the use of data to track change and validate strategy. If relevant data is used in strategic decision making and set off a sequence of activities using this information it may drive actual system change. This information is also essential in the evaluation of the effectiveness of the actions actually taken, including to explore risk and make assessment before and after an initiative, as described in prospective risk analysis. Based on motivation theory, there is a need for a sense of achievement and ownership to make the change worthwhile. For this feedback and reports on progress is essential.

Agency relates to the need for both internal and external change agents as well as the need for developing an agency between research and practice to reach practical relevance and leverage in complex research and development issues. The need for both knowledge from research and practice is needed in successful change. However not enough applied successful research has been made applicable in practice and researchers will never gain enough domain knowledge from the outside and many industries lack research departments on the inside.

Figure 1. The MASCA model including dimensions such as Strategy, Process and Competence as well as Agency, Integration and Analysis. In the center is a change process to assess, plan, implement and evaluate change.

The MASCA Change Management System, CMS is designed to support the integrated change management capability for a company. Several components have been developed for the CMS: The *SCOPE* process modeling tool which is a direct development of the HILAS OPM /KSM tool; The Learning (Modules), Training (Courses) and Mentoring framework (*LTM*) developed to enable relevant competence building at various organizational levels. An *evaluation framework* will enable assessments of progress and results of various stages, as well as directing and initiating change. *Serious games* are developed as a simulation capability for facilitating understanding change of states in an organization and for learning and training specific concepts that are difficult to understand in regular training. An *information support framework*, consolidated from existing technical tools may give enhanced opportunity to gather and analyse aviation data, especially with new theory applied. The CMS will be further developed, implemented and evaluated in actual, real life, case studies of change in the industrial partners.

2.4.1 SCOPE

The SCOPE model represents causal influences on the process. It is designed so that it can link to the real operational data and reports representing antecedents, activities and outcomes of the process.

Supply in the model relates to resources and affordances, that humans receive from the system (socio-technical) that they are operating in i.e. information, other people (teams) and materials they need to run the **Process** that produce the outcome. Other contributing support is linked to the **Context**, such as work environment, physical and cognitive interfaces to technology. **Organizing** relates influences of the operational process by means of competence, procedures, co-ordination, feedback, planning, and allocation of task and resources. SCOPE provides a picture how these dimensions combined influence the performance of the process in delivering the **Effect**, output.

Human factors may be treated similarly to other models or lists in the SCOPE categories but the main innovation is that much of the combined human performance and process outcome depends on the support of the process not the individuals per se. Together they make up factors that shape the process, a process that make sense or "that make us smart" to travesty Norman's "things that make us smart". Indicators can be derived per dimension, which define critical areas which may not be covered in standard approaches to performance measurement.

In the literature review it is clear that this research is not a trivial task and that development is needed in various areas in relation to become applicable in practice. The theoretical framework needs to be applied to the development of indicators and relevant training of operational personnel and managers, both relevant for change and safety. Another identified gap is that existing management systems and financial governance and approaches such as lean is not enough for safety or managing safety performance. Aviation has enormous amount of data. The question remains how to use the available data after risks are identified, classified, prioritized or when safety outcome is reported not sufficient. The real challenge is to manage the system outcome by letting the indicators drive change. A process needs to be developed of a full cycle in the performance management loop to prove a capability to manage change so that cost-benefit analysis, more powerful than graphs of historical data trends may support strategic decisions.

3   METHOD

MASCA is an action research project with several partners in aviation industry, technology firms and universities. There are multiple objectives of iteratively developing research as well as applying models in industrial cases studies and by that further develop and validate them by supporting partners to improve their systems according to their needs. To identify industrial partners' needs for change field research with our two airports and two airlines has been conducted. Initial result (Leva et al., 2011) led to definition of a set of case studies tailored for each partner. The case study reported on here is a major airline and the main team is the safety department. This particular case study may be seen as a proof of concept in a smaller scale test bench, since performance management is at the same time an essential part of the methodology for system change that is concurrently under development.

Identifying industry needs and objectives in relation to MASCA. Field research has so far included identification of needs around the three dimensions in the MASCA model: Strategy, Competence and Process. This was done through: Structured interviews and observations with key stakeholders at various levels in the organization; Documentation analysis; Focus Groups workshops were partners were introduced to the MASCA project, method and model. The WS also aimed at building an internal MASCA improvement team. Identification of needs around other aspects of the MASCA model such as Agency, Analysis and Integration has been conducted in follow up field research, including deeper interviews, plenary meeting with industrial partners, and analysis of their current PM.

This particular case study has focus on (safety) performance management in an airline. Literature reviews have been made on *Safety performance management state of the art including authors' prior research.*

Gap analysis from existing state to a predictive safety performance management system is made combining initial findings on existing and missing qualities in the airline's PM framework. The result guides planned future research.

4 RESULT

The vision of the Airline is to the "The obvious choice" in its market with the values SAFETY, PUNCTUALITY and SERVICE. The airline has developed parallel CORE strategies to meet these challenges. A long-term way called the "Cultural journey" and a short term way called the "Road to rationalisation". The first immediate step is a cost-cutting mode and the airline has gone from 25 000 to 15 000 employees which is an enormous change and today they work more closely. Today the airline is Europe's most punctual airline and a good example of many of their safety activities. A Production optimisation Process (PoP) as well as a LEAN approach has been chosen as methodology to reduce waste in terms of resources and increase efficiency which is believed to reduce costs. The initial LEAN phase is a top-down initiative. The long-term goal is to implement lean as a bottom-up initiative including developing and changing the into a continuous improvement culture. When lean has matured on a lean maturity scale the short term cost-driven changes is believed to provide a basis to build further on the way for the cultural journey. Existing support for this change is mainly the Lean initiatives structure, Business Intelligence Portal, Training Management System, Sentinel reporting system, and the new VisionMonitor A. Lean training for continuous improvement with 10 specific lean experts support departments doing a lean cycle for 12 weeks and then it is training by doing.

The safety department is developing a more advanced SMS (Ulfvengren et al, 2009). In HILAS a prototype framework was depicted in a simple matrix of structuring indicators at strategic, tactical and operational organizational levels and in relation to processes managing change, risk and performance (Ulfvengren, 2009). This simple model was adopted and translated and further developed. The model was chosen because it is built on state of the art research and it is communicable. It helped internal discussions and to broaden the use of SPI's beyond the traditional safety outcomes. A new PM methodology was introduced using 13 spreadsheets and a data base and Visual Basic for Applications a program to get data from various sources. For risk assessment the ARMS framework (Nisula, 2008) has recently been adopted for occurrence reports. ARMS is taking into account high frequency as a risk value not only severity of one event. This led to development of a new safety performance management system application, VM A, for presenting and managing data as well to give an overview of the current safety measures in the organization. An organizational structure and processes to support strategic decisions based on the new PM was also developed and implemented around a Flight and Safety Quality Board, FSQB.

Initial studies were conducted already in HILAS. A 5-step methodology for airlines (Rignér et al., 2009) was proposed as a start to explore ways of better utilizing data already in their organization by the use of existing tools. The logic assumed was that the safety plateau is partly due to the fact that the systems already in place are not fully explored. Without a clear strategy on how to get more effect out of the current system the airline may lose focus and revert to simply measuring what is readily available, but potentially less meaningful. By following the systematic 5 step methodology, regardless of result, should give an increased internal knowledge and awareness related to the organizations data utilization capabilities. Benefits should include: an increased understanding about the effectiveness of current activities, a better capability to allocate resources to where they are more effective. Working with the computer systems and databases containing data often require data to be extracted in ways that the systems were not designed for which can be very troublesome requiring investments not only in human resources but possibly also in i.e. new hardware. Some systems may have been developed in-house for a specific purpose and may then have been given new features and add-ons and fixes. The traditional department "silos" may lead to a fragmented data picture and there might be reluctance to allow harmonization of data in interest to keep consistency.

The steps were: 1) Select Direct Safety Performance Indicators, 2) Identify contributory factors stated in or concluded from safety or voluntary reports, 3) Identify contributory factors through external sources, 4) List selected DSPIs and identified contributory factors relations. 5) Select other relevant (non-reporting) data.

Direct Safety Performance Indicators were many times the same as the data collected from occurrence reports. Some difficulties were where to draw the line between operational outcomes and potential related contributory factors and to have data available in a user friendly format. It was clear that contributory factors are not always identified by information given in reports or in the subsequent investigation. The terminology used can be vague or very broad, which make it difficult also to categorise and aggregate data. The task requires domain knowledge, safety and human factors expertise. A basic spreadsheet was started with the data available and filled out with derived logic of dependencies and relations. However, this methodology lack support of: a functional analysis, process model, a HF model and theoretical framework for assessing influences on the process. Understanding trends requires more sufficient, robust and detailed data as well as an interpretive capacity. If sufficiently detailed process maps not are available, it makes it difficult to relate factors in a format suitable for relationship or data correlation.

4.1 Existing PM

Two main recent improvements in the SMS and PM framework briefly described here are the FSQB, equivalent with a safety review board (AMC-OR GEN 200a) and the VisionMonitor Airline application. The FSQB process is as follows:

1. All available sources for data are used to collect safety related data.
2. All production areas collect, assess and analyze available safety data. Local activities are initiated and issues requiring coordination at tactical level is prepared for the Safety Action Group (SAG).
   a. Hazard identification and conclusions are made within respective area of responsibility.
3. All SPI's: are collected and presented in a formal report. The presentation includes present values and trends. Additional subjective information is added in order to conclude hazard identification and conclusions of risk exposure.
4. SAG reviews all SPI's. Participants brief about assessments and activities. Issues requiring coordination between the organizations are discussed and tactical decisions are taken.
5. A formal report is prepared from SAG to FSQB including data requiring FSQB attention. The structure is formalized and contains selected SPI's and explanatory text.
6. FSQB receives SPI's /issues requiring strategic decisions. Conclusions of risk exposure and mitigations to reduce risk. The information is analyzed and background data is attached. A cluster of SPI's may result in a KPI at this level, meaning a single high level objective indicator.
7. At the Core airline level a summary of risk exposure and mitigations to reduce risk is presented.
8. The board receives an executive summary of risk and exposures and mitigations.

The airline has cooperated with VisionMonitor Software Aviation solutions to develop a tool for gathering, monitoring, analyzing, visualizing and aggregating data. The main enhanced capability from prior systems was to connect data originating from a large number of different sources. The VM A application presents data in numbers and graphically in a Radar Graph Dashboard. Data may be looked at from several views, a 30-day rolling trend, risk index per department and to show trigger thresholds busts as well as effectiveness of corrective actions taken. VM A is a system that supports identification of where attention needs to be directed in terms of safety related issues in the production areas. With VM A key personnel are immediately made aware of any safety metrics that begin to exhibit undesirable trends thus it enables authorized users to drill down into the underlying data to see trends and measures. This may drive and fosters a continuous improvement process. The data and underlying structure is the same as before but VM helps communicate and visualize a common picture on performance. Data sources are for example Sentinel, Reliability, MR, ACMS (Flight Data Monitoring), CSM (Cabin Safety Monitoring) and AO costs (AOC).

5   DISCUSSION

The airline has come a long way and owns many of the qualities proposed in the MASCA approach already. As an example of the new HF model of change the airline did change their process and structure of the SPI work using VM A and the FSQB process. With initial resistance among stakeholders the new integrated approach enabling cross-departmental overview was making sense and the attitude and behaviours changed, which relates to integration. In relation to agency a continuous collaboration with external researchers has contributed to the last years of developments. In relation to analysis and in accordance to recommendations in PRA the airline has developed tools for gathering and analysing operational data.

There is still a need for developing a systematic process for change to support the FSQB process. These things combined are essential to become predictive and enable PRA including data modeling and analysis of process influences so that risks may be explored and change initiatives monitored and evaluated. One of the key issues here is to define the logic behind data management of existing current and historical data and particularly how data can be combined to provide a prospective view of future risk. Although the airline has a SPI structure it is not yet in a structure that allows quantative analysis of data from various production areas i.e combining KPI's and SPI's in order to integrate lean and safety. A great challenge remains also to validate aggregated pictures of risk, composed of several risk and performance metrics. The scope of indicators is the integral of the safety perspective or safety model in practice and says a lot of how management approaches safety. The current SPI structure needs to be validated in accordance to an agreed safety model. This involves the classification of risk data, contributory factors for risk, and the prudent use of safety performance indicators, using those which will drive most beneficial system change.

5.1   Future research

In parallel work (Ulfvengren, et al. 2012) on this case study a performance management framework was proposed to be integrated with existing tools and processes as well as implementing MASCA CMS where applicable. It included a prototype model over the structural, typological performance measurement framework and a procedural, step-by-step processes part of the overall PM process. Essential parts of the framework need to be developed and improved in order for the PM process to be functioning, despite implementing the MASCA CMS. Both the structural and procedural parts of the framework need to be further developed. A main contribution will be implementing SCOPE as a modeling tool and methodology for analysis as well as for knowledge transformation process in a process of change in the overall PM process. This includes studies such as:

- Revision and evaluation of current SPI and safety data framework.
- New HF theoretical framework-review applied to the airline.
- Improvements of current structural framework.
- Integration of existing processes and procedures like Lean and Safety.

MASCA methodology for change and the CMS will complement what has been identified missing and contribute to further system change with respect to what influences operational processes that delivers outcome and do it in a way that makes sense, at all levels, including support of the organizational processes delivering improvements.

ACKNOWLEDGMENTS

The above-mentioned research has received funding from the European Commission's Seventh Framework Programme FP7/2007-2013 under grant agreement 266423 "MASCA".

REFERENCES

Amalberti, R. (2001) The paradoxes of almost totally safe transportation systems. Safety Science 37 (2001) 109-126.

Ax, C., Johansson, C. and Kulvén, H. (2009) Den nya Ekonomistyrningen, 4th ed, Liber Förlag, Sweden.

Beauchamp-Akatova, E. (2008) Supporting decision on change. Balancing Safety issues with operational and commercial Decision in an Airline company. In proceedings of 3rd Resilience engineering symposium, Oct. 28-30, Antibes, France.

Bunderath, M., McDonald, N., Grommes, P. and Morrison, R. (2008). The operational Impact to the Maintainer. In proceedings of IET Conference, London.

Cooke, M. and McDonald, N. (2009) Prospective Risk Analysis. HILAS deliverable. Project Number 516181. EC - 6th FP.

Dent, E. & Goldberg, S. (1999). Challenging "resistance to change." Journal of Applied Behavioral Science 25-41.

European Aviation Safety Agency (2011-1). Retrieved from: European Commercial Aviation Safety Team (ECAST).

European Commission (2011): Setting up an Aviation Safety Management System for Europe. COMMUNICATION FROM THE COMMISSION TO THE COUNCIL AND THE EUR. PARLIAMENT, Brussels, 25.10.2011 COM(2011) 670 final

Harms-Ringdahl, L., Dimensions in safety indicators, Safety Sci. (2008), doi:10.1016/j.ssci.2008.07.019

HILAS (2005 – 2009) Human Integration Into the Lifecycle of Aviation Systems. Priority No: 1.4 - Aeronautics and Space Research Area 3. Priority Title: Improving Aircraft Safety and Security, IP8. Proposal/Contract No: 516181. Brussels: EC.

Hollnagel, E. (1998). CREAM: Cognitive reliability and error analysis method. Oxford: Elsevier.

Hollnagel, E., Woods, D. D., Leveson, N. (Eds.) (2006). Resilience engineering – Concepts and Precepts. Ashgate.

Hopkins, A., Reply to comments, Safety Sci. (2008), doi:10.1016/j.ssci.2008.07.020

International Civil Aviation Organisation (2008) Safety management manual (SMM), 2nd ed., Montreal , Canada: Doc 9859

Learmont, D. (2009) Year in review. In keynote speak at EASS 2009, the Eur. Aviation Safety Seminar, March, Nicosia, Cyprus.

Leva, C., Ulfvengren, P., Corrigan, Zon, R., Baranzini, D., McDonald, N., Licata, V (2011) Deliverable: D 2.1 Review of requirements for change Deliverable to EC as part of MASCA project FP 7-AAT-2010-4.3-4.: Grant agreement n° 266423

Leva M.C Cahill J Kay A Losa G. Mc Donald N. (2010) The Advancement of a New Human Factors Report – 'The Unique Report' Ergonomics Feb;53(2):164-83.

Mawdsley, D (2010) Measuring safety. IATA IOSA SMS workshop 22/10.

McDonald N. (ed.) (1999). Human-centred management for aircraft maintenance ADAMS-WP4A-D2 (April 1999). Deliverable to the European Com., ADAMS project. Dpt. of Psychology, Trinity College Dublin

McDonald N. (2001 a). Human systems and aircraft maintenance. Air and Space Europe, 3 (3/4), 221-224.

McDonald, N. (2001 b). The development and evaluation of SCARF. Paper at 'Managing the risk of the flight-ground safety interface' conference. Australian Aviation Psychology Ass., Melbourne, Oct 2001.

McDonald N., Corrigan S., Cromie S., Daly C. (2000). Safety management systems and safety culture in aviation maintenance organisations. Safety Science, 34, 151-176.

McDonald, N. (2006 a) Organizational resilience and industrial risk, in Resilience engineering – Concepts and Precepts. Ashgate

McDonald, N. (2006 b). Human Integration in the Lifecycle of Aviation Systems, HILAS report. Proj. No 516181. EC 6th FP.

McDonald, N. (2009). Transformation: Human Factors and Organisational Change. HILAS report, deliverable and draft manuscript to book. Project No. 516181. Funded by European Commission - 6th FP.

Morrison, R., (2009) Operational Process Modell / Knowledge Space Model HILAS deliverable and draft manuscript to book. Project Number 516181. Funded by European Commission - 6th FP.

Nisula, J. (2008) Operational Risk Management, Work by the ARMS WG, Delivery report, ECAST Dec.2010.

Norman, D.A. (1993) Things That Make Us Smart. Addison-Wesley, New York, 1993

Norman, D. A. (1986) Cognitive Engineering, Ch. 3 in Draper S.W. and Norman, D.A. (Editors) UCSD, User-Centered System Design, Lawrence Erlbaum Ass. ISBN 0-89859-872-9 (PB).

Perrow, C. (1999). Normal accidents - Living with High-Risk Technologies. Princeton, NJ: Princeton University Press.

Rasmussen, Pejtersen, Goodstein (1994) Cognitive systems engineering, Wiley and sons.

Reason, J. (1990). Human Error. New York: Cambridge University Press.

Reason, J. (1997) Managing the risks of organizational accidents. Ashgate Publishing Ltd: Aldershot, UK.

Rignér, J., Ulfvengren, P., Cooke, M., Leva, C. and Kay, A. (2009) Study of safety performance indicators and contributory factors as part of an airline systemic safety risk data model. In proc. of the 17th IEA 2009, Bejing 9-14 Augusti 2009.

Ulfvengren P., Rignér J., Ydalus M., Mårtensson L. (2009). SAS HILAS Participation and Experiences.HILAS report, deliverable and draft manuscript to book in preparation. Project Number 516181. Funded by EC- Sixth Framework Programme.

Ulfvengren, P., McDonald, N. and Ydalus, M (2012, in press) A methodology for managing system change- SAS airline PM development. In proc. in PSAM11/ESREL2012, Helsinki, Finland.

Ward, M., McDonald, N., Morrison, R., Gaynor, D. and Nugent, T. (2010) A performance improvement case study in aircraft maintenance and its implications for hazard identification. Ergonomics, 53:2, 247-267.

Womack, J. P. and Jones, D.T. (2003) Lean Thinking, Banish Waste and Create Wealth in Your Organisation. 2nd ed. Free Press.

Wreathall, J., Leading? Lagging? Whatever!, Safety Sci. (2008), doi:10.1016/j.ssci.2008.07.031

CHAPTER 50

Power Relations and Safety in Commercial Aviation

Professional Identity and Occupational Clashes

Maria Papanikou, Ulke Veersma

University of Greenwich
London, UK
M.Papanikou@gre.ac.uk

ABSTRACT

This paper focuses on safety, the potential for intervention with accidents or near accidents and the determinants of safety. At the organisational level the interaction and division of labour between employees may have a critical impact. Trust in the employment relationship and training on procedures is required for an adequate safety culture. Earlier research suggests that pilots are employees with substantial bargaining power compared to other professions and most of their skills are a scarce resource and only limited transferable. However, our research shows that fleet standardization as a cost cutting tool creates experts on an aircraft type with possibly less job discretion and more specialized, and therefore fewer transferable skills. The degradation of pilot skills seems to result from the business environment where managerial control dominates. This fosters a culture of blaming with respect to safety culture impacting the type of reporting practices in the organisations regarding safety actions or inactions. Early evidence collected shows a clash between identities of various occupational groups – pilots, managers and cabin crew. Power relations and managerial control play a major role determining the level of discretion within jobs and the reporting culture.

Keywords: safety management, safety culture, professional identity, occupational culture, pilots

1 INTRODUCTION

Safety and the incidence of (near-) accidents are most crucial on almost a day-to-day basis for the operating of airlines. Other relevant factors contributing to safety are the technology in use and organisational aspects. The structuring and functioning of organisations is again dependent on strategies of airlines. Like Reason (2000) suggests the systems theory can be more fruitful as a theoretical framework than approaches, in which errors are looked at as consequences of human nature and intervention by individual actions. Social system thinking is more focused on systemic failures rather than on human nature. Our approach is in line with this type of thinking, where system aspects, including organisational processes and the design of workplaces may give rise to errors which may impact the safety on airplanes. Our research focuses on conditions under which people work rather than human nature and characteristics of individuals and their actions. Other determinants looked at are the various strategies of airlines, which have been identified as the background of accidents (Herrera et al 2009). The business objective of many airlines of a high labour productivity seems now to have been succeeded by a relative decrease of wage-levels and a higher workload and longer working hours as the crew have longer flying hours and shorter rest periods between flights, which leads generally to an intensification of the job. This is even more the case as crew members are increasingly more responsible for cleaning of the aircrafts (Alamdari & Morrell 1997). In addition, the organisational culture is affected by the adoption of various business models emerging and/or changing either due to the preconditions, like the gap with airlines adopting the low cost model, other external reasons, like the recession, high competition and terrorism, or as part of the incumbents' reactions. Hopkins' research (2006) on rail safety found, similarly to what takes place in airlines, obsessive rule making in organisations and a typical punctuality focus leading to a high level of risk blindness.

1.1 Environmental and Business Strategy Factors

Beside airline deregulation and privatisation socioeconomic phenomena, such as the economic recession, intense competition and a more demanding customer orientation, account as conditions leading to cost cutting pressures (Doganis 2002; Bamber et al, 2009), which may be affecting flight safety. Changes at the organisational level accompanying such trends have been identified as an explanation for accidents (Herrera et al 2009). In addition, this context may cause increasing insecurity as unemployed people may have to move or change jobs, therefore limiting their discretion. Thus, current phenomena in the environment of airlines, and their response to them may affect the safety culture in airlines. In addition, the political economy of the industry set some airlines to take either high road or low road accounts for survival (Boyd, 2001).

Doganis (2002) states that Low Cost Carriers (LCC) compete on the basis of profit maximization, including higher seating density, higher daily aircraft utilisation, direct operating costs being dependent on a single type of aircraft, as well as flexible staff undertaking two or more functions. Bamber (2007), however, argues that aircraft productivity stems from longer block hours, the number of seats available per mile and the advantages for legacy airlines due to large aircraft fleet/long haul routes. However, the growth of LCCs has affected working conditions impacts on the position of both LCCs and legacy airlines since personnel of LCCs works longer hours and is paid lower wages, while a certain percentage of LCC flight crew has professional experience from Full Services Carriers (FSCs) and now works under different terms and conditions, such as longer hours and lean staffing (Hunter 2006; Bamber 2007). Accordingly, with Neilson (2007) it may be concluded that, in Marx's terms, both formal and real subordination of the proletariat employees to capital is driven by competition intensive industries and environments. In other words, in much more competitive, highly liberalised environments and industries, to which organisations are exposed to, the subordination of employees thus results in low salaries in terms of real wages, long working hours, disciplinary managerial actions and retaliation.

1.2 Organisational culture and occupations as sociological factors

Cultures that are driven by rule making, stringent job role divisions and focusing on procedures are characterised by the existence of explicit rules for 'right' behaviours (Trice 1993). There may be competing cultures and subcultures, where one dominates the other, but there may also be unintended consequences having reverse effects on safety and the prevention of failures.

Gillen & Gados (2008) suggest that pilots are employees with substantial bargaining power compared to other professions and that they are characterised by low substitutability and less skill transferability. However, the standardisation of fleets as a cost cutting tool creates experts with regard to the specific aircraft type, which may be a rather transferable skill. Thus, the culture the pilots are transferring along with their skills may create industry level sub-standards. At the same time, as becomes clear in Easyjet founder's statement, in which LCCs are compared to a bus service, implies a lowered status of pilots as they are compared with bus drivers rather than professionals. At the same time, other research carried out in 2001 among BA pilots found that they felt their status being devalued and that management dictated many decisions that should lie with the captain (Harvey & Turnbull, 2010). Complying pilots may be good in the perception of management, but not for safety per se, as is also reflected in the complaints of some pilots against flying for many hours (Bennett, 2010). Thus behaviours promote a norm, as culture evolves from employee behaviours bottom up rather than being imposed by organisational norms dictating employee behaviours.

Increasingly, professionals work in corporations, subject to the decisions of company managers. However, the more stratified the organisational structure the more the conflict (Form 1987). De skilling includes reducing the complexity of tasks and therefore reducing the control over tasks and the status by and of the occupier (Trice 1993). Thus, it can be argued that the deregulation and the simplification of the airline business model into a more flexible and low cost structure has affected the latter point. However, it can be argued that the claimed degradation of pilot skills can be due to the job security issues driven by economic downturn and cost cutting mechanisms to increase profitability in a high cost industry. Moreover, Sennett (1998 cited in Ackroyd et al 2005) indicates that power relations have shifted as the structure of organisations evolved and changed, but yet are not obsolete. More specifically, power relations between groups of employees and management changed to fit the decentralised structure and the new form of control is imposed by performance auditing and customer-focused culture (Harrison 1994; Sennett 1998 cited in Ackroyd et al 2005).

1.3 Safety Culture and Crew Resource Management (CRM) as Risk minimization mechanisms

A lot of the existing literature is focused on the training component of Crew Resource Management (CRM), or stress factors related to fatigue and a high workload leading to human errors which are beyond training deficiencies. Nevertheless, limitations shown in studies already conducted, such as the variable of cross-cultural generality, is acknowledged and the impact of culture is highlighted (Helmreich et al 1999). However, organisational practices embedded in the safety culture may be more impacting than formal training, as reporting of errors may be inhibited by fear of being fired (Zhao & Olivera 2006). In this way the positive results of CRM may be reduced.

Safety culture includes proactive commitment to ensure safety and reducing errors (Merritt & Helmreich 1997 cited in Helmreich et al 2001). Therefore, since it can be argued that the culture can positively nurture attitudes towards a safety climate and promote the respective actions from its members, it can be argued that the organisation may be promoting negative attitudes and actions as well (Helmreich et al 2001). On the other hand, norms set by occupational groups are a constraint for organisations as they inhibit flexibility by their default set characteristics and ideologies, while they provide a shelter or a security net for professionals (Freidson 1994). Therefore, there is an inherent conflict within the organisation and between the professionals. However, the power structure may shift or may have already been shifted as pre-conditions change. In other words, the power and the respective contingencies may have shifted from professionals to the organisation. Accordingly, in previous research occupational identity of captains (Aschcraft 2007) questions the deployment and practicality of CRM: there have been accidents in the past because of poor cockpit communication in regards with the authority of the captain; the assumption here is that the organisation itself may be reinforcing such perceptions and sub cultures. If the organisation is not reinforcing the latter subculture, then another issue may be power and status in the organisation which affects the crew interaction, discretion and power. However, occupational commitment of pilots for their professional status may conflict with the economy and job security, while management retaliation may include scheduling pilots on lower paid routes or aircrafts that are not popular or are to be withdrawn from fleets.

2 METHODS

Drawing on interviews with 15 company flight crew members of different status, experience and airlines across Europe, this study examines how organisational factors affect the crew's discretion on their work. Thus, the levels of consistency deviance and conformity are explored. In addition, the reasons underlying the levels and the implications on flight safety are addressed.

2.1 Study Context

We find important to understand and investigate both types of action according to Weberian methodology and their different derivatives. Since action is identified by Hindess (1977) as a natural event, in the research topic the action of the crew and their inherent decisions is the event-assuming that the decision results into pilot error. Moreover, it is rigour to use qualitative interviews in order to identify both the actions-events as well as the actions-behaviours of pilots. There are several methods to analyse interviews, varying from focus on meaning, focus on language, bricolage –using different techniques. Bricolage is identified as an analysis of data using a combination of tools which allows us to overcome limitations embedded in the single method analysis (Kincheloe 2001; Chao et al 2007; Denzin & Lincoln 2008). At the same time, words are becoming categories when they share meaning (Cavanagh 1997 cited in Elo & Kyngaes 2007). Therefore, bricolage-ing can be argued to include a combination of content analysis with some hermeneutical elements.

2.2 Participants & Control factors

The purpose of the interviews was to gather initial perspectives from different airlines and sectors –public and private-, as well as from both flight deck and cabin crew members. The research therefore explored experiences in order to address the industrial organisational and safety culture among occupational groups. The research required the participation and purposive selection of a specific group with the respective skills with knowledge that Marshall (1996) has identified as key informants. In addition, key informants include those with differing and wide range of views. Those included the respective rank of the interviewee (cabin crew rank and flight crew, i.e. pilot's rank), experience, and number of total employers in their career (see control factors).

2.3 Themes
Profession, Training & CRM

The theme of profession emphasized on interviewees reasoning regarding their choice of occupation which included feelings, perceived benefits and training investment on their behalf. Organisational training and CRM were used as a theme in order to measure consistency between occupational groups, their knowledge deriving from the training and their perception of its significance.

The dimension of safety culture included the identification of safety mechanisms for each occupational group, including managerial directives and actions to promote the culture. Reporting was the main focus which included a discussion on regulations and actual practices on the organizational level. We included the predominant element of hierarchy in the occupations within airlines in order to observe perceptions of the usefulness of strict ranking statuses in daily practices. Moreover, hierarchy was tested against the safety culture mechanisms.

Workload theme included questions on both the regulatory conditions set by Aviation Authorities and whether those were adhered to in the organizational level. Therefore, the organisation's perspective on safety was measured. Because the research focuses on safety, the interviewees were asked to narrate any lived experience on unsafe actions during or before flights. Moreover, their observations regarding the safety on flights with regards to the deployment of safe practices or not were recorded.

2.4. Data analysis

We pre coded the interview themes and sub themes before data collection according to our assumptions. Therefore, the coding included a system that captured crew perceptions on safety and organizational culture; the level of consistency in practices between occupational groups; the level of importance those practices have on crew. It was deemed important to pre code ambiguous responses as a conflict was pre assumed. However, re-reading the interview transcripts, we identified one more pattern; the repetition of key words in relation to culture, as well as the fact that the interviewees' may not have been willing to indicate actual events on the record. The last part of the analysis involves linking the codes with the control factors and identifying the respective patterns (see table 2).

Control factors of interviewees

Rank: perceptions among flight deck and cabin crew, as well as perceptions among captain and first officer (co-pilot)

Experience: assumption that more experienced crew is more likely to have observed or lived incidents and/or accidents; length of the service where key informants have worked for only 1 air carrier may affect responses and sharing information

Number of airlines as employers: perceptions may vary or be richer in context due to various organizational cultures lived

Airlines: The choice on airline selection is not justified explicitly as it was not purposive and was in line with the requirement that there will be no mentioning of the air carriers employing or having employed participants.

3 RESULTS & DISCUSSION

Profession: Both flight and cabin responses regarding their motivation included mainly emotions of 'enjoy', 'like the feeling', 'love to fly' to justify the choice of the profession. Status was identified and repeated by pilots, using prestige and status interchangeably. However, cabin crew included the 'myth' around the profession and the life-threatening element ('our life is in danger'). The latter indicates negative aspects of the occupation which counteract the positive feelings or indicates how strong those feelings are in order to sustain life threatening situations. This is a very important finding as it has direct effect on the safety culture and the willingness to comply with managerial procedures. Moreover, comparing captains and copilots responses, we found some asymmetry in terms of the power of CPTs, and reasoning in terms of little experience for FO who highlighted the issue of competitiveness and 'desirable candidates'. Thus, further research will include the issue of pilot's experience.

Training and CRM: In general responses were positive on the usefulness of CRM, but not consistent in terms of organisational training. CRM was identified as highly important, despite its absence in one of the airlines for part-time crew. Therefore, there are two contradicting perspectives; that of the importance of CRM training and that of actual deployment of CRM in the training sessions. Copilots' responses on CRM included a definition of CRM rather than answering the interviewer's question. That point triggered the researcher to explore more the case of FO and reach the point that overall there was a tension to give politically correct answers. This observation in relation with FO request to the researcher to exclude from the transcripts lived experiences, makes seeming that further research focused on co-pilots as well. At the same time CPTs were eager to provide detailed information. Moreover, different perspectives on CRM were evident amongst cabin and flight deck crew, while CRM is provided to enhance teamwork amongst crew.

Safety culture: There were varying perspectives among cabin and flight deck crew, as SCCMs identified no safety reporting procedures, whereas CPTs and FOs did. However, what was common in their responses was the norm of unofficial reporting of safety issues to those 'trusted', had a 'good relations with', yet that they would not report a 'colleague'. Furthermore, the findings were ambiguous, as responses from the same interviewees included 'is safe' and '..but it is not safe'. Accordingly, ambiguity on safety is observed in responses of less experienced FOs which include expressions such as 'it depends on your position in the company'. The latter further advocates the power relations finding.

Hierarchy: The mode of flight - whether it is under normal operations or under an emergency- was identified by all interviewees as the distinguishing element that acts as a fine line in hierarchy. Moreover, the importance of hierarchy welcomed or not, was noted in terms of use and misuse or abuse of power. More specifically and depending on the flight mode, CPTs stated that there 'should be no democracy' (emergency mode). SCCMs stated there is 'no meritocracy' (normal operation mode). While both the FOs and SCCMs stressed the importance of hierarchy in terms of decision making (emergency mode). Less experienced FOs suggested 'finding a way to say what you want to pass on', indicating that there are some obstacles in interaction and communication (power). However, due to the politically correct stance of less experienced FOs, we researcher question whether such suggestions do not reflect a lived experience.

Workload: The issue of regulations governing flight and duty limitations was addressed and it was stated that they are officially met, although the interviewees' observations are different. The crew was flying fatigued, sleepless, or even drunk. However, what appears to be a common denominator on the interviewees' perspective on the rigidness of duty limitations and workload is the fact that it is ultimately rests upon personal choices. The responses of the pilots included that safety is not endangered per se, but that flying as such is 'less safe' rather than 'unsafe'. Less experienced FOs responses were again ambiguous, as FO1 stated that 'manufacturer knows what is safe and what is unsafe', whereas at some other point they found manufacturer manuals to be inefficient in reflecting and accounting organisational culture.

Lived experiences: Most interviewees narrated both a lived incident and observations with the exception of the less experienced FO. Accordingly, FO1 may had or not a lived incident experience, however most of the discussion was conducted off the record. 'Too many' incidents which were covered up until the airline 'blamed' someone to illustrate intolerance. CPTs further identified costs as reasons for incidents observed. Cost issues were seemingly identified as causes of incidents or near incidents from SCCMs as well, along with 'pressure' and 'competition'. Competition among airlines -especially between those privately and publicly held- has been identified frequently in the pilot study, yet because of length limitations this will not be analysed in this part of the study. Competition and sectors have, however, been noted and will be in the interest of further research.

Moreover, CPTs concluded with discussing about trade unions' power to protect pilots from 'being fired' and a blaming culture.

Table 2. Summary of findings

	Pilots	Cabin Crew
Rank effect on disclosure of information	High	Moderate
Ambiguity in responses regarding common practices	High	Moderate
Consistency of practices	Low	Low
Sub cultures	High	Moderate
Sub mechanisms	High	High

4 LIMITATIONS, STRENGTHS AND FUTURE DIRECTIONS

The research focused on airline crew and how organisational factors are affecting the safety culture and levels. However, generalizations are to be avoided as the sample of the still in depth interviews is not large enough. Moreover, analysis could use discourse analysis and pre coding should be avoided as it limits the factors and lurking variables to a predetermined mindset on what constitutes power. Therefore, at this stage we cannot confirm that skill degradation has a negative impact on safety actions. Nevertheless, our future direction is towards exploring in specific power relations of the pilots' professional elite and management in both full and low cost sectors. Power relations seem also to interfere with the quality of human intervention to prevent accidents and incidents that may happen. The lack of cohesion in policies and training leads to compromised situations where crew resource management should take effect. The reporting of incidents or unsafe acts/observations was not deployed or depended on ranks. In addition, a fine line between necessity (emergency modes) and abuse of power (normal flight modes) is observed. Occupational culture is in conflict with the organisational culture, whereas there is conflict between organisational rationality and professional identity. Thus, organisational factors include the increasing insecurity and flexibility for employees. Under such conditions the discretion in the job for intervention may be decreased, which could impact more directly safety on board of airplanes.

ACKNOWLEDGMENTS

We wish to thank all the participants of this study for their insightful contribution and Dr. Graham Symon for his input and support throughout the project.

REFERENCES

Ackroyd, S. et al 2005 "The oxford handbook of work and organisation", Oxford University Press, UK

Alamdari, F. E. & Morell, P. 1997 "Airline labor cost reduction: post-liberalisation experience in the USA and Europe", Journal of Air Transport Management, 3(2), pp.53-66

Ashkraft, K. L. 2007 "Appreciating the 'work' of discourse: occupational identity and difference as organizing mechanisms in the case of commercial airline pilots", Discourse and Communication, 1(1), pp. 9-36

Bamber, J. G. et al 2009 "Up in the air: how airlines can improve performance by engaging their employees", Cornell University Press, USA

Bennett, S. 2010 "Contemporary issues: fatigue impacts of employee commutes", Parliamentary Advisory Council for Transportation Safety, UK

Boyd, C. 2001 "HRM in the airline industry: strategies and outcomes", Personnel Review, 30(4), pp. 438-453

Chao, W. et al 2007 "University of British Columbia & Simon Fraser University-The Bricolage", IEEE Symposium on Visual Analytics Science and Technology, USA

De Groote, P. 2005 "The success story of European Low-Cost Carriers in a changing airworld", GaWC Research Bulletin 174

Denzin, N. K. & Lincoln, Y. S. 2008 "Collecting & interpreting qualitative material", Sage Publications, UK

Doganis, R. 2002 "Flying off course: the economics of international airlines", 3rd edition, Routledge, UK

Elo, S. & Kyngaes, H. 2008 "The qualitative content analysis process", JAN Research Methodology, 1, pp.107-115

Form, W. 1987 "On the degradation of skills", Annual Review of Sociology, 13, pp. 29-47

Freidson, E. 1994 "Professionalism reborn: theory, prophecy and policy", University of Chicago Press, USA

Gillen, D. & Gados, A. 2008 "Airlines within airlines: assessing the vulnerabilities of mixing business models", Research in Transportation Economics, 24(1), pp. 25

Harvey, G. & Turnbull, P. 2010 "On the Go: walking the high road at a low cost airline", International Journal of Human Resource Management, 21(2), pp. 230-241

Helmreich, R. L. et al 1999 "Models of threat, error, and CRM in flight operations", Proceedings of the Tenth International Symposium on Aviation Psychology, USA

Helmreich, R. L. et al 2001 "Culture, error, and crew resource management" in E. Salas, C.A.Bowers &E. Edens (Eds.), Applying resource management in organisations: A guide for professionals, Hillsdale, USA

Herrera, I. A. et al 2009 "Aviation safety and maintenance under major organisational changes, investigating non-existing accidents", Accident Analysis & Prevention, 41(6), pp. 1155-1163

Hindess, B. 1977 "Philosophy and methodology in social sciences", Humanities Press, USA

Hopkins, A. 2006 "Studying organisational cultures and their effects on safety", Safety Science, 44(10), pp. 875-889

Huckaby, M. F. 2010 "Researcher/researched: relations of vulnerability/relations of power", International Journal of Qualitative Studies in Education, 24(2), pp. 165-183

Hunter, L. 2006 "Low Cost Airlines: Business Model and Employment Relations", European Management Journal, 24(5), pp. 315-321

Kincheloe, J. L. 2001 "Describing the bricolage: conceptualising a new rigor in qualitative research", Qualitative Inquiry, 7(6), pp. 679-692

Kvale, S. 2007 "Doing interviews", The Sage Qualitative Research Kit, Sage Publications, UK

Marshall, M. N. 1996 "The key informant technique", Oxford University Press, 13(1), pp. 92-97

Neilson, D. 2007 "Formal and real subordination and the contemporary proletariat: Re-coupling Marxist class theory and labour-process analysis", Capital and Class, 31(1), pp. 89-123

Reason, J. 2000 "Human error: models and management", BMJ, 320, PP. 768-770

Schein, E. 2008 "Organisational Culture and leadership", 2nd edition, Jossey-Bass Psychology Series

Trice, H. M. 1993 "Occupational subcultures in the workplace", Cornell International Industrial and Labor Relations 26, USA

Zhao, B. & Olivera, F. 2006 "Error reporting in organizations", Academy of Management Review, 31, pp. 1012-1030

Section VIII

Cognition and Workload

CHAPTER 51

Functional Near Infrared Spectroscopy: Watching the Brain in Flight

Angela Harrivel, Tristan Hearn

NASA Glenn Research Center
Cleveland, OH, USA
angela.r.harrivel@nasa.gov

ABSTRACT

Functional Near Infrared Spectroscopy (fNIRS) is an emerging neurological sensing technique applicable to optimizing human performance in transportation operations, such as commercial aviation. Cognitive state can be determined via pattern classification of functional activations measured with fNIRS. Operational application calls for further development of algorithms and filters for dynamic artifact removal. The concept of using the frequency domain phase shift signal to tune a Kalman filter is introduced to improve the quality of fNIRS signals in real-time. Hemoglobin concentration and phase shift traces were simulated for four different types of motion artifact to demonstrate the filter. Unwanted signal was reduced by at least 43%, and the contrast of the filtered oxygenated hemoglobin signal was increased by more than 100% overall. This filtering method is a good candidate for qualifying fNIRS signals in real time without auxiliary sensors.

Keywords: cognition, fNIRS, in-task monitoring, Kalman filter, real time, signal processing

1   INTRODUCTION

Functional Near Infrared Spectroscopy (fNIRS) is an emerging neurological sensing technique applicable to optimizing human performance in transportation operations, such as commercial aviation. FNIRS quantifies hemoglobin concentration ([Hb]) changes in the brain based on optical intensity measurements. Hemodynamic activations can be detected across the whole head. These ambulatory, non-invasive measurements allow in-task monitoring. Cognitive state is determined via pattern classification of the functional activations. Thus, we can watch the activations within the brain of a pilot during the safety-critical task of flying. Importantly, this is distinct from the use of vigilance tests which the pilot could undergo only while breaking from the task at hand. Such "fit-to-fly" tests may be passed due to short-term increases in attentional effort (Sarter, 2006) which may not be sustainable for long durations. During monitoring, information regarding cognitive state could be used to trigger appropriate risk mitigations in real time via changes in flight automation and information display.

Our work continues to focus on improving techniques for applying fNIRS to in-task operator characterization. Optical sensing may improve upon the existing use of other modalities for operator characterization, especially through combination with them. Examples include electroencephalography and other physiological measures used for Augmented Cognition (Schmorrow, 2007), operator performance research (Schnell, 2004), and crew cognition research (Pope, 1995).

FNIRS quantifies [Hb] changes with time based on optical intensity measurements of light that has scattered through the outer layers of the cortex beneath the optical probe. FNIRS measures the same hemodynamic changes as the functional Magnetic Resonance Imaging (fMRI) Blood Oxygen Level Dependent signal, with lower spatial but improved temporal resolution. fNIRS provides a direct connection to the wealth of information in the fMRI literature. We believe this improves the outlook for sensing more complicated states.

2   MOTIVATION

FNIRS works well in the laboratory, and it has been used in many neuroscientific research studies, including real-time classification of state using an extended Kalman filter and known stimulus timing (Abdelnour, 2009). However, implementation for monitoring outside the laboratory requires techniques that do not rely on known stimulus timing or block averaging. Also, no standard for the removal of physiological noise and motion artifact yet exists. Many opportunities remain for real-time applications (Zhang 2009) and for improving robustness and reliability. Motion artifact can be significant, and is likely to occur in operational environments. Field application calls for further development of algorithms and filters for the automation of bad channel detection and dynamic artifact removal. Such advances in operational use will benefit use in clinical outpatient scenarios.

Frequency domain (FD) instruments for fNIRS measure radio-frequency-modulated signal intensity amplitude and offset, plus phase shift of the detected optical intensity signal relative to that of the source. The phase shift data provide a direct indication of the coupling noise associated with signal detection and thus the quantification of [Hb] at that time for that channel. Thus we explore the use of this phase measurement to drive an adaptive filter for the removal of motion artifact in real-time.

Principally we are interested in avoiding human error to improve commercial aviation safety by identifying the loss of attentional engagement using fNIRS signals. The attentional monitoring system under development (Harrivel, 2010) could be adapted for the detection of other cognitive states by changing the location of the optodes and adjusting the classifier training data and parameters. Our previous work has demonstrated improvements in the comfort of optical head probes to enable monitoring for over an hour (Harrivel, 2009), and has shown on average 70% accuracy for real time classification of attentional state using Support Vector Machines with training data free of artifact (Harrivel, 2011). To improve hemodynamic activation measurements and classification accuracy, we continue to implement existing cutting edge techniques while exploring novel headgear and data processing methods to provide such clean data in real time during operational use in the field. Here we introduce a filter based on the FD phase measurement for real-time signal quality detection and improvement. This new filtering technique is well-suited to improving the quality of fNIRS signals in any application where motion is an inherent part of the task to be performed during monitoring.

3   RATIONALE

3.1   Optical rationale

Optical intensity is attenuated by absorption and scattering along the optical path length (the light's path through the head). The phase is sensitive to motion because of its dependence on the source-detector separation and the index of refraction of the tissue through which the light travels. Both affect the optical path length. An increase in the actual optical path length causes greater absorption. This can lead to overestimation of the change in [Hb] if the path length is assumed to be constant. High variability of the FD phase shift can indicate inconsistent coupling at the optode-scalp interface, which changes the separation distance and possibly exposes the detector to ambient light (which has no consistent phase shift with respect to the source). Shifts in the phase also can indicate motion of a probe along the scalp surface. As the probe senses different volumes of tissue, changes in intensity are caused by changes in the optical scattering and absorptive properties of the tissue. This also contributes to probe relocation errors and inter-subject variability.

If the phase and [Hb] changes are correlated, it is likely the [Hb] changes are due to motion-induced changes in path length or absorption. This is because the increased phase indicates a larger path length, which causes more absorption and lowers the detected intensity, which finally increases the calculated [Hb]. If the phase is steady, it is likely the calculated [Hb] changes are due to activations of interest, systemic physiology or detector gain changes. Thus, the phase can be monitored to automate signal removal or alert the user to reset the probe.

This filter assumes the phase does not change significantly with physiological activation. Neuron swelling or the influx of new hematocytes may affect the scattering and thus the optical path length and phase, but not on time scales accessible to FD instrumentation.

This filtering method is most appropriate for the Frequency Domain Multiple Distance (FDMD) method where FD instrumentation is used to calculate optical absorption and scattering properties so absolute [Hb] values can be determined (Fantini, 1999). In this case, the filter would be applied to the measured optical intensity, and anti-correlation between the phase and the intensity could be used. This can be explored in future work. Alternately, in cases where relative measures suffice while probe footprints on the head must be small and the signals must be reliable, it can be used to filter the relative [Hb] changes calculated using the Modified Beer Lambert Law. The phase signal used should be for the wavelength that matches the hemoglobin species of the trace being filtered. This is the case presented here.

3.2   The Kalman filter and its suitability

Kalman filtering (Kalman, 1960) is a technique for estimating the state of a linear discrete dynamical system. The Kalman filtering implementation used for this work assumes that the unknown true value x_k and measurement value z_k of a system at time k is determined by the model

$$x_k = Ax_{k-1} + w_{k-1}$$

$$z_k = Hx_k + v_k,$$

where A and H are known linear operators. A and H were each assumed to be an identity operator for this work. In this way, we are assuming that the process under study is simply the contamination of the data by additive noise. The variables w_k and v_k are process and measurement noise, respectively. The distributions of these two noise sources are assumed to be

$$w_k \sim N(0, Q) \text{ and } v_k \sim N(0, R),$$

where $N(\mu, \sigma^2)$ is a normal probability distribution with mean μ and variance σ^2. The Kalman filter produces estimates, by means of the explicit recurrence relations, of the data

$$\hat{x}_k = \hat{x}_{k-1} + \frac{p_{k-1} + Q}{p_{k-1} + Q + R}(z_k - \hat{x}_{k-1})$$

and the variance of the filter

$$p_k = \left(1 - \frac{p_{k-1} + Q}{p_{k-1} + Q + R}\right)(p_{k-1} + Q),$$

beginning with the initial condition $x\hat{}_0 = z_0$ and $p_0 = 1$. Because of the explicit nature of these computations, no iterative solver or optimization procedures are required for filter implementation. Furthermore, the computed estimates $x\hat{}_k$ are optimal in the mean squared error sense. More specifically,

$$\hat{x}_k = \text{argmin } E(|x_k - \hat{x}_k|^2)$$

for each time k; where E is the mathematical expectation operator. The optimal properties of the Kalman filter estimates along with the simplicity of its implementation (Zarchan, 2005) make it an excellent choice for the smoothing of fNIRS data traces.

Values for the variances of the process and measurement noise (Q and R) must be determined before the Kalman filter can be implemented; since they are required in the calculation of each recurrence relation (each k). These values can be interpreted as tuning parameters for the Kalman filter. A larger R weights the data estimate and distrusts the measurement to produce a smoother output, while a smaller R follows the measured signal more closely. Q can be increased to allow more dynamic changes in the estimate of the system.

We propose that the phase channel φ associated with a [Hb] channel x^ is a natural choice of information for computing an appropriate value for R at each time.

4   METHODS

Synthetic data were created to mimic [Hb] and phase shift traces with a data rate of 6.25Hz for 400s to demonstrate the feasibility of this filter. Four [Hb] time series were simulated: relative changes in both oxygenated and deoxygenated (reduced) hemoglobin species concentration ([HbO] and [HbR]) for two source-detector separations, which we label close and far. A close source can be used to measure systemic physiological contributions to the signal of interest (biological "noise") on a per-sensor basis (Zhang, 2009) by interrogating shallow tissue (not cortex). The far source interrogates both superficial and cortical tissue.

The input trace to be filtered was created by adding non-dynamic physiological signals for cardiac at 1Hz, respiration at 0.25 Hz, and Meyer waves at 0.1Hz to the close source traces. Functional activations, and smaller physiological contributions, were added to the far source traces. Random noise was added to all traces. In figure 3, the input [Hb] trace is shown (top, black) after within-species subtraction of the close source signal from the far source signal, then subtraction of [HbR] from [HbO]. Functional activations were simulated with gamma-variant functions as in Abdelnour, et al. (2009) eq. 3, with the [HbR] amplitude being -1/4 of [HbO].

Four different types of motion artifact were simulated. These are outlined in Table 1. No motion artifact is simulated from 0s to 50s. Artifact from probe slip (across skin with no air gap) was simulated between 50s and 100s by allowing the phase to vary randomly over tens of degrees. Artifact from decoupling (an air gap at the detector interface) was simulated between 100s and 200s with phase varying over hundreds of degrees. Artifact from a probe bump was simulated with a spike in the phase at 225s, while artifact from a bump and relocation was simulated with a spike and a step increase of 10 degrees at 325s. Phase changes on these orders of magnitude do occur in real data. [Hb] signal which covaried with the phase was generated for the slip and the relocation bump artifacts.

In the case where a noisy signal could appear to be a [Hb] change (50s – 200s), filter performance is quantified as percentage of signal mean reduction. In the cases where motion noise obscures the activation of interest, performance is quantified as contrast-to-noise ratio (CNR) improvement between the unfiltered and the filtered signal. CNR is defined here as mean activation signal minus mean reference signal (0s to 50s), divided by the square root of the sum of the variances in those signals.

4.1   Mathematical Implementation of the Kalman filter

To implement the Kalman filter, the measurement noise variance, R, can be set to a function, possibly non-linear, of the variance in the phase over a window of time prior to that instance. The use of the phase differentiates our method from other implementations of Kalman filtering for motion artifact removal (Izzetoglu, 2010). Here, the tuning parameter R was set to the variance of the phase in a rolling window of 1s prior to the current instance, plus the correlation of the phase with the [Hb] trace being filtered during 20s prior (which turns on at 20s). The contributions of the variance and correlation were linearly scaled. The absolute value of the correlation was added if it was greater than a threshold of 0.3. Thus only high correlations impact R. This threshold has not been optimized, and can be raised for less aggressive filtering.

4.2   Parameter Selection

Parameters were selected empirically for this pilot study to maximize signal retention and artifact rejection. Different scale factors and rolling time window sizes were explored for both the variance and the correlation. The scale factors were selected to bring the magnitude of R into the useful range of 0 to 5. R should be small (~0) to allow desired changes in signal to be retained, and large to remove artifact. R on the order of 2 or more was found to remove artifact effectively (as in figure 3, 100s – 150 s).

Increasing window size increases signal plateau due to filter turn off delay, as seen in figure 3 at 340s. It is due to correlation contributing to R even after it has passed in time. This may be tolerable for conservative investigators who wish to err on the side of signal removal. Shorter windows may be less able to detect artifact. Figures 1 and 2 show improved rejection (signal mean reduction) of the probe slip (50s - 100s) and the probe relocation (325s) artifacts attained by allowing the correlation to contribute to R.

Q was set to 1e-4 or 1e-5 as noted in the figures. If larger, high frequency spikes are not removed. If smaller, desired signal changes are attenuated. The initial state estimate was set to $z_0 = 0$, and the initial variance of the filter was set to $p_0 = 1$. All of these parameters are available for adjustment and optimization, with the objective of maximizing the rejection of artifact and the retention of good signal.

In real application, optimal parameters may depend on source-detector separation. For example, the signal to noise ratio of the phase itself depends on the source-detector separation (Dehaes, 2011). Ideally, the [HbO] close and far, and [HbR] close and far traces should be filtered with their own parameters.

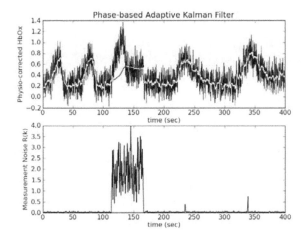

Figure 1 Calculated R (bottom) used to filter the [HbO] trace (top, filter output in negative). The phase variance from a rolling window looking back 1s is used to calculate R. Q was set to 1e-4.

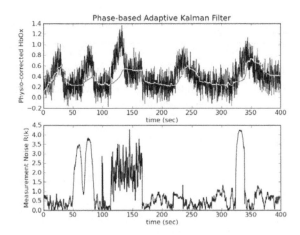

Figure 2 Calculated R (bottom) used to filter the [HbO] trace (top, filter output in negative). Both the phase variance from a rolling window looking back 1s, and the correlation from a rolling window looking back 20s, are used to calculate R. Q was set to 1e-4.

5   RESULTS

After filtering, the mean of the unwanted peak between 50s and 100s is reduced by 43%. That between 100s and 200s is completely removed. The reference signal mean is reduced by 9%. CNR is increased by more than 100% overall. The results are listed in Table 1. The filter output trace is shown in figure 3 (negative, top).

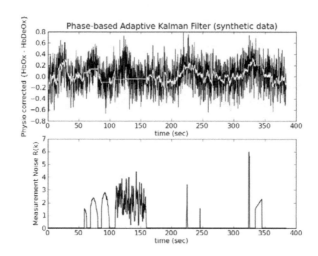

Figure 3 Calculated R (bottom) used to filter the [Hb] trace (top, filter output in negative). Both the phase variance and the correlation are used to calculate R. Q was set to 1e-5.

Table 1 Phase-based adaptive Kalman filter performance by motion type

Time	Motion simulated	Phase shift signal feature	Filter Performance Result
0s-50s	none (reference)	none (reference)	• CNR increase: 125% • signal mean reduction: 9%
50s-100s	probe slip	slow changes correlated with [Hb]	• unwanted signal mean reduction: 43%
100s-200s	probe decoupling	very high variance	• unwanted signal mean reduction: 100%
200s-300s	transient probe bump	spike at 225s	• CNR increase: 144% • signal mean reduction: 12%
300s-400s	probe bump, relocation	spike and step increase at 325s, correlated with [Hb]	• CNR increase: 204% • signal mean increase:11% • late filter turn off extends activation by 10s

6   DISCUSSION

An advantage of this filtering method is that it only comes on when needed according to actual conditions without adding a channel or auxiliary sensor. The phase provides an objective indication of whether [Hb] changes are due to motion without relying only on the [Hb] trace. Also, the phase can simply be used to indicate data quality for complete removal during post-processing.

A disadvantage is the introduction of time delays. The filter can only change the estimate so quickly, and delays can be on the order of seconds when the filter is on (high R). Filter turnoff delay is also a problem, as discussed in section 4.2 and as evidenced by the artificial increase in signal mean between 300s and 400s. Q may be increased and time windows shortened to reduce delays.

Other methods for real time artifact removal include simply smoothing, using other signal correlations, or using a moving standard deviation. With smoothing, unwanted signal is not rejected, and the temporal resolution advantage of fNIRS is eroded in all instances. The Kalman filter can be adjusted to quickly follow non-noisy data by instance. However, smoothing does not introduce delay.

Cui, et al. (2010) present a successful method for removing motion artifact from fNIRS signals which takes advantage of the anti-correlation between [HbO] and [HbR] inherent in functional activation. However, the phase-based filter removes motion artifact from both channels intended to measure functional activations and channels for systemic physiology, which lack neural activation.

A moving standard deviation measurement on the [Hb] signal itself has been used to detect spikes for removal (Scholkman, 2010). However, mechanical changes can degrade the signal via poor coupling resulting in a relatively flat [Hb] signal. This would not trigger a moving standard deviation detection algorithm.

7   CONCLUSIONS AND FUTURE WORK

The phase is a good candidate for qualifying fNIRS signals in real time, enabling automated bad-channel detection and motion artifact removal without auxiliary sensors. Successful future work would increase the amount of motion which can be tolerated while obtaining useful fNIRS signals. This real-time filtering technique relies upon frequency-domain instrumentation, but is well-suited to improving the quality of systemic physiology and functional activation measurements with fNIRS in any operational environment where motion is an inherent part of the task being performed.

Further parametric study is needed to minimize delays and maximize the performance of the filter for data collected with human subjects. The investigation of non-linear functions of the phase variance and the correlation between the phase and the [Hb] is planned. The filter should be tried on the measured optical intensity with the FDMD method. Human subject studies to quantify the motion of the probe with respect to the head during the phase shift measurement should be undertaken. Ultimately, cognitive state classification could be performed with this filter to determine the effect on classification accuracy.

ACKNOWLEDGMENTS

The authors would like to acknowledge NASA's Aviation Safety Program and the Biomedical Engineering Department at the University of Michigan.

REFERENCES

Abdelnour, F., Huppert, T. 2009. Real-time imaging of human brain function by near-infrared spectroscopy using an adaptive general linear model. *NeuroImage,* 46: 133-143.

Cui, X., et al. 2010. Functional near infrared spectroscopy (NIRS) signal improvement based on negative correlation between oxygenated and deoxygenated hemoglobin dynamics. *NeuroImage,* 49: 3039-3046.

Dehaes, M., Selb, J., et al. 2011. Assessment of the frequency-domain multi-distance method to evaluate the brain optical properties: Monte Carlo simulations from neonate to adult. *Biomedical Optics Express*, 2(3): 552-567.

Fantini, S., Hueber D., et al. 1999. Non-invasive optical monitoring of the newborn piglet brain using continuous-wave and frequency-domain spectroscopy. *Physics in Medicine and Biology,* 44(6): 1543-1563.

Harrivel, A., et al. 2009. Toward Improved Headgear for Monitoring with Functional Near Infrared Spectroscopy, *15th Annual Meeting of the Organization for Human Brain Mapping*, July, 2009, San Francisco, CA. *NeuroImage,* 47: S141.

Harrivel, A., et al. 2010. A System for Attentional State Detection with Functional Near Infrared Spectroscopy. *Human Computer Interaction - Aerospace*, November 2010, Cape Canaveral, Florida.

Harrivel, A. et al. 2011. Monitoring attentional state using functional near infrared spectroscopy: A pilot study. *17th Annual Meeting of the Organization for Human Brain Mapping*, June, 2011. Quebec City, Canada, poster #519WTh.

Izzetoglu, M, et al. 2010. Motion artifact cancellation in NIR spectroscopy using discrete Kalman filtering. *Biomed. Eng. Online*, 9: 16.

Kalman, R.E., 1960. A new approach to linear filtering and prediction problems. *Journal of Basic Engineering*, 82: 35-45.

Pope, A., Bogart, E.H., Bartolome, D.S. 1995. Biocybernetic system validates index of operator engagement in automated task. *Biological Psychology*, 40: 187-195.

Sarter, Gehring, Kozak. 2006. More Attention must be paid: The neurobiology of attentional effort. *Brain Res. Reviews,* 51: 145-160.

Schmorrow, D.D. and L.M. Reeves. 2007. 21(st) century human-system computing: Augmented cognition for improved human performance. *Aviation Space and Environmental Medicine*, 78(5): B7-B11.

Schnell, T., et al. 2004. Improved Flight Technical Performance in Flight Decks Equipped with Synthetic Vision Information System Displays. *International Journal of Aviation Psychology*, 14(1).

Scholkman, F., et al. 2010. How to detect and reduce movement artifacts in near-infrared imaging using moving standard deviation and spline interpolation', *Physiological Measurement*, 31: 649-662.

Zarchan, P and Musoff, H. 2005. Fundamentals of Kalman Filtering: A Practical Approach. *AIAA Progress in Astronautics and Aeronautics*, 232.

Zhang, Q., et al. 2009. Adaptive filtering for global interference cancellation and real-time recovery of evoked brain activity: a Monte Carlo simulation study. *Journal of Biomedical Optics*, 12(4): 044014-1-12.

CHAPTER 52

An Integrated Experimental Platform for the Crew Workload Measurement

Dayong Dong, Shan Fu

Shanghai Jiao Tong University
Shanghai, CHINA
Dongdayong@sjtu.edu.cn;dy_dong@163.com

ABSTRACT

The measurement of crew work load is an important part in the evaluation of the airworthiness of human factors. In clause 25.1523 of air worthiness' regulation on minimum flight crew, it is explicitly required that crew work load be put into consideration. The applicant must demonstrate in a rational manner the appropriate level of the crew that is to be given certification in order to ensure the safety of the aircraft.

The evaluation of crew workload is to be done in an objective way so as to offset the uncertainty of subjective evaluations. The research of this paper constructs an integrated simulator experiment platform to collect and analyze the physiological data of pilots during the experiment, to achieve a quantitative measurement of crew work load. The integrated experiment platform mainly comprises of two parts: firstly, a flight simulation system that can simulate various flight task scenarios; the other is a pilot physiological data collection and analysis system, where objective measurement of the flight crew is achieved by recording and analyzing the relationship between physiological and flight data of the tested pilots. In this research, experiments of approaching and landing are done and through data recording and analysis, the feasibility of the integrated experiment platform is tested.

Keywords: Airworthiness, simulator, crew workload, human factors, physiological

1   INTRODUCTION

Airworthiness certification is issued for an aircraft by the national aviation authority in the state in which the aircraft is registered. The main purpose of airworthiness certification is to insure the aircraft must be in a condition for safe operation. The human factors have been widely recognized as critical to aviation safety and effectiveness (FAA, 1993). Human error has been documented as a primary contributor to more than 70 percent of commercial airplane hull-loss accidents. (Boeing) As a result, in the type certification processes of air crafts, human factors must be considered as a key point of evaluation.

The current FAA (Federal Aviation Administration) and CAAC (Civil Aviation Administration of China) airworthiness standards (25.1523) explicit requirement on crew workload:" The minimum flight crew must be established so that it is sufficient for safe operation, considering the workload on individual crewmembers." Measurement and evaluation of light crew workload is the main work of essential flight crew certification. There are mainly four ways of flight crew workload measurement and evaluation. (Cowin workload, 1989): subjective workload measurement, physiological workload measurement, performance workload measurement, and analytical assessment technique of Timeline Analysis (TLA).Applicant must use an appropriate workload measurement method to demonstrate that the crew workload in normal, non-normal or emergency situation and could be maintained at an acceptable level. Though subjective evaluation is still the prevalent workload measurement method. However, as subjective method relies too much on the pilots' subjective opinions, during its application inconsistency of experiment results might result from different understandings of questionnaires. Physiological measurement as an objective measurement method is becoming increasingly accepted, with many researchers trying to measure the workload of pilots with various physiological parameters (Malcolm, 2002) (Roscoe, A. H., 1992). (Wilson, G. F., 2001, 2002)Thus technologies of objective workload measurement will become an inevitable trend.

This research constructs an integrated simulation experiment platform that performs real time collection of physiological parameters during simulated flights and uses the parameters for workload measurement and evaluation. The platform comprises of mainly two parts. One is a flight simulation system that simulates task scenarios of certain models aircrafts. It includes aircraft dynamics simulation, virtual meters, operational and control devices and external view. Tested pilots could undergo flight task experiments of various scenarios in this simulation system.

The other part of the experiment platform is the pilot physiological data collection and analysis system, where real time collection could be performed during simulated flight experiments. The data are then integrated via TCP/IP network, then gathered and analyzed as a work station. At the current stage, the data that can be collected mainly include: eye movement data, heart rate data, pilot behavioral video data and flight performance data. The system is open, in that by expanding data ports data other than aforementioned could be collected. Through analysis of the collected data, objective measurement and evaluation of flight crew workload could be achieved. In this research, confirmatory experiments are done on the workload measurement during approaching and landing. The results show that the platform is capable of collecting and processing physiological and performance data from simulated flight tasks, and thus is a technology to achieve objective evaluation of workload.

2 SYSTEM PLATFORM FRAMEWORK

Of the MOC (means of compliance) for air worthiness evaluation, testing through simulator experiments is one of the means. It is also one major method used in related researches on human factors. Common processes of simulator experiment are that pilots conduct simulated flight tasks in simulators and fill in questionnaires, both in-flight and post-flight, which are referenced against rating scales, to evaluate crew workload. In order to collect physiological data of subject pilots and to measure and evaluate crew workload objectively, this research constructs an integrated experiment platform for flight simulation and physiological data collection and analysis. The platform comprises of two parts, to achieve flight simulation and collection of physiological data of subject pilots. See figure 1 for the framework of the system platform.

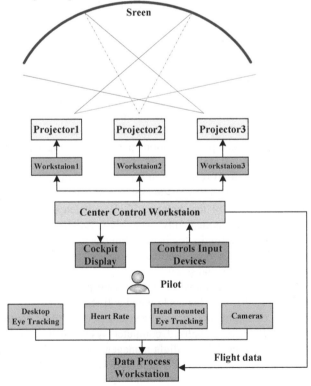

Figure 1 framework of integrated experiment platform

2.1   Flight Simulator

Flight simulator mainly constructs a virtual flight environment, where pilots could undergo experiments of various tasks of flight simulation. It is made up of flight deck display system, visual scenery system, flight control system, control devices, etc. The platform adopts a module design, and simulations of different scenarios could be performed by adjusting the corresponding modules.

Flight Deck Display System

Flight deck display system is the most important human-machine interface in the cockpit and its design is of critical importance to crew workload. In this research, VAPS is chosen as development tool, and flight deck display system simulation is designed successfully. The systems is made up of: Principle Flight Display (PFD), Multi-Function Display (MFD), Engine Indicating and Crew Alerting System (EICAS) (yet unwritten), mode control panel, overhead panel and control display unit, etc.

Among the listed, PFD is a modern aircraft instrument dedicated to flight information. It is located right in front of the pilot, provides an accurate virtual horizon, and displays angle of pitch of the plane, roll angle, and other information concerning the attitude of the plane; displays flight altitude, speed, Mach number and lifting speed and other flight parameters; also displays heading, flight track, Flight Director and other navigation information.

MFD is a screen in an aircraft that can be used to display information to the pilot in numerous configurable ways, allowing the pilot to display their navigation route, moving map, weather radar, (ground proximity warning system)GPWS, (Traffic collision avoidance system)TCAS and airport information.

Engine Indicating and Crew Alerting System (EICAS) is an integrated system to provide aircraft crew with aircraft engines and other systems instrumentation and crew annunciations. EICAS typically includes instrumentation of various engine parameters, including for example revolutions per minute, temperature values, fuel flow and quantity, oil pressure etc.

Mode Control Panel (MCP) is located at the middle of the glare shield. It is an instrument panel that controls advanced autopilot and related systems. It contains controls that allow the crew of the aircraft to select which parts of the aircraft's flight are to be controlled automatically. The MCP can be used to instruct the autopilot to hold a specific altitude, to change altitudes at a specific rate, to hold a specific heading, to turn to a new heading, to follow the directions of a flight management computer (FMC), and so on.

Overhead Panel is installed above the pilot, made up of many switches which control lighting, fire protection of the cockpit and anti-ice, etc.

CDU is the core of Flight Management System (FMs). Pilots control FMS by controlling CDU. FMS is a specialized computer system that automates a wide variety of in-flight tasks, especially in-flight management of the flight plan.

Design of the display and control system followed relevant requirements on "electronic cockpit display" issued by FAA (06/21/2007).

Visual Scenery System

The hardware of the visual scenery system is made up of three-channel projector and large dome screen projection. Vega Prime software is used to realize combination of visual scenery projection and three channel data mixing.

Visual Scenery based on Vega Prime is mainly made up of three parts: construction of model of the visual scenery, graphic interface design of LynX Prime, and visual scenery program design. The construction of visual scenery model mainly means using Creator to build static visual scenery model pool and using Creator Terrain Studio to generate large scale terrain visual scenery model pool. Graphic interface design of LynX Prime includes settings of the basic environment, model initial location, and frequently used weather and so on. Visual scenery simulation program is designed to finish to whole process of simulation to provide users with real time visual scenery simulation results. Drawing from the model pool generated during visual scenery construction using API functions from Vega Prime, and documents generated during LynX Prime's graphic interface design, the program drives the entire simulation.

Modeling of virtual three-dimensional entities in simulated scenarios is done in Creator; and the three-dimensional models are generated into virtual scenarios in Vega Prime, and the scenarios are controlled through Vega Prime's cross functions.

Flight Dynamics Model

Flight dynamics simulation model is one of the core software systems of flight simulator. Its main function is to calculate aircraft atmospheric environment parameters, aircraft engine parameters, aerodynamic equation and kinetic equation and attitude parameters. It is the data hub of the flight simulator. It takes flight environment and pilots' operational instruction signals to levers, helm, throttle lever and cockpit equipment as input parameters, and calculates in real time the attitude, location, speed, altitude and other flight parameters, which will all be transferred to other sub-systems as driving instructions and computing input parameters, also transferred via network to visual scenery system to render external visual scenery.

In this research, the commonly used 6-degrees-of-freedom kinetic equation is adopted to solve the aircraft's kinematical equation and dynamic equation and then transfer to the plane's attitude, which is then delivered to the visual scenery system. In the development of the software, open data ports are designed for aircraft aerodynamic parameters and engine parameters, so as to achieve dynamic simulation of various types of aircrafts.

Control Devices

Control devices mainly include steering wheel/lever, pedal, throttle, flap handle and other control input devices. To give the system better compatibility, all control devices all use standard USB ports. For cockpit layout, steering wheel style layout of Boeing and side stick style layout of Airbus are referred to, and switch between the two can be easily made, to simulate various design plans.

Fidelity

Flight simulation should satisfy requirements of human factor experiment fidelity. In system development processes, "General Specification for Airplane Simulator" should be followed to finish design of the simulation system. In this way the validity of data gathered from experiments can be ensured.

2.2 Physiological Data Collection

To evaluate crew workload with multi physiological parameters, multi-channel physiological data collection system is constructed. This research chooses heart rate and eye movement parameters that are effective for work load evaluation, together with pilot behavioral analysis based on videos recorded, considering both physiological parameters and behavioral features, to construct flight crew workload evaluation model.

Heart Rate

In this research, we adopt POLAR's RS800CXmultiterm movement heart rate watch as heart rate parameter collection system. In the experiments, subjects wear heart rate sensors to collect heart rate data in real time. And via a wireless transmission module, the collected heart rate data will be to the data analysis software at a terminal workstation.

Eye Movement

According to statistics, over 60% of all information input is through vision. Eyes, as the main organ for visual information input, displays features during information obtaining corresponding to mental workload during task performing.

Pilot's workload during flying can be shown in data of eye movement. Obtaining of eye movement data is through a SmartEye desktop eye tracking system. By recording and analyzing distribution of eye focus and pupil data during the experiments, a relationship between workload and eye movement data can be determined.

Operation Behavior

Pilot's control of the aircrafts is ultimately through the interaction between the limbs and the display control interface. So through analyzing pilots' behavioral features in flight tasks, pilot performance can be evaluated. When combined with eye movement data, pilot's eye-hand coordination modes in executing operations can be analyzed as well. Thus further studies on the relationship between information obtaining in the interaction process, information processing and information execution can be done.

For behavioral data collection, general video collection modules are used. Videos collected via camera during performing tasks will be transferred to terminal workstation for analysis with data processing software.

Data Summarization and Analysis

For data summarization and processing, a workstation is linked to all physiological collection devices to collect and store all data from experiments. And after finishing experiments, integrated data analysis software synchronizes the data for analysis.

3   EXPERIMENT

To verify the validity of the experiment platform, subject students went through many experiments involving approaching and landing. And through the simulation, various parameters were collected and processed.

Initial status of experiment task: altitude 2000 ft, speed 250 knots, distance from runway 6 miles, flying level straight towards the runway.

Basic operations:
1. Descend to 1000 ft, decelerate to 200 knots
2. Extend gear
3. Descend to 700 ft. decelerate to 150 knots, extend flap to 20 degrees
4. Descend to 400 ft. decelerate to 130 knots, extend flap to full
5. Flatter out, land gear down
6. Brake

6 subjects in total, experimenting 2-3 times per day, all data recorded, and analyzed with software.

Figure 2 Experiment

4   RESULTS

During experiments, various parameters of the aircraft during the entire flight are recorded as well, including space coordinates (longitude, latitude, altitude), attitude, (angle of pitch, roll angle, orientation), status of various parts of the aircraft (status of engine, flaps, ailerons, elevator, rudder), etc.

Table 1 Experiment Data

Physiological and Mental Measurement Equipment	Form of Recorded Data	Content Information
Polar Heart Rate Watch	1 text file	Heart Rate
SmartEye Desktop Tracking System	1 video file	Video of scenario in front of the subject
	1 text file	Coordinates of focus Number of gazing Number of blinking Number of glancing
SMI Eye Tracking System	1 video file	Video of scenario in front of the subject
	1 text file	Coordinates of focus Diameter of pupils Duration of gaze
Hand Movement Capturing Camera	1 video file	Operational behaviors of the subject
Body Movement Measurement Camera	1 video file	Range of body movement

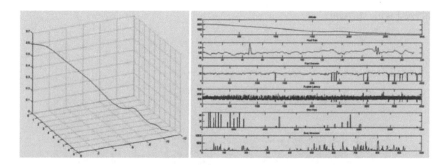

Figure 3 Experiment Data

5   CONCLUSION

Through analysis of the experiment data, the following results can be gathered:
1. During approaching, as the plane descends, the subject's heart rate fluctuates but the range of fluctuations tends to gradually lessen after gradually increasing. And heart rate rises more obviously near landing than when ascending.
2. When plane altitude changes intensely, pilot's operation movements are more frequent, range of movements larger and amount of movement larger.
3. Heart rate and movement range are rather sensitive to changes in plane altitude.
4. When the simulator has just gone into initial status, the subject has a higher frequency of blinking. And when the plane is descending, frequency of blinking decreases, but rises again when approaching the ground.
5. Relationship of pupil diameter of subjects and duration of gazing, to plane altitude is not clear.

This research constructs a rather complete human factor experiment platform that could undergo pilot-in-the-loop flight experiments. And through collection and processing of experiment data during flights, objective analysis of parameters, pilot performance and workload could be achieved. In follow-up studies to come, more experiments will be conducted using this platform, for deeper research into the methods for objective analysis of human factors in aviation.

ACKNOWLEDGMENTS

The authors would like to acknowledge the support of 973 Program (No. 2010CB734103)

REFERENCES

Boeing Aero Magazine No.8. Human factors. http://www.boeing.com/commercial/aeromagazine/aero_08/human_textonly.html

Corwin, William H. (1989) Assessment of Crew Workload Measurement Methods, Techniques and Procedures. ADA217699, WRDC-TR-89-7006. Volume I and Volume II

Eggemeier, F. T., & Wilson, G. F. (1991). Performance and subjective measures of workload in multitask environments. In D. Damos (Ed.), Multiple-task performance (pp. 217–278). London: Taylor & Francis.

FAA. (1993). 9550.8 - Human Factors Policy

Malcolm A. Bonner & Glenn F. Wilson (2002): Heart Rate Measures of Flight Test and Evaluation. The International Journal of Aviation Psychology, 12:1, 63-77

Roscoe, A. H. (1992). Assessing pilot workload: Why measure heart rate, HRV, and respiration? Biological Psychology, 34, 259–288.

Wilson, G. F. (2001). In-flight psychophysiological monitoring. In F. Fahrenberg & M. Myrtek (Eds.), Progress in ambulatory monitoring (pp. 435–454). Seattle, WA: Hogrefe & Huber.

Wilson, G. F. (2002). Psychophysiological test methods and procedures. In S. G. Charlton & T. G.O'Brien (Eds.), Handbook of human factors testing and evaluation (2nd ed., pp. 127–156). Mahwah, NJ: Lawrence Erlbaum Associates, Inc.

CHAPTER 53

The Relationship between Pilots' Manual Flying Skills and Their Visual Behavior: A Flight Simulator Study Using Eye Tracking

Andreas Haslbeck[1], Ekkehart Schubert[2], Patrick Gontar[1], Klaus Bengler[1]

[1]Institute of Ergonomics, Technische Universität München
Munich, Germany
haslbeck@tum.de

[2]Section Flight Guidance and Air Transportation, Technische Universität Berlin
Berlin, Germany

ABSTRACT

This paper presents an experimental evaluation of pilots' ability to support their manual flying skills through visual behavior. To this end, two groups of pilots with different levels of practice and training are compared in a full flight simulator. Different visual information acquisition strategies are used during the flight phases. In flight, pilots must direct their attention towards monitoring, while in a manual flying phase (approach and landing), a more frequent and accurate panel scan is imperative. The gaze data collected during this high-taskload flight period makes it possible to detect the differences between these two groups.

Keywords: manual flying, eye tracking

1   INTRODUCTION

With highly automated 'fly-by-wire' aircraft and the introduction of glass cockpits (e.g. types A320 and A340 from Airbus, which were chosen for this experiment), pilots' management duties in the cockpit have begun to squeeze out manual flying tasks from the cockpits. In general, this process should not be considered negative. Long-haul flights, for example, are very monotonous and require a very low level of action from the pilots. So automation can support pilots in their monitoring tasks during such uneventful flying phases. For critical phases of a flight, however, an active pilot is necessary.

Manual flying does not just mean "flying by hand." When defined as the opposite of automated flight, manual flying covers cognitive processes from information acquisition to cognitive processing, ending with response execution (information implementation). Information is acquired via sensory perception and includes processes such as visual, auditory, and vestibular perception. Verbal communication defines a meta-level of some auditory information. In the next step, all of the information is processed. A broad variety of models exists for this core element of cognition. One frequently used example of these models was created by Rasmussen (1983). The information is implemented via eye-hand coordinative activities, such as trajectory control via side sticks or yokes, and the speed is set by the thrust levers. Another output is also communication.

There is some evidence that today's pilots have lost a significant degree of their manual flying skills. Haslbeck et al. (2012) have discussed these degradation effects and introduced an experimental design to measure such effects. They presented core requirements to find variance in pilots' manual flying abilities within a realistic but difficult landing scenario.

This paper introduces the results of the aforementioned work with a focus on the relationship between pilots' manual flying skills and their visual behavior. In this experiment, pilots had to perform a landing scenario with eye tracking. The measurements were taken during different approach phases under different automation conditions and taskload levels.

2   PAST AND PRESENT RESEARCH

There are only a small number of studies analyzing visual behavior in modern glass cockpits.

Anders (2001) analyzed eye movements as well as head movements of pilots during flight trials in a full flight simulator (Airbus A330). All in all, 16 pilots and thus eight crews took part in this experiment, flying up to four standard missed-approach scenarios. Anders worked out the average attention time allocated to different areas of interest. Thereby he found out, that the gaze allocations on the several indicators differ. The attitude indicator, for example, is fixated more often than speed and altitude indicator. In contrast, they could not find any outstanding differences between the four approaches on the miscellaneous airports or between the different pilots.

Another experiment conducted by Diez et al. (2001), asked five Boeing B777 pilots to perform two flight scenarios on a Boeing B747-400 desktop simulator (single-pilot operation). One scenario involved the takeoff, climb, cruise, descent and approach phases, while a second scenario involved the descent and approach phases. The authors found that the proportion of the gaze allocation to the different areas of interest depended on the flight phase and on specific environmental events. Furthermore, they came to the conclusion that the pilots especially failed to remember the current FMA modes.

Sarter, Mumaw, and Wickens (2007) asked ten volunteer Boeing B747-400 flight crews to fly their scenario. The study's focus was on analyzing mode awareness and appropriate behavior when "wrong" modes were selected. Their results clearly show that a large number of pilots failed at this: only 47% of all required FMA checks were conducted. Furthermore, they found "that such failures were attributed, at least in part, to inappropriate or incomplete knowledge."

Frische, Osterloh, and Lüdtke (2011) conducted a study in an experimental flight simulator cockpit according to the Airbus A350 XWB layout. The tasks' focus for 15 German airline pilots was on planning und making decisions with a newly built flight management system. The analysis of gaze distributions showed a significant change in the importance of the primary flight display (PFD) among all areas of interest (AOIs). The newly presented flight management system became the most important AOI for several flight tasks. Only during final approach did the PDF remain the most frequently observed AOI.

None of these studies referred to manual flight in detail. So there remains a need to link visual behavior on the cognitive processes' input side with manual flying activities on the output side.

3   EXPERIMENTAL DESIGN AND METHOD

The experiment presented in this paper was motivated by the question of whether the level of practice and training influences manual flying skills. The independent variable, the level of practice and training, was varied by using two different groups of German commercial airline pilots. With a low occurrence of this variable, 27 long-haul captains (CPTs) participated. Representing a high level of practice and training, 30 short-haul first officers (FOs) took part in the experiment. All of the subject pilots were analyzed during their regular duties, so participation was not voluntary. The participants were the pilots responsible for flying (pilot flying). For pilot monitoring, the same FO/CPT was always on duty and played a rather passive role – but did not provoke any errors. Manual flight performance was measured as a dependent variable. A full flight simulator objectively measured the actions according to information implementation. Here, deviations from the ideal localizer, glide slope during approach and from the ideal touchdown point, vertical speed, and g-forces while touching down were recorded. A DIKABLIS eye tracking system recorded visual behavior (Lange, Spies, Bubb, and Bengler 2010). In combination with audio records, this allows important elements of the pilots' information acquisition to be analyzed. The participants had been asked for some explanatory variables such as perceived tiredness and workload. Haslbeck et al. (2012) gave a comprehensive overview of the experimental design.

The flight scenario started about half an hour before the scheduled touchdown. After an uneventful approach, the participants were forced to go around because of a strong emerging tailwind after passing an altitude of 1.000ft. After performing the missed approach, an automation defect made a manually flown landing necessary.

4   RESULTS

4.1   GAZE ALLOCATION

For the four different flight phases listed in table 1, relevant AOIs were defined and gaze allocation was analyzed afterwards. The primary flight display shown on figure 1 was separated into five different AOIs: attitude indicator (ATT) with flight director (FD) when engaged, airspeed indicator (SPD), heading (HDG), flight mode annunciator (FMA), and altitude (ALT).

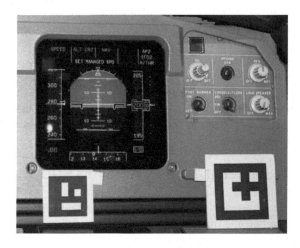

Figure 1   Airbus A340 primary flight display with markers for the DIKABLIS eye tracking system

Table 1   Overview of different flight phases

Phase	Content	Start / End
I	Automated descent on runway 08L EDDM: autopilot (AP) and flight director (FD) engaged	Start: ATC instruction HDG 110 (vectored ILS approach on RWY 08L) End: PF disconnects the AP
II	Continuous manual flown approach with FD engaged	Start: PF disconnects the AP End: call-out "go-around"
III	Go-around due to strong tail wind component while ATC changes landing direction	Start: call-out "go-around" End: ATC instruction HDG 080 (vectored ILS approach on RWY 26R)
IV	Manual landing (raw data approach) without AP and without FD engaged (RWY 26R)	Start: malfunction (disconnect of AP and FD) End: touch-down

It was not possible to distinguish between the five different areas of the FMA, nor between altitude and vertical speed, due to the resolution capabilities of the eye tracker.

Table 2 Gaze allocations for CPTs and FOs , flight phase I - IV

flight phase	pilots	ATT	SPD	ALT	FMA	HDG	chi-square test
I	CPTs	.177	.104	.084	.052	.009	$X^2(4)=49.45$; p<.001
	FOs	.178	.124	.069	.048	.006	
II	CPTs	.312	.027	.003	.011	.000	$X^2(4)=82.89$; p<.001
	FOs	.495	.078	.027	.016	.001	
III	CPTs	.261	.121	.040	.034	.011	$X^2(4)=240.38$; p<.001
	FOs	.233	.160	.056	.044	.002	
IV	CPTs	.425	.089	.045	.028	.070	$X^2(4)=710.15$; p<.001
	FOs	.367	.123	.139	.024	.108	

Table 2 shows general gaze allocations for the PFD. All glance durations at the PFD were counted for all pilots, standardized to 100%, and finally the median was calculated for the two groups. The chi-square test shows differences between the flight phases regarding the number of glances at the various AOIs (CPTs: $X^2(12)=566.57$; p<.001. FOs: $X^2(12)=2078.49$; p<.001). So in general, it can be said that the pilots' visual behavior differs according to the level of automation and the flight task.

Further comparison of the CPTs' and the FOs' visual behavior for each flight phase indicates differences in gaze allocations between the two groups. The CPTs' eyes are more often caught by the ATT or the FD. This even happens in flight phase IV, where the ATT does not provide the FD. The CPTs lapse into an accustomed gaze behavior. Consequently, the gaze allocation varies between pilots with different levels of training and practice.

4.2   FMA AWARENESS

Situation awareness is a frequently mentioned but rather abstract concept nowadays. Its measurement and quantification appear difficult. When this construction is broken down into several convenient units, it becomes easier to control in an empirical sense. This paper addresses three different aspects of situation awareness: FMA, speed, and wind awareness.

Because modern glass cockpits are equipped with different auto flight systems, involving direct interactions with the flight commands involved, it is necessary to know which systems are active, engaged or armed; in short, the pilots need to maintain a broad mode awareness. These different system statuses are indicated by 'flight modes'. The FMA displays the current selected flight modes in five columns and three rows within the PFD. The first three columns show the thrust, the vertical modes such as the glide slope, and the lateral modes such as the localizer. The fourth column indicates the selected decision altitude and the approach capabilities, while the last column shows the autopilot and flight director status. A separate analysis of these five different areas by the eye tracking system is not reliable and is therefore not considered separately. During the entire flight scenario, several mode changes were conducted. All of these mode changes had to be cross-checked by the participants (pilot flying). Mode awareness is thus given when this cross-check is done within five seconds after the actual mode change happened. This time interval of five seconds for each check was generated in agreement with aviation experts as well.

The participating CPTs performed 43.3% (SD 13.79%, n=20) of the FMA checks while the FOs did 53.4% (SD 21.47%, n=26) of all FMA checks. So on average, all pilots conducted just under half of the necessary FMA checks. Furthermore, it seems that the CPTs check the FMA even less reliably than the FOs do. In fact, this difference is statistically significant (t(44)=1.96; p=.028). Because FMA checks are mandatory for the pilots, standard operating procedures are violated in about half of the cases, and this behavior will not support a broad FMA awareness. It should be asked, however, whether all mandatory FMA checks are really necessary.

4.3 WIND AWARENESS

When passing an altitude of 1.000ft during a first approach in the flight scenario, the wind picks up and turns into a strong tailwind. Landing with this tailwind component of about 16kts is prohibited. Interestingly, the standard operating procedures (SOPs) of the aviation partner distinguish between the two aircraft types. In the A340, the SOPs ask pilots to recognize possible shear potential and/or significant cross-wind or tailwind components after the final altitude check (about 1.250ft). In contrast, the A320 SOPs require this check at an altitude of 500ft. Because the wind changes at an altitude of 1000ft in this example, the FOs can be expected to have a better awareness of the changed wind situation shortly before going around again due to this increasing tailwind. If the CPTs were to only follow their SOP instructions, they would not recognize this situation, while the FOs in contrast would.

The mandatory wind check at 1.250ft was done by 30% of the CPTs, while 64% of the FOs performed their wind check at 500ft. 60% of the CPTs and 57% of the FOs did two or more wind checks after final altitude, and thus finally became aware of the changed wind situation. But in contrast, about one-third of the pilots in both groups did no wind check at all, violating the SOP at this point. They would have most likely performed a landing without satisfying the wind awareness requirement – in the case of tailwind, resulting in an unsafe landing. In this case the monitoring pilot was instructed to go around at 70ft if the flying pilot did not. It should be said that the wind strength measured by the tower is binding, not the indicated wind speed provided by the aircraft. Nevertheless, it is absolutely mandatory to check the wind regularly during the approach.

4.4   SPEED AWARENESS

Despite a huge taskload during the approach, some aviation experts advice to check speed and vertical speed at threshold overfly on the runway. This final check provides energy information about the aircraft in question and offers a last opportunity to regulate power. When flying with a visual reference, the pilots do not need to check vertical speed once they can focus on the end of the runway. When there is a great distance to an object, humans are able to estimate angles and thus angular rates very well (Chapman, 1968). In the case of a landing, this would be the angular rate between the aircraft and the end of the runway, which implicitly provides the vertical speed information. But in the flight scenario here, the pilots were not able to see the end of the runway due to weather conditions. So they had no opportunity to estimate their vertical speed based on environmental information. In measuring speed checks, two different areas within the FMA are interesting: the SPD indicator on the left side and the vertical speed / sink rate indicator located on the very right side, to the right of the ALT indicator.

Between 100ft and touchdown, 19% of the CPTs and 64% of the FOs checked the SPD. The vertical speed was checked by 10% of the CPTs and 60% of the FOs. These numbers indicate that FOs maintain a higher awareness of speed and sink rate in the last moments before touching down.

5   CONCLUSIONS AND OUTLOOK

The general analysis of pilots' gaze distribution has shown that the pilot's taskload influences visual behavior. An in-depth analysis also showed differences between pilots depending on their level of practice and training. These differences also appear when considering various aspects of situation awareness. FOs performed significantly more mandatory FMA checks than their elder colleagues. The same result was found when looking at speed checks. Only the analysis of wind checks failed to deliver any relevant differences in the performance of both groups of participants. So the main hypothesis of this study, that the level of training and practice has an influence on manual flying skills, has been proven.

A second relevant finding was that today's pilots fall short in one important area of manual flying skills: visual behavior / information acquisition. Half of the mandatory FMA checks are not performed, and one-third of all pilots ultimately did not check the wind at all to ensure a safe landing and speed; the sink rate was also neglected according to the pilots' level of practice and training. At first glance, these results are negative. However, the question now is whether this visual behavior is still at an acceptable level or not. Here, the main focus of interest is on the SOPs concerning mandatory FMA checks. Is every check necessary, or can some checks be eliminated from the SOPs, in part to reduce the pilot's taskload without increasing general risk? Criticisms of pilots' professional behavior and their training programs are directed toward the missing wind checks. Even when the binding wind information is based on the tower wind, it has a latency of several minutes and cannot support scenarios where the wind suddenly picks up. Here, pilots are in charge of reducing the risk of landings with a high tailwind, which is an important contributing factor in hard landings and runway excursion incidents. Another safety aspect, but also relating to passengers' comfort, is the speed awareness shortly before touching down in low visibility conditions. In particular, the participating captains checked neither the speed nor the sink rate. If there is another way to control g-forces while touching down, this was not measured in this study.

These considerations also show the next steps to be taken. Speed awareness must be compared to g-forces while touching down in order to determine whether speed awareness reduces g-forces while landing. Another remaining question addresses all kinds of missing checks (regarding FMA, speed, wind): does visual behavior worsen according to the perception of an increased workload (visual tunneling)?

REFERENCES

Anders, G., 2001. Erfassung und Verarbeitung digitaler Blickdaten im Cockpit eines Airbus A330 Flugsimulators. Dissertation. Technische Universität Berlin.

Chapman, S., 1968. Catching a Baseball. American Journal of Physics, 36 (10), 868–870.

Diez, M., Boehm-Davis, D. A., Holt, R. W., Pinney, M. E., Hansberger, J. T., and Schoppek, W., 2001. Tracking pilot interactions with flight management systems through eye movements. Fairfax, VA: George Mason University.

Frische, F., Osterloh, J.-P., and Lüdtke, A., 2011. Modelling and Validating Pilots' Visual Attention Allocation During the Interaction with an Advanced Flight Management System. In: P.C. Cacciabue, et al., eds. Human Modelling in Assisted Transportation: Models, Tools and Risk Methods. Milan: Springer, 165–172.

Haslbeck, A., Schubert, E., Onnasch, L., Hüttig, G., Bubb, H., and Bengler, K., 2012. Manual flying skills under the influence of performance shaping factors. Work: A Journal of Prevention, Assessment and Rehabilitation, 41 (Supplement 1/2012), 178–183.

Lange, C., Spies, R., Bubb, H., and Bengler, K., 2010. Automated Analysis of Eye-Tracking Data for the Evaluation of Driver Information Systems According to EN ISO 15007-1 and ISO/TS 15007-2. In: W. Karwowski, ed. 3rd International Conference on Applied Human Factors and Ergonomics (AHFE): Conference Proceedings.

Rasmussen, J., 1983. Skills, Rules, and Knowledge; Signals, Signs, and Symbols, and Other Distinctions in Human Performance Models. IEEE Transactions on Systems, Man, and Cybernetics, SMC-13 (3), 257–266.

Sarter, N.B., Mumaw, R.J., and Wickens, C.D., 2007. Pilots' Monitoring Strategies and Performance on Automated Flight Decks: An Empirical Study Combining Behavioral and Eye-Tracking Data. Human Factors, 49 (3), 347–357.

CHAPTER 54

Air Traffic Controllers' Control Strategies in the Terminal Area under Off-Nominal Conditions

Lynne Martin, Joey Mercer, Todd Callantine, Michael Kupfer &
Christopher Cabrall

San Jose State University
NASA Ames Research Center
Moffett Field, California
Lynne.Martin@nasa.gov, Joey.Mercer@nasa.gov, Todd.Callantine@nasa.gov,
Michael.Kupfer@nasa.gov, Christopher.D.Cabrall@nasa.gov

ABSTRACT

A human-in-the-loop simulation investigated the robustness of a schedule-based terminal-area air traffic management concept, and its supporting controller tools, to off-nominal events – events that led to situations in which runway arrival schedules required adjustments and controllers could no longer use speed control alone to impose the necessary delays. The main research question was exploratory: to assess whether controllers could safely resolve and control the traffic during off-nominal events. A focus was the role of the supervisor – how he managed the schedules, how he assisted the controllers, what strategies he used, and which combinations of tools he used. Observations and questionnaire responses revealed supervisor strategies for resolving events followed a similar pattern: a standard approach specific to each type of event often resolved to a smooth conclusion. However, due to the range of factors influencing the event (e.g., environmental conditions, aircraft density on the schedule, etc.), sometimes the plan required revision and actions had a wide-ranging effect.

Keywords: CMS, ATC, traffic management strategy, collaborative work

1   INTRODUCTION

In today's air traffic system, control in a busy terminal area can include multiple speed changes, altitude level-offs, and/or heading vectors. In periods of heavy demand, this very safe system works hard to maintain high throughput, but does so at the expense of flight efficiency. That efficiency is commonly traded for positive control, so that controllers can quickly deliver aircraft from one sector to the next with proper spacing that allows aircraft to safely descend for landing. When demand is low, there is more space between aircraft and therefore less reason to be concerned with compression; less controller intervention is needed and aircraft can more easily descend along an efficient descent profile.

The Super-Density Operations research under NASA's Airspace Systems Program aims to safely sustain high runway throughput while still accommodating fuel-efficient operations. Advanced scheduling capabilities create schedules at the runway to enable Optimized Profile Descents (OPDs) along Area Navigation (RNAV) routes. Assuming en route controllers feed the Terminal Radar Approach Control (TRACON) with reasonable schedule errors, TRACON controllers can avoid costly altitude and heading maneuvers and instead rely primarily on speed adjustments to minimize runway schedule conformance errors (Isaacson, Robinson, Swenson, & Denery, 2010).

Controller-Managed Spacing (CMS) research in the Airspace Operations Laboratory (AOL; Prevôt, et al., 2010) has used a series of real-time human-in-the-loop simulations to investigate specific controller decision support tools (DSTs) for such operations. With relatively straight-forward display enhancements, TRACON controllers were able to manage dense arrival flows that followed OPDs along RNAV routes and met runway schedule times (see Kupfer, Callantine, Martin, Mercer & Palmer, 2011; Callantine, Palmer & Kupfer, 2010).

Having tested the CMS concept and tools under conditions with only speed variances, in 2011 research examined how robust the concept was to significant disturbances and off-nominal conditions. Accepting that large disturbances to operations in the TRACON are an eventuality, the response and recovery to an off-nominal event also requires investigation. In particular, the research sought to investigate whether the CMS concept and tools can support the response to and recovery from a disturbance (see Callantine, Cabrall, Kupfer, Martin, Mercer & Palmer, 2011).

This paper will focus on the role of the supervisor, who managed the runway schedule as part of the controller team's response to scripted off-nominal events. The paper will describe in more detail the operations tested, the CMS tools, and the roles and responsibilities of the participants.

2   METHODS

The 2011 simulation built on the previous studies using some of the same elements (e.g., the airspace), extending the investigation of other elements (e.g., winds), and adding consideration of off-nominal events. Of interest were the supervisor's strategies and how the TRACON team implemented resolutions to these off-nominal events.

2.1   The Simulation: Route Structure, Scenarios and Winds

The airspace simulated was the terminal area around the Los Angeles International Airport (LAX). Figure 1 shows a map of the airspace displaying the

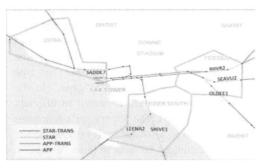

routes, waypoints, and sector boundaries for west-flow operations (based on current sectors in the Southern California TRACON). The airspace was comprised of three feeder sectors: Zuma, Feeder and Feeder South, and two final sectors: Stadium and Downe. Aircraft in the simulation flew OPDs on merging RNAV routes to runways LAX24R and

Figure 1. Test sectors in the LAX airspace created for the simulation

LAX25L. Path options in the form of named RNAV arrival/approach transitions were made available inside the TRACON to absorb large delays. RNAV go-around routes were designed to enable controllers to use the CMS tools to reinsert go-arounds into the arrival flows to the runways.

Three one-hour scenarios were developed for the simulation with a planned mean throughput of 31 arrivals to each runway. The scenarios were built under the assumption that aircraft had been delivered to the TRACON meter fixes by en route control with nominal schedule errors between 60s early and 30s late. However, due to wind forecast errors, these errors between the estimated time of arrival (ETA) and scheduled time of arrival (STA) at TRACON entry differed from that range. In addition to the standard wake spacing distances an additional buffer of 0.5nmi (Nautical Miles) was added into the scheduler.

Winds were a headwind aligned with the landing runways from 265° a third of the time, and the rest of the time were from 45° north or south of the runway. Below 1,500ft the forecast wind profile matched the actual wind profile, but above this altitude there were two wind-forecast-error conditions where the actual wind differed from the forecast wind by either 7 or 13 knots. This had an effect on the accuracy of the higher-level tools (see below) because their calculations took the forecast winds into account.

Four types of off-nominal event were planned: on-board medical emergencies, radio outages (NORDO or "no radio"), pilot-initiated go-arounds (e.g., due to gear malfunctions), and tower-initiated go-arounds (e.g., due to another aircraft on the runway). Two of these events, of different types, were scheduled in each run.

The study was run in the Airspace Operations Laboratory (AOL) at the NASA

Ames Research Center using Multi Aircraft Control System (MACS) software (Prevôt, et al., 2010). Simulated aircraft were assumed to be Flight Management System and Automatic Dependent Surveillance-Broadcast-out equipped.

2.2 CMS Tools

Controller participants worked with a MACS emulation of the Standard Terminal Area Replacement System (STARS) onto which the CMS tools were added. As the Supervisor's role was to manage the traffic schedule, his four tools were located on two timelines (Figure 2). The two timelines showed schedules to the two arrival runways with aircraft ETAs on the left side of the time tape and their STAs on the right. The ETA computations and the tools were based on aircraft trajectories being predicted through the forecast winds. Green bars in the time tape indicated excess spacing, or "gaps," in the schedule and red bars indicated insufficient spacing (overlap). The Supervisor's four tools were located in a row of buttons near the top of each timeline. He was able to "assign" a particular STA to an aircraft by dragging its current STA to a new desired time; "swap" the STAs of two aircraft, i.e., the STA of aircraft b became the STA of aircraft a and vice versa; "reset" all aircraft after a specified time which re-scheduled aircraft according to the current ETA information; and, "move" multiple aircraft STAs forward or backward on the schedule by a constant amount. These tools are based on aircraft trajectory predictions computed using the forecast winds. The timelines were located on a display ranged out to show the entire TRACON as well as about 40nmi into the Center airspace (approx 100nmi radius).

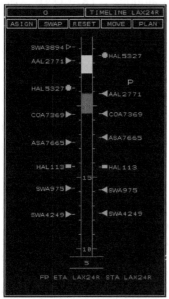

Figure 2. Supervisor timeline with scheduling tools

The controllers also had a suite of four tools. These differ from the Supervisor's tools as they are focused on individual aircraft rather than the schedule. The CMS controller tools are designed to provide a temporal and spatial awareness of each aircraft's progress relative to its STA, and speeds a controller could issue to correct schedule errors (see Kupfer, et al., 2011 for a detailed description).

2.3 Participants and Their Tasks

Eight retired air traffic controllers participated in this simulation. Three staffed the feeder positions and two the final positions. The sixth participant, who served as the Supervisor, was recruited specifically for his professional experience as a terminal-area traffic manager. Two controllers staffed supporting confederate "ghost" and tower positions, and general aviation students and pilots ran eight simulation pilot stations to control the simulated aircraft.

Controllers were asked to manipulate the traffic using only speed and the pre-defined path options, if possible, to bring the aircraft to land at the runways on schedule, although vectoring was still a valid option. The Supervisor was asked to manage the schedule and to try to maintain high runway utilization by working the schedule only as far out as the first aircraft that was outside the freeze-horizon (about 80nmi out from the runways). How the controllers and the supervisor used the tools to achieve this and how they coordinated to formulate off-nominal recovery plans was a point of research interested and left open for them to determine.

Participants took part in five days of training during shakedowns prior to the study to ensure their familiarity with the concept, tools, and procedures. The study ran for two weeks (non-consecutive) during which 46 one-hour runs took place – 42 runs were unique and four were repeats. Each run had two scripted off-nominal events that occurred in three base scenarios under varying wind conditions. The run matrix was randomized to reduce controller learning effects. All but one scripted event initiated as planned, yielding 91 off-nominal event examples.

Data were recorded for each run through the MACS' data collection logs, including aircraft and tool states, as well as controller and pilot actions. Screen and voice recordings were also collected. Following each run, the participants completed an online questionnaire that included questions about their strategies and problem solving. In addition, detailed observer notes were compiled from each experimental run and debriefing. The following section describes results from analyses conducted thus far.

3   RESULTS

The Supervisor's role was to manage the schedule manipulations that were required during off nominal events. As this was an exploratory position, the aim of the study was to observe the strategies the Supervisor developed and how they were executed, and to record his interactions with the rest of the team, who were controlling the traffic in the TRACON.

3.1   Supervisor Strategies

The Supervisor developed four basic strategies that he often used to begin handling the four different types of off nominal event. When a NORDO aircraft was identified, the Supervisor typically began by setting its STA to match its ETA (using his "assign" tool). He then assessed the schedule and how aircraft ahead or behind the NORDO aircraft could be affected. He usually consulted with the feeder controller(s) to determine whether to swap STAs, and in some cases devised a contingency plan for having the aircraft ahead of the NORDO go around if safe spacing was lost. For medical emergencies, the Supervisor coordinated with the controllers to expedite the emergency aircraft if possible; the plan could include schedule swaps and delaying other aircraft. When go-arounds were declared, the controllers needed to formulate a plan for climbing the aircraft to a safe altitude and assigning the desired RNAV route to re-enter the aircraft into the flow. The Supervisor informed to this route choice because he considered the schedule and adjustments of neighboring aircraft in the planned sequence. In some cases, the Supervisor organized the aircraft to go around to the other arrival runway if its schedule had more space. For tower-initiated go-arounds, less time was spent developing a plan because the event was announced much later on an aircraft's final approach than for pilot-initiated go-arounds.

These differences in strategies were reflected in the work that the Supervisor estimated he did for each type of off-nominal event (Table 1). For example, he worked three times as long, and twice as hard, on NORDO problems as on medical events because for medical events he was able to set up a plan and let it play out, whereas in a NORDO event the actions of the aircraft were unknown, even though they were expected to follow the charted procedure, which resulted in NORDOs entailing higher levels of monitoring.

Table 1. Supervisor's workload during four types of off-nominal events

	Medical Emergency	Pilot initiated go-around	Tower initiated go-around	NORDO
Mean estimated task time	2 min 46 s	5 min 15 s	5 min 15 s	9 min
Mean estimated task mental load	1.76	2.87	2.9	3.62
Mean estimated task time pressure	1.66	2.41	2.85	3

3.2 Variations to a Plan

Unsurprisingly, not every strategy played out the way the Supervisor intended. Some plans fell victim to wind changes or to other actions that impacted them. For example, in Trial 25 the wind shift to a direct headwind pushed the ETAs of all aircraft back, making them late with respect to the schedule; while in Trial 45 a favorable wind change helped an aircraft advance into its slot marker at an earlier waypoint. Some strategies simply followed a different course than the one the Supervisor predicted. For example, in Trial 36 a NORDO aircraft kept its speed up, forcing the Supervisor to abandon his plan of schedule 'tweaks' and take an aircraft around to the other runway.

The Supervisor reported he made adjustments to his original plan about 60% of the time but these were most often 'a few adjustments' and were 'large revisions' for less than 10% of the events. Usually he did not adjust a medical event plan (median =1 out of 7: no adjustments) but he had to make 'a few adjustments' to plans for NORDO events (median =3). In a companion question, the Supervisor was asked what he would have done differently in hindsight. He reported he would not

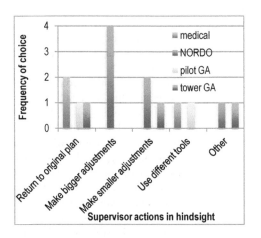

Figure 3. Supervisor's reports of different plans in hindsight

have done anything differently in 73% of the events, but he would have made changes on 15 occasions (17%). Half of these reports were after NORDO events and, although the Supervisor would usually have done more (Figure 3), on one or two occasions he thought a better course of action would have been to do less: "I would have let the NORDO run and monitored the situation," (Trial 36). On four occasions he felt his initial plan would have worked and he should not have made the additional adjustments that he did. For example, in Trial 17, after initially saying the aircraft with the medical emergency should stay at its assigned STA, the Supervisor decided to swap it with the aircraft in front. To achieve this, Stadium had to issue multiple vectors to both aircraft – a large increase in workload.

The larger number of adjustments and plan changes (with hindsight) that the Supervisor made during the NORDO events highlight that the NORDOs, in general, were the least flexible of the four event types. Not only was a NORDO inherently less predictable, because the pilot could not be contacted, it was also less manipulable because the controllers could not redirect the problem aircraft and the Supervisor had to adjust the schedule around it.

3.3   Collaboration

As the team-member with an overall view of the TRACON traffic and the schedule, the Supervisor was the natural role to initiate recovery plans for off-nominal events. Plans were sometimes formed by the Supervisor alone but often, if there was time, he would discuss the situation with the controllers to generate a plan that was informed by their views. For example, in Trial 29 the Supervisor considered a swap but as part of a discussion with Downe he demonstrated this action and Downe said that he would not be able to meet the new STAs. Due to this, they began to work on other options. This strategy of consulting the team was usually worthwhile because the controllers had to be able to execute the plan that the Supervisor created. However, controllers often did far more than merely "execute" the plan. In approximately a third of the events the plan would not have had the successful outcome that it did if a controller had not paid additional attention to the key aircraft and creatively manipulated them into the right place. Sometimes these actions were to use the path extension routes or extra speed control but at other times the controller solution reflected years of experience; in Trial 28

Downe vectored one aircraft off the final in a U-shaped detour that allowed another aircraft to overtake without having to swap them.

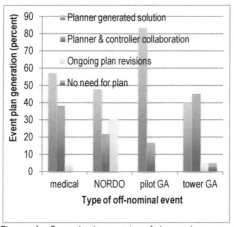

Figure 4. Supervisor's reports of how plans were generated

Sometimes a plan was determined or requested by a controller. Although the Supervisor estimated that he generated the plan 60% of the time (Figure 4) and he collaborated 30% of the time, these were occasions where the controller suggested a distinctly different option (not the creative execution discussed above). Joint plans were created most often for tower-initiated go-arounds probably because the initial actions for this event – to break an aircraft out of its final descent and assign it the go-around route – had to happen quickly. Final controllers often assigned a route and then asked the Supervisor if it would work which began an exchange of ideas.

In the handful of cases where there needed to be a new plan controllers were involved half the time in the Supervisor's revisions. This included asking the Supervisor for additional help because the STA could not be met, reporting that a plan would not work for their position, and working through options to find a viable solution.

3.4   Interactions of actions

Given that the Supervisor's role was a new component of the research and that he didn't have clearly defined procedures to follow, there was a certain degree of experimentation as his role organically evolved and he became more comfortable. While this was a necessary part of the learning process, it also came with some interesting side effects.

The supervisor tools were all related to adjusting the runway schedule; his actions directly changed the ground system's STA of an aircraft. At the same time however, the controllers' tools were driven by aircraft STAs. Consequently, DSTs on the controller displays fluctuated as a direct result of the supervisor's actions. This was usually a good thing: e.g., an aircraft needed to be delayed to make room for the re-insertion of a go-around aircraft. In such a case the controller tools helped to understand the current state of the aircraft relative to meeting its goal of the new STA. On the other hand, there were situations when formulating a plan was difficult for the Supervisor and required him to "try out" a few potential ideas before finalizing the plan. To do this, the Supervisor simply adjusted the STAs of the affected aircraft in a particular way and examined the result (sometimes with and sometimes without controller feedback). At that point, the supervisor could have kept the plan, reverted to the original state, or tried adjusting the STAs of the affected aircraft in a different manner. Even if a plan formulation like this took less than a minute, during that time the controller DSTs could change dramatically in response to the supervisor's actions. In multiple instances, a controller saw the change to their aircraft's STA and immediately began working toward the new goal, only to discover that it was a "test" rather than the finalized goal.

A more extreme example shows how the interaction of the Supervisor's planning process and the controllers' efforts to manage their aircraft can cause undesired outcomes. In Trial 35, the NORDO aircraft was a little ahead of schedule, and the Supervisor planned to swap the NORDO with the aircraft scheduled just ahead (flight HAL5327). After the swap, HAL5327 was ahead of its STA and the Zuma controller working the flight issued it a path extension as well as a speed reduction. Three minutes went by while the Supervisor examined the changes he made to the schedule, after which he determined this plan would not work, and swapped the two aircraft back. This now put HAL5327 behind its STA, so much so that Zuma informed the Supervisor that even with speed increases the STA could not be met. The Supervisor adjusted the schedule a third time, entailing moving the HAL5327 STA back again, as well as delaying several other aircraft STAs. Zuma slowed HAL5327 again, and was somewhat frustrated at having issued multiple contradictory speed clearances. He then spent the rest of Trial 35 absorbing sizable delays for most of the aircraft flying through his sector.

In this example, the Supervisor's delay in solidifying a plan without coordination, combined with the actions already taken by the controllers, complicated the problem unnecessarily, and resulted in an inefficient solution. Interestingly, Trial 35 was repeated as Trial 46, and in Trial 46 the supervisor did not attempt to swap the NORDO aircraft with HAL5327, yielding a much smoother outcome. Figure 5 shows the lateral tracks of aircraft that flew through the Zuma and Stadium sectors during Trials 35 and 46.

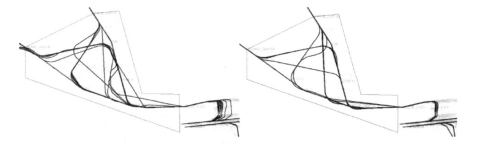

Figure 5. Lateral tracks of aircraft flown through Zuma and Stadium sectors in Trial 35 (left) and Trial 46 (right). Compared to Trial 46, the tracks in Trial 35 show increased vectoring in both the Zuma and Stadium sectors, stemming from the Supervisor's inefficient plan.

4   DISCUSSION

While prior studies of nominal operations (Kupfer, et al., 2011) illustrated little need for schedule manipulations, this study of off nominal events clearly shows the importance of the role that the Supervisor played. Although his responsibilities were pre-defined as monitoring and managing the schedules, the role of 'strategist' and the procedures for managing events were specifically not defined. During the study, the need for a single coordinator role emerged. As the position with an overview of the TRACON the Supervisor was able to identify options – such as swapping aircraft to a different runway, especially in the go-around cases – that the sector controllers were unable to see. In this sense, he became the team strategist as well as the schedule manager. During the debrief, when asked who should be in charge of creating the plan, the controllers echoed the idea the supervisor was in the best position to do so, stating "I knew [the supervisor was] the only one with the big picture. I learned that [lesson] early; that I've got the small picture."

Over the course of the two-week study, the TRACON team developed initial strategies for each of the four event types that gave them a going-in position that they modified to meet the specific conditions of each trial. These strategies could become the basis for developing specific procedural guidance for managing these particular types of off-nominal events. Notwithstanding this, the trials in this study provide case-studies for comparing the different approaches the team tried and illustrate how some approaches were more effective than others at bringing the schedule back to a nominal state. Further research is needed to ensure off-nominal recovery procedures are specified at the correct level of abstraction and conditions for applying particular procedures are clearly defined (Callantine, 2011). It is interesting that although the same basic event was initiated multiple times, they rarely played out in the same way.

The functioning of the controllers and the Supervisor as a team unit were highlighted by the off-nominal events. During nominal operations there was little need for the team to coordinate because the CMS tools provided the required information. However, off-nominal events illustrated that supervisor-controller communication was key to many of the successful solutions the team identified. This is most definitely not a one-way channel; controller input was valuable to the Supervisor in many cases. Not only do the controllers have a detailed view and understanding of the situation that a supervisor does not have but in this study, he needed their feedback to assess whether he had solved a problem or just shifted it. A "ripple" effect was observed for some Supervisor actions where workload increased in sectors that were not involved with the off-nominal problem (usually the East Feeder).

As seemingly small actions on the schedule can have large and often unforeseeable consequences, off-line planning tools are key. Problems were observed that were complicated by the Supervisor trying a solution to look at its effects and reversing his actions but not before the controllers had begun to issue clearances based on the change. The team tried to develop a workaround where the Supervisor announced when he was planning and when he had set the schedule, but this frustrated the controllers as their tools reset and they had to wait to make adjustments. The Supervisor needed a 'schedule trial-planning' function so he could assess the implications of his plans before he set them as changes into the ground automation. A pre-existing MACS schedule-adjustment functionality could be streamlined and extended to enable schedule trial-planning to meet this requirement in future studies. Further analyses to determine whether any of the Supervisor's tools could or should be automated also need to be conducted.

5   CONCLUSION

This study investigated the robustness of a schedule-based arrival-management concept using controller tools and introduced a new supervisory position to manage the schedule under off-nominal events. The controller team successfully managed most of the trial cases during a two-week study, suggesting that this concept has potential to demonstrate consistently robust performance even during off-nominal events. The results indicate recovery from off-nominal events is most efficient when rescheduling is accomplished reasonably quickly and the TRACON team is able to use the tools to make the required adjustments. The larger number of adjustments that the Supervisor made during the NORDO events and the greater number of actions he would have changed with hindsight illustrate that the NORDO events, in general, were the hardest, and most workload-intensive to manage. The case studies have provided rich information about possible strategies for development of off-nominal recovery procedures, and have shown where improvements to the supervisor tools are required and where the study can be improved for future investigations. While the research is a first step toward establishing the necessary safety case for CMS operations, the results are promising, and should help pave the way for future development of controller tools, procedures, and simulation-capabilities.

ACKNOWLEDGEMENTS

Thank you to the AOL development team and the controller subject matter experts without whom this study could not have been completed.

REFERENCES

Callantine, T. 2011. Modeling off-nominal recovery in NextGen terminal-area operations, AIAA-2011-6537. *Proceedings of the AIAA Modeling and Simulation Technologies Conference,* Portland, OR.
Callantine, T., C. Cabrall, M. Kupfer, L. Martin, J. Mercer & E. Palmer. 2011. Investigating the impact of off-nominal events on high-density "green" arrivals. *30th Digital Avionics Systems Conference*, October 17-20th, Seattle, WA.

580

Callantine, T., E. Palmer & M. Kupfer. 2010. Human-in-the-loop simulation of trajectory based terminal-area operations. *27ʰ International Congress of the Aeronautical Sciences (ICAS)*, Nice, France.

Isaacson, D., J. Robinson III, H. Swenson & D. Denery. 2010. A concept for robust, high density terminal air traffic operations. *Proceedings of the AIAA Aviation Technology, Integration, and Operations (ATIO) Conference*, Fort Worth, TX.

Kupfer, M., T. Callantine, L. Martin, J. Mercer & E. Palmer. 2011. Controller support tools for schedule-based terminal-area operations. *Proceedings of the Ninth USA/Europe Air Traffic Management Research and Development Seminar*, June, Berlin, Germany.

Prevôt, T., P. Lee, T. Callantine, J. Mercer, J. Homola, N. Smith & E. Palmer. 2010. Human-in-the-loop evaluation of NextGen concepts in the Airspace Operations Laboratory, AIAA 2010-7609. *American Institute of Aeronautics and Astronautics*, Reston, VA.

Section IX

Experiments and Evaluation

CHAPTER 55

Assessing Wake Vortex Hazards Using Pilot/Vehicle Analyses

Ronald A. Hess

Dept. of Mechanical and Aerospace Engineering
University of California
Davis, CA 95616
rahess@ucdavis.edu

ABSTRACT

The development of a methodology that will allow the assessment of hazards associated with wake-vortex encounters by fixed and rotary-wing aircraft is presented. The methodology utilizes pilot/vehicle analyses and is intended to serve as a prelude to high-fidelity pilot-in-the-loop flight simulation. The pilot model, itself, has been introduced in the literature for flight simulator fidelity investigations. The analysis procedure provides an economical and time-saving tool for preliminary assessment of wake vortex hazards. The approach is exercised using a model of a small business jet as an in-trail aircraft interacting with a relatively simple vortex model assume generated by a large transport aircraft. Hazard metrics are proposed and employed in the analysis. The possibility of wake vortex-induced upsets providing triggering events for pilot-induced oscillations is emphasized.

Keywords: Aircraft wake vortices, analytical pilot models, aviation   safety

1   INTRODUCTION

The danger of aircraft encounters with trailing vortices has been known for over 40 years (Carten, 1971). Computer simulations of such encounters have been utilized for nearly as long, e.g., (Iversen and Bernstein, 1972). The negative impact of wake turbulence upon air traffic capacity has been well documented (Anon., 2008a), the dangers to both fixed and rotary-wing aircraft outlined (Turner,

Padfield, and Harris, 2002). Only recently has emphasis been placed upon the importance of modeling the pilot/aircraft system in such encounters, (Luckner and Amelsberg, 2010).

2   APPROACH

2.1   Background Research

The groundwork for a pilot modeling approach to wake vortex hazard assessment is in place, e.g., (Hess, 2010a). The modeling approach was developed in research devoted to flight simulator fidelity assessment (Hess and Marchesi, 2009). In addition, past research by the author has focused upon modeling atmospheric disturbances on aircraft and rotorcraft, e.g., (Hess, 2005), and on aircraft handling qualities, e.g., (Hess, 2010b).

2.2   A Vortex Model

Consider the wake vortex velocity model shown below described in terms of the lateral (v_g) and vertical (w_g) components of a frozen turbulence field. The model is similar to one previously proposed (Tatnall, 1995). Figure 1 defines the variables shown in Eq. 1.

$$v_g = \frac{\Gamma_\infty (h - h_0)}{2\pi\pi[(Y + d/2)^2 + (h - h_0)^2 + r_c^2]} - \frac{\Gamma_\infty (h - h_0)}{2\pi\pi[(Y - d/2)^2 + (h - h_0)^2 + r_c^2]}$$

$$(1)$$

$$w_g = \frac{-\Gamma_\infty (Y + d/2)}{2\pi[(Y + d/2)^2 + (h - h_0)^2 + r_c^2]} + \frac{\Gamma_\infty (Y - d/2)}{2\pi[(Y - d/2)^2 + (h - h_0)^2 + r_c^2]}$$

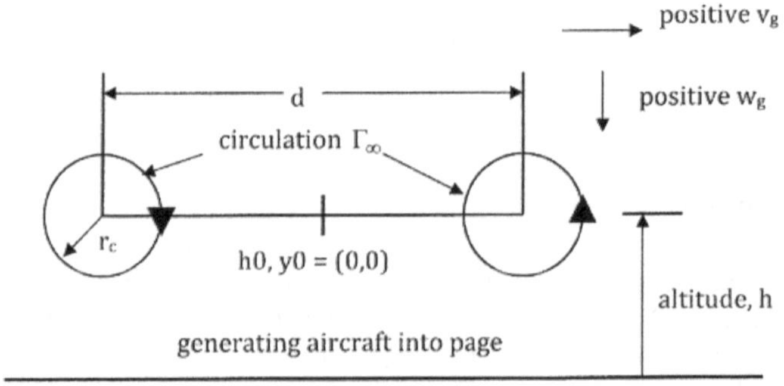

Figure 1 Geometry of vortex model.

For the purposes of this study, the following wake vortex parameters were used:

$$d = 195.68 \text{ ft}, \quad \Gamma_\infty = 6{,}587.5 \text{ ft}^2/\text{sec} \quad r_c = 10.5 \text{ ft} \quad (2)$$

The values in the equation above are based upon parameters cited in the literature (Vaughan, Brown, Constant, Eacock, and Foord, 1996), and here consider the generating aircraft to be a Boeing 747. The age of the vortex was assumed to be approximately 100 sec. Using the concept of a frozen turbulence field, "gust" inputs w_g, v_g, p_g, and q_g (vertical, lateral, roll and pitch atmospheric velocities) were generated based upon the position of the in-trail aircraft (Hess, 2005).

2.3   Candidate Hazard Metrics

Candidate metrics for wake vortex encounter hazards can be proposed based upon closed-loop pilot vehicle responses. These would include the magnitudes of roll/pitch attitude excursions and cockpit accelerations, and whether control surface actuator amplitude/rate limiting occurred in the encounter. As a point of departure consider the following limits, the violation of which would denote a hazard to flight safety.

maximum pitch/roll attitudes = ± 30 deg;
maximum lateral acceleration at cockpit = ± 0.5 g's
maximum vertical acceleration at cockpit = ± 1.0 g;
time spent with amplitude/rate saturated elevator or aileron actuators = 0 sec.

2.4   In-Trail Aircraft

The in-trail aircraft was selected as the Jetstar business jet as described in (Heffley and Jewell, 1972) and shown in Fig. 2. The vehicle model was chosen because of its size and the ready availability of an aerodynamic model. The flight condition selected was a trim velocity U0 = 257 ft/sec at sea level. The controls available to the pilot were elevator through column inputs, aileron through wheel inputs and rudder, through pedal inputs. Second-order amplitude and rate-limited actuator models were included for each of the control surfaces. Finally, a limited-authority yaw-damper was included as the only stability and command system on the aircraft.

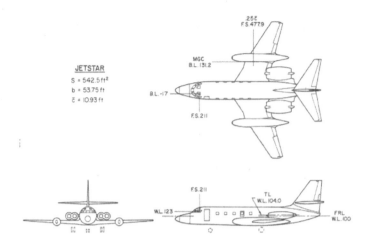

Figure 2 The Jetstar aircraft.

2.5 The Pilot Model

A multi-axis pilot model was created for this aircraft. Figure 3 shows the genera feedback structure. Pilot cues included visual, proprioceptive and vestibular. An additional heading control loop was closed around the roll-loop in the pilot model. Finally, the possibility of coordinated pedal and wheel inputs were considered. In Figure 3, the pilot model elements Yp-θ and Yp-ϕ included proprioceptive feedback from the cockpit inceptors (column and wheel). Details of pilot model parameter selection can be found in (Hess and Marchesi, 2009).

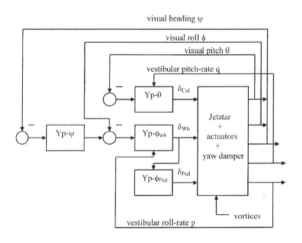

Figure 3 The pilot model feedback structure.

3   COMPUTER SIMULATION OF A VORTEX ENCOUNTER

3.1   Slanting Encounter with hinitial = h0 = 0 ft

The encounter to be simulated first involves the Jetstar with a nominal flight path making an angle of 5 deg with the longitudinal axes of the vortices, i.e., close to a parallel encounter with $h_{initial} = h_0 = 0$. The simulation begins with the Jetstar 1000 ft to the left of the longitudinal axis of the left trailing vortex of the generating aircraft. Note that because of the vortex disturbances, the Jetstar will necessarily depart from this nominal path. It is of some interest to demonstrate the vertical vortex velocity w_g that would be experienced at the Jetstar center of gravity were the aircraft to remain on the nominal trajectory. Figure 4 shows the w_g component. The sign convention for w_g is negative "up", i.e., a negative w_g is an updraft. Due to the vortex symmetry v_g is zero in this encounter. Figure 5 shows v_g for h = 20 ft.

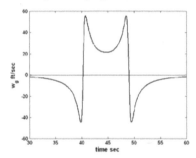

Figure 4 Vortex velocity component w_g; In-trail aircraft constrained on path with 5 deg heading path relative to longitudinal axes of vortex pair, h = h_0 = 0 ft.

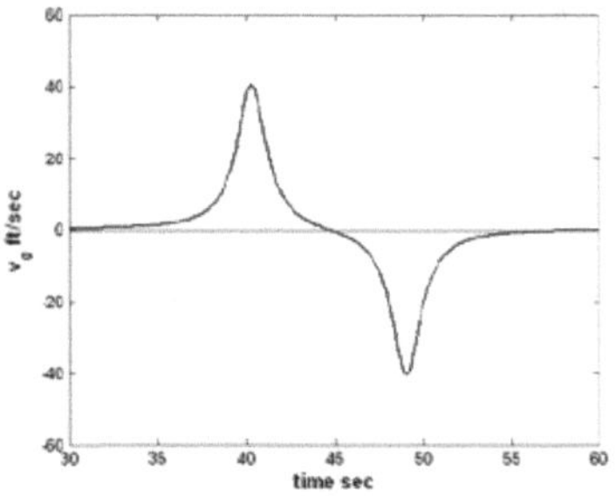

Figure 5 Vortex Velocity component v_g; In-trail aircraft constrained on path with 5 deg heading relative to longitudinal axes of vortex pair, h = 20 ft, h_0 = 0 ft.

Figures 6-10 show the pilot/vehicle responses for the slanting encounter in which the initial aircraft altitude was $h_{initial} = h_0 = 0$ ft. Of interest here is (1) the nearly 30 deg negative roll attitude (Fig. 6), (2) the large h and y deviations from the nominal path (Fig. 7), (3) the lateral acceleration at the cockpit exceeding 1 g, (Fig. 8), (4) the aileron actuator amplitude limiting (Fig. 9) and rate limiting (Fig. 10). The large negative roll excursion at t ≈ 47.5 sec in Fig. 6 is attributable to the fact that the left trailing vortex effectively "tosses" the aircraft into the right trailing vortex. Note the y excursion in Fig. 7.

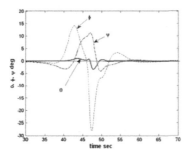

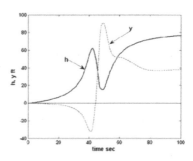

Figure 6 Pitch (θ), roll (φ) and heading (ψ) disturbances with Jetstar traversing vortex pair; Nominal path is 5 deg heading relative to longitudinal axes of vortex pair, $h_{initial}$ = h_0 = 0 ft.

Figure 7 Vertical (h) and lateral (y) deviations relative to nominal path with Jetstar traversing vortex pair; Nominal path is 5 deg heading relative to longitudinal axes of vortex pair, $h_{initial}$ = h_0 = 0

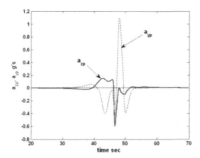

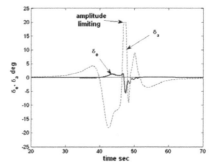

Figure 8 Vertical (h) and lateral (y) deviations relative to nominal path with Jetstar traversing vortex pair; Nominal path is 5 deg heading relative to longitudinal axes of vortex pair, $h_{initial}$ = h_0 = 0 ft.

Figure 9 Elevator and aileron deflections with Jetstar traversing vortex pair; Nominal path is 5 deg heading relative to longitudinal axes of vortex pair, $h_{initial}$ = h_0 = 0 ft.

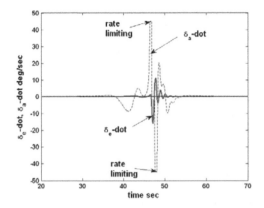

Figure 10 Elevator and aileron deflection rates with Jetstar traversing vortex pair; Nominal path is 5 deg heading relative to longitudinal axes of vortex pair, $h_{initial} = h_0 = 0$ ft.

3.2 Slanting Encounter with Coordinated Pedal and Wheel Inputs

The slanting encounter of Section 3.1 was revisited. Now coordinated pedal and wheel inputs by the pilot model were allowed at the occurrence of the large negative roll angle in Fig. 6 at t $\approx$ 47.5 sec. Figure 11 shows the pilot inputs δ_{Wh} (wheel inputs controlling aileron), δ_{Ped} (pedal inputs controlling rudder) and roll-attitude ϕ. As can be seen, a steady oscillation occurs and the roll attitude and wheel/rudder inputs are 180 deg out of phase. The frequency of oscillation is approximately 0.25 Hz (1.57 rad/sec).

In 2008 an Air Canada A319 encountered the trailing vortex of a United Airlines Boeing 747-400 10.7 nm ahead (Anon., 2008b). Analysis of Digital Flight Data Recorder (DFDR) data indicated that, during the event, vertical accelerations reached peak values of +1.57g and -0.77g. Lateral accelerations reached peak values of +0.49g (right) and 0.46g (left) during four oscillations. The rear vertical stabilizer attachment fitting was subjected to loads of 129 per cent of limit load, and the rear fuselage fitting, 121 per cent of limit load. It must be emphasized that these loads were not a result of the vortex, per se, but rather the pilot's response to the encounter. The oscillations that ensued lasted 18 sec, and were primarily attributable to the large, oscillatory control inputs of the pilot. The 180 deg phase differences shown in Fig. 11 were also in evidence in the Air Canada incident.

590

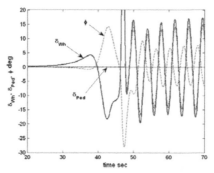

Figure 11 Pilot wheel and pedal inputs and roll attitude with Jetstar traversing vortex pair;

For the sake of completeness, a normal encounter was simulated with the nominal path making an angle of 90 deg with the longitudinal axes of the vortex pair. The symmetry of the encounter meant that only the Jetstar's longitudinal modes were excited. Figures 12-15 show the pertinent results. The normal acceleration at the cockpit is seen to reach 1.5 g's making this a hazardous encounter given the metrics of Section 2.3. However, the responses are considerably less severe than those of the slanting encounter with $h_{intial} = 0$. The maximum angle of attack excursion in this encounter was +3.5 deg.

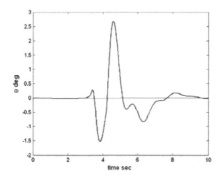

Figure 12 Pitch (θ) disturbance with Jetstar traversing vortex pair; Nominal path is 90 deg heading relative to longitudinal axes of vortex pair, $h_{initial} = 0$ ft, $h_0 = 0$ ft.

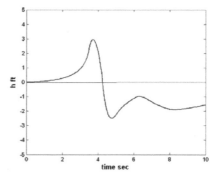

Figure 13 Vertical (h) deviations relative to nominal path with Jetstar traversing vortex pair; Nominal path is 90 deg heading relative to longitudinal axes of vortex pair, $h_{initial} = h_0 = 0$ ft.

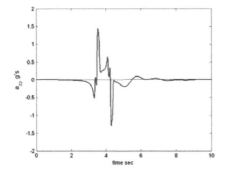

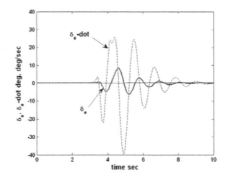

Figure 14 Vertical acceleration (a_{zp}) at pilot station with Jetstar traversing vortex pair; Nominal path is 90 deg heading relative to longitudinal axes of vortex pair, $h_{initial} = h_0 = 0$ ft.

Figure 15 Elevator and elevator rate inputs with Jetstar traversing vortex pair; Nominal path is 90 deg heading relative to longitudinal axes of vortex pair, $h_{initial} = 0$ ft, $h_0 = 0$ ft.

4   CONCLUSIONS

Based upon the research summarized herein, the following brief conclusions can be drawn:

(1) An established pilot modeling procedure can be applied to the problem of wake vortex interaction.

(2) The resulting computer simulation can serve as economical and time-saving tool for preliminary assessment of wake vortex hazards.

(3) The modeling procedure addresses findings and recommendations cited by international bodies concerned with the impact of wake vortices on air-traffic capacity.

REFERENCES

A Report of the Committee to Conduct and Independent Assessment of the Nation's Wake Turbulence Research and Development Program, National Research Council, National Academies Press, Washington, DC.

Anon., 2008b, Aviation Investigation Report, Encounter with Wake Turbulence, Air Canada, Airbus A319-114 C-GBHZ, Washington State, United States , Transportation Safety Board of Canada, 10 January 2008, Report Number AO8W0007.

Carten, A. S., Jr., 1971. Aircraft Wake Turbulence – An Interesting Phenomenon Turned Killer, Air University Review.

Hahn, K.-U, Schwarz, C, Friehmelt, H., 2004, A Simplified Hazard Area Prediction (SHAPE) Model for Wake Vortex Encounter Avoidance, 24th International Congress of the Aeronautical Sciences, ICAS 2004.

Heffley, R. K., and Jewell, W. F., 1972, Aircraft Handling Qualities Data, NASA CR-2144.

Hess, R. A., 2005, Response to Atmospheric Disturbances, *The Engineering Handbook*, 2nd Ed., CRC Press, Boca Raton, Fl., 2005, Chap. 194.

Hess, R. A., and Marchesi, F., 2009, Analytical Assessment of Flight Simulator Fidelity Using Pilot Models, *Journal of Guidance, Control and Dynamics*, 32(3) pp. 760-770.

Hess, R. A., 2010a, "Multi-Axis Pilot Modeling," Report of the WakeNet3-Europe Specific Workshop on Models and Methods for Wake Vortex Encounter Simulations, Technical University of Berlin, Berlin, Germany.

Hess, R. A., 2010b, "Fixed-Wing Control and Handling Qualities," *Encyclopedia of Aerospace Engineering*, Eds: R. Blockly and W. Shyy, John Wiley & Sons Ltd, Chichester UK, pp 2717-2726.

Iversen, J. D. and Bernstein, S., 1972. Dynamic Simulation of an Aircraft Under the Effect of Vortex Wake Turbulence, Annals de L'Association Intenationale Pour Le Calcul Analogique – No. 3, pp. 136-144.

Luckner, R. and Amelsberg, S., 2010, Report of the WakeNet3-Europe Specific Workshop on Models and Methods for Wake Vortex Encounter Simulations, 1st and 2nd June, 2010, Technical University of Berlin, Berlin, Germany, http://www.wakenet3-europe.eu/index.php?id=168.

Tatnall, C.R., 1995, A Proposed Methodology for Determining Wake-Vortex Imposed Aircraft Separation Constraints, Master of Science Thesis, The School of Engineering and Applied Science of The George Washington University, Washington, DC.

Turner, G. P., Padfield, G D., and Harris, M., 2002, Encounters with Aircraft Vortex Wakes: The Impact on Helicopter Handling Qualities, *Journal of Aircraft*, 39(5), pp. 839-849.

Vaughan, J. M., Brown, D. W., Constant, G., Eacock, J. R., and Foord, R., 1996, Structure, Charcterisation and Modification of Wakes from Lifting Vehicles in Fluids, AGARD CP-584.

CHAPTER 56

An Experimental Investigation on Air Quality inside WindJet Aircraft

Catalisano D.*, Giaconia C.**, Grillo C.**, Montano F.**

*WindJet S.p.A.
**University of Palermo, Italy
caterina.grillo@unipa.it

ABSTRACT

In order to improve the passengers and crew comfort during the flight, the aim of the present paper is a study of the cabin air quality through experimental measures on randomly selected flight segments of an Italian airline company, WindJet.

Carbon dioxide, ultrafine particles, temperature and relative humidity have been measured by using low cost high efficiency instrumental equipments.

Exploring ways to improve aircraft cabin air quality, WindJet and University of Palermo are investigating equipment, filters and components of the ventilation system.

In this paper, after description of both ventilation systems for aircrafts of WindJet and the instrumental equipments used to measure environmental characteristics on board, obtained results are shown.

Keywords: passengers and crew comfort, airline, experimental measures

1   INTRODUCTION

As aircraft operators have sought to substantially reduce propulsion fuel cost by flying at higher altitudes, the energy cost of providing adequate outside air for ventilation has increased.

Fundamental problems are: low ventilation rate (less than 5 l/s per person); reduced partial pressure of oxygen and its effect on susceptible people; very low relative humidity (10–20 per cent); ozone (sometimes over 100 ppb); cosmic radiation, which increases with altitude and latitude; fumes from breakdown products of leaks of lubricants and hydraulic fluids.

In the USA, credible scientific investigations of cabin air quality have been conducted by the National Academy of Sciences, United States Department of Transportation (DOT),National Institute for Occupational Safety and Health(NIOSH), independent research groups and airplane manufacturers.

In the context of the EU 5th framework project the ASD-STAN European preStandard prEN4618 for aeronautical air quality and comfort has been published. The standard is intended for use in design, manufacturing, maintenance and normal operation of commercial aircraft. It specifies requirements and determination methods for indoor air quality and thermal comfort, therefore it distinguishes between safety, health and comfort conditions for passengers and crew under a variety of phases of flight.

In the UK, many research activities are carrying out about the air cabin quality, so, British Airways contracted with the British Research Establishment to conduct environmental monitoring and a health and comfort survey of cabin crews flying.

Because of in Italy, nowadays, there aren't similar activities, an Italian airline company, WindJet, and the Energy Department of the University of Palermo are carrying out an experimental research activity by measuring carbon monoxide, carbon dioxide, temperature and humidity on randomly selected flight, to evaluate and improve the passengers and crew comfort during the flight.

The present paper shows the results obtained during the first campaign of tests.

2   PASSENGER COMFORT

The new European cabin air standard significantly contribute to enhance the already high comfort level onboard of commercial passenger aircraft. Although the most recent studies have shown that changing one cabin parameter may have an influence on other parameters, at the present the European Standards "Aircraft integrated air quality and pressure standards, criteria and determination methods" set values for the following cabin parameters: pressure conditions, air quality, thermal conditions, humidity conditions, noise and vibration.

In particular pressure conditions are related to either rates of change of cabin Air Pressure or Absolute cabin Air Pressure altitude. This one must be of not more than 2438 m at the maximum operating altitude of the aircraft under normal operating condition.

The rate of change of cabin air altitude should be as low as possible and as constant as possible during climb or descent. Therefore it is limited to 2.5 m/s (sea level equivalent) for decreasing air pressure and to 1.5 m/s (sea level equivalent) for increasing air pressure. Air quality specifications lead to limits for carbon monoxide, carbon dioxide, ozone, ultra fine particles, etc.

Thermal conditions set that, for Comfort the cabin air Temperature must be $21°C < t_a < 25°C$ while $20°C < t_{a,01} < 25°C$ and $\Delta t_a = |t_{a,11} - t_{a,01}| < 3$ K, with t_a

ambient temperature ta,01 ambient temperature at ankle height (0.1 m) ta,11 ambient temperature at head height (1.1 m). Besides the Minimum contact temperature must be 15°C and the Maximum surface temperature must be 40°C. Moreover local air velocity is limited v < 0.2 m/s at draft sensitive bare body parts: ankles and neck, v(a) < 0.36 m/s otherwise.

No specific comfort limits are imposed on humidity conditions.

Obviously, Standard set limits for the above mentioned cabin parameters regarding either safety, or health and comfort.

In the present paper of course, the measured parameters are related to Comfort limits.

3   ECS ON WINDJET AIRCRAFT

3.1   AIR CONDITIONING

The air conditioning system maintains the air in the pressurized fuselage zones at the correct levels of temperature, freshness and pressure. Air supply comes from the pneumatic system.

The outside air is taken by ventilation from the turbo compressor. This air is filtered, treated and diffused into the cabin. In general, there are at least two twins plants powered by two different engines.

Then, air is regulated in temperature by the temperature regulation sub-system. Proper ventilation of the avionics equipment is ensured by the avionics ventilation sub-system. The pressurization sub-system ensures a cabin altitude compatible with crew and passenger comfort. A special outlet valve regulates the internal overpressure in order to maintain, in the worst case, approximately 2500 m.

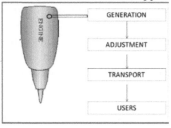

Figure 1 Air conditioning function scheme

The air coming from the pneumatic system is controlled in flow before reaching two air conditioning packs which ensure basic temperature regulation.

Air delivered by the packs is mixed with recalculated air from the cabin zones. Fine temperature adjustment of air distributed in the pressurized zones is obtained by controlling the amount of hot air added to the air coming from the mixer unit. Correct pressurization is obtained by controlling the conditioned air discharge through one outflow valve.

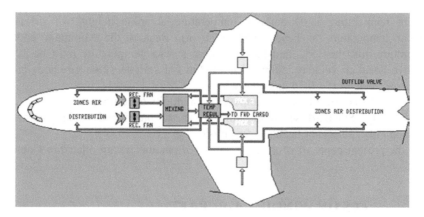

Figure 2 Air conditioning system scheme

3.2 ZONE TEMPERATURE CONTROL SYSTEM

Hot air flow coming from the air bleed system is regulated before entering the packs in order to be temperature regulated. Hot air pressure is maintained above the cabin pressure allowing the hot air flow to join the pack air supply when necessary.

A part of cabin air is recirculated to decrease air supply demand.

The air flow from the air bleed system is regulated by two pack flow control valves. Then two independent packs provide regulated temperature air to the mixer unit. Both packs provide air at the same temperature.

A mixer unit mixes regulated temperature air from the packs with part of the cabin air supplied by recirculation fans.

Hot air tapped upstream of the packs supplies the trim air valves through a hot air pressure regulating valve. This valve regulates the downstream pressure above the cabin pressure.

The air trim valve is associated with each zone optimizes the temperature by adding hot air to the cold air coming from the mixer unit.

Overall the conditioned air is distributed to three main zone: cockpit, forward cabin and aft cabin. Normally the mixer unit allows the cockpit to be supplied from pack1 and FWD and AFT cabins from pack 2.

The temperature regulation is automatic and controlled and optimized by the corresponding pack controller which in turn is controlled by the zone controller. Each zone and pack controller consists of one primary channel and one electrically independent secondary channel. The secondary computer is used in case of failure.

The control of basal temperature and the flow regulations was performed by the two packs in accordance with demand signals from the zone controller.

The zone controller optimizes the temperature regulation and flow by means of trim air valves to obtain the selected ambient temperature in the related zone.

The pack flow selector permits selection of pack valve flow according to number of passengers and ambient condition. In particular we have two selections: LO, if passengers are below 81, and HI, for abnormal hot and humid condition.

When the LO is selected and heating or cooling cannot be carried out, the zone controller automatically provides a normal flow and if necessary increases the engine power.

Whatever the selection, the zone controller automatically provides high flow in case of bleed air taken from the APU or single pack operation.

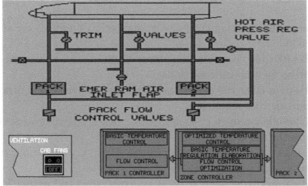

Figure 3 Zone temperature control system

It is possible to manage either pressure or temperature for specific cabin sectors.

3.3   PRESSURIZATION SYSTEM

The pressurization system ensures a cabin altitude safe and compatible with crew and passenger comfort. Pressurization is performed by controlling the amount of air discharged overboard through one outflow valve.

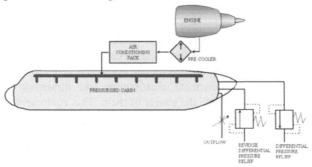

Figure 4 Pressurization system scheme

The system has two identical and independent automatic controllers. Only one controller operates at a time. In normal condition the system is fully automatic.

The automatic cabin pressure control operation is dependent on programmed control laws and information form FMGCs and ADIRUs. In case of failure of both, manual mode is permitted.

Two safety valves are installed on the aft pressure bulkhead to prevent excessive positive and negative differential pressure. The two safety valves are installed above the aircraft floatation line.

4   EQUIPMENTS

Equipment we used during the experimental campaign inside the aircraft cabin was composed by:
- **BABUC/M**: portable equipment to survey, visualize, memorize and elaborate ambient conditions;
- **Lastem BSO 103 probe**: equipment to analyze air and measure carbon dioxide concentration;
- **Globethermometer**: equipment to measure heating exchanges due to irradiation;
- **Fluke 983**: equipment to survey the concentration of particles suspended in air;
- **iButton®**: equipment to measure temperature and relative humidity.

BABUC/M permits to evaluate, through different kinds of probes, many ambient characteristics i. e.: temperature, relative humidity, airspeed, gas concentration and rotation speed. It effords either to visualize in real time the measures, or to memorize elements that could be tansferred to PC for further elaborations.

The Lastem BSO 103 probe is connected to the previous equipment. It takes an imposed air quantity, convoy it in an internal room in which it measures the concentration through an infrared sensor.

To measure the concentration is important because it permits to evaluate the air quality and through this measures it is possible to understand if the air changing is appropriate. This is possible making a relationship between the air characteristic and the internal production of pollutants.

The Globethermometer is composed by a thin copper sphere inside which there is a temperature sensor. The sphere color is black and it has an emittance equal to 0.95 W/m^2. This particular shape is used because it is the shape that better approximate the surface/volume ratio of human body. It is connected to BABUC/M.

Fluke 983 permits to determine distribution and dimensions of particles transferred in air or to individuate the source of them. This equipment measures and visualize at the same time, through six channels, the particles dimensions, temperature and humidity. It can measure particles dimensions until 0.3 μm.

A problem we had during experiment was the electric supply. In fact, while Fluke has high endurance battery (till 8 hours), BABUC/M has not battery packages but should be connected to the airplane electricity network but, for safety reasons, it's impossible to connect anything not certified to the airplane electricity network. This problem was solved connecting an external battery to the equipment. This battery guaranteed a limited autonomy (till 5 hours) to the equipment. It was however sufficient to make measure during all the time of flight.

iButton® were used for temperature and humidity during all flight phases. We used 13 equipments and located them in many parts of the cabin to divide it in five sections. In each section we put one of them on the top of the section and others on the opposite sides in a lower position. Besides we put them on voluntary passengers. To individuate the exact positions of the buttons we used the Olensen formula:

$$T_{skin}=0.5 \, T_{torax} + 0.14 \, T_{elbow} + 0.36 \, T_{calf}$$

So for each volountary passenger we use three buttons: one put on the thorax, one put on the elbow and one put on the calf.

5   RESULTS

By dividing aircraft cabin into 3 sections it has been possible to obtain the dew point Temperature T_{mr}, and the operative Temperature T_{op}

$$T_{mr} = \sqrt[4]{(T_g + 273)^4 + 0.4 \cdot 10^8 \cdot (T_g - T_{bs}) \cdot \sqrt[4]{|T_g - T_{bs}|}} - 273$$

$$v_a < 0.1 \text{ m/s}$$

$$T_{op} = \frac{(T_{bs} + T_{mr})}{2}$$

$$v_a < 0.2 \text{ m/s}$$

The obtained results are shown in Figures from 5 to 11. These state that during cruise there is an high comfort level. In fact only 17% of the measurements is out of the comfort zone. Most of the operative temperature values are between 25 and 26 °C. Such a values are considered optimal values for comfort in ASHRAE Standards.Relative humidity ranges from 10% to 20% .

To obtain a reference cabin values, average values of the measured relative humidity and temperature have been evaluated and a comparison has been made between average values and measured values.

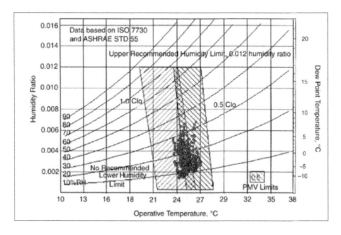

Figure 5 Measured Temperature and Relative Humidity

600

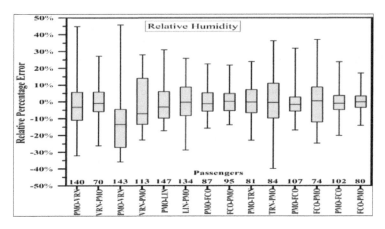

Figure 6 Relative humidity results

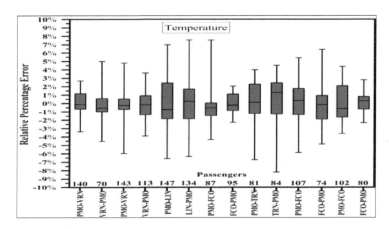

Figure 7 Temperature results

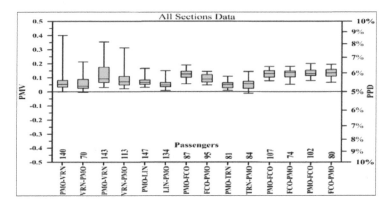

Figure 8 PMV e PPD results

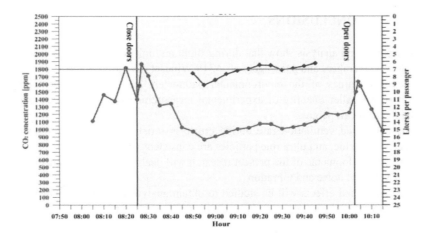

Figure 9 CO2 concentration and ventilation rate in flight Palermo-Verona (Sept 8th 2012)

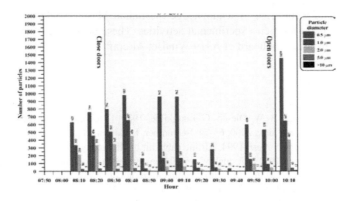

Figure 10 Particles from 0.5 μm to >10 μm in flight Palermo-Verona (Sept 8th 2012)

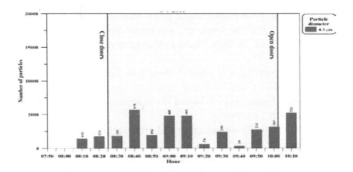

Figure 11 Particles ≤ 0.3 μm in flight Palermo-Verona (Sept 8th 2012)

6   CONCLUSIONS

The carried on analysis show that during flight an high comfort level is ensured to both crew members and passengers of A319 WindJet Aircraft.

Measured values of the environmental parameters are regular in the whole aircraft so a smaller quantity of experimental measurements will be implemented during the next campaigns.

The calculated ventilation rate during cruise is satisfactory Concentrations of both carbon dioxide, and ultra fine particles are consistent with Standards.

Further developments of the present research will deal with the determination of local air velocity, noise and vibration.

Also combined effects will be studied to obtain analytical correlations between cabin environmental parameters and passengers or crew comfort.

ACKNOWLEDGMENTS

The authors would like to acknowledge the Italian Master Students M. Ariolo and D. Giurintano for their experimental activities. These ones have carried out the campaign of measures onboard of A319 WindJet Aircraft.

REFERENCES

Fanger P.O., L. Bànhidi, B. W. Olesen, G. Langkilde, 1980. Discomfort caused by overhead radiation. Proc. of Clima 2000, 65-74. September, Budapest, Hungary.

Air Transport Association. April 1994. Airline cabin air quality study.

ASHRAE Journal. April 1991. Air quality, ventilation, temperature and humidity in aircraft.

National Academy Press. 1986. The airliner cabin environment: Air quality and safety.

Report No. DOT-P-15-89-5. December 1989. Airliner cabin environment: Contaminant measurements, health risks, and mitigation options

Wolff C., 2009. ICE and European preStandard on health and cabin environment. International Cabin Environment (ICE)-International Conference

ASD-STAN ,(2005) European preStandard prEN4666

Hunt E. H., Space D. R.,1994 The Airplane Cabin Environment: Issues Pertaining to Flight Attendant. International In-flight Service Management Organization Conference, Montreal, Canada.

RITE-ACER, (2010) Report No. CoE-2010-1 Report to the FAA on the Airliner Cabin Environment

Grün G., (2009) Cabin climate and its impact on passengers. ICE International Aviation Conference , Munich ,Germany

Evans T., (2009) ICAO activities on passenger and crew health. ICE International Aviation Conference , Munich ,Germany

Fox R., (2009)   A US Perspective on Cabin Air Quality Standard Development ICE International Aviation Conference , Munich ,Germany

CHAPTER 57

Initial Evaluation of Video Game Technologies for Future Unmanned Air Vehicle Interfaces

Marc Gacy[1], Gloria Calhoun[2], Mark Draper[2]

[1]Alion Science & Technology, Boulder, CO
[2]Air Force Research Laboratory, Dayton OH
agacy@alionscience.com; [gloria.calhoun][mark.draper]@wpafb.af.mil

ABSTRACT

This paper describes an initial evaluation of several gaming technologies and details their applicability to future control concepts for multiple unmanned air vehicles (UAVs) systems. Current gaming technologies evaluated included Real-Time Strategy games, Massively Multiplayer Online Role Playing Games, and arcade-style games. Accordingly, these technologies are then compared to current UAV control concepts, noting several key similarities and differences, to provide direction for future application.

Keywords: unmanned aerial vehicles, video games, situation awareness

1.0 INTRODUCTION

Unmanned aerial vehicles (UAVs) are becoming an increasingly critical aspect of military operations. With these successes comes an Air Force directed vision for more capable systems, including the ability of a *single operator* to control *multiple platforms,* and the desire for UAV accomplishment of more complex, dynamic missions in close collaboration with other assets. Achieving this vision is predicated on a thorough understanding and proper design of the human-automation interface. Thus, there is a need to determine the most effective, over-arching architecture for implementing the automation-operator interface. The architecture also establishes the rules for the transfer of control between the automation and the operator, as well as communication protocols, interaction styles, and cognitive strategies for reasoning with adjustable autonomy in operational contexts. An example of such an approach is the operator serving as a supervisor and delegating the division of labor after determining what, when, and how the automated systems will function.

In many ways, this automation architecture is similar to the approach taken by many current Real-Time Strategy (RTS) video games, where the player acts as a commander, delegating actions to the units, which are essentially autonomous agents in the computerized "world." First-Person Shooter (FPS) games have also become more sophisticated, integrating interactions with virtual team members. Understanding what capabilities the video game commander or team leader has may indicate capabilities that are missing or need to be augmented in real UAV systems.

2.0 OBJECTIVE AND PROCEDURES

The objective was to evaluate gaming approaches to determine their applicability to future multi-UAV control. In particular, we addressed how these systems complement and augment advances in traditional autonomy. The survey was codified into a standalone database program that is searchable and editable. The majority of the games fell into one of several broad categories, and only a few of these provided more than a cursory insight into multi-system control. As expected, RTS games provided many examples of both interface concepts and actual control concepts. FPS games, especially with multiplayer interaction options, also provided valuable insight into what is and what is not considered successful. Arcade-style games often provided "pure" examples of interface and control concepts, uncluttered by game complexity and graphics. Finally, turn-based games provided the greatest level of time luxury, and thus most accurately represent the UAV planning stages. From this initial evaluation of how gaming systems do or do not show promise for application to multi-UAV autonomy, several research requirements were also identified.

3.0 CONTROL CONCEPTS EXEMPLIFIED IN GAMES

In the following sections, first a control concept will be described, along with a game that demonstrates that concept. Finally, a potential research area and how it relates to multi-UAV control will be discussed.

3.1 Operator Overload– *Mission Command* and *Tetris*

In single player, level-based video games (not story-driven) there is often a point at which the game becomes too difficult and the players will subsequently lose the game. For many games this is a gradual decline, where the loss of lives often occurs gradually over the course of the game, with the final loss indicating the end. There are several games, however, where the losses can come more quickly than suggested by the incremental increase in difficulty. These games represent a catastrophic failure on the part of the game player, with a sudden shift in behavior arising from a small change in circumstance. *Missile Command* and *Tetris* are good examples of this behavior and represent ways in which such behavior can arise.

In *Missile Command*, the operator is using three independent missile silos to fire at and destroy incoming missiles (see Figure 1). The operator selects which silo is best to fire from based on the location of the incoming missiles. The object of the game is to keep the missiles from destroying both the missile silos and the cities on the ground during an attack wave. When players achieve a certain level of proficiency, they can often play for many levels without losing a "life," only to lose all or nearly all of the cities on the next level. Usually this occurs because the player lets a silo get destroyed, or they lose their "cadence" with placement of missile hit sites, which results in a number of enemy missiles getting through. *Tetris* has a similar effect; if the operator is unable to correctly place a game piece or makes a mistake in placement, subsequent placement of game pieces is difficult because of the reduced space available and an inability to remove the blockage (Figure 2).

The popularity of these games may be attributed to the sense of accomplishment for keeping the cadence and staying ahead of the failure. Both games highlight this kind of effect in error propagation, where making a mistake makes subsequent mistakes more likely. This kind of behavior can lead to catastrophic failure whereas simple probabilistic errors rarely achieve such a collapse in performance.

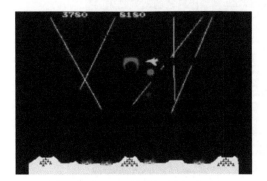

Figure 1. Missile Command screen capture.
(From URL: http://www.hollywood.com/news/Missile _Command_Coming_To_The_Big_Screen/774789)

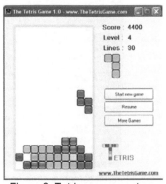

Figure 2. Tetris screen capture.
(From 4Tetris, URL: http://www.thetetisgame. com. Figures 1 & 2 Accessed May 4, 2011.)

A potential research area is whether or not the UAV system is prone to increasing error probabilities. If it is, can mitigation strategies, much like *Tetris'* presentation of the next piece, alleviate some or all of that behavior? To continue the *Tetris* example, if the system detects that an error occurred, for the next few drops it may suggest placement options or show the next two pieces to come, until a certain number of correct placements occur. Thus, this research area may be relevant to issues of control with a very large number of fairly autonomous UAVs where the operator is issuing simple commands, but very frequently.

3.2 Awareness of Mode and Plan Progress of Units – *Starcraft*

The game genre that best represents the high-level control of multiple unmanned systems is the RTS genre, of which *Starcraft* is an early and successful example. The ability to manage a large number of units is facilitated by its interface with a summary of unit types and states with more detailed information upon selection. The units are also given rudimentary, single point commands, such as patrol (move back and forth between two points), move (from current position to specified position), attack, build, or repair. The system provides auditory (verbal) feedback that the unit has been properly selected and that it has received the input command. However, it does not do a good job of indicating that the units are following the command or that they have finished the task. Such behavior might be present with UAVs moving through cluttered or restricted space with the potential for unintentional rerouting around the spaces. The units also do not give any indication that they are finished with a task. Usually that determination is left up to the player by observing movement (or lack of it) on the small mini-map (lower left Figure 3).

Figure 3. Starcraft screen capture and manual portion, showing minimap (lower left), selected units or individual unit information in the center and command icons on the right. (From URL: http://www.gossipgamers.com/starcraft-course-opened-for-college-earn-credits/. Accessed May 4, 2011.)

A potential research area is how to best indicate that units have finished their task, or more importantly, somehow deviated from their task. Auditory alerts are used to great effect in real-life combat scenarios to indicate when units are under attack; thus, plan deviations (another class of "unexpected behavior") might also be well suited to auditory alerting. In order to reduce auditory clutter, a visual indicator might be better suited for the initial selection and tasking of the unit.

It is also important to point out which unit is selected and that the selected unit is given the correct task. Many times, as in *Starcraft*, only the unit type is referred to in the feedback; any identifier, special designation, or grouping is not used, furthering the potential for confusion. This is particularly likely to be a problem for an operator controlling a number of UAV systems with mostly identical capabilities. Subtle indicators, such as vehicle type or available/unavailable commands, do not exist, leaving the operator to rely on name and memory of the UAV's position and mission alone for distinction.

It should be noted that a likely explanation for *not* including this behavior in games is that it has the potential to remove the operator as an active participant, such that they are simply responding to the various cues they receive as opposed to actually controlling the situation. This lack of engagement can often be translated to a loss of situation awareness (SA), which will be discussed in more detail later.

3.3 Interface Concepts – *Mech Commander*

The game *Mech Commander* has a similar interface approach, but with more user control over the presentation. In *Mech Commander*, the player has the option of selecting Map, Data, Briefing, and Salvage, where Data is information about the selected Mech, Briefing is a summary of the mission, and Salvage is a record of picked up goods (inventory). The player may select any of these to be in the information window (upper left in Figure 4). The player may also choose to minimize the information window, so more of the screen area is devoted to the main "battle" view. The definite advantage of the *Mech Commander* interface is that there is a large area for individual unit information, which can take up the entire information area, whereas in the static display, it is confined to a smaller area.

Figure 4. Mech Commander screen with information area in upper left corner. (From: URL: http://games.softpedia.com/progScreenshots/Mech-Commander-Gold-No-Cd-Patch-Screenshot-53836.html. Accessed May 4, 2011.)

This attention to unit status is also evident in the rectangular "traveling status bar" that accompanies each unit. Similarly, the selection aspect of the game makes use of text overlays to provide information about the selected object, in addition to the color change among neutral, friendly, or enemy targets. These enhancements, present in many subsequent games, are examples of augmented reality. While in a video game it is easy to keep the overlay information correctly registered with the player's view, in a real situation, especially with assets and entities adding clutter, small errors in the location/registration of the overlay information may confuse the operator instead of providing immediate status information.

"Attaching" information directly to the units, through a form of augmented reality, is certainly an attractive concept based on the evolution of RTS games. As described above, however, there is the potential for confusion with multiple real assets, particularly on a live map. Research is needed to address the minimum level of "persistent" or always available information to the operator, regardless if the UAV is in the operator's current map view. Some questions that could be addressed are whether the alerting and representation behavior for those UAVs "off the map" should somehow be different for those UAVs that are "on the map," have status information immediately visible, and be readily selectable. Such research could address attempts to minimize the "out of sight, out of mind" effect for non-visible UAVs, and how to reduce the time to address a situation with "off the map" UAVs.

3.4 Teaming – *World of Warcraft*

World of Warcraft is the most popular Massively Multiplayer On-line Role Playing Game (MMORPG) and as such has incorporated many display control concepts from other games over the years. One area that is a key consideration for MMORPGs is teaming (with other human players). *World of Warcraft* has specific rules for teaming at two different levels: parties (small groups) and guilds (many members). A party can be viewed as essentially an ad-hoc, task-based team structure where one leader invites others to join a party. Parties are usually created for a single adventure and the game structure is very specific about who the leader is, how communication is handled, and how "loot" is distributed. A guild is a larger, more permanent organization, with a stratified hierarchy and more stringent rules. A raid is a special subset of a party, which is a large collection of players working toward a common goal. The complexities of keeping such a large number of individual players focused and efficient is interesting to study as the actual interaction rules for the structures are fairly loose, being primarily in the form of communication filters (party only or guild only) and some additional status indicators (see the right side of the screen in Figure 5).

Furthermore, *World of Warcraft* has one of the richest display environments, with many interface concepts being present. First, it has the bottom status bar, with a centralized mini-map. Several team features are provided as overlays, with raid members' status in green and a target list in blue (on the right in Figure 5). Color coding is also heavily used as both a primary (squares) and secondary (names list) indicator of information. Character names/groups are included as transparent words that move with the character, providing an "augmented reality" overlay to the view.

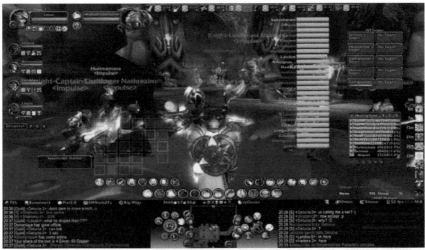

Figure 5. World of Warcraft screen capture with an information bar and myriad overlays. (From URL: http://log.archpaper.com/wordpress/archives/date/2009/04/page/3. Accessed May 4, 2011.)

World of Warcraft (and other MMORPGs) raise several potential research topics. First, the wide variety of interface concepts and the ability to change concepts as desired provides the level of control over the interface needed for any comparative study. A good example of this type of add-on is called "Decursive." that provides small translucent squares on the left side of the screen (see Figure 6).

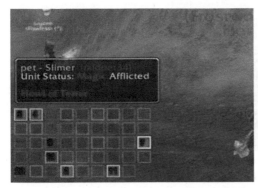

Figure 6. "Decursive" overlay screen capture showing colors and transparent overlay. (From URL: http://www.2072productions.com/to/decursive.php. and URL: http://wow.joystiq.com/2008/04/18/addon-spotlight-decursive-2-0. Accessed May 4, 2011.)

These squares are color coded to provide relevant status information and are organized in descending order of importance: self, group, and raid group. These squares represent a form of augmented reality, in that they automatically assess the relevant status information of all the team members, and they also represent a limited form of autonomy, by automatically preparing a response to adverse status conditions. This allows the player to take on the role (at least for this function) of a supervisor authorizing actions. This approach could be pushed slightly in either direction to a more or less autonomous mode, by only indicating the status without the suggestion (less autonomy), and by automatically performing the requested action by default after a predetermined time period (more autonomy). Furthermore, instead of just defaulting to hierarchical importance, the physical proximity to the character and/or the severity of the condition could also play a role. For control of multiple UAVs, these differences in autonomy level could change from well-coordinated missions to either unacceptable slow missions or losses of SA. The effect compounded over many choices with many units can highlight the less optimal solutions. Using *World of Warcraft* as a research tool may be different as there is a current rule and built-in limitation against purely automated behaviors, which encourages including the human element in the game.

3.5 Situation Awareness – *Supreme Commander*

The geographic scale of multi-UAV control is well beyond that of most games. Due to the larger area covered by the UAVs, it may be expected that maintaining SA through graphic cues would be more difficult for UAV control than in video games. In many games, the "action" takes place in a well-defined area that often is indicated by a small overview map (described above). While in most cases this is true, there are a few games that take advantage of a much larger area. *Supreme Commander* has such a larger area of operations than most games, claiming to be theater-based, as opposed to battle-based. The player interacts with the map in a very fluid manner, being able to zoom in at any level from single-vehicle, third person close-ups, all the way out to the entire theater space (see Figure 7).

Figure 7.
Supreme Commander screen capture zoomed out to theater level. (From URL: http://pc.ign.com/articles/929/929907p 1.html. Accessed May 4, 2011.)

The intuitive and easy to use controls of the continuous zoom feature go a long way to maintaining SA by allowing the player to visually map the changes in zoom level. Furthermore, the player may choose to have anywhere from one to three maps, all independently zoomable, including a split screen and a small overview map. Each is independently controllable and can provide both top-down or angled views (much like Google Earth). What is missing from the split screen maps is a way to easily relate them to each other, especially if there is no area of overlap. Ideally, one could argue that the maps should have a common overview map linking them for proper SA and registration.

3.6 Real-Time Strategy/First-Person Shooter – *Raven Squad*

The combination of a RTS game with a (FPS) game would provide one of the most thorough overviews of system capabilities. However, this combination is rarely attempted, and when it is, the result is often a poorly received game. *Raven Squad* seems to be one of the first video games to really combine both RTS and FPS aspects into the main gameplay (Figures 8 and 9). Several other games, *Iron Grip: Warlord*, *Savage 2*, *Battlezone 2*, and *Natural Selection* also combine these two elements, although usually either one of the two aspects heavily dominates, or the game-play is very sequential. *Raven Squad*, however, allows a player to shift back and forth between the two aspects rather seamlessly, but the FPS aspect seems more engaging and the player often feels more in control.

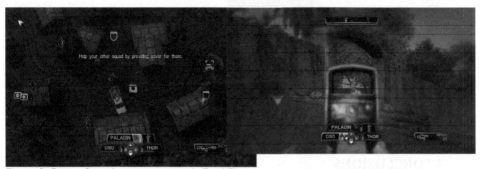

Figure 8. Raven Squad screen capture in Real-Time Strategy Mode. (From URL: http://www.destructoid.com/review-raven-squad-147718.phtml. Accessed May 4, 2011)

Figure 9 Raven Squad screen capture in First-Person Shooter Mode. (From URL: http://www.destructoid.com/review-raven-squad-147718.phtml. Accessed May 4, 2011.)

As suggested, the "confusion" that results from switching between modes is a real problem both in this game and in the real world. In *Raven Squad*, even though it is fairly simple to switch, there are a limited number of links to facilitate the transition. Developing tools to ease the switch, by providing appropriate information for the transition and indicating which units are now being controlled (possibly in a temporal highlighting fashion), should help prevent the loss of SA.

3.7 Turn-Based Planning – *Strategic Conquest*

Turn-based games are not nearly as popular amongst video gamers as they used to be. A large part of this is the instant and continuous immersion and involvement capability with increased speed and graphics processors in computers and consoles. This does however lead to a preponderance of emphasis on immediate interactions and split-second decision-making. While these are necessary skills for the control of military systems, there is also value in planning and having a proper tool to facilitate such planning. *Strategic Conquest* is an old game that stresses detailed planning using a variety of different assets with widely different capabilities (see

Figure 10). It can take many game turns to produce and then use. Similarly, since movement takes a long time, desired units can easily be "out of position" when either planning or responding to an attack.

Strategic Conquest is unique in terms of its planning and general risk assessment as the main hallmarks of game-play, with the "fog of uncertainty," or unexplored regions. This game may be the best representation of supervisory control of multiple UAVs, and the game alerts the operator (e.g., flashing icons and auto-jump to attack a unit), to improve the SA of the player. Players also need to keep track of fuel consumption and the location of units. The techniques a player uses and the success of their own SA building strategies are a ripe area for multi-UAV research.

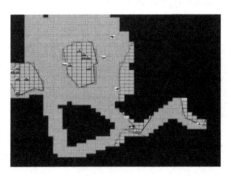

Figure 10. Strategic Conquest screen capture showing active battles in red, and unexplored regions in black. (From URL: http://www.answers.com/topic/strategic-conquest. Accessed May 4, 2011.)

4 CONCLUSIONS

While this assessment of video games identified potential research areas, the similarities among many games, and the shortage of games that address some of the key problems with unmanned systems (mainly switching between the supervisory role and the direct control role), made this effort less fruitful than originally anticipated. A probable reason relates to the differing primary goals of games and real unmanned system control. Games are designed to maximize the user's sense of engagement, rather than providing the most efficient means for accomplishing a given mission. Hence, the interface is optimized for continuity at the expense of "mission performance". This is contrasted with unmanned system control where the goal is successful mission accomplishment, while reducing the operator's workload and preserving SA. Periods of inactivity are *not* penalized in control of real, unmanned systems, and delayed responses are expected (e.g., images collected are not always fed directly back into a subsequent mission). Also, real systems require a fairly fluid transition among UAV control states, whereas games usually concentrate on one "style" of play (e.g., direct control or supervised control). These fundamental differences complicate translating successful concepts from one domain to the other. Admittedly, this may reflect our superficial study of a large number of games. A more targeted and thorough analysis that includes data from experienced gamers may help identify candidate UAV applications of video gaming features and determine additional future research directions.

CHAPTER 58

Human-in-the-Loop Simulations of Surface Trajectory-Based Operations

Emily K. Stelzer, Constance E. Morgan, Kathleen A. McGarry, and Kathryn A. Shepley

The MITRE Corporation
McLean, VA, USA
estelzer@mitre.org

ABSTRACT

The Federal Aviation Administration's Next Generation Air Transportation System (NextGen) concept proposes a suite of decision support tools for use in the air traffic control tower to support safe and efficient operations. This paper describes the proposed set of surface automation capabilities to support ground and local controller activities in the NextGen mid-term, including automated decision support tools to generate taxi routes and monitor pilot conformance to an assigned taxi route. The MITRE Corporation's Center for Advanced Aviation System Development (CAASD) conducted two human-in-the-loop experiments to evaluate ground controller performance with these decision support tools. These simulations specifically examined capabilities provided in surface automation designed by Mosaic ATM. This paper describes the results of those two experiments to refine and validate surface decision support tool concepts.

Keywords: air traffic control, surface automation, taxi routing

1   SURFACE TRAJECTORY-BASED OPERATIONS

The Next Generation Air Transportation System (NextGen) is the Federal Aviation Administration's (FAA's) concept for modernizing the National Airspace

System. High-density airports in the NextGen mid-term timeframe are expected to operate with a set of integrated automation decision support tools (DSTs), which provide access to timely and dynamically updated information; a means for collaboration with stakeholders; maximization of airport capacity; and improved efficiency of airport and surface operations (Air Traffic Organization, 2010). The DSTs described in the NextGen concept are the basis for an operational concept known as Surface Trajectory-Based Operations (STBO) that will manage traffic flows and resources on the airport surface and will enable surface trajectory-based operations in the far-term.

1.1   STBO Concept Overview

The need for STBO is derived from a set of identified shortfalls in the current operations, including departure delays and long runway queues (Nene, Morgan, Diffenderfer, and Colavito, 2010). While current surface operations tend to be tactical and reactive in nature, operations under NextGen will need to be strategic and predictive to reduce these issues.

STBO is expected to increase airport surface efficiencies through shared situation awareness and local collaboration across all stakeholders in surface operations, allowing operators to work proactively to manage high surface demand and dynamic surface and airspace constraints. These capabilities are expected to support better decision-making through local information sharing and automated DSTs (Audenaerd, Burr, Diffenderfer, and Morgan, 2010).

1.2   STBO Capabilities in the NextGen Mid-Term

The STBO concept elements will be implemented as automated DSTs that support surface operations. These automated DSTs are envisioned to support (1) airport configuration management, (2) runway assignment, (3) scheduling and sequencing, and (4) taxi routing. While departure routing can also be considered within the scope of surface DSTs under STBO, this DST will not be described in the context of this paper.

Airport Configuration Management

Airport configuration management capabilities will assist air traffic control in planning and decision-making for airport configuration changes to improve the utilization of runways. These automated capabilities will be based on arrival and departure demand, with consideration of external factors, such as weather. The prescribed airport configuration will include runways for arrivals, departures, or mixed use; taxiway segments, whether open or closed; and standard operating procedures for the airport.

Runway Assignment

Runway assignment capabilities will assist controllers in early planning of flight-specific departure runway assignment based on factors such as the filed departure route, the airport configuration, aircraft type, projected runway loading, and flight operator preferences, requirements, and limitations (Morgan, 2010). Departure runway assignments for a flight and any updates will be shared with flight operators and available to NAS domains. The automation capabilities will display runway assignment recommendations to ground control, who will be able to enter or modify a runway assignment for a flight.

Scheduling and Sequencing

Scheduling and sequencing capabilities will assist controllers in managing the surface schedule and runway sequences for both arrivals and departures and will support collaboration with flight operators. Sequencing support provides guidance on the suggested order of aircraft in a queue, while scheduling support involves time-based recommendations. The surface schedule and sequence will satisfy traffic management constraints and controlled departure times and will optimize the use of surface resources to meet the demand.

Taxi Routing

Taxi routing capabilities address both taxi route generation and surface conformance monitoring. Taxi route generation capabilities will consider current aircraft position, aircraft surface destination (e.g., the assigned departure runway), and the airport configuration (runway use and taxiway or runway closures), user preferences for gate and taxiway, and other relevant factors to propose to the controller an appropriate taxi route for a flight. To enhance safety, a hold-short will be inserted automatically when a taxi route crosses a runway. In addition, taxi route information will be stored and shared electronically with other DSTs or systems.

Surface conformance monitoring capabilities will assist the controllers in monitoring the aircraft's conformance with its assigned taxi instructions. This capability will be reliant on surface surveillance capabilities. The automation will compare the aircraft's current location with its assigned taxi route and will determine if the aircraft is in conformance or out of conformance. Out-of-conformance conditions that relate to aircraft velocity will also be defined. One example includes an aircraft failing to move when instructed to move, such as an aircraft that fails to begin the take-off roll when cleared for take-off. When the aircraft is out of conformance, alerting will be based on severity of the impact of the nonconformance. The concept envisions a visual alert to the controller and an additional aural alert for safety-related surface conformance issues.

2   RESEARCH QUESTIONS

Extensive work has been conducted defining the concept of use for these new technologies (Morgan and Diffenderfer, 2010; Morgan, 2010). Despite these accomplishments, a series of research questions need to be addressed before these technologies can be operationally implemented. These questions address both the measurable benefit associated with the taxi routing and conformance monitoring decision support tool and controllers' acceptance of these capabilities:

- Do controllers assign automation-generated taxi routes to departure aircraft?
- Do surface conformance monitoring capabilities improve controllers' response to nonconformance events?

- Do controllers trust the automation?

3   HUMAN-IN-THE-LOOP SIMULATIONS

This paper details the methods and results from two simulations, which took place during 2010. Since that time, an additional simulation has been conducted and is documented in Stelzer and Stanley (2011). The described simulations were designed to examine the human performance implications associated with the use of initial prototype taxi route generation capabilities, including the surface conformance monitoring tools that are projected to be implemented in the mid-term plan for NextGen.

3.1   Simulation 1

Methods

Participants: Thirteen FAA employees, who reported an average of 22.5 years of experience controlling traffic, participated in the study.   At the time of the simulation, seven individuals were serving in roles at FAA Headquarters, while the remaining six participants came from tower facilities in the United States. None of the participants had worked at Dallas Fort Worth International Airport (DFW), which served as the simulated site for the study.

Simulation Environment: The simulation was conducted at the MITRE CAASD's Aviation Integration Demonstration and Experimentation for Aeronautics (IDEA) laboratory.   CAASD's air traffic control tower simulator was used to generate a high fidelity representation of DFW during a standard south flow operation (see Figure 1).

Figure 1   MITRE air traffic control tower simulator, located in MITRE's Integrated Demonstration and Experimentation Aviation Laboratory.

Configurable workstations were located directly below the out-the-window scene and were outfitted with an electronic flight strip display and a surface surface map display that were developed by Mosaic ATM. These displays are shown in Figure 2.

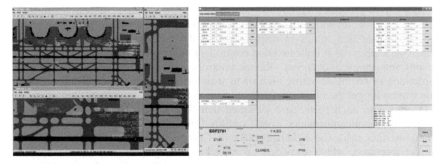

Figure 2   Surface maps and electronic flight strip display, developed by Mosaic ATM.

Procedure: Participants served as the ground controller in the tower, directing departure traffic from the ramp area to the runway edge. They were responsible for verbally assigning a taxi route to pilots, monitoring pilots' conformance to that route, and revising the route when necessary. Participants were given a half day of training on the airport environment and surface automation, one practice trial, and four 45-minute experimental trials. After each trial, participants completed questionnaires to capture subjective ratings of their trust in and acceptance of the automated tools.

Four nonconformance events were randomly distributed within each scenario. These events were comprised of lateral deviations that occurred from erroneous turns or from pilots' failure to turn at a correct intersection. When participants detected a pilot deviation, they were asked to press a button on the electronic flight strip display and verbally respond to the pilot to correct the error. All pilot deviations created an inconvenience to the controllers, though note that none of these deviations produced a direct safety concern.

Experimental Design:The presence of automation was manipulated in a within-subjects design. When no automation was present, participants were provided with the displays shown in Figure 2; however, no taxi routing or conformance monitoring capabilities were provided. When automation was present, decision support tools were used to generate taxi routes based on the assigned departure fix and alert the controller to a pilot deviation. When a nonconformance event was identified, the electronic flight strip for the aircraft was highlightged in red and an aural alert was presented. While the aural alert was terminated after five seconds, the visual alert persisted until the aircraft had rejoined the route or until a new route had been manually entered on the strip. All conditions were counterbalanced across participants.

Simulation 1 Results

The following sections outline participants' use and subjective assessments of the taxi route generation and surface conformance monitoring capabilities. A more detailed review of these findings is also available (Klein, Stelzer, Nelson, Brinton, and Lent, 2010; Stelzer, 2010).

Taxi Route Generation: An analysis was conducted to examine the frequency with which participants manually modified routes that had been generated by the automation. This analysis indicated that, in 100% of cases, controllers used the automation-generated taxi routes and never modified these routes. To further explore this issue, an additional analysis was performed to examine how frequently participants modified routes in response to a pilot's deviation from the assigned route. This analysis also indicated that controllers never modified taxi routes when a nonconformance event occurred. Rather, controllers always provided corrective taxi instructions to return deviating aircraft to their original route, even if the original route was no longer the most efficient.

Debrief with each participant indicated that they were often reluctant to make changes when nonconformance events occurred because characteristics of the route editing interface made these changes difficult. Comments from participants indicated that the interface, which provided an alphabetical keyboard for input, required significant modifications before being suitable for implementation in an operational environment. To understand this behavior at a deeper level, questionnaire items were added for Simulation 2 to collect data on the use of specific interface features.

In addition to examining participants' trust of the taxi routing capabilities, participants were also asked to provide subjective ratings of trust in the tool. On average, participants rated their trust of this decision support tool as 8.58 (out of 10.0), which is high for a prototype system that has not been extensively used, but not necessarily surprising given that the performance of the tool was perfect.

Surface Conformance Monitoring: Statistical analysis indicated that the presence of surface conformance monitoring automation strongly influenced participants' performance ($F_{(1, 11)} = 3.41$, p = 0.001). While participants detected less than half of events when no automation was provided (M = 46.7%), all events were detected with the aid of the nonconformance alerts. Average response time was found to be 23.1 s, though automation was not found to improve the speed with which participants responded (p > 0.10). Variations in traffic load were not found to affect these measures.

While the reported value of trust was relatively high, additional analysis indicated that trust was largely impacted by the performance of the automated alerts. The simulation was designed to examine perfect alerting performance, though there were documented cases where the logic of the algorithms, when paired with certain actions or states in the simulation environment, produced false alerts. A correlation was performed to examine the relationship between the number of false alerts to which a given participant was exposed with subjective ratings of trust on that trial. This analysis indicated a significant negative correlation between false alarm frequency and subjective ratings of trust ($r = -0.35$, $p = 0.10$). This relationship is shown in Figure 3.

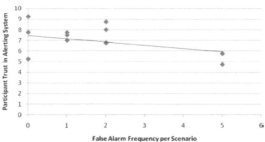

Figure 3   Correlation of false alarm frequency and trust.

3.2   Simulation 2

Methods

Participants: Twelve participants took part in the simulation were working at FAA Headquarters or at an air traffic control tower facility at the time of the study. The participants ranged in age from 27 to 51 years (M = 43.3 years). On average, participants reported 17.2 years of experience controlling traffic. As reported in Simulation 1, no participants had experience working DFW traffic.

Simulation Environment and Procedure: The second simulation used the same environment and procedures described for Simulation 1, with the exception of a subset of the implemented pilot deviations. As in Simulation 1, nonconformance events included of lateral deviations that occurred from erroneous turns or from pilots' failure to turn at a correct intersection. In addition to lateral path errors, hold short deviations occurred in each scenario, during which a pilot failed to hold short at either a taxiway or a runway crossing. The path deviations and the hold short deviation at the taxiway crossing were non-safety critical events. The hold short deviation at the runway crossing was a safety critical event that resulted in a runway incursion but not one that would have generated an ASDE-X alert. The simulation confederate local controller held departing traffic around the time of the safety critical nonconformance event to insure there was no potential for a collision. When participants detected a pilot deviation, they were asked to press a button on the electronic flight strip display and verbally respond to the pilot to correct the error.

Experimental Design: As in Simulation 1, automation presence was manipulated as a within-subjects variable. The order of scenario presentation was randomly selected across the thirteen participants from the full set of permutations associated with a completely counterbalanced design.

Simulation 2 Results

Key results from the simulation are summarized in the section below. A full review of these findings is available in a published technical report (McGarry and Kerns, 2010).

Taxi Route Generation: After each of the automation scenarios, participants responded to six questionnaire items that addressed whether and when they used system-generated taxi routing. They also responded to questions about taxi route modifications.

Over half of the participants (54%) indicated that they used the automated route generation function in at least one of the automation scenarios. Note that this finding indicates less reliance on the automated taxi routing capabilities than were found in Simulation 1. This difference may be attributed to changes that were made in the route editing interface for Simulation 2. In this simulation, the keyboard used in Simulation 1 was augmented with a graphical editor interface that allowed editing on a map-based display by clicking on key taxiway intersections. These two route editing capabilities were available regardless of the automation condition.

Participants were asked to describe and assess their use of taxi route amendments to change the initial taxi route or to reestablish route conformance following a nonconformance alert. According to the responses, less than a quarter of the participants modified a taxi route in any scenario, for any reason, regardless of whether the route was self-generated or generated by the automated system. Responses further indicated that only 19% of the participants used the keyboard while only 12% used the graphical interface to amend taxi routes. Thus, while we observed additional cases in which edits were made to the taxi routes, these cases were still rare, regardless of the editing methods used.

In fact, participant assessments of the graphical editor were moderately negative, indicating that they did not find the new capability to be particularly helpful. Participants indicated a moderate preference for using the keyboard over the graphical editor for modifying routes. Commentary about the graphical editor capability expressed concern over the extensive head-down effort involved. These results indicate that the graphical editor interface should be modified to make it easier for the controllers to input both initial and modified taxi routes. One possible method is to only show the area of interest (taxiways and runways) specific to the selected aircraft. As the taxi route is selected, the algorithm can expand the area if needed.

Surface Conformance Monitoring: An analysis of variance (ANOVA) was conducted to examine the impact of both automation and traffic load on nonconformance event detection accuracy. This analysis revealed a significant main effect for automation ($F_{(1, 8)} = 6.65$, p = 0.03), indicating that participants detected less than half of events when no automation was provided for support (M = 44.2%) and a majority of events when alerting was provided (M = 86.44%). This finding was comparable to the results described for Simulation 1, where the detection rate in the no automation condition was 46.7%. In Simulation 1, however, the automation resulted in an increase in the detection rate to 100%, while here the increase was to 86.44%. This difference in Simulation 2 can be attributed to participants who failed to indicate that they had detected the nonconformance event by pressing an indicated key on the display interface. However, the data in both simulations indicate a clear benefit of automation to detection accuracy.

To further examine participants' detection performance, analyses were conducted on nonconformance detection response times. Paired t-tests were conducted to evaluate the impact of automation. As with detection accuracy, automation had a significant impact on detection response time ($t_5 = 2.93$, p = 0.03), indicating participants detected nonconformance events faster under the automated condition (M = 8.1 s) than under the manual condition (M = 22.9). T-tests also revealed a significant difference in response times between the two types of nonconformance events, with response time to the safety critical events shorter ($M = 9.3s$) than response times to the other events ($M = 27.4s$; $t_3 = -3.53$, $p = 0.04$), though this effect was only present in the manual condition. This relationship is shown in Figure 4.

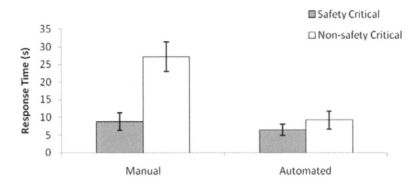

Figure 4   Detection response time to safety critical and non-safety critical nonconformance events.

4   SUMMARY AND CONCLUSIONS

Under STBO, decision support tool capabilities will provide controllers with recommended taxi routes to move departure aircraft from the ramp area to the runway. The two reported studies indicated that participants used the automation-generated taxi routes in nearly all cases. While participants did indicate that it was difficult to edit taxi routes, which may have attributed to their reluctance to readily make changes to these routes, these participants also indicated that the automation generating these routes was reliable and trustworthy. Collectively, these findings indicate the viability of the route generation concept, while simultaneously pointing to the need to develop intuitive and low workload methods for revising taxi routes on an electronic flight strip interface. While these simulations provide a basic evaluation of the taxi route generation tool concept, additional research is still needed to closely examine taxi route assignment under more dynamic and complex scenarios.

Results from the present studies indicate that providing surface conformance monitoring automation results in nearly perfect detection of pilot deviations, doubling the detection rates found when automation was not present. Results also point to the finding that controllers detect these events more quickly with automated support, at least for non-safety critical deviations, which are typically detected very slowly. While this latter finding may not be associated with a direct safety benefit, it can be tentatively concluded that more rapid detection of non-safety critical events is likely to reduce the likelihood that non-safety critical deviations become safety critical errors (e.g., a lateral deviation results in an aircraft approaching a runway entrance).

Participants also indicated that the surface conformance monitoring automation was trustworthy in both simulations. Finally, research from Simulation 1 provided some initial evidence that false alerts, which will exist in the operational world when surveillance data is imperfect, impacts trust and may be likely to affect controllers' use of these tools (Klein et al., 2010).

To fully exercise these technologies, additional research will be needed to examine a broader range of nonconformance events (e.g., taking off without clearance) and an assessment of multi-level alerting, in which different alerting methodologies are used for different categories of severity. Research is also necessary to evaluate the use of conformance monitoring for taxi routes related to performing scheduling and sequencing tasks.

ACKNOWLEDGMENTS

The authors would like to thank the participants who provided invaluable feedback, Paul Diffenderfer for providing subject matter expertise, Dave Tuomey for developing MITRE's simulation environment, and Mosaic ATM for software development efforts on the surface automation prototype.

NOTICE

This work was produced for the U.S. Government under Contract DTFAWA-10-C-00080 and is subject to Federal Aviation Administration Acquisition Management System Clause 3.5-13, Rights In Data-General, Alt. III and Alt. IV (Oct. 1996).

The contents of this document reflect the views of the authors and The MITRE Corporation and do not necessarily reflect the views of the FAA or the Department of Transportation (DOT). Neither the FAA nor the DOT makes any warranty, guarantee, expressed or implied, concerning the content or accuracy of these views. This work has been publically released under case 11-0190.

REFERENCES

Air Traffic Organization, NextGen and Operations Planning, Research & Technology Development, 2010. *Air Traffic Systems Concept Development, NextGen Mid-Term Concept of Operations for the National Airspace System*, Version 2. Washington, DC: Federal Aviation Administration.

Audenaerd, L, F., C. S. Burr, P. A. Diffenderfer, and C. E. Morgan, 2010. *Surface Trajectory-Based Operations (STBO) Mid-Term Concept of Operations Overview and Scenarios*, MP090230R1. McLean, VA: The MITRE Corporation.

Klein, K. A., E. K. Stelzer, E. T. Nelson, C. Brinton, S. Lent, 2010. *Improving Efficiency with Surface Trajectory-Based Operations and Conformance Monitoring*, Proceedings from the Digital Avionics Systems Conference, Salt Lake City, UT.

McGarry, K. A. and K. Kerns, 2010. *Results of a Second (Controller) Human-in-the-Loop Simulation Study of Automated Capabilities Supporting Surface Trajectory-Based Operations*, MTR100483. McLean, VA: The MITRE Corporation.

Morgan, C.E., 2010. *Mid-Term Surface Trajectory-Based Operations (STBO) Concepts of Use: Departure Runway Assignment*, MTR100269V4. McLean, VA: The MITRE Corporation.

Morgan, C.E. and P. A. Diffenderfer, 2010. *Mid-Term Surface Trajectory-Based Operations (STBO) Concepts of Use: Summary*, MTR100269V1. McLean, VA: The MITRE Corporation.

Nene, V., C. E. Morgan, P. A. Diffenderfer, and A. P. Colavito, 2010. *A Mid-Term Concept of Operations for a Tower Flight Data Manager (TFDM)*, MP090169R1. McLean, VA: The MITRE Corporation.

Stelzer, E. K., 2010, *Human-in-the-Loop Concept Evaluation of Surface Conformance Monitoring Automation*, MTR100188. McLean, VA: The MITRE Corporation.

Stelzer, E K. and Stanley, R. 2010. *Examination of Air Traffic Controller Use of Surface Trajectory-Based Operations Decision Support Tools*, MTR110344. McLean, VA: The MITRE Corporation.

CHAPTER 59

An Investigation into the Potential Sources of Interference in SA Probe Techniques

Corey A. Morgan, Dan Chiappe, Joshua Kraut, Thomas Z. Strybel, & Kim-Phuong L. Vu

California State University, Long Beach
Long Beach, CA
coreyandrewmorgan@gmail.com

ABSTRACT

Currently there is much disagreement on the best tools for measuring situation awareness (SA) as well as the nature of the processes that underlie it. The present project sought to assess the appropriateness of two commonly used techniques, SAGAT and SPAM, by examining the intrusiveness of each to operator SA and workload. This was done by manipulating whether or not scenarios were paused and operator displays blanked during probe presentation—two of the key features distinguishing SPAM and SAGAT. The results indicated that both of these factors have an effect on SA and workload. We also found that the intrusiveness associated with blanking and not pausing scenarios depended on the type of information being queried. In particular, making information more difficult to access from external displays had a more negative effect on the ability to answer queries pertaining to information regarding specific aircraft as opposed to general scenario characteristics. These results support the Situated approach to SA, which holds that operators often off-load information to their environment to limit what they have to store internally.

Keywords: situation awareness, SPAM, SAGAT

1   INTRODUCTION

Situation Awareness (SA) is the understanding required for an operator to safely and efficiently manage a complex, dynamic system (Endsley, 1995a, Chiappe, Strybel, & Vu, in press; Stanton, Salmon, Walker, & Jenkins, 2010). Despite having been a topic of investigation for years, there remains considerable disagreement on the best tools for measuring SA (e.g., Jeannot, Kelly & Thompson, 2003; Salmon, Stanton, Walker & Green, 2006), and on the nature of the processes that underlie it (Stanton et al., 2010). This paper explores the use of probe techniques to measure SA in the context of an air traffic control (ATC) task. These techniques differ in important respects, including whether scenarios are paused or not paused, and whether the scenarios remain visible or not visible during query presentation (Durso & Dattel, 2004; Endsley, 1995b). We examine the effects of scenario pause and visibility on operator SA and workload, and argue that these effects can be used to illuminate the processes by which operators acquire and maintain SA.

The most widely cited theory of SA is Endsley's (1995a) Three-level model. It holds that SA is made up of knowledge pertaining to the *perception* of system elements, the *comprehension* of their status in light of current operational goals, and the *projection* of future states of the system. This information is integrated into a detailed, stable, consciously reportable, internal representation referred to as a "situation model" (Endsley, 1995a). This theory is presupposed by Endsley's (1995b) measurement tool, the Situation Awareness Global Assessment Technique (SAGAT). It involves pausing scenarios at random intervals and blanking the display. Once this occurs, a battery of SA probe queries is presented, with accuracy serving as the dependent variable. SAGAT is claimed to have a minimal impact on operator workload and SA (Endsley, 1995b). Pausing prevents an increase in workload that would arise as a result of operators having to process probe queries while carrying out their primary task. Such an increase in workload would likely compromise their ability to maintain SA and their performance. Furthermore, blanking displays is done to avoid inflating estimates of SA. Leaving displays available would allow operators to answer a query by looking at a screen in case they did not represent the answer directly in memory. Thus, SAGAT is consistent with the assumption of Endsley's Three-level Model that for information to count as SA it must be represented internally in conscious, working memory.

A very different approach to SA measurement is the Situation Present Assessment Method (SPAM; Durso & Dattel, 2004). In SPAM, scenarios are not paused, and displays remain visible during the presentation of probe queries. This is done so as not to disrupt the way operators normally carry out their primary tasks. Although scenarios are not paused, Durso and Dattel (2004) claim that SPAM does not significantly increase workload or affect performance. This is because it presents operators with a "Ready" prompt prior to queries being presented. Operators are instructed to indicate they are ready to receive a query only when their workload permits. Once they do so, an SA probe is presented. Accuracy and RT are both dependent variables. Furthermore, SPAM keeps displays visible because removing them can lead to an underestimation of operator SA. Operators should be free to consult them because they do not hold a complete picture of the task environment internally. For example, an air traffic controller (ATCo) may not store an aircraft (AC) callsign internally, but may remember where to locate this information on their display, which also counts as good SA (Durso & Dattel, 2004).

SPAM is therefore consistent with the Situated SA theory (Chiappe et al., in press). This theory holds operators often "off-load" information onto their external displays so as to minimize the amount of information they have to hold internally. Operators do not create the detailed, stable, internal representations of situations envisioned by Endsley's model. Rather, they create only minimal representations, relying instead on frequent interactions with their task environment to access information on an as-needed basis. According to Situated SA, then, much of our understanding of a situation may involve not storing information internally, but rather knowing where to locate this information in the environment when we need it. In fact, because it collects RT along with accuracy, SPAM can be used to examine whether information is represented internally, or off-loaded onto a display. If the information is stored internally, the RT to answer a probe query will be fast; if it is represented externally and the operator knows where to find it, RT will be slower. It will be slowest still in cases where the operator does not even know where in the environment to find the information (Durso & Dattel, 2004).

The Situated SA position outlines several possible factors that could determine whether information is represented internally in the mind of the operator, or off-loaded onto a display (Chiappe et al., in press). This study examined two factors in the context of an ATC simulation. These are the generality of the information and its priority. *General* information queries referred to the general trends of many or all AC within the sector. *Specific* information queries referred to a single identifiable AC within the sector. It is possible that operators rely on their displays to store information that is specific, while only storing general information about their task environment internally. The latter representations would help operators know where in the display to access specific information when it is needed. It is also possible operators store high priority information internally, while low priority information is stored externally. Having high priority information stored internally would ensure it is promptly acted upon, without having to spend time accessing the information on the display. Storing low priority information externally, on the other hand, would serve to limit the amount of information that needs to be held in working memory.

To summarize, SPAM and SAGAT probe metrics make different theoretical assumptions about the nature of SA. Proponents of these techniques claim that their measures have a minimal impact on SA and workload. Although several studies have compared these two probe techniques (e.g., Durso, Bleckley & Dattel, 2006; Endsley, Sollenberger, & Stein, 2000), these have largely compared them intact. In this study, we separated the core aspects of each (scenario pause and visibility) to examine whether scenario visibility or pause condition are intrusive to SA and workload. We also examined whether the two factors interact, so that intrusiveness is greatest under certain combinations of visibility and pause. To this end, this experiment measured accuracy and RT to answer SA probe queries and workload under four conditions: (scenario visible/paused, scenario visible/not paused, scenario not visible/paused and scenario not visible/not paused). We predicted that workload would be greater when the scenarios are not visible and when they are not paused, with the greatest workload for the not-visible, not-paused condition.

Manipulating the conditions under which SA probes were administered also allowed us to test whether information generality and priority affect whether it is represented internally, or off-loaded onto displays. This was achieved through the creation of probe queries that differed in specificity (general vs. specific) and priority (high vs. low). This resulted in four query categories (See Table 1).

Table 1   Four query categories and sample queries

	High Priority	Low Priority
General	Will any conflicts be present if no further action is taken for the next 2 minutes?	In the past 5 minutes, did most AC enter into your sector from the East?
Specific	Will any conflicts be present with AAL123 if no further action is taken in the next 2 minutes?	In the past 5 minutes, did AAL123 enter into your sector from the East?

The Situated SA approach predicts that general queries should be answered faster than specific queries, and that high priority queries should be answered faster than low priority queries. This is because in the latter cases, individuals have to access information on the display, rather than internally. Furthermore, it predicts that SA query conditions that make it more difficult to access information in the environment (i.e., when the scenario is not visible, and when it is not paused) will have a negative effect on operators' ability to answer queries for information that is stored externally, i.e., low priority information, and specific information. However, information stored internally should be relatively unaffected by these factors. In contrast, the Three-level model predicts that these different types of information should not be differentially affected by scenario visibility or pause, because all task relevant information is stored internally. Making it more difficult to access information externally should have no effect on this view.

2   METHOD

2.1   Participants

Participants were 14 students enrolled in an FAA certified Collegiate Training Initiative program at Mt. San Antonio College, who are studying to become air traffic controllers. They were compensated $80.00 for completing the entire, 6-hr study. All participants had previous experience with the simulation software, sector characteristics, and tasks required of them from a previous 16-week ATCo radar internship hosted in the Center for Human Factors in Advanced Aeronautics Technologies (CHAAT) at CSULB.

2.2   Apparatus and Scenarios

The simulation was conducted using the Multi Aircraft Control System (MACS), a medium-fidelity simulation environment (Prevot, 2002). Each participant was seated in front of two computers. One displayed an ATCo radar scope that simulated Indianapolis Center Sector 91 (ZID 91), and one was a touch screen computer used for presenting and answering probe queries.

Four experimental scenarios were run, each lasting 50 minutes. Each scenario featured mixed equipage and was balanced in terms of difficulty. Mixed equipage means that approximately 50% of the AC were NextGen equipped. As such, ATCos had several NextGen tools available, including integrated Datalink, Conflict Alerts and Probes, and trial planners. All equipped AC were managed primarily through NextGen tools while all non-equipped AC were managed manually, using current-day tools. Scenario difficulty was balanced by ensuring there was an approximately equal number of AC present within the sector at any given time and that an equal number of AC occurred in each of the four experimental scenarios.

2.3   Design

The study used a 2 (visibility: visible vs. not visible) x 2 (pause: paused vs. not paused) x 2 (specificity: general vs. specific) x 2 (priority: high vs. low) within-subjects design. The order that SA measurement conditions were presented (i.e., visible/paused, not-visible/paused, visible/not-paused, not-visible/not-paused), was counterbalanced across participants. The order of probe queries was also counterbalanced across participants. Dependent variables included metrics of ATCo performance safety and efficiency, and measures of SA and workload. However, this paper presents results only for the latter two.

2.4   Procedure

Each participant received three 20-minute training trials. The first was used to re-familiarize participants with the MACS software. The second and third were used to familiarize participants with the conditions in which their display would be not visible, and the scenario would either be paused or not paused. At the beginning of each experimental scenario the participants were informed as to which measurement condition they would receive. If the condition called for the display to not be visible, a custom Visual Basic program overlaid the ATCo display with a black screen for the duration of the query presentation. If the condition called for the scenario to be paused, the simulation manager did so for the duration of the query presentation. Scenarios were paused or blanked for 30 seconds. In all, 16 probe queries were asked during each 50 minute trial. Four were Air Traffic Workload Input Technique-like (ATWIT-like) probes where participants rated workload on a scale from 1 to 7. SA Probe queries and workload probes were presented on a touch screen display adjacent to the ATCo radar display. A SPAM-like methodology was employed for all conditions in that participants were first presented with a ready prompt, and presentation of the query occurred only when they indicated they were ready to receive it. Probes were presented every three minutes, with the first query presented three minutes from the start of the trial.

3   RESULTS

3.1   Situation Awareness

SA data were analyzed using a 2 (specificity: general vs. specific) x 2 (priority: high vs. low) x 2 (visibility: visible vs. not visible) x 2 (pause: paused vs. not paused) repeated measures ANOVA for probe query accuracy and probe RT.

3.1.1 Probe Accuracy

The results revealed a main effect of pause condition, $F(1, 13) = 4.54, p = .053$. Probe accuracy was 5% higher when the scenarios were paused than when they were not paused. There was also a significant main effect of visibility, $F(1, 13) = 19.08, p = .001$, with accuracy being 19% higher when the display remained visible than when it was not visible. In terms of the type of information queried, we found no overall difference in accuracy between high and low priority information, $p > .10$, and no overall difference between general and specific information, $p > .10$

However, further results suggest the effects of pausing and visibility depend on the type of information being queried. The results revealed a significant interaction between visibility and specificity, $F(1, 13) = 7.50, p = .017$. When the display was visible, there was no difference in accuracy for general vs. specific information. However, when the display was not visible, accuracy for general queries was about 14% higher than for queries regarding specific information. Thus, blanking the displays had a more negative effect on student controllers' ability to provide specific information about the sector they were controlling, but had less of an effect on their ability to provide general information about their sector.

With respect to the effect of scenario pause, we found a significant three-way interaction between pause, specificity, and priority, $F(1, 13) = 8.81, p = .011$ (See Figure 1). This interaction reflects the fact that for the high priority information, general queries were answered more accurately than specific queries. However, for low priority information, there was no difference in accuracy between general and specific information. In short, not-pausing the scenarios while probe queries were presented had the greatest negative effect on specific information about the scenarios, provided that information was also of high priority.

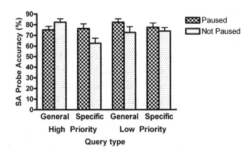

Figure 1   SA Probe Accuracy as a function of Pause, Specificity, and Priority.

3.1.2   SA Probe Response Time

Only RTs for queries that were answered correctly were analyzed. Analyses were conducted on the log transformations of the data, but we report the results in the original units (in seconds). We found a significant effect of visibility, $F(1, 9) = 36.96, p < .001$. When the display was not visible queries were answered almost 2s faster than when the display was visible. This is likely because when they were not visible participants had to respond based only on what they remember, and did not spend time searching through a display for the needed information. There was no main effect of pause condition. In terms of types of information queried, there was a significant effect of specificity, $F(1, 9) = 25.71, p = .001$. Queries regarding general information were answered about 1s faster than queries regarding specific information. There was no effect of priority.

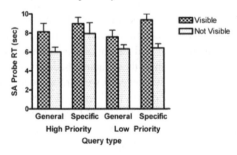

Figure 2   SA Probe RT as a function of Visibility, Specificity, and Priority.

The analysis also revealed that the effect of visibility depended on the type of information queried, with an interaction between visibility, specificity, and priority, $F(1, 9) = 5.76, p = .04$ (See Figure 2). This interaction reflects the fact that when the display was not visible queries regarding general information were answered faster than queries regarding specific information, provided the information was of high priority. However, when the display was visible, the specificity and priority of information did not interact.

3.2   Workload

Average ATWIT ratings were calculated for each participant within each measurement condition and analyzed using a 2 (visibility: visible vs. not visible) x 2 (pause: paused vs. not paused) repeated measures ANOVA. There was a marginally significant effect of pause, with workload tending to be lower when the scenario was paused than when it was not paused, $F(1, 13) = 3.97$, $p = .068$. There was no main effect of visibility, $p > .10$. However, there was a marginally significant interaction between pause and visibility, $F(1, 13) = 3.21$, $p = .097$ (See Figure 3). When the display was not visible, participants tended to experience less workload when scenarios were paused than when they were not paused.

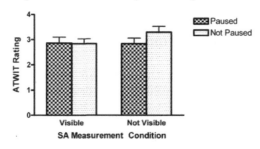

Figure 3   Mean ATWIT Rating as a function of Visibility and Pause conditions.

4   DISCUSSION

The present study examined two key factors distinguishing SA probe techniques—whether or not scenarios are paused, and whether or not displays remain visible during query presentation. SAGAT involves pausing scenarios and removing the displays, while SPAM does not pause scenarios, allowing displays to remain visible. We found that both of these factors had an influence on SA. In terms of pause condition, participants answered fewer probe queries correctly when the scenarios were not paused, likely due to the increased workload associated with the non-paused condition. Fortunately for the SPAM technique, however, we found an increased workload resulting from not-pausing the scenarios only when the participants were not able to see the displays. In terms of visibility condition, we found that participants answered more SA probe queries correctly and did so more quickly, when displays were visible.

Should we interpret the increased accuracy under visible conditions as an inflation of SA as proponents of SAGAT maintain, or does it represent a more accurate assessment of operator SA than under non-visible conditions, as proponents of SPAM maintain? This depends on whether one accepts the Situated SA approach or Endsley's Three-level model. According to the Three-level model, information counts as SA only if it is represented in the conscious mind of an operator. For the Situated SA approach, however, operators often rely on external representations to limit what they have to hold internally. This approach holds that people eschew detailed, stable world modeling, and create instead minimal internal representations, exploiting the structure of the physical world to store and manipulate information (Chiappe et al. in press). For information stored externally to count as SA, operators need to know where to access it in a timely manner, incorporating this information fluidly into their activities. We examined whether information specificity and priority determine whether operators store information in the mind or externally.

We found that participants were just as accurate in answering probe queries for high and low priority information, and for general and specific information. The RT results, however, are of greater theoretical relevance. We found that participants were faster to answer probes for general information than for specific information. This is consistent with the Situated SA approach because it holds that specific information is likely off-loaded (and thus has to be accessed on the display), while general scenario characteristics are represented internally (and are therefore answered more quickly). Storing internally general scenario information limits the load on working memory, while at the same time providing operators with the information that they need in order to know where in the display to access more specific information, should it be required.

However, failing to find an effect of priority on accuracy and probe RT suggests that our participants, being student controllers, were not as sensitive to this manipulation as more experienced controllers might be. Indeed, part of ATCo expertise is learning what information needs to be dealt with, and what information can safely be ignored. Thus, future research should examine whether expert controllers are more likely to off-load low priority information and store high priority information internally. Storing high priority information internally would ensure that the information is acted on promptly, without taking the time to access it in the environment. In contrast, storing low priority information externally would limit the load on working memory.

Also consistent with the Situated SA theory are the effects that display visibility and scenario pause had on the ability to answer SA probe queries. The theory predicts that making information more difficult to access from the environment should only affect information that is off-loaded, a prediction not shared with the Three-level model. Nonetheless, we found that removing the displays affected people's ability to answer probe queries about specific scenario information, but not queries about general scenario characteristics. This was also revealed to some extent in the RT analysis, which showed that when the displays were not visible, people were faster at answering queries regarding general information than specific information, though this only held for high priority information. A similar effect was found for the pause condition. When the scenarios were not paused (making it more challenging to access information), participants were more accurate in answering probe queries for general information than for specific information, again provided that the information was of high priority.

To conclude, the present study examined the consequences of scenario pause and display visibility on operator SA and workload. We found that pausing scenarios (as done during the SAGAT technique) can increase SA and decrease workload. Similarly, allowing displays to remain visible (as done during the SPAM technique) can increase the SA of operators, though its effect on workload depends on whether or not the scenarios are paused. We also found support for the Situated approach to SA, which holds that much of SA involves knowing where in the environment to find key information, rather than relying exclusively on internal representations. We found operators off-load specific information to their displays, retaining internally general information about scenarios they are controlling, relying instead on frequent interactions with external props and tools to maintain their understanding of a situation.

ACKNOWLEDGEMENTS

This research was supported by the NASA cooperative agreement NNX09AU66A, *Group 5 University Research Center: Center for Human Factors in Advanced Aeronautics Technologies* (Brenda Collins, Technical Monitor).

REFERENCES

Chiappe, D., T. Z. Strybel, & K.-P. L. Vu. in press. Mechanisms for the acquisition of Situation Awareness in situated agents. *Theoretical Issues in Ergonomic Sciences.*

Durso, F. T. & A. Dattel. SPAM: The real-time assessment of SA. In. *A Cognitive Approach to Situation Awareness: Theory, Measures and Application,* eds. S. Banbury & S. Trembley (pp. 137–154). New York: Aldershot.

Durso, F., K. Bleckley, & A. Dattel. 2006. Does situation awareness add to the validity of cognitive tests? *Human Factors* 48: 721-733.

Endsley, M. R. 1995a. Toward a theory of situation awareness in dynamic systems. *Human Factors* 37: 32-64.

Endsley, M. R. 1995b. Measurement of situation awareness in dynamic systems. *Human Factors* 37: 65-84.

Endsley, M. R., R. Sollenberger, & E. Stein. (2000). Situation awareness: A comparison of measures. *Proceedings of the Human Performance, Situation Awareness and Automation: User Centered Design for the New Millennium Conference.* Savannah, GA.

Jeannot, E., C. Kelly, & D. Thompson. 2003. The development of situation awareness measures in ATM systems (HRS/HSP-005-REP-01). Brussels, Belgium: EUROCONTROL.

Prevot, T. 2002. Exploring the many perspectives of distributed air traffic management: The Multi Aircraft Control System MACS. In. *Proceedings of the HCI-Aero 2002,* eds. S. Chatty, J. Hansman, & G. Boy. Menlo Park, CA: AAAI Press.

Salmon, P. M., N. A. Stanton, G. Walker, & D. Green. (2006). Situation awareness measurement: A review of applicability for C4i environments. *Applied Ergonomics* 37: 225–238.

Stanton, N., P. Salmon, G. H. Walker, & D. P. Jenkins. 2010. Is situation awareness all in the mind? *Theoretical Issues in Ergonomics Science* 11: 29-40.

CHAPTER 60

Effect of Force Feedback on an Aimed Movement Task

R. Conrad Rorie, Hugo Bertolotti, Thomas Strybel, Kim-Phuong L. Vu,
Panadda Marayong, Jose Robles

California State University, Long Beach
Long Beach, CA
robert_rorie@yahoo.com

ABSTRACT

This study examined the effect of force feedback on target acquisition in a simulated cockpit display of traffic information (CDTI) task, using a computer mouse and the Novint Falcon with and without force-feedback information. Participants were asked to select targets that varied in size, distance, and angular direction from the start location. Results showed significant differences between the movement times for the various input devices. Movement times with the mouse and the Falcon with force feedback were significantly faster than the Falcon without force feedback. Force feedback was found to reduce the effect of target size, but not the effect of target distance. Force feedback also produced faster movement times when the direction of movement was in two dimensions. The findings suggest that force feedback may be a useful method for assisting pilots in future CDTIs.

Keywords: input device, movement time, Fitts' Law, force feedback, movement direction, CDTI

1   INTRODUCTION

In response to a projected increase in overall air traffic, the Federal Aviation Administration is engaged in a series of improvements and changes in the National Airspace System (NAS). This new approach, known as the Next Generation Air Transportation System (NextGen), should significantly transform the way that pilots and air traffic controllers (ATC) manage their airspace. One concept of operation within the NextGen is Trajectory Based Operations (TBO). TBO have been designed to address the operator capacity limitations that characterize our current system, such as operator workload and imprecise navigation. TBO will provide greater flexibility and autonomy to pilots when modifying flight plans mid-flight (JPDO, 2010). Pilots will interact with new onboard avionics to alter their own flight plan, which will use GPS coordinates rather than radar and pre-existing waypoints. This will eliminate the need for ATC to provide pilots with longer deviations and clearances around traffic or weather, reducing controller workload considerably.

To accommodate a shift to TBO in the cockpit, new onboard technologies, are needed. Most importantly, pilots will require information that allows them to maintain their awareness of surrounding traffic, terrain and weather (e.g., Alexander and Wickens, 2001). One such device, the Cockpit Display of Traffic Information (CDTI), presents pilots with the information necessary for effective route planning and separation assurance (e.g. information regarding the speed and direction of ownship, location of nearby traffic, weather, and terrain). CDTIs could potentially include conflict detection and resolution tools, automated route modification tools, and precise spacing functions.

Although CDTIs have been tested in simulated cockpits, there are constraints on the use and implementation of CDTIs in current day cockpits. For example, the limited space in the cockpit may require that the CDTI be kept small, which would restrict the workspace of operators when performing operations on the CDTI. Moreover, the unstable nature of a cockpit environment would further increase the difficulty of interacting with the CDTI, making accurate selection difficult and increasing task completion time. Lastly, if 3D CDTIs such as the one being developed by NASA's Flight Deck Research Group are implemented, pilots would require an input device that would allow them to manipulate objects in three dimensions (Granada et al., 2005).

These constraints have implications for the design of CDTI input devices. The present study examined whether providing force-feedback would benefit pilot input procedures, specifically movement times in a point and click task. Force-feedback has been suggested as a potential way of improving operator performance on tasks with high degrees of difficulty (e.g., Griffiths and Gillespie, 2005). As the first step in the evaluation of force feedback for cockpit applications, we examined movement times for simple point and click tasks with and without force feedback.

1.1   Research on Force-Feedback

Several studies have examined the effects of tactile and force-feedback on operator performance, particularly in the domain of human-computer interaction (e.g. Dannerlein and Yang, 2001; Tornil and Baptiste-Jessel, 2005; Akamatsu and MacKenzie, 1996; Ahlstrom, 2005). Tactile feedback refers to the presentation of information to the operator through the stimulation of mechanoreceptors on the skin (He and Agah, 2001). The most common type of tactile feedback is vibration, and it is used most often with the intent of reducing burden on the visual channel. Force feedback, on the other hand, stimulates mechanoreceptors in the joints and muscles to effectively assist or resist movement. Force feedback has been evaluated as a method of integrating human-automation control sharing. Griffiths and Gillespie (2005) showed that automated torques applied to the steering wheel in a driving task significantly improved lane following and reduced visual demands made on the driver. Lee, Adelstein and Choi (2008) described a weather display that enables users to feel the weather at the geographic locality indicated by the position of the interface cursor. Lee et al. suggested that tactile feedback would be beneficial to the operators; however, no performance measures were reported for that study.

Most of the research on tactile and force feedback employ Fitts' Law tasks using two-dimensional input devices (like the computer mouse, trackball, joystick) within relatively constrained directions of motion. In some experiments, force feedback has been found to decrease target selection time by 20-25% when employing an attractive force around a target region (e.g., Eberhardt, Neverov, West, and Sanders, 1997); however, the benefits of the force feedback depend on the device and tasks used. Akamatsu and MacKenzie (1996), for example, examined the effects of tactile and force-feedback on a modified computer mouse. They found that the additional types of feedback worked to reduce target selection time. Moreover, the impact of tactile and force feedback was on the final stages of movement, when the cursor entered and clicked on the target. Tactile and force feedback were found to decrease the amount of time the participants spent within a target before selecting it, especially for small targets. Force feedback was responsible for reducing the number of errors. On the other hand, Tornil and Baptiste-Jessel (2005) obtained no improvement with a force feedback mouse compared with no feedback in a point and click task containing several target locations.

Most of these experiments on simple movement times analyzed force feedback using a mouse as the input device. A mouse, however, is impractical in cockpit applications. The current study examined movement times with and without force feedback on a simple point and click task using a nontraditional input device, the Novint Falcon. Although the Falcon is also unsuitable for cockpits, it has similar characteristics with cockpit input devices. It is also inexpensive and can be easily programmed. These features made the Falcon an attractive device for evaluating force feedback in point-and-click tasks. This study also used a wider range of target directions than is typically found in studies employing Fitts' Law tasks in order to better approximate the movements pilots will be required to make on a CDTI.

2   METHOD

2.1   Subjects

Twelve students from California State University, Long Beach participated in the experiment. They were paid 60 dollars at the conclusion of the study for six, one-hour sessions over six days. Eight of the subjects were female and four of the subjects were male. All subjects were right handed and had extensive experience with a standard computer mouse; no participants had previously used the Novint Falcon.

2.2   Apparatus

The experiment was conducted with a Logitech laser mouse and the Novint Falcon input device. The Falcon is capable of position sensing and force feedback in three dimensions, with an operational workspace of 4" x 4" x 4". Since the task was two-dimensional in nature, a virtual unidirectional plane was programmed to limit the range of movement for the Falcon to a planar surface. While the Falcon is typically oriented to accommodate hand motion in the plane parallel to the orientation of the visual display, in the present study, the device was turned 90 degrees to allow hand movement in the horizontal plane (i.e. perpendicular to the visual display; see Figure 1). This was done to maintain a similar form of manipulation relative to the mouse condition as well as to improve participant comfort while using the device. Buttons located on the Falcon's interchangeable grip allowed participants to select targets as they would with the mouse. The height of the grip was set at the level of the participant's armrest.

Force feedback was added to the Falcon using Newton's gravitational law equation to generate an attractive force that guided the user toward the center of the object based on the square of the distance between the cursor's position and the center of the target. A constant force feedback gain was selected to yield a perceivable and stable force display (for more details on the force model, see Robles et al., 2012).

The start and target icons were displayed on a screen shot of a CDTI with no traffic present. The CDTI display was 8" x 8", presented on a 17" x 11" computer monitor with 1680-pixel x 1050-pixel resolution (see Figure 2). Participants sat roughly 20 in. from the computer monitor. The worktable was aligned with the height of the chair's armrests, so that movements in all conditions could be accomplished with the arm resting on the chair. An experimenter remained in the room with the participants for the duration of the experiment in order to field any questions and activate the program between trials.

Figure 1 Novint Falcon oriented horizontally

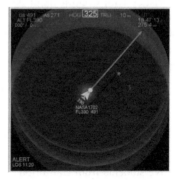

Figure 2 Sample screen shot of the task

2.3   Procedure

Participants performed a simple target selection task with all input devices: the computer mouse, the Falcon without force feedback, and the Falcon with force feedback. At the beginning of each trial the start position and target were displayed. Participants started a trial by clicking a green start circle; they then moved the cursor to the target as quickly as possible and clicked the target to end the trial. Movement times were recorded for each target selection. After the target was clicked, the start and target stimuli for the next trial were presented. The task was therefore self-paced, as the trial did not begin until the start location was clicked.

The start icon was always 0.25in. and remained at the center of the display throughout the duration of the experiment. Targets varied in size, distance, and angular direction. Two target sizes, 0.17in. and 0.25in., and two target distances, 0.83in. and 2.50 in., were used. Targets appeared at one of six angular directions relative to the starting position (with 0° corresponding to 12:00 clock position): 0°, 60°, 120°, 180°, 240°, and 300°. Each size, distance, and direction combination was presented to the participant five times, resulting in 120 trials per experimental block.

Each participant completed a practice block on the first day of testing with a new device. The practice block consisted of two trials of each combination of variables. Each experimental block took an average of five minutes to complete. Upon completion of a block, participants were provided with a brief rest period before starting the next block. Participants completed four experimental blocks per day in a single session lasting about 40 minutes. Participants completed eight blocks per condition. All blocks for one device were completed before moving to a new device.

2.4   Design

The experiment was a 2 x 2 x 3 x 6 repeated measures design, as shown in Table 1. The order of presentation for the device variable was counter-balanced between participants. The order of presentation for the target variables (i.e. size, distance, and direction) were randomly generated for each participant.

The dependent variable was movement time, recorded to the nearest millisecond. In order to compare the efficiency of each device with respect to target size and distance, the index of difficulty (ID) for each combination of size and distance was calculated. The four IDs were 2.74, 3.32, 4.32, and 4.91 bits.

Table 1 Levels of each independent variable

Variable	Levels
Target Size	0.17 and 0.25 in.
Target Distance	0.83 and 2.50 in.
Device Type	mouse, Falcon with force, Falcon without force
Target Direction	0°, 60°, 120°, 180°, 240° and 300°

3   RESULTS

A 2 (target size) x 2 (target distance) x 3 (device) x 6 (target direction) repeated measures ANOVA was conducted with movement time as the dependent variable. All main effects were significant. A main effect of device type was found, $F(2, 22)$ = 120.44, $p < .001$, where movement times using the Falcon with force feedback and the mouse were significantly faster than the average movement time of the Falcon without force feedback, p's $< .001$. Differences in movement time between the Falcon with force feedback and the mouse were not significantly different ($p >$ .05). Not surprisingly, main effects of target size $F(1, 11) = 386.68$, $p < .001$ and target distance, $F(1, 11) = 652.86$, $p < .001$, were found. Movement times were significantly faster with the large target size and short target distance.

The effect of target size was modified by a significant interaction between device type and target size, $F(2, 22) = 204.82$, $p < .001$, as shown in Figure 3. Simple effects analyses demonstrated that when participants selected the small target, movement times using the Falcon with force feedback were 82 ms faster, on average, than the mouse condition ($p < .01$). For large targets, the difference in movement time between the Falcon with force feedback and the mouse (MD=10.5 ms) was nonsignficant ($p > .05$). When simple effects of target size were analyzed for each device, significant effects were found for all three devices (p's $< .001$). However, the reduction in movement time with the large target size was minimal with force feedback (average reduction = 38 ms) compared with the effect of target size on the mouse (average reduction = 110 ms) and the Falcon without force feedback (average reduction = 209 ms).

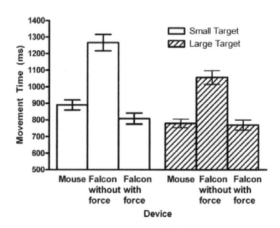

Figure 3 Movement time as a function of target size for each input device.

The main effect of target distance also was modified by a significant interaction between target distance and device type, $F(2, 22) = 20.51$, $p < .001$. As shown in Figure 4, movement times with the Falcon with force feedback were 69.2 ms faster, on average, than with the mouse for the near target, $p < .01$. However, when participants were selecting the far target, the difference in movement times (MD=23.5 ms) between the mouse and Falcon with force feedback was nonsignficant, $p > .10$, indicating that force feedback significantly improved performance with the near target, but not the far target. However, the increase in movement time with a large distance was greater for the Falcon with force (M=256 ms) compared with the mouse (M=210 ms). Movement times were again the highest for the Falcon without force.

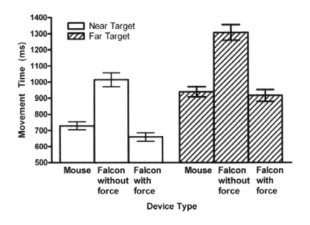

Figure 4 Movement time as a function of target distance for each input device.

A significant main effect of target direction was also found, $F(5, 55) = 9.32$, $p < .001$. The target directions of 120° and 300° resulted in the fastest movement times, and were significantly faster than target directions 0° and 240°, p's $< .05$, as shown in Table 2. Target directions of 60° and 180° did not differ significantly from the other four targets, p's $> .05$. An interaction between device type and target direction was also found, $F(10, 110) = 13.98$, $p < .001$. The source of the interaction was in the differences between the Falcon with force feedback and mouse conditions at each direction of movement. As shown in Figure 5, movement times for the Falcon with force feedback at target directions of 60°, 120°, 240°, and 300° were significantly faster when compared to the mouse at those same angles ($p < .05$).

Differences in movement times between the Falcon with force feedback and mouse conditions at target directions of 0° and 180° were nonsignificant. In other words, the Falcon with force feedback produced superior performance compared with the mouse for all directions except for straight forward (0°) and straight back (180°). Movement times for the Falcon without force condition were significantly longer than all other devices, at all six target directions, p's $< .001$).

To determine the efficiency of each input device, movement times were regressed against all four target IDs. The fit of these equations was quite low (mouse: $R^2 = .41$; Falcon with force feedback: $R^2 = .34$; Falcon without force feedback: $R^2 = .42$.), most likely because Fitts' Law was not intended to take more

Table 2 Mean (SE) movement times (ms) by device and target angular direction

Device Type	0°	60°	120°	180°	240°	300°
Mouse	790.75	851.24	853.66	809.67	862.56	839.48
	(31.09)	(32.16)	(30.31)	(24.84)	(26.20)	(27.31)
Falcon	1218.00	1161.55	1091.58	1200.33	1194.52	1100.41
w/o Force	(50.03)	(38.57)	(49.87)	(49.99)	(50.91)	(43.84)
Falcon	826.80	753.64	772.16	807.74	802.63	766.16
w/ Force	(35.15)	(31.42)	(32.39)	(31.81)	(31.51)	(30.32)
Average	945.19	922.14	905.80	939.25	953.24	902.02
	(35.11)	(30.87)	(34.49)	(32.31)	(33.27)	(30.79)

than one dimension of movement into account (Cha and Myung, 2010). When movement times were regressed against the four IDs for each device and angular direction separately for each participant, however, all but five values of R^2 exceeded .50 and over 60% of the values exceeded .80.

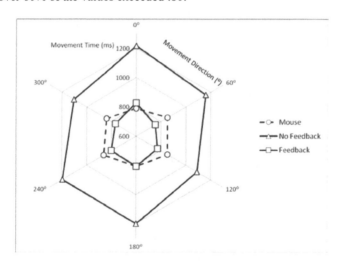

Figure 5 Movement time at each angular direction for each input device.

To determine whether performance with the input devices varied in efficiency, a repeated measures ANOVA was run on the individual slopes with the factors device and movement direction. Significant main effects of device type and direction were obtained. The effect of device showed that the slopes of the mouse (M=204 ms/bit) and Falcon with force feedback (M=207 ms/bit) were significantly lower than the Falcon without force feedback (M=309 ms/bit). Direction of movement also affected the slope of Fitts' equation. The slopes were lowest at directions of 0° and 180°.

4   DISCUSSION

This study compared movement times for simple point-and-click tasks between the mouse and a relatively new device, the Novint Falcon with and without force feedback. Previous research examining force feedback typically compared movement times within a single input device, the standard computer mouse. As mentioned earlier, the mouse is not well suited for cockpit applications. Although the mouse was shown to be superior to the Novint Falcon without force feedback, when force feedback was added, performance on the Falcon was equivalent, and in some cases superior to performance with a standard mouse. In other words, the addition of force feedback improved performance on a novel input device to a level comparable to the mouse.

Superior performance with force feedback was found in two areas. First, the Falcon with force feedback produced faster movement times for target directions in two dimensions (60°, 120°, 240° and 300°) as shown in Figure 3. The mouse and Falcon with force produced equivalent movement times only when movements were made in one dimension (0° and 180°). Whisenand and Emurian (1996) showed that mouse movements were generally faster with movements in one dimension, (vertically or horizontally), compared with movements in two dimensions (diagonally). Therefore, force feedback in our experiment eliminated the limitation seen with mouse movements. Second, force-feedback was most effective when participants were selecting the small target size. As shown in Figure 1, decreasing target size produced an average increase of only 38 ms with force feedback, and 110 ms when the mouse was used (and 300 ms for the Falcon without force feedback). While force feedback did improve participant performance relative to the mouse for the near target, differences between all three devices on the effect of target distance were minimal. When distance was tripled, movement times increased 256 ms for the Falcon with force feedback and 210 ms for the mouse.

The advantage of the Falcon with force feedback for small targets combined with the larger effect of increased target distance is consistent with Akamatsu & MacKenzie (1996). They found that movement times can be decomposed into three stages: approach time, stopping time, and clicking time. Akamatsu and MacKenzie found that tactile and force feedback had the most effect on the final stages of movement: the stopping time (time between cursor entering the target and the cursor stopping) and the clicking time (time between cursor being stopped and target clicked). This indicates that force feedback should minimize the effects of a small target, but not a large distance between the start and the target region.

These results suggest that force feedback for CDTI input devices may be a promising method for overcoming some of the constraints on their use in the cockpit, such as a small display size and limited space for movement. The CDTI will require relatively difficult target selection tasks, which will be made increasingly difficult by the unstable nature of a cockpit environment. While the optimal configuration of force-feedback within the CDTI is still under development, the findings reported here encourage further evaluation of this possibility.

ACKNOWLEDGEMENTS

This research was supported by the NASA cooperative agreement NNX09AU66A, *Group 5 University Research Center: Center for Human Factors in Advanced Aeronautics Technologies* (Brenda Collins, Technical Monitor).

REFERENCES

Ahlstrom, D. 2005. Modeling and improving selection in cascading pull-down menus using Fitts' law, the steering law and force fields. *Proceedings of the Conference on Human Factors in Computing Systems*. Portland, OR.

Akamatsu,M. and I.S. MacKenzie 1996. Movement characteristics using a mouse with tactile and force feedback. *International Journal of Human-Computer Studies* 45:483-493.

Alexander, A.L. and C.D. Wickens. 2001. Cockpit display of traffic information: The effects of traffic load, dimensionality, and vertical profile orientation. *Proceedings of the 45th Annual Meeting of the Human Factors and Ergonomics Society*. Santa Monica, CA.

Cha, Y. & R. Myung. 2010. Extended Fitts' law in three-dimensional pointing tasks. *Proceedings of the Human Factors and Ergonomics Society, 54th Annual Meeting*. San Francisco, CA.

Dannerlein,J.T. and M.C. Yang. 2001. Haptic force-feedback devices for the office computer: Performance and musculoskeletal loading issues. *Human Factors* 43:278–286.

Eberhardt, S., M. Neverov, T. West, and C. Sanders. 1997. Force reflection for wimps: A button acquisition experiment. *Proceedings of the 6th Annual Symposium on Haptic Interfaces, International Mechanical Engineering and Exposition*. Dallas, TX.

Granada, S., A.Q. Dao, D. Wong, W.W. Johnson, and V. Battiste. 2005. Development and integration of a human-centered volumetric cockpit display for distributed air-ground operations. *Proceedings of the 12th International Symposium on Aviation Psychology*. Oklahoma City, OK.

Griffiths, P. and R. B. Gillespie. 2005. Sharing control between human and automation using haptic interfaces: Primary and secondary task performance benefits. *Human Factors* 47: 574-590.

He, F. and A. Agah. 2001. Multi-modal human interactions with an intelligent interface utilizing images, sounds, and force feedback. *Journal of Intelligent and Robotic Systems* 32: 171-190.

Joint Planning and Development Office (JPDO). 2010. *Concept of operations for the Next Generation Air Transportation System version 3.2*. Accessed January 25, 2012, http://jpe.jpdo.gov/ee/docs/conops/NextGen_ConOps_v3_2.pdf.

Lee, C., B.D. Adelstein, and S. Choi. 2008. Haptic weather. *Proceedings of the Symposium on Haptic Interfaces for Virtual Environments and Teleoperator Systems*. Reno, NV.

Robles, J., M. Sguerri, R.C. Rorie, K.-P. Vu, T. Strybel, and P. Marayong. 2012. Integration framework for NASA NextGen volumetric cockpit situation display with haptic feedback. Proceedings of the International Conference on Robotics and Automation. Saint Paul, MN.

Tornil, B. and N. Baptiste-Jessel. 2005. Force feedback and performance in a point-and-click task. *Proceedings of International Conference on Multimodal Interaction (ICMI)*. Trento, Italy.

Whisenand, T.G. and H.H. Emurian. 1996. Effect of angle of approach on cursor movement with a mouse: Consideration of Fitts' law. *Computers in Human Behavior* 12: 481-495.

T - #0215 - 101024 - C0 - 234/156/35 [37] - CB - 9781439871164 - Gloss Lamination